Waste Management for Sustainable Environment

Waste Management for Sustainable Environment

Edited by **Victor Bonn**

SYRAWOOD
PUBLISHING HOUSE
New York

Published by Syrawood Publishing House,
750 Third Avenue, 9th Floor,
New York, NY 10017, USA
www.syrawoodpublishinghouse.com

Waste Management for Sustainable Environment
Edited by Victor Bonn

International Standard Book Number: 978-1-68286-319-0 (Hardback)

Printed in China.

Contents

Preface

This book aims to highlight the current researches and provides a platform to further the scope of innovations in this area. This book is a product of the combined efforts of many researchers and scientists from different parts of the world. The objective of this book is to provide the readers with the latest information in the field.

This book focuses on the various measures and techniques for the collection, recycling, treatment of waste and environmental assessment. It explores all the important aspects of waste management and environmental conservation with the help of topics such as green energy, waste water treatment, assessment of heavy metals, etc. in the present scenario. It includes state-of-the-art inputs and evaluations by eminent experts from around the globe, in this field. It aims to serve as a reference for students and experts alike and contribute to the growth of the discipline.

I would like to express my sincere thanks to the authors for their dedicated efforts in the completion of this book. I acknowledge the efforts of the publisher for providing constant support. Lastly, I would like to thank my family for their support in all academic endeavors.

Editor

Spatial variation of volatile organic compounds and carbon monoxide in Blantyre City, Malawi

Mapoma, H. W. T.[1], Zimba J. J.[2], Utembe W.[1] and Kosamu I. B. M.[1]

[1]The Polytechnic, University of Malawi, Private Bag 303, Blantyre 3, Malawi.
[2]Dairibord Malawi Limited, P.O. Box 30647, Chichiri, Blantyre 3, Malawi.

This study assessed variations of ambient volatile organic compounds and carbon monoxide (CO) levels in Blantyre City, Malawi. Volatile organic compounds and carbon monoxide measurement was done using the Multi-Gas Monitor between August and December, 2010. Rapid urbanization and lack of standard enforcement on air pollution as well as lack of data on pollutant levels in the city prompted this study. The study focused on methane volatile organic compounds and non-methane volatile compounds besides carbon monoxide. Results of the analysis showed that non-methane volatile compound levels were highest at public bus terminuses with 12.40 ± 0.83 ppm, and lowest for industrial locations (2.16 ± 0.09 ppm). However, all points registered zero (0.00) ppm for methane volatile organic compounds. Similarly, mean levels of carbon monoxide were highest at public bus terminuses with 10.92 ± 0.74 ppm, and lowest for industrial locations (1.11 ± 0.13 ppm). With the exception of Ginnery Corner, all the sampling points had levels of both parameters within the air quality standards. The study revealed a spatial and diurnal variation of the pollutants in all sampling points. It further revealed that the sources of these pollutants are diverse with vehicular emissions being the main source for Masauko Chipembere highway.

Key words: Air pollution, ambient, permissible limit, urban, vehicular emission.

INTRODUCTION

According to the World Health Organization, about one-third of Africa's disease burden can be attributed to environmental hazards and, among these, air pollution is a significant contributor (Prüss-Üstün and Corvalán, 2006). Rapid urbanization in developing countries has led to increased air pollution due to increased vehicular and industrial emissions (Fenger, 1999; Akimoto, 2003). Examples of atmospheric pollutants include volatile organic compounds (VOCs) and carbon monoxide (CO). The use of fossil fuels (for example, gasoline and coal) is the dominant activity aspect that introduces abundant amounts of VOCs and CO into urban atmospheres (Chow and Chan, 2003; USEPA, 2010; Chen et al., 2011).

Several studies on urban air pollution have been carried out elsewhere (Mendoza et al., 2009; Makra et al., 2010) and impacts of such pollution on human health (Peters et al., 2000; WHO, 2003; Lu et al., 2004) and ecosystem (IPCC, 2007; Fowler et al., 2009) are on record. However, there is no published data on such pollution in Blantyre City and air pollution monitoring is not routinely done.

Although, complete control of air pollution is almost impossible, it is reasonable to expect control of emissions to the lowest possible levels consistent with available technology and within a reasonable cost (Khonje, 2004). Malawi has air quality standards although air quality

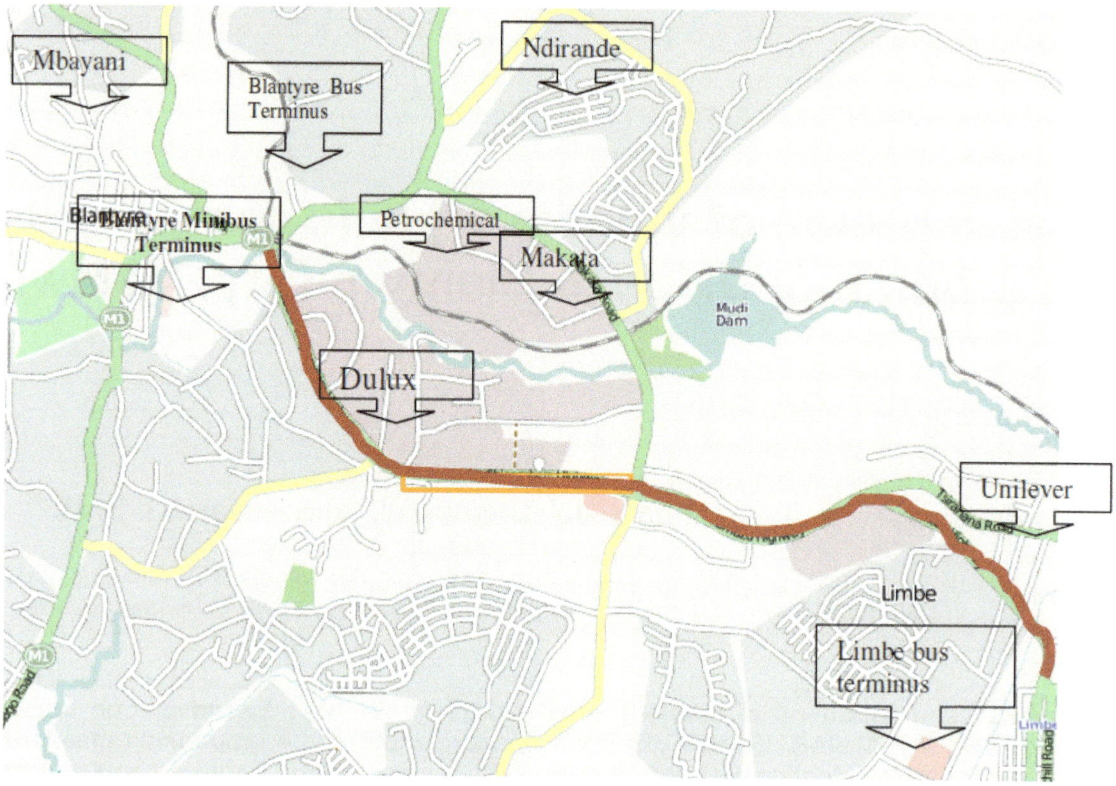

Figure 1. Main sampling points. Red line shows Masauko Chipembere Highway between Blantyre business district and Limbe (adapted and modified from openstreetmap (online)).

Table 1. Summary of meteorological variables obtained from the stationary monitoring system used by the Department of Physics and Biochemical Sciences located along the Masauko Chipembere Highway.

Meteorological variable	Aug	Sep	Oct	Nov	Dec
Temperature (°C)	20	23	25	25	24
Precipitation (mm)			21	83	125
Wind speed (m/s)	6.3	6.6	6.0	5.7	4.1
Relative humidity (%)	56	56	48	59	76

monitoring is rarely done. Besides, short-term air quality forecasting, air pollution monitoring and trend-predictions are very crucial because they help in anticipating preventive and evasive action during episodes of elevated air pollution (Makra et al., 2010). Blantyre City is one of such cities where such pollution needs scientific evaluation and monitoring. This study was therefore conducted in order to assess the extent of air pollution in Blantyre City.

MATERIALS AND METHODS

Study area: Blantyre City

The assessment study took place in major pollution prone areas within the city, namely: Masauko Chipembere highway, bus/minibus terminus, busiest residential areas and selected industries (Figure 1). The choice of sampling sites was based on anticipated source of emissions (vehicles) and large population. Blantyre City is located in the southern region of Malawi (15°47′10″S and 35°0′21″E). Data collection was done between August and December, 2010 (5 months).

During the study period, temperature and wind speed averages were measured by a stationary Air Monitoring System at the polytechnic as indicated in Table 1.

Sampling design and data collection

Data collection was done using an active mobile multi-gas monitor (Drager X-am 7000 (Online www.draeger.co.uk)) equipped with electrochemical sensors that measured continuous concentrations of VOCs and CO per 1 h 30 min session during the sampling period. In all the locations, the equipment was placed at 1 m from the ground to capture maximum emissions from both vehicles and industrial sources. The distance from the anticipated emission source was 1.5 m. Calibration of the instrument was done in fresh air which was free of gases to be measured and suspected to have interfering gases. During data collection, 10 min consecutive average measurements were recorded during morning (09:00 - 10:30), midday (11:30 - 13:00) and afternoon (15:30 - 17:00), providing 10 sets of values per session per day. This was aimed at ensuring accuracy and reliability of results.

Data analysis and statistical methods

Results were recorded, processed and analyzed using Microsoft Excel 2007 and SPSS version 16 for windows and results were tabulated. Multiple data collection as well as statistical quality control ensured that data was as accurate as possible. Graphs were used to illustrate spatial and diurnal variations as well as regression relationship in VOCs and CO concentrations.

Landau and Everitt (2004) elaborate the statistical procedures employed in this study. To verify whether there were significant spatial and temporal differences in VOCs and CO levels in the study areas, one way ANOVA at 0.05 significance level was used. The independent T-test was used to evaluate differences in levels of NMVOCs and CO between the two residential areas (Ndirande and Mbayani). In a special case, one sample T- test at 95% confidence level was performed to ascertain whether the 2010 carbon monoxide mean values for Masauko Chipembere highway were higher than those of 2004 (Khonje, 2004) and also whether they were higher than permissible levels.

RESULTS AND DISCUSSION

Levels of VOCs and CO

Table 2 presents a summary of NMVOCs and CO concentration levels. The results indicate mean day values split into morning, midday and afternoon. The standard error of mean is included to indicate the reliability of results. However, the summary excludes values of methane simply because all study locations did not register any methane VOCs.

Diurnal and spatial variations in VOCs and CO levels in main locations

Observations from Table 2 indicate a remarkable variation in diurnal mean values for all sampling locations with a general overview of highest mean values observed for midday period. The outcome of ANOVA analysis indicates a general significant spatial and temporal (diurnal) differences in mean concentrations of NMVOCs and CO in all sampled locations ($p < 0.05$, $F = 11.350$, $F = 11.600$, respectively). A least square deviation analysis however showed that there were no significant differen-ces in mean values between morning and afternoon periods of the day in all sampled locations ($p > 0.05$). On the other hand, midday periods for Masauko Chipembere Highway had the highest mean values for both parameters. The phenomenon can be attributed to high traffic intensity that emits large quantities of NMVOCs and CO (Baumbach et al., 1995). Figure 2 presents a picture of the diurnal mean variations in both parameters observed for all sampling locations. Clearly, the midday values are higher than morning and afternoon for bus terminus and Masauko Chipembere Highway.

According to Figure 2, the minibus and bus terminals had the highest concentration of NMVOCs and CO when compared with the rest of the sites except for midday rush hour due to high intensity of traffic in the main highway. The lowest recorded mean values for both NMVOCs and CO levels were observed for industries along the highway followed by Mbayani area. Diurnal variations are distinctly clear for Bus terminals and highway (Figure 2) unlike the other sites.

In general, for these locations, morning concentrations were lower than midday and afternoon (Figure 2). This exposes the fact that NMVOCs and CO concentrations may be accumulating in the atmosphere, depending on the time of the day. On the other hand, the decrease in concentration in the afternoon as compared to mid-day concentrations is indicative of the atmospheric auto-epuration capacity. A special observation is made for the industries along the highway where concentration levels indicate highest mean value in the afternoon period as compared to mid-day and morning session (Table 2 and Figure 2). This is simply due to the operation mode of industries since most of them operate in continuous production mode peaking close to mid-day and resume after lunch.

The independent T-test confirmed that the mean NMVOCs and CO levels for Ndirande and Mbayani residential areas were significantly different ($p < 0.05$). In general, Mbayani residential area had higher morning and afternoon mean NMVOCs record as compared to Ndirande while, the mean CO values were found to be significantly higher in Ndirande as compared to Mbayani residential area.

Diurnal and spatial variations in VOCs and CO levels in Masauko Chipembere Highway

Figure 3 is a map of Masauko Chipembere Highway stretching from Limbe to Blantyre business district, while Figure 4 illustrates the diurnal and spatial variations in mean NMVOCs and CO concentrations in the highway.

Sampling point CH4 (Ginnery Corner) had extreme values according to Figure 4. Ginnery corner is located close to the main referral hospital (Figure 3), a situation that is alarming if these emissions drift and find their way into the wards especially during lunch hour. This may pose a serious health hazard. The highway is the central dual carriage route where most of the vehicles pass through from Limbe to Blantyre and vice versa making it the route with the highest density of traffic passing through during morning, lunch-time and afternoon hours. As such, the areas prone to highest concentration of vehicular emissions are intersections (especially roundabouts) and where small minibuses stop to pick up passengers commuting between Limbe and Blantyre. It is thus reasonable to attribute the concentration of VOCs and CO to vehicular emissions and that the variations are due to traffic stopping and starting (passenger pickup points) and slowing down into an almost traffic jam at the roundabouts which are not wide enough to provide room for current large volumes of vehicles in the city.

Sampling point CH8 was observed to have the lowest concentrations of both NMVOCs and CO because the roundabout offers a very wide passage way for vehicles. Vehicles do not normally experience traffic jams at this point, as such, cruising through minimizes the rate of emissions in a given time unlike other sampling points. Between CH1 and CH3, both directions experience fast

Table 2. Summary results of NMVOCs and CO levels in Blantyre City. The results indicate mean levels and their standard error of mean.

Sampling point	NMVOC (ppm)			CO (ppm)		
	Morning	Midday	Afternoon	Morning	Midday	Afternoon
BT-Limbe (minibuses)	12.65 ± 2.02	8.40 ± 0.81	7.20 ± 0.81	9.90 ± 1.22	8.20 ± 0.84	14.90 ± 2.37
BT-Limbe (buses)	7.00 ± 1.09	20.30 ± 4.40	6.30 ± 0.88	6.30 ± 0.82	23.10 ± 4.71	4.30 ± 0.34
BT-Mibawa (minibuses)	6.55 ± 0.94	10.40 ± 0.97	7.70 ± 1.32	6.45 ± 1.07	3.90 ± 0.38	10.20 ± 1.06
BT-Wenela (buses)	5.85 ± 0.84	11.15 ± 1.14	10.20 ± 0.89	6.30 ± 0.73	11.00 ± 1.21	11.00 ± 1.21
IN-Petrochemical	0.95 ± 0.12	1.70 ± 0.32	1.50 ± 0.28	0.00 ± 0.00	0.20 ± 0.20	0.00 ± 0.00
IN-Dulux	2.25 ± 0.49	1.25 ± 0.13	1.55 ± 0.19	0.30 ± 0.21	0.20 ± 0.20	1.00 ± 0.25
IN-Unilever	0.95 ± 0.35	1.15 ± 0.20	3.50 ± 0.34	0.70 ± 0.23	0.20 ± 0.20	3.35 ± 0.30
IN-Makata	0.95 ± 0.27	1.15 ± 0.15	2.15 ± 0.27	0.40 ± 0.27	0.20 ± 0.20	1.80 ± 0.42
Mb-Market	5.40 ± 0.58	2.90 ± 0.33	4.45 ± 0.56	4.90 ± 0.67	2.05 ± 0.41	4.20 ± 0.51
Mb-Ndalama	3.55 ± 0.35	1.75 ± 0.15	3.65 ± 0.45	1.80 ± 0.44	0.65 ± 0.28	2.40 ± 0.45
Mb-Zandeya	2.10 ± 0.43	0.90 ± 0.24	1.85 ± 0.30	0.45 ± 0.24	0.35 ± 0.21	0.70 ± 0.37
Mb-Gayisi	3.40 ± 0.19	3.75 ± 0.48	2.95 ± 0.42	1.30 ± 0.37	3.20 ± 0.33	1.30 ± 0.45
Mb-Anyamata	2.40 ± 0.22	2.55 ± 0.38	3.75 ± 0.54	0.30 ± 0.21	1.85 ± 0.43	4.20 ± 0.70
Nd-Market	5.40 ± 0.31	6.10 ± 0.74	5.35 ± 0.56	6.70 ± 0.92	6.70 ± 0.92	6.30 ± 0.84
Nd-Newlines	4.30 ± 0.63	0.75 ± 0.19	2.05 ± 0.24	0.20 ± 0.20	0.30 ± 0.21	2.45 ± 0.16
Nd-Chinseu	1.80 ± 0.13	5.15 ± 0.55	4.40 ± 0.83	4.00 ± 0.75	4.90 ± 0.62	5.70 ± 0.84
Nd-Zion	1.10 ± 0.23	1.20 ± 0.19	1.60 ± 0.24	0.30 ± 0.21	0.20 ± 0.20	0.00 ± 0.00
CH-Clocktower (CH1)	4.80 ± 0.34	15.55 ± 2.38	6.65 ± 0.95	4.00 ± 0.37	13.60 ± 2.00	3.95 ± 0.32
CH-Kristwick (CH2)	5.95 ± 0.53	5.45 ± 0.76	3.65 ± 0.35	5.70 ± 0.65	5.20 ± 0.80	3.30 ± 0.37
CH-QECH Roundabout (CH3)	4.75 ± 0.47	6.00 ± 0.83	5.60 ± 0.42	4.50 ± 0.40	5.80 ± 0.84	5.85 ± 0.42
CH-Ginnery Corner (CH4)	4.20 ± 0.60	64.20 ± 12.52	16.65 ± 5.07	3.90 ± 0.48	52.40 ± 12.52	9.10 ± 2.21
CH-Polytechnic (CH5)	7.80 ± 0.93	10.05 ± 1.08	5.35 ± 0.59	7.40 ± 0.93	10.05 ± 1.08	5.50 ± 0.69
CH-Shoprite bus Stage CH6)	11.70 ± 2.34	13.90 ± 1.11	5.20 ± 0.52	9.90 ± 1.88	12.90 ± 1.35	4.35 ± 0.51
CH-Stadium Roundabout (CH7)	4.85 ± 0.38	15.35 ± 3.39	9.05 ± 1.20	4.30 ± 0.42	12.80 ± 2.74	8.30 ± 1.30
CH-Maselema Roundabout (CH8)	4.40 ± 0.40	4.35 ± 0.25	5.30 ± 0.26	4.20 ± 0.39	4.00 ± 0.26	5.30 ± 0.28
CH-Yiannakis Roundabout (CH9)	10.15 ± 1.14	7.00 ± 0.87	16.50 ± 1.79	8.70 ± 0.94	6.75 ± 0.93	15.60 ± 1.55
CH-Hardware Stage (CH9)	5.40 ± 0.60	6.55 ± 0.69	9.10 ± 1.43	5.50 ± 0.62	5.95 ± 0.80	8.60 ± 1.54

Abbreviations preceding the sample site stand for main sampling locations: BT = Bus Terminus, IN = Industries, Mb = Mbayani residences, Nd = Ndirande residences and CH = Masauko Chipembere Highway. Morning = 09:00-10:30, midday = 11:30 – 13:00 and afternoon = 15:30 – 17:00.

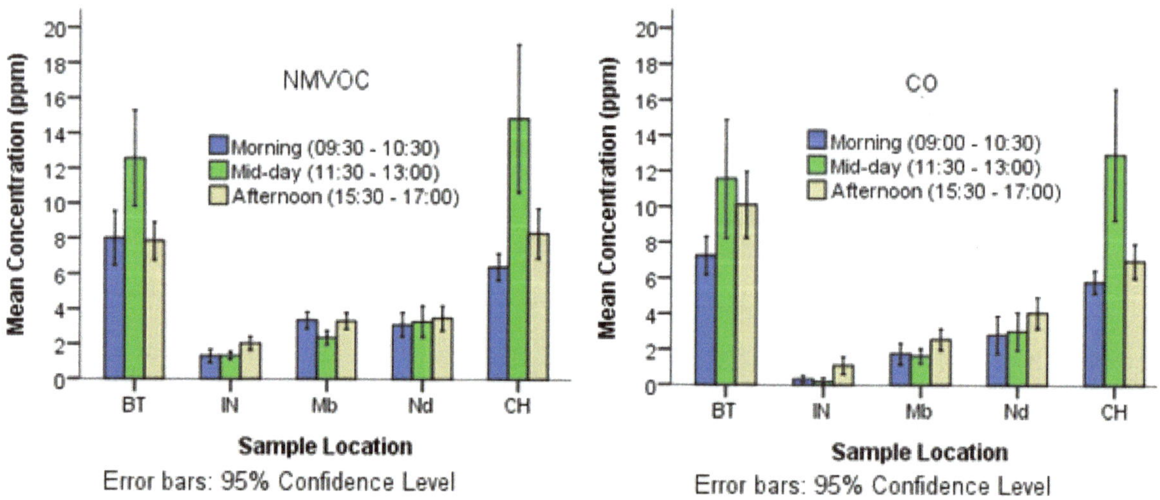

Figure 2. Plot of overall mean values of NMVOCs and CO levels in the main sampling locations. The graphs show diurnal mean variations for each sampling location. BT = Bus terminus, IN = Industries, Mb = Mbayani residential areas, Nd = Ndirande residential area and CH = Masauko Chipembere Highway.

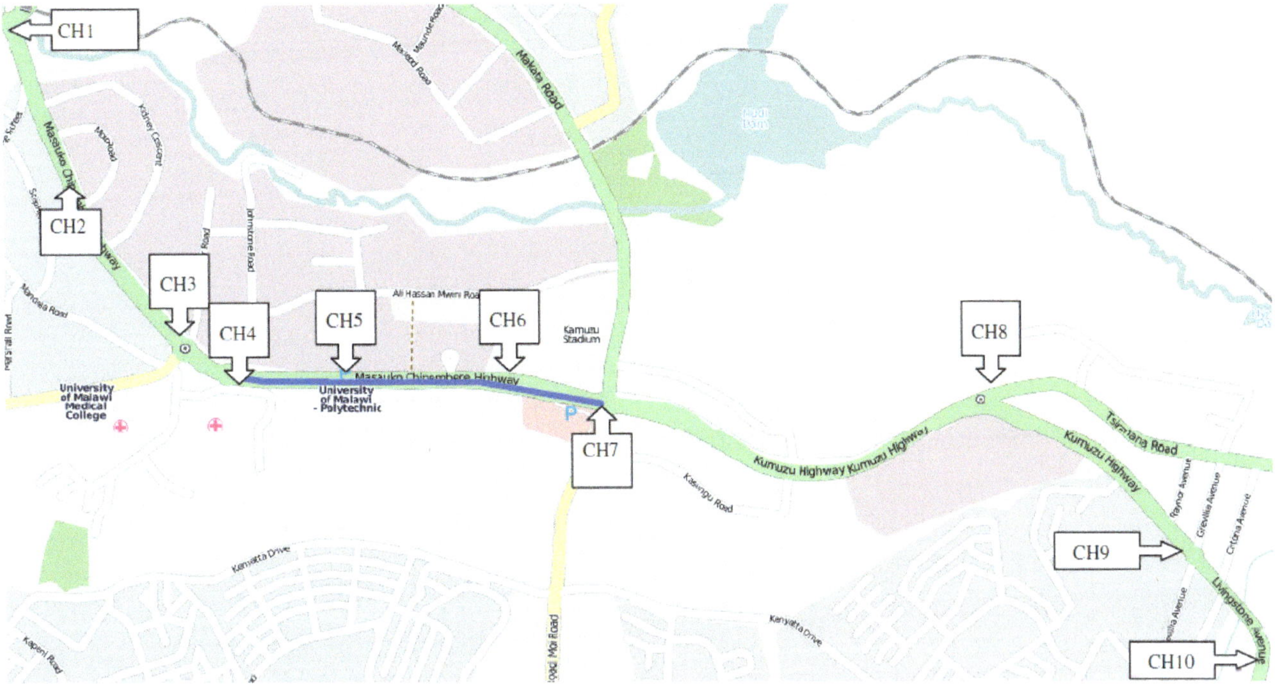

Figure 3. Sampling points along Masauko Chipembere Highway. CH1 = Clocktower roundabout, CH2 = Kristwick, CH3 = Queen Elizabeth Central Hospital Roundabout, CH4 = Ginnery Corner roundabout, CH5 = Polytechnic minibus stage, CH6 = Shoprite shopping mall stage, CH7 = Kamuzu Stadium Roundabout, CH8 = Maselema Roundabout, CH9 =Yiannakis Roundabout and CH10 = Hardware minibus stage (adapted and modified from www.openstreetmap.org).

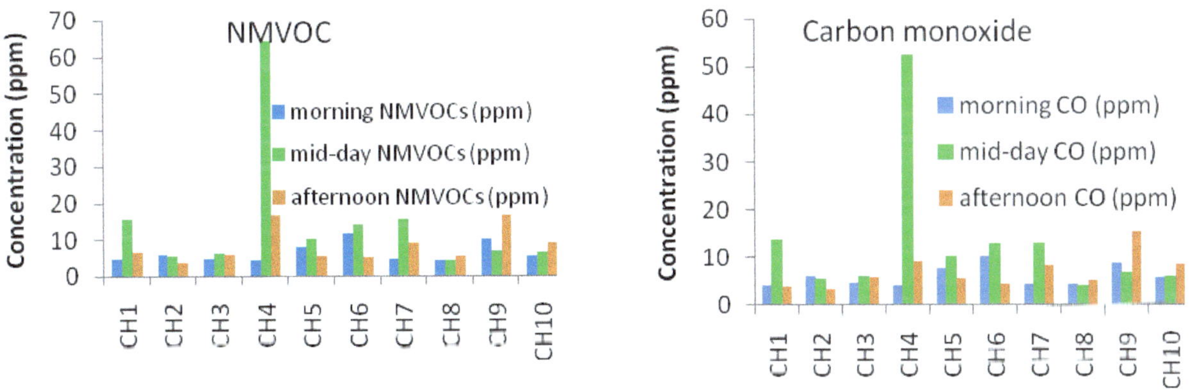

Figure 4. Graphs showing spatial and diurnal variations in NMVOCs and CO levels along the highway from Clock Tower to Hardware Minibus Stage (Limbe). CH1 = Clocktower roundabout, CH2 = Kristwick, CH3 = Queen Elizabeth Central Hospital Roundabout, CH4 = Ginnery Corner roundabout, CH5 = Polytechnic minibus stage, CH6 = Shoprite shopping mall stage, CH7 = Kamuzu Stadium Roundabout, CH8 = Maselema Roundabout, CH9 =Yiannakis Roundabout and CH10 = Hardware minibus stage.

moving vehicles since the volume of vehicles point to Blantyre or Limbe are reduced in density considering the fact that there are several roads that vehicles can take when moving from Blantyre to Limbe and rejoin the highway at CH3. On the other hand, between CH4 and CH8, the volume of vehicles increases due to joining in of more roadways into the highway at points CH3, CH4 and CH7. Between CH8 and CH10, concentrations are high due to the narrowness of the road since up to date the stretch is not yet developed into a dual carriage way.

Relationship between NMVOCs and CO levels

Figure 5 shows the relationship between NMVOC and CO concentration levels and elucidate whether the emissions

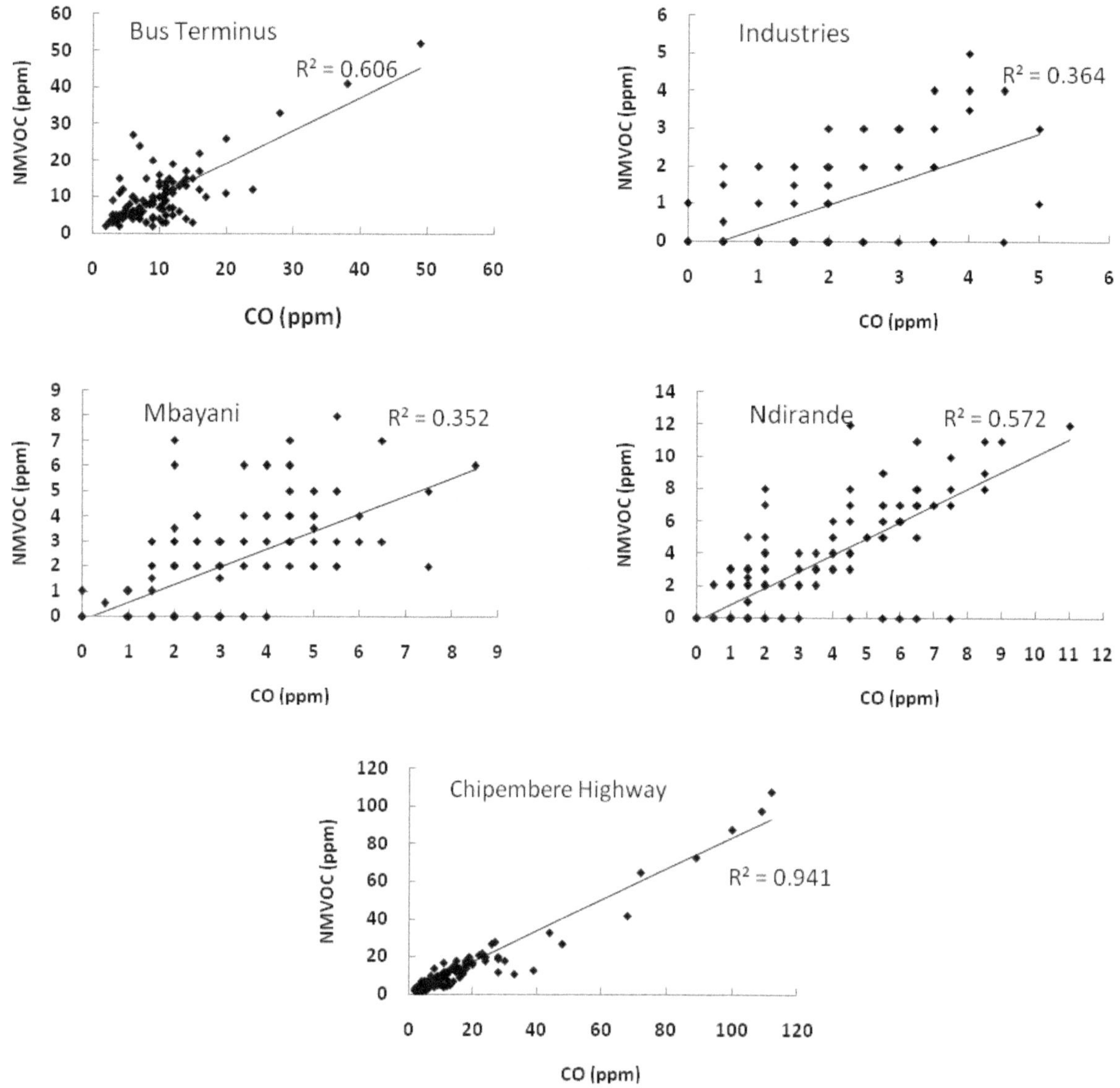

Figure 5. Relationship between VOCs and CO levels in the main sampling locations.

observed are coming from the same sources. In general, the sources, according to the figures, are different. However, for the Bus Terminus and Masauko Chipembere Highway, the cause is vehicular emissions considering the positive relationship exibited by the regressions (R^2 = 0.61 and 0.94, respectively). This is consistent with the fact that such pollutants are emitted from mobile sources everyday (Lee et al., 2002).

The source of emissions in the industrial area is the manufacturing processes that involve coal burning, solvent utilization and chemical manufacturing. This is why the regression coefficient is low for these sampling points which is a significant contribution to the fact that the sources in these industries depend on the process with coal burning contributing more of the CO while solvents and petrochemical product lines contributing more of the fugitive VOCs. Despite that, vehicular emissions also contribute to the concentration by wind movement since the industries are located in areas where vehicles pass through abait in lower volumes unlike that of the highway. Mbayani residential area has a very low vehicle traffic and is strictly a high density low housing area where you expect very few vehicles to contribute significantly to the emission concentrations. The main source of these emissions are various human activities and small scale business ventures that utilize charcoal and firewood as fuel. On the other hand, Ndirande residential area, despite having low volumes of vehicles, is a thriving business area where various small scale manufacturing, garages and furniture manufacturing is done. Besides, it is the area that has the fuel storage tank for the southern region where most of the incoming fuel (petrol, diessel and parafin) is stocked prior to distribution. As such, these activities together contribute to the levels of NMVOCs and CO. Despite a positive relationship, the sources are not necessarily the same in Ndirande.

NMVOC and CO levels compared to Standard Permissible Levels and 2004 Study

The comparison was made only for CO due to the fact that, of the two parameters, only the standards for CO exist in Malawi and also that the previous study concentrated on CO levels in Masauko Chipembere Highway. It was observed, however, that the mean CO values for all sites except Ginnery Corner were significantly lower ($p < 0.05$) than the standard 8 h permissible levels stipulated by Malawi Bureau of Standards (Malawi Bureau of Standards, 2005) and WHO (Colls, 1997).

Morning, mid-day and afternoon carbon monoxide mean levels for Masauko Chipembere Highway (5.81, 12.95 and 6.99 ppm, respectively) were then compared with the values (5.77, 5.24 and 4.95 ppm, respectively) obtained for the same study period in 2004 (Khonje, 2004). During the previous study, only CO levels were recorded and analyzed. The current CO levels are found to be significantly higher ($p < 0.05$) than those of 2004. Since, according to this study, CO levels in the highway are strongly correlated to NMVOC, same conclusion can be assumed. However, further research is required. This is a serious warning that there is need for consideration of means to reduce emissions in the city as well as safeguard citizens who utilize public transport from the harmful long effects of VOCs and CO. Evidence points to vehicular emissions as that consideration of reducing traffic jams and highways would be a very significant direction in minimizing pollution. Some areas close to hospitals require stringent measures to abate significant emissions.

Conclusion

This study did not discuss the influence of meteorological variables because at the time of the study, the facility to measure such variables on site where not available. However, it is anticipated that in the ongoing studies, such parameters will be budgeted for and incorporated into the analysis. Despite the fact that the study reveals a spatial and diurnal variation of atmospheric NMVOCs and CO levels in the busy areas of Blantyre City, vehicular emissions are the main sources of NMVOCs and CO levels in the Masauko Chipembere Highway, while in the rest areas is a combination of activities including vehicles. The study further reveals that vehicular emission of VOCs and CO in Blantyre City is increasing in the Masauko Chipembere Highway although the levels are still within the Malawi Bureau of Standards and World Health Organization, 8 h mean permissible limit of 10 ppm. With the current demand for car ownership and lessening usage of public transport in Malawi, it is time to put in place transport policies that will minimize traffic-related air pollution and safeguard public health.

ACKNOWLEDGEMENTS

This research was supported by the Department of Physics and Biochemical Sciences at the University of Malawi. The research was made possible with the donation of the multi-gas monitor by Environmental Affairs Department (Malawi). The authors extend their deepest appreciation to Mr. Albert Maliro for the technical support. The anonymous reviewers' input that greatly improved the scientific outlook of this paper is highly appreciated.

REFERENCES

Akimoto H (2003). Global air quality and pollution. Science 302 (5651): 1716 – 1719.

Baumbach G, Vogt U, Hein KRG, Oluwole AF, Ogunsola OJ, Olaniyi HB, Akeredolu FA (1995). Air pollution in a large tropical city with a high traffic density – results of measurements in Lagos, Nigeria. Sci. Total Environ. 169(1-3): 25 – 31.

Chen R, Pan G, Zhang Y, Xu Q, Zeng G, Xu X, Chen B, Kan H (2011). Ambient carbon monoxide and daily mortality in three Chinese cities: The China Air Pollution and Health Effects Study (CAPES). Sci. Total Environ. 409(23): 4923–4928.

Chow WK, Chan MY (2003). Field measurement on transient carbon monoxide levels in vehicular tunnels. Build. Environ. 38(2): 227 – 236.

Colls J (1997). Air Pollution: An Introduction. E & Fn Spon, London. p. 341.

Fenger J (1999). Urban air quality. Atmos. Environ. 33(29): 4877–4900.

Fowler D, Pilegaard K, Sutton M, Ambus P, Raivonen M, Duyzer J, Simpson D, Fagerli H, Fuzzi S, Schjoerring JK, Granier C, Neftel A, Isaken ISA, Laj P, Maione M, Monks PS, Burkhardt J, Daemmgen U, Neirynck J, Peronne E, Wickink-Kruit R, Butterbach-Bahl K, Flechard C, Tuovinen JP, Coyle M, Gerosa G, Loubet B, Altimir N, Gruenhage L, Ammann C, Cieslik S, Paoletti E, Mikkelsen TN, Ro-Poulsen H, Cellier P, Cape JN, Horváth L, Loreto F, Niinemets Ü, Palmer PI, Rinne J, Misztal P, Nemitz E, Nilsson D, Pryor S, Gallagher MW, Vesala T, Skiba U, Brüggemann N, Zechmeister-Boltenstern S, Williams J, O'Dowd C, Facchini MC, de Leeuw G, Flossman A, Chaumerliac N, Erisman JW (2009) Atmospheric composition change: Ecosystems-atmosphere interactions Atmos. Environ. 43(33): 5193–5267.

IPCC (2007). Climate Change. The physical science basis. Contribution of working group I to the fourth assessment report of the Intergovernmental Panel on Climate Change. Chapter 1, Historical overview of climate change. Cambridge University Press, Cambridge.

Khonje T (2004). Determination of Carbon monoxide levels in Masauko Chipembere Highway in Blantyre. BSc Thesis, University of Malawi, The Polytechic, Blantyre, Malawi.

Landau S, Everitt BR (2004). A Handbook of Statistical Analyses using SPSS, Chapman & Hall/CRC Press LLC

Lee SC, Chiu MY, Ho KF, Zou SC, Wang X (2002). Volatile organic compounds (VOCs) in urban atmosphere of Hong Kong. Chemosphere 48: 375–382

Lu W, Wang W, Wang X, Yan S, Lam JC (2004). Potential assessment of a neural network model with PCA/RBF approach for forecasting pollutant trends in Mong Kok urban air, Hong Kong. Environ. Res. 96: 79–87

Makra L, Mayer H, Mika J, Sánta T, Holst J (2010). Variations of traffic related air pollution on different time scales in Szeged, Hungary and Freiburg, Germany. Physics and Chemistry of the Earth, Parts A/B/C 35(1-2): 85–94

Malawi Bureau of Standards (MBS) (2005). Industrial emissions from mobile and stationary Sources- Specification. MS 737:2005. Malawi Standards Board, Blantyre.

Mendoza A, Gutiérrez AA, Pardo EI (2009). Volatile organic compounds in the downtown area of Mexicali, México during the spring of 2005: analysis of ambient data and source-receptor modeling. Atmósfera 22(2): 195-217

Peters A, Liu E, Verrier RL, Schwartz J, Gold DR, Mittleman M, et al. (2000). Air pollution and risk of cardiac arrhythmia. Epidemiology 11:11–7.

Prüss-Üstün A, Corvalán C (2006). Preventing disease through healthy environments. World Health Organization. Geneva (Online) http://www.who.int/quantifying_ehimpacts/publications/en/ accessed 20/04/2013

USEPA (2010). "Methane". Methane and Nitrous Oxide Emissions From Natural Sources. [Online] http://www.epa.gov/methane/sources.html (accessed 16th October, 2012).

WHO (2003). Health Aspects of Air Polution with Particulate Matter, Ozone and Nitrogen Dioxide (http://www.euro.who.int/document/e79097.pdf), pp. 98, EUR/03/5042688, World Health Organisation, Bonn, Germany.

URLhttp://www.draeger.co.uk/sites/en_uk/Pages/Industry/Draeger-X-am-7000.aspx.Accessed18.04.2013Openstreetmap. http://www.openstreetmap.org/?box=yes&bbox=35.02187%2C-15.80184%2C35.03575%2C-15.80085. Retrieved on 20-04-2013.

2

Ecosystem services and drivers of change in Nyando floodplain wetland, Kenya

P. J. K. Rongoei[1,4], J. Kipkemboi[1], J. B. Okeyo-Owuor[2] and A. A. van Dam[3]

[1]Egerton University, Department of Environmental Sciences, P.O. Box 536-20115, Egerton, Kenya.
[2]Chepkoilel University College, Constituent College of Moi University, P.O. Box 1125, Eldoret, Kenya.
[3]UNESCO-IHE Institute for Water Education, Department of Water Science and Engineering, P.O. Box 3015, 2601 Da Delft, Netherlands.

Papyrus wetlands in East Africa play a vital role in supporting livelihoods of people living around them. Although, subject to natural fluctuations and threats by anthropogenic activities, little is known about historical changes in wetland functions and services, or their present status. We focused on Nyando wetland on the eastern shores of Lake Victoria, Kenya. Three sites in the wetland were identified for assessment of history and current status. Changes during the past fifty years were assessed through participatory exercises with local communities and a review of published work. To establish the current status, we used field surveys and transect walks. Results showed that the wetland is important for hydrological and also ecological functions, which depend on the connectivity of the wetland with river and lake. The major direct drivers of change were hydrological regimes and livelihood activities. The main indirect driver of change was population growth, which leads to more pressure on wetland resources. Provisioning services are important in Nyando wetland but are generated at the expense of regulating services. Hydrology and livelihoods are strongly interlinked as flooding limits access to the wetland. Understanding the historical changes in wetland functions and services is important for rural communities, policy makers and for wetland managers in guiding, planning and wetland management.

Key words: Papyrus wetland, wetland ecosystem services, drivers of change, community perception, Nyando wetland.

INTRODUCTION

The benefits that humans derive from the use of ecosystems are now commonly referred to as ecosystem services (de Groot, 1992; MEA, 2005). A distinction can be made between ecosystem functions, defined as the capacity of natural processes and components to provide goods and services (de Groot, 1992) and ecosystem services, defined as the benefits that people derive from using the ecosystem functions (MEA, 2005; TEEB, 2010). Maltby (2009) defined three categories of ecosystem functions: hydrological, ecological and biogeochemical functions. The Millennium Assessment distinguished provisioning (food and materials provided by the ecosystem), regulating (water purification, flood protection), supporting and cultural (education, spiritual and recreation) ecosystem services. Growing pressure by human exploitation to maximize provisioning services of wetlands has led to the degradation of wetland functions and regulating ecosystem services (MEA, 2005).

Wetlands are important because of their position in the landscape between terrestrial and aquatic environments and their high productivity. The *inter alia* provide food and other materials, store water, improve water quality, sequester carbon and support biodiversity (Mitsch and Gosselink, 2007; Maltby, 2009). Despite their importance, wetlands are threatened by conversion to agriculture, industrialisation and human settlement. They face hydrological modification and pollution from human activities, including climate change. Although, the use of wetlands has contributed to human well-being and economic development, over-exploitation has degraded wetland functions, leading

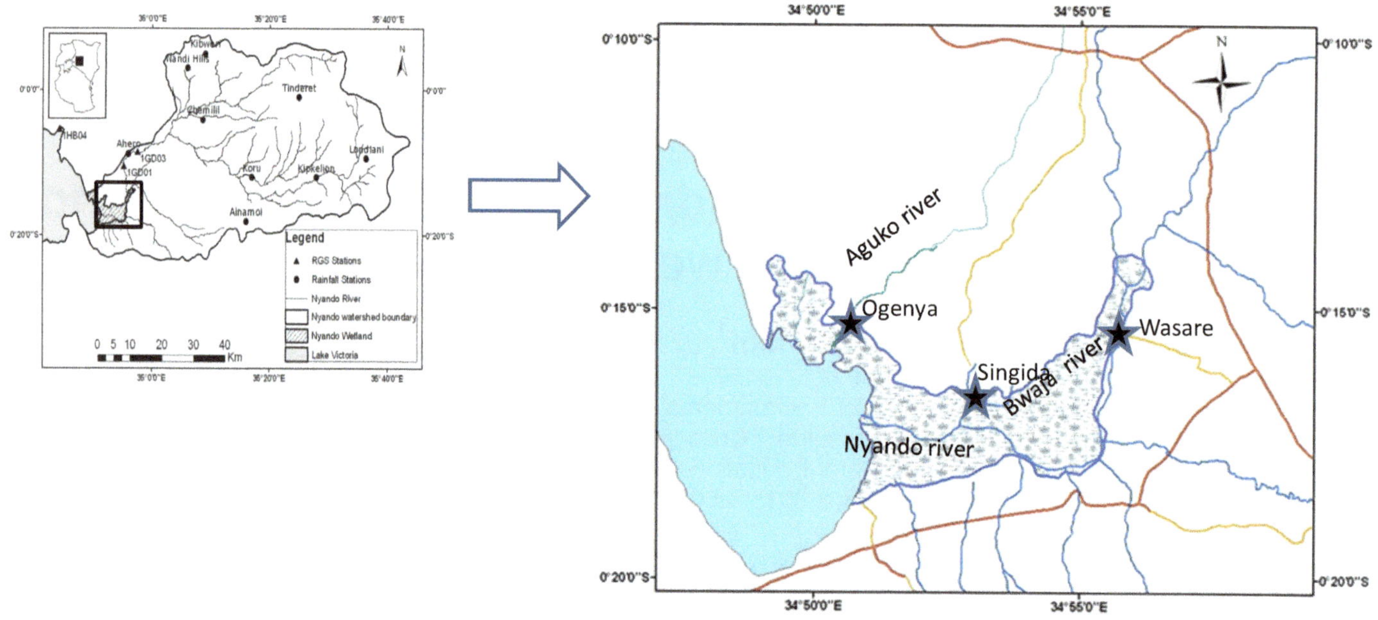

Figure 1. Location of Nyando floodplain wetland in relation to Lake Victoria in East Africa. The transect sites are shown with a black star- Ogenya, Singida and Wasare, while the dotted part is the wetland and the blue colour is the Nyanza Gulf of lake Victoria (Courtesy of P. Khisa).

to increased poverty and loss of biodiversity (Wetlands International, 2009). Improving sustainable exploitation, conservation and "wise use" of wetland ecosystems is needed (MEA, 2005; Finlayson, 2012).

Cyperus papyrus (common name: papyrus) is an emergent macrophyte that occurs in riverine, lacustrine and floodplain wetlands across mid-latitude from the Okavango Delta in southern Africa through central and eastern Africa with isolated occurrences in Egypt, Israel and southern Italy. Papyrus is a tall sedge growing to five meters on average and is known for its rapid growth and dense monoculture stands (Muthuri et al., 1989; Saunders et al., 2007). Papyrus wetlands support livelihoods of millions, providing food, construction materials, clean water and other benefits (Kipkemboi et al., 2006; Swallow et al., 2009). Large areas of papyrus wetlands occur in the Lake Victoria basin in East Africa, and the littoral and floodplain wetlands of the lake are dominated by papyrus. The papyrus vegetation either occurs as rooted in the seasonally flooded areas on the landward side of the wetlands or as floating fringes at the edge of the lake. In the past 50 years, wetland area around Lake Victoria has undergone tremendous change owing to considerably lake level changes due to uncertainty of rainfall events, human encroachment and intensified land use (Owino and Ryan, 2007).

The research project "Ecology of Livelihoods in East African Papyrus Wetlands" (ECOLIVE) integrated studies of hydrology, ecology and livelihoods in the Nyando wetland, aiming at the development of a framework for sustainable wetland management (van Dam et al., 2011). While the study reported here focused on ecosystem functions and services, two related studies focused on the hydrology (Khisa et al., in prep.), livelihoods and governance (Nasongo et al., in prep.) of Nyando wetland. This paper aims at describing and reviewing past and current status of ecosystem functions and services in Nyando wetland in preparation for more in-depth studies. Specific objectives were to: (1) describe the current ecosystem functions and services of Nyando wetland and changes of these during the past 50 years; (2) identify the key drivers of change in the Nyando wetland; and (3) define the main gradients of hydrology and exploitation in the wetland and identify transects for further research.

MATERIALS AND METHODS

Study site location and description

Nyando wetland is located in Nyanza Province, western Kenya, and at the mouth of Nyando River at Nyanza Gulf (on the eastern part of Lake Victoria, East Africa). Administratively, the wetland is part of Nyando, Nyakach and Kisumu East districts which according to the new constitution of Kenya fall in Kisumu County. It stretches from 0°11'- 0°19'S to 34°47'- 34°57' E (Figure 1). Reports of its size vary from 3000 ha in the dry season to 5000 ha in the wet season (Kipkemboi, 2006; Mwakubo and Obare, 2009). In the 2008 dry season, it was estimated at approximately 4212 ha (Khisa et al., in prep.).

The mean annual rainfall for the Nyando wetland is 1184 mm with an estimated mean temperature of 23°C (Muthusi et al., 2005; World Agroforestry Centre, 2006). Rain is divided into March-May ("long rains") and October-December ("short rains") periods. During these wet seasons, flooding of Nyando River creates connectivity between river, wetland and lake. The soil in this area is poorly drained and rich in organic matter (Jaetzold and Schmidt, 1982) which influences crops grown and the distribution of natural vegetation. The human communities living in the floodplain is predominantly the Luo ethnic group who practice subsistence farming, livestock herding and fishing (Kipkemboi et al., 2006).

For this research, three sites were identified based on hydrological characteristics and rural livelihoods activities. Singida (0°16'S and 34°53'E), about 3 km east of Ogenya (Figure 1), is influenced by the Bwaja stream that joins Nyando River before entering the lake. This site becomes inundated during the wet season, with water levels up to 79 cm above the soil surface, but dries up in the dry season. The vegetation is dominated by *C. papyrus* mainly rooted and *Ipomoea wrightii* with *Vossia cuspidata* along the edges of the Bwaja River. On the landward side, natural vegetation has been replaced with crops.

Ogenya (0°16'S and 34°51'E) at the shore of Lake Victoria is influenced directly by wind related positive water balance (seiche effect) that inundate the littoral zone. The drying and wetting of the papyrus vegetation at the edge of the wetland makes this site different from the others. Papyrus vegetation comprises of the rooted zone towards the terrestrial zone and a floating mat at the lake fringe. Water levels above the soil surface fluctuate on a daily basis between 20 and 65 cm depending on the magnitude in the seiche effect associated with the wind action blowing in the lake surface. There is a distinctive zonation of plants from floating *Eichhornia crassipes*, *Pistia stratiotes* and *Nymphaea sp.* to emergent *Vossia cuspidata* at the edge of the lake and along Aguko channel. *C. papyrus* stands float at the lake edge and are rooted at the landward side in association with patches of *Phragmites* sp. and *Typha* sp., including a strip of *Sesbania* spp. and *Triumffeta* sp. The wetland-terrestrial interface and abandoned farms are characterized by occurrence of *Cyperus latifolius* and *Cynodon dactylon*. The spatial extent of the permanently and seasonally flooded zones varies with dry and wet seasons.

Wasare (0°15'S and 34°55'E), located approximately 15 km east from the edge of Lake Victoria, is influenced by annual flooding of Nyando River. Unlike the other two sites, the gradient rises rapidly on the landward site restricting the river into the floodplain. The water depth above the soil surface ranges between 0 and 45 cm. The floodplain is dominated by trees such as *Acacia seyal, Balanites aegyptiaca, Euphorbia* sp. *Thevetia peruviana* and *Eucalyptus* sp. In the wet season, *Cyperus papyrus* dominates the wetland in association with *Ipomoea aquatica*, *Vossia cuspidata* and *Polygonum* sp. However, plant composition changes in the dry season to species like *Leersia hexandra*, *Amaranthus spinosa* and *Cyphostema babusoti* after clearing and burning of the wetland ecosystem. Since burning does not affect the rhizomes, papyrus will regenerate at the onset of rain enabling it to dominate again.

Data collection

Informal and formal sources of knowledge were used to describe the ecosystem services and their main drivers of change. Semi-structured community interviews were used to solicit local knowledge on the wetland and its changes while open-ended questions were asked to recall knowledge on changes that had occurred in the wetland within the past 50 years. Guided questions were used to obtain information on livelihood activities, hydrological regimes and biodiversity (flora and fauna). The number of persons interviewed was 15, 18 and 25 in Singida, Ogenya and Wasare, respectively. Both men and women between 30 and 80 years old were interviewed.

Information obtained during the interviews was supplemented and verified with results from research published in government reports and the scientific literature (CBS, 1999; Gichuki et al., 2001; Kipkemboi et al., 2006; Owino and Ryan, 2007; KNSB, 2009; unpublished reports of the Ministry of Agriculture for the years 2007-2011). Data on historical rainfall, river discharge and lake water level changes were obtained from Kenya Meteorological Department (KMD) through the Water Resource Management Authority (WRMA). Demographic information was obtained from CBS (1999) and KNBS (2009) census data reports of The Republic of Kenya. Based on the interviews and literature review, a list of functions and services and their changes in the last 50 years was established. To classify ecosystem functions, the approach used by Maltby (2009), which includes hydrological, biogeochemical and ecological functions was used. Ecosystem services were identified according to MEA (2005). Because of the overlap of regulating and supporting ecosystem services (MEA, 2005) with Maltby's ecosystem functions, only provisioning and cultural services were identified. The assessment of ecosystem functions and services was qualitative based on own observations and literature review. No attempt was made to quantify indicators or scores for ecosystem services as this was beyond the scope of the present study. This paper therefore focuses on historical to present knowledge on wetland resource use and provisioning ecosystem services.

Community members were asked to list livelihood activities in the wetland that were important to their well-being and rank them in terms of impact on their daily lives using a pair-wise ranking matrix (Jackson and Ingles, 1998; Asia Forest Network, 2002). After ranking, community members scored the three most important livelihood activities in their area during the last 30 years (1980-2010) the year in which these activities intensified in the wetland.

To define the hydrological and exploitation gradients, transect walks were done across each study site in the dry (March 2010) and wet seasons (May 2010) with the help of community members (Lightfoot et al., 1992). At each site, soil types, vegetation and crops, and livelihood activities were observed, photographed and recorded. The results were summarized in resource transect diagrams.

RESULTS

According to the community members, Nyando wetland historically was restricted to the narrow strip along the shores of Lake Victoria and along the river channels. However, with the heavy rains experienced in 1962-1963, the wetland expanded landward resulting in loss of croplands, settlements and even lives. To date, some members of the local community still believe that their land is submerged in water and part of the wetland, which they occasionally clear and grow their crops during severe drought. People's livelihood activities at the time included fishing mainly in the wet season and livestock herding and seasonal wetland cultivation in the dry season.

Ecosystem functions and services of Nyando wetland

Table 1 provides an overview of the current ecosystem functions and services of Nyando wetland. Of the hydrological functions, sediment retention with respect to the river load is important although there has not been quantification of this function in Nyando wetland. Due to its position in the lower floodplain extending to the river mouth, flood water retention and groundwater recharge

Table 1. Description and status of functions and services of Nyando floodplain wetland and the sources of information.

Category	Function/Service	Description and status	Source
Hydrological functions	Flood water detention	Wetland hold flood water and release it slowly to groundwater. However, frequent floods occur causing destruction	Muthusi et al., 2005; Ogutu et al., 2007; Awange et al., 2008b
	Ground water recharge and discharge	Some studies have shown that this wetland discharges water during dry season to maintain river flow. However, more studies are required to prove the extent of this function in Nyando.	Karicho, 2010
	Sediment retention	Papyrus can trap sediments and retain them before entering the lake. However, it is not clear if the wetland still has the capacity to retain as more sediments get to Lake Victoria.	ISRIC, 1997; World Agroforestry Centre 2006; Swallow et al., 2009.
Biogeochemical functions	Nutrient retention	Studies have shown that papyrus wetland is able to retain nutrients in plants and soil reducing nitrogen that cause eutrophication of the lake water. This function is still unclear for the Nyando floodplain wetland which is characterized by seasonal connectivity with River Nyando and Lake Victoria.	Mwaura and Widdowson, 1992; Boar et al., 1999; van Dam et al., 2007; Kelderman et al., 2007; Kansiime et al., 2007
	Insitu carbon retention	Evidence from other studies done on carbon retention shows that papyrus wetlands can be good in retaining carbon.	Gichuki et al., 2005; Saunders et al., 2007
	Trace element storage and export	Wetland may not function well in retaining trace elements including heavy metals as studies have shown that some of these elements were found in lake sediment, water column and fish tissues of Lake Victoria.	Muwanga and Barifaijo, 2006; Ongeri et al., 2009; Oyoo-Okoth et al., 2011
	Organic carbon concentration control	Evidence from studies show that papyrus wetlands are a source of organic carbon to the lake	Gichuki et al., 2005; Loiselle et al., 2008
Ecological functions	Habitat for biodiversity	Papyrus wetland is known for protecting various endemic bird species that have been known to be globally threatened or rare. They also are refuge areas to fish species, reptiles, amphibians and macroinvertebrates	Birdlife International, 2004; Maclean et al., 2006; Owino and Ryan, 2007; Gichuki et al., 2001
	Production, biomass and detritus for food web support	Papyrus wetlands are known to be productive systems in terms of biomass but also provide food for other organisms through high decomposition rate. The litter from plants also supports soil formation that enhances soil fertility of the wetland.	Boar et al., 2006; Mnaya et al., 2007; Terer et al., 2012
Provisioning services	Food	People use a variety of food products from the wetland such as papyrus rhizomes and *Typha* species' stems but also fish, wild vegetables and animal meat. The wetland is used to produce food crops, livestock grazing that give products like milk, meat, skin and dung to the rural communities	Geheb and Binns, 1997; Kipkemboi et al., 2006, 2007; Kibwage et al., 2008; Rongoei, pers. obs.

Table 1. Contd.

	Fresh water	Fresh water for both human and livestock use is gotten from wells, rivers and lake	Rongoei, pers. obs.
	Fibre, fuel and other construction materials	Papyrus culms are used to make mats, furniture, wall hangings, ropes, fish traps and roofing. Papyrus rhizome is used as fuel wood and umbel used as a broom	Kipkemboi et al., 2007; Kibwage et al., 2008; Osumba et al., 2010; Rongoei, pers. obs.
	Biochemical materials	Biochemical materials like medicinal plants for preventing illnesses are collected from the wetland. *Thevedia peruviana* is used as a repellent for mosquitoes in the homesteads. Not much is known about their properties	Kipkemboi et al., 2007; Kuga Ambiso, pers. comm.; Opiyo Yimbo, pers. comm.; Samuel Andega, pers. comm.
	Genetic materials	These resources have not been quantified in Nyando wetland though has potential in adding to the genetic diversity which is currently unknown.	
Cultural services	Spiritual and inspirational	Papyrus wetlands are used for spiritual cleansing and cultural rituals such as driving away bad spirits.	Kipkemboi et al., 2006; Kibwage et al., 2008
	Recreational and aesthetics	Swimming and fishing for locals occur. Wetland ecosystem and the surrounding has the potential for exploiting ecotourism industry because of its support to endemic and rare species that depend on papyrus vegetation	Kibwage et al., 2008
	Educational	Studies have been done in the wetland by universities, government departments and NGOs	www.unesco-ihe.org/ecolive

may be less important as there are no human communities downstream of this part of the river. Nutrient and carbon retention is important as shown by generally high vegetation biomass and peat formation in the permanent wetland zone (Jaetzold and Schmidt, 1982). The importance of the hydrological and biogeochemical functions depends strongly on the degree of connectivity of the wetland with the river. Connectivity seems to be limited for the Nyando wetland, which is flooded once or twice annually, and only for a period of approximately two weeks (Muthusi et al., 2005). It is not clear if the short period in which the river interacts with the floodplain allows for significant sediment retention in the Nyando wetland. This phenomenon is aggravated by the recent construction of dikes on both sides of River Nyando to restrict overtopping of the river. At the time of this study, most river sediment loads were deposited at the river mouth.

In terms of ecological functions, Nyando wetland is a habitat for a variety of flora and fauna. Thirty macrophyte plant species have been identified in the Nyando floodplain wetland (Rongoei et al., in review). These macrophytes form a habitat to the rare and endangered antelope Sitatunga (*Tragelaphus spekei*), and to papyrus endemic bird species such as the white-winged warbler (*Bradypterus carpalis)* and the papyrus gonolek (*Laniarus mufumbiri*) (Birdlife International, 2004). Papyrus biomass is high and can be over 8000 g m^{-2} in Ogenya wetland (Osumba et al., 2010).

Provisioning ecosystem services include food, water, fuelwood, building materials and herbs. Food products that are directly extracted from the wetland are wild vegetables, including *Solanum nigrum*; fish (main species include the lungfish *Protopterus aethiopicus,* local name "kamongo"; catfish *Clarias gariepinus,* local name "mumi"; and *Schilbe intermedius,* local name "sire"); wild meat from Sitatunga *(Tragelaphus spekei*), hippopotamus, birds and insects. In addition, there is farming in the seasonally flooded wetland sugarcane growing for commercial purposes mainly sold to sugar processing industries has been expanding into the macrophyte zone over the recent years. Papyrus biomass is harvested for making mats, furniture, wall hangings, ropes, brooms, fish traps,

Table 2. List of plant and animal species lost or reduced: year of change and their uses in Nyando floodplain wetland.

Scientific name	Common name	Local name (Dholuo)	Year lost	Reasons for loss/reduction	Use
Plants					
Aspilia pluriseta		Get	1960s	Change in climatic conditions: long dry spell	Firewood, medicine
Triumfetta cordifolia.	Burweed	Owich	1963	Flooding after "Uhuru" rains	Firewood, medicine, making robes
Sesbania sesban	Common sesban	Omburi	1980s	The demand for use in rowing boats, fencing, charcoal burning, firewood	Firewood, poles, posts for fencing, charcoal and boat rowing pole
Euphorbia tirucalli	Pencil plant	Ojuok	1990s	Replaced with fast growing *Thevedia sp.*	Live fences around homesteads
Aloë sp.	True aloe	Okaka	1997	El Niño rains	Medicinal, live fences
Ficus sycamore.	Sycamore tree	Ng'ou	2000	Tree felling for timber made easy by use of power saw	Furniture, firewood, building, fencing
Nymphaea sp.	Water lily	Oyungu	2002	Draining of water pans that used to occur at the edges of the wetland	Making straw for taking local brew "busaa"
Animals					
Crocodylus niloticus	Nile crocodile	Nyang'	1948	Flooding after "Uhuru" rains	Hunted for skin for making leather bags and belts
Redunca redunca	Bohor reedbuck	Mwanda	1980	Hunting and habitat encroachment	Meat
Python sebae	African rock python	Ng'ielo	1994	Drought which led to burning and clearing of wetland	Believed to bring rain and bumper harvests
Hylochoerus meinertzhageni meinertzhageni	Giant hog	Mbidhi	1997	El Niño	Meat
Ptemistis squamatus	Scaly francolin	Aywer	1997	El Niño	Meat
Ephippiorhynchus senegalensis	Saddle-billed stork	Rabala	1997	Habitat encroachment	Cultural

fencing and thatching. Other emergent macrophytes harvested for use by local people include *Phragmites* sp. (fish traps and fencing), *Vossia cuspidata* (livestock fodder) and *Cyperus* sp. (thatching). Medicinal plants collected from the wetland are used to treat and prevent ailments, including snake bites. These include plants such as *Cyphostema suaveolens, Solanum nigrum* (herb and vegetable), *Senna* sp. and *Sphaeranthus* sp. while *Thevetia peruviana* is used as a repellent for mosquitoes in the homesteads. However, there is little information on the biochemical characteristics of these plants some of which are known to have medicinal properties. In addition, Nyando wetland is important for domestic water supply and irrigation either through direct abstraction from the river and pools or shallow wells dug in the wetland periphery.

Nyando wetland plays an important role in the tradition and culture of the Luo community, who use the wetland for spiritual cleansing and rituals of sending away evil spirits (Samuel Andega-resident, pers. comm.). The wetland is also home to a reptile used for indigenous weather pattern prediction among the Nyakach community. The famous "omieri" python is mythically associated with flooding and bumper harvests in the region (Peter Guga - resident, pers. comm.). The local community uses the wetland for recreational fishing and swimming but tourism potential remains unexploited.

Table 2 presents a list of wetland plant and animal species considered by the local community to be reduced or lost between 1960 and 2010. The majority of changes were attributed to hydrology, notably rainfall patterns (El Niño); and to human encroachment into the wetland like

Table 3. Fish species perceived by community members as lost or reduced in number and the period when change was detected.

Scientific name	Common name	Local name	Year
Mormyrus kannume		Suma	1950s
Bagrus bajad	Bayad/Lisi	Seu	1950s
**Bagrus docmak*	Semutundu/Sew	Seu	1950s
**Barbus artianalis*	Ripon barbel	Fwani	1952
Brycinus sadleri		Osoga	1960s
**Brycinus jacksonii*	Victoria robber	Osoga	1960s
Mastercembelus frenatus		Okunga	1990s

* These species have been seen in the satellite lakes and fringing wetlands of Lake Victoria.

burning of wetland vegetation, farming in the wetland, hunting for wild animals and livestock herding. Increased encroachment is caused by the expansion of the local population but could also be attributed to the increase in people migrating to the floodplain from upland areas.

Overtime, plant species composition in the wetland has changed. Alongside the loss of native species, new plant species were introduced. One of the dominant of these species is *Thevedia peruviana* (local name: chamama), which is not a wetland plant but dominates areas around homesteads. This plant is used for live fences and, according to respondents, serves as an insect repellent. Other plants that respondents reported to have increased with time include *Eucalyptus* sp. (local name "mbao"), *Azadiractai indica* (local name "mwarubaini") and *Schefflera actinophylla* (common name: umbrella tree), which are used for construction (housing, boats), fencing, furniture, firewood and charcoal production. These plants are exotic and are not wetland plants but grown near homesteads and at farm boundaries. However, there are also introduced species like *Eichhornia crassipes*, *Pistia stratoites* and *Azolla* sp. which dominate open water lake, pools and river channels. Some new animal species have also been noticed, including African hare, *Lepus* sp. (local name: "apwoyo"; since 1994) and squirrel *Paraxerus* sp (local name: "ayidha"; since 2005).

The Luo community traditionally depended on the lake and wetland fisheries for protein, employment and income. However, respondents reported the reduction and disappearance of a number of fish species from the lake, some of which were delicacies for special occasions (Table 3).

Categories of drivers of change

Three main categories of drivers of change that influence ecosystem functions and services in the Nyando wetland were identified: hydrology, demographic changes and associated livelihoods activities in the wetland. The major hydrological event that shaped Nyando wetland was the El Niño event known as the "Uhuru rains" of 1962-1963 (Figure 2a). The Uhuru (Swahili word for independence) rains referred to as the heavy downpours that affected most parts of the Lake Victoria basin at the time when the Republic of Kenya was gaining its independence. This rain led to extensive inundation along the lake and in the river mouth floodplain and consequently the shift and expansion of once narrow papyrus vegetation to extend into a landward direction, forming the current Lake Victoria fringing wetlands. The raging floods resulted in destruction of property, loss of life and the migration of many people to higher areas in Miwani, Lambwe, Koru, Kibigori and Muhoroni. The respondents identified historical rainfall and drought events over the years which they believed to have shaped the wetland ecosystem. They suggested that before 1961, rainfall patterns were easy to predict as they fell between the months of March and May for long rains and in October to December for short rains. However, abnormal floods were experienced in the years 1962-1963, 1970, 1977-1978, 1984 and 1997-1998 while droughts occurred in the years 1961, 1980, 1985, 1993, 1997, 2003, 2004-2005 and 2007 (Figure 2a).

A result of the heavy rainfall was a change in the flow of River Nyando (Figure 2b) but it generally maintained two discharge peaks, in May and December. However, recently the river has shown a more irregular behaviour, e.g. with a discharge peak observed in the field in September 2010. Apart from rainfall and river discharge, Lake Victoria water level fluctuations (Figure 2c) have an important influence on the Nyando wetland, both on a daily and seasonal scale. On a daily basis, a rise in water level in the littoral zone floods the papyrus vegetation along the edges of the lake, resulting in a permanently flooded wetland zone. This pattern is due to tilting in the orientation of the lake volume in the lake level in the afternoons as a result of change in wind direction blowing over the lake. On a seasonal scale, the lake level is influenced by rainfall and river flow regimes in the Lake Victoria basin.

The second driver identified is the demographic patterns which triggers change in the wetland ecosystem services. Generally, the Kenyan population has increased by over

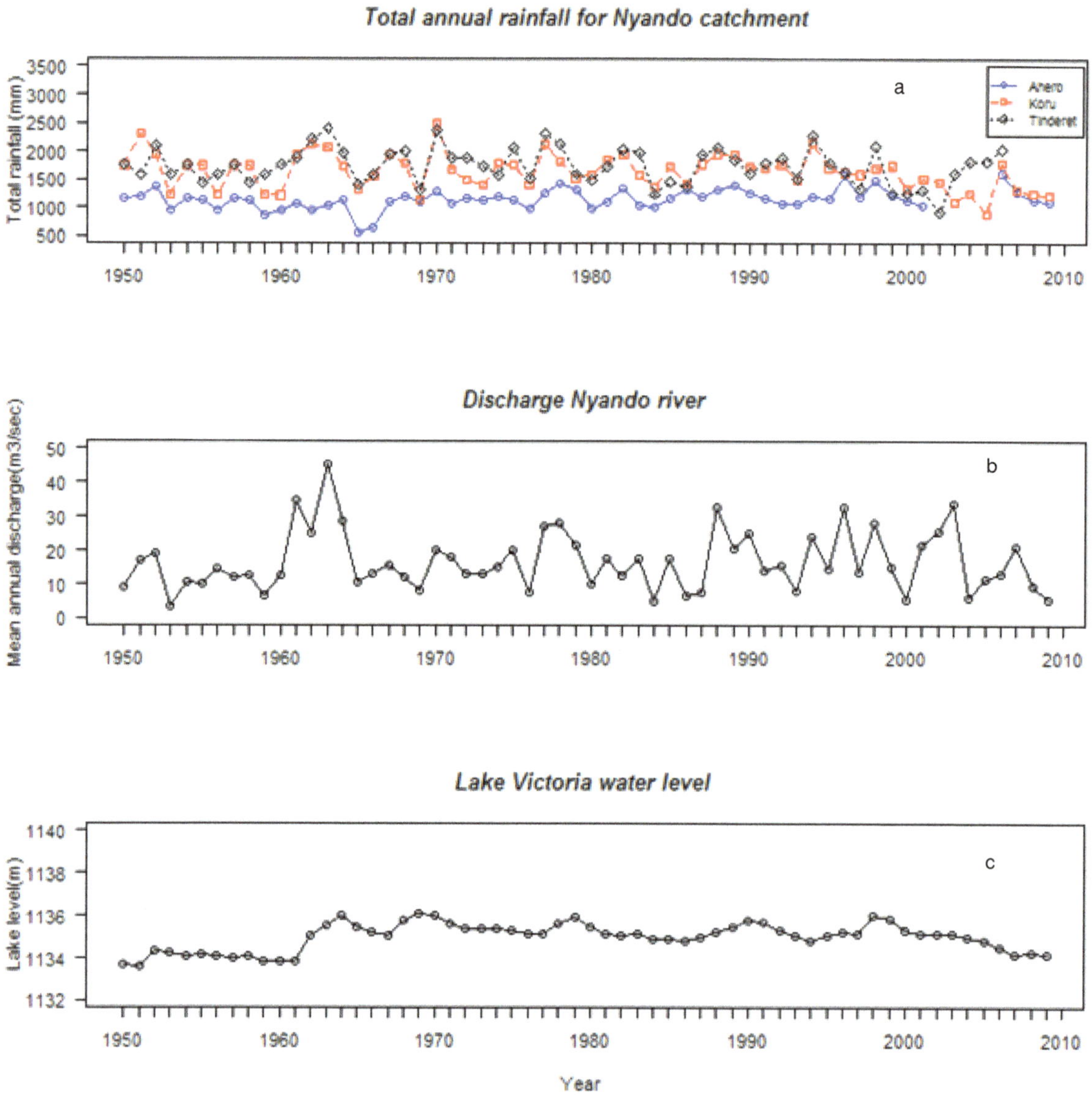

Figure 2. Hydrological regimes in the Nyando River Basin since 1950 – 2009; (a) total annual rainfall for Ahero, Koru and Tinderet stations in Nyando catchment from 1950-2009 measured on a daily basis, (b) River Nyando discharge at Ahero data showing mean of hourly measurements and (c) mean annual lake water level fluctuations at Kisumu. Data points represent annual totals for rainfall and annual mean for river discharge and lake water level.

10 million people by 1969 and reached over 38 million in the year 2009 with a population growth rate of 2.7% per year between 1999 and 2009 (KNBS, 2009). The Kenyan side of Lake Victoria comprises about 40% of the country's population (Odada et al., 2006). Nyando wetland has an average population density of 979 persons per km^2 with a high level of poverty (CBS, 1999). Between 1989 and 2009, human population increased steadily (Table 4). In addition, cattle population also increased steadily in the three transects between 2007 and 2011 (Table 5). The Luo rural communities still value their culture where cattle is turned into cash to pay school fees or dowries and buy food and other household goods. More so, due to increase in population and loss of income opportunities, more people have migrated to the wetland area in search of grazing land (Joel Ouma- resident Wasare, pers. comm.).

The last major category of drivers of change is related to livelihood activities. Table 6 presents the livelihoods activities identified by respondents in the three research sites, ranked in order of importance. Farming in wetland ranked first in Singida and was in the top three of livelihood activities in all sites. Papyrus harvesting and firewood collection were important in Wasare, livestock herding and fishing in Singida, and irrigation and livestock herding in Ogenya. A variety of other livelihoods activities

Table 4. Human population size in the study districts of Nyando floodplain wetland between 1989 and 2009 (before 1999 census, Kisumu East and Nyakach were administratively part of another district; therefore data was not included as it may not be representative).

Location	1989	1999	2009
Kisumu East	-	64,883	473,649
Nyakach	-	196,233	299,930
Nyando	48,914	64,311	350,353

Source: CBS (1999) and KNBS (2009).

Table 5. Cattle population trend in three districts within Nyando floodplain wetland between 2007 and 2011.

Location	Type	2007	2008	2009	2010	2011
Kisumu East	Dairy	8,598	9,728	9,894	10,440	2,123
	Zebu	54,000	57,348	65,000	70,147	49,000
Nyakach	Dairy	2,200	2,300	2,368	2,415	2,467
	Zebu	47,800	49,000	51,400	52,430	137,842
Nyando	Dairy	-	188	285	300	340
	Zebu	-	32,600	35,100	36,600	39,000
Total	Dairy	10,798	12,216	12,547	13,155	4,930
	Zebu	101,800	138,948	151,500	159,177	225,842
	All	112,598	151,164	164,047	172,332	230,772

Source: District Agricultural and Livestock Development Annual Report (unpublished).

Table 6. Pair wise ranking of livelihood activities in the study transects. "Score" is the number of times the activity scored "more important" than the other activities. "Rank" indicates the ranking of an activity within each transect. Overall ranks indicate the mean of the scores in the three transects; and the coefficient of variation (CV) of the means.

Activity	Singida		Wasare		Ogenya		Overall Rank	
	Score	Rank	Score	Rank	Score	Rank	Avg	CV
Wetland farming	7	1	9	3	7	2	2.0	50
Livestock herding	6	2	8	4	5	4	3.3	35
Water for irrigation	4	4	7	5	8	1	3.3	62
Firewood collection	1	7	10	2	6	3	4.0	66
Papyrus harvest	3	5	11	1	3	6	4.0	66
Fishing	5	3	5	7	4	5	5.0	40
Medicinal plants harvest	-	-	5	6	-	-	6.0	-
Fish farming		-	4	8	-	-	8.0	-
Sand harvest	-	-	3	9	2	7	8.0	18
Thatching grass harvest	2	6	1	11	1	8	8.3	30
Hunting	0*	8	0*	12	0*	9	9.7	22
Liquor brewing	-	-	3	10	-	-	10.0	-

*The activity was mentioned by communities in all transects but was not important to the livelihoods of the community. These activities were not mentioned as important in Ogenya and Singida transects.

were reported for all sites (Table 6). In the last 30 years, the contribution of these activities to the wetland community livelihoods increased (Figure 3). Farming is a common livelihood activity in all the three transects which gained its importance since 1995 when drought was experienced. As described previously in the hydrology, rainfall patterns

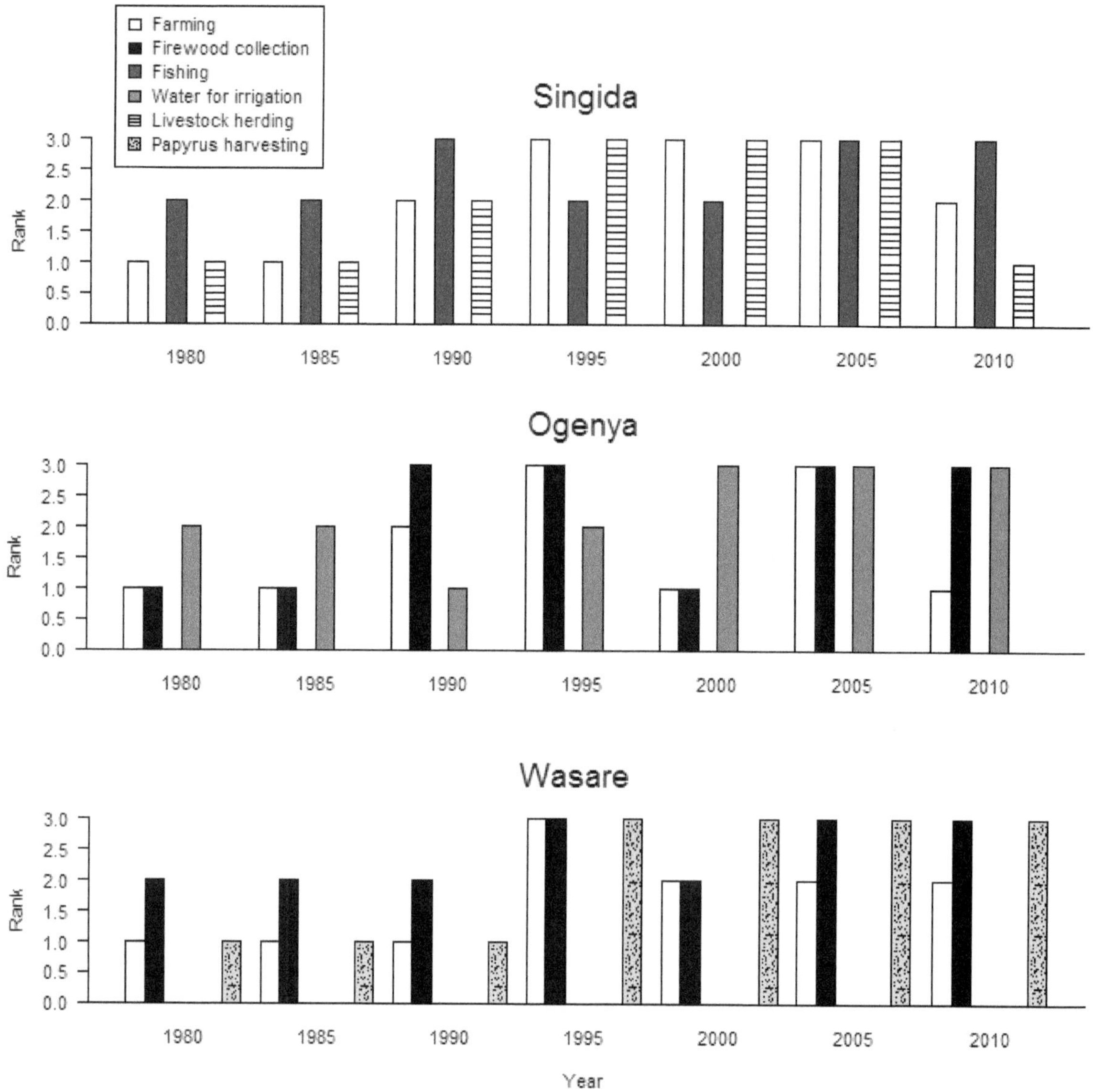

Figure 3. Trend analysis of the three major livelihood activities in Singida, Wasare and Ogenya (data covering the period between 1980 and 2010). White bars indicate farming in wetland which was common in all transects. The scores given are 1- low, 2- moderate and 3- high in importance in contributing to the livelihoods of the rural people. The legend abbreviations of fire_col stand for firewood collection; Irri_water stands for irrigation water while pap_harv stands for papyrus harvesting.

influence the trend of livelihood activities where high water level makes the farms in wetland inaccessible but dry condition opens an opportunity for people to extent their croplands into the wetland where there is enough soil moisture. Firewood collection was common in Ogenya and Wasare and gained its importance from 1990s. This was the time the respondents perceived that more trees had been cleared from the homesteads. Consequently, rural women had to look for firewood in the wetland where there were still intact shrubs and trees. Other activities like livestock herding, fishing and water for irrigation increased too as a result of increase in human population that led to demand for wetland resources. Papyrus harvesting was important too in Wasare as people discovered more uses of papyrus plant such as mat-making, furniture and wall hangings for income purposes.

Of the three transects, Singida had the highest flood inundation water levels in September with an average of 66 cm as compared to Ogenya (20 cm) and Wasare (33 cm) over the six month study period (Figure 4). These changes were influenced by the position of each transect in relation to the water body. The Singida transect is about 1700 m long and starts from the last homestead to Bwaja river running across farms and seasonally flooded areas and finally to the wetland. The land is divided into dry land, seasonally flooded land and a permanently

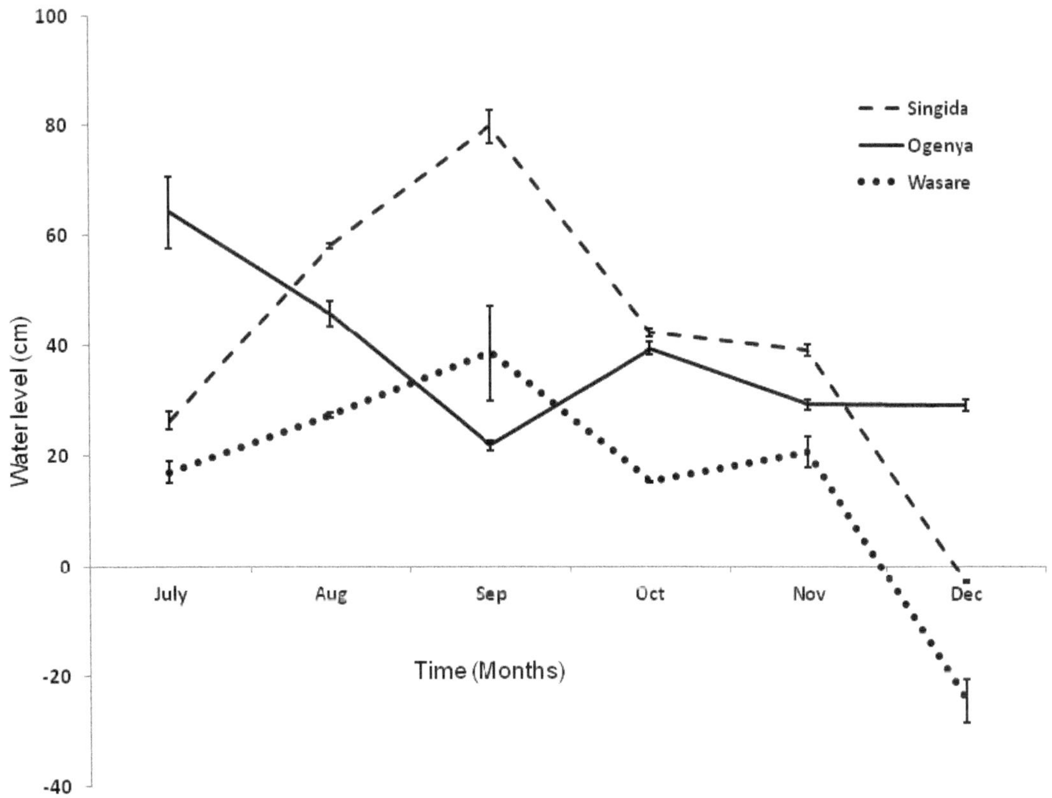

Figure 4. Line graph showing monthly average water depth measurements from July to December 2010 in the three study sites. Measurements were done above the soil surface in centimetres. Data points are the mean of three replicate measurements in the permanently flooded zone at each site. Error bars represent ± S.E.M.

flooded area (Figure 5). In the dry season, the dry land is characterized by a few livelihood activities such as mat making, thatching and trade, while in the wet season, there is farming of crops such as maize, sorghum and vegetables. The average farm size in the wetland is 2 ha, while that of dry land is 5 ha and the average household size was 8 persons. People keep domestic animals such as cattle, goats, sheep and chicken which provide them with meat, skin, milk and cash for the education of their children. In the seasonally flooded area, the dry season is characterized by maize and sugarcane farming, livestock herding, thatch and fodder grass harvesting, and fishing along Bwaja stream. In the wet season, flood waters associated with the onset of rainfall season in the highlands inundates the floodplain destroying maize and vegetable crops while sugarcane can tolerate flooding to some extent. The farms become inaccessible and farm activities are curtailed. However, fishing activity was observed within the wetland farms where fishermen set nets to catch species such as *Xenoclarias* sp. (local name - ndhira), *Synodontis afrofischeri* (local name - okoko rateng'), *Clarias gariepinus* (local name - mumi) and *Protopterus aethiopicus* (local name - kamongo).

The Ogenya transect is influenced by the daily changes in lake water levels and based on this study, extends for about 2 km, starting in the last homestead towards Ogenya beach and ending in Ogenya point on the lake edge. This transect is divided into permanently and seasonally saturated zones and the extent of inundation varies with season (Figure 6).

The dominant livelihood activities are papyrus harvesting, firewood collection, water for irrigation, farming, livestock herding and fishing and depend on the seasonal changes in the wetland. In the dry season, farmers expand their farms towards the wetland to benefit from the moist soil. Farm average size here is less than 1 ha and main crops are maize, sugarcane, cassava, sweet potatoes and vegetables, while in the wet areas of the wetland, rice and arrowroots are planted. In addition, papyrus harvesting occurs at the lake edge in both dry and wet seasons, using boats to access the sites in the flooded conditions. In the wet season, it was observed that farm areas are flooded, destroying crops such as vegetables and maize. During this time, fishing becomes important in the lake, in Aguko channel and in pools within the farms. Fish species commonly harvested in this transect include *Barbus cercops* (local name: adel), *Rastrineobola argentea* (local name: omena), *Oreochromis niloticus* (local name: nyamami), *Clarias gariepinus*, *Protopterus aethiopicus* and juveniles of *Lates niloticus*

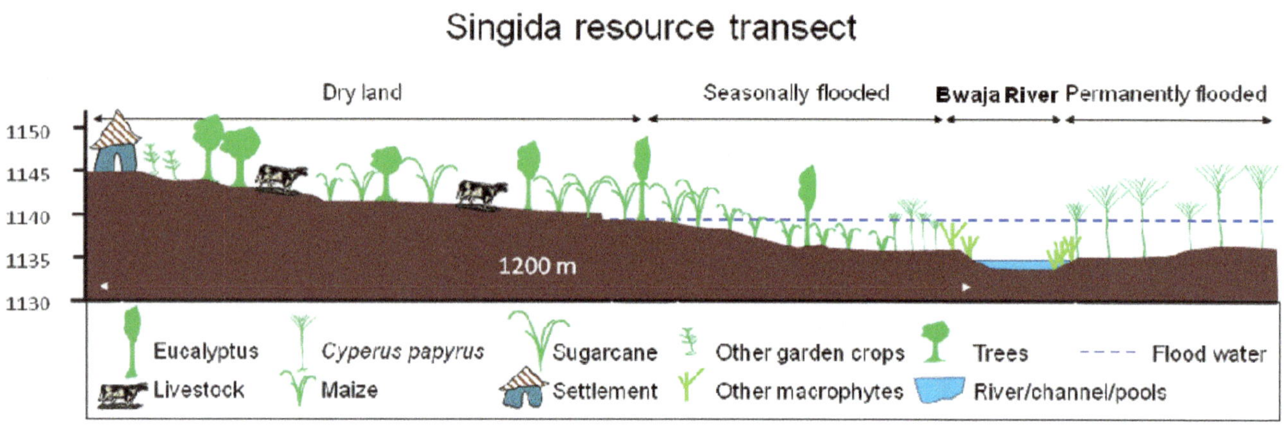

HGMU	Dry land	Seasonally flooded wetland	Permanently flooded wetland
Soil type	Black cotton soil and sandy soil	Black cotton soil	Black cotton soil / clay
Vegetation	Grass, *Makhamie* sp., *Thevedia peruviana*, *Eucaplytus* sp.	*Cyperus latifolia*, shrubs, grass, *Eucalyptus* sp., *Phragmites* sp., *Cyperus papyrus*, *Seabania* sp.	*Cyperus papyrus*, *Cyperus* sp., *Sesbania* sp. *Ipomoea whightii*, *Azolla* sp., *Pistia stratiotes*, *Phragmites* sp., *Vossia cuspidata*
Crops	Sorghum, maize, beans cow peas cassava, sweet potatoes	Sugarcane (wet), cowpeas and maize (dry)	-
Avg. farm size (ha)	5	0.1	
Avg. house-hold size (P/HH)	8	-	-
Permanent water source	Borehole	Borehole	Shallow wells, Bwaja river
Domestic animals	Cattle goats, sheep, chickens, donkeys	Cattle	-
Livelihoods activities	Farming, thatching, mat making, trade	Farming, thatch grass harvesting, thatching, fodder grass harvesting, fishing	Fishing

Figure 5. Resource transect map of Singida from the last homestead to River Bwaja. HGMU-Hydro-geomorphic unit depict different zones from dry land to wetland or open-water. Different characteristics are described for each zone where permanently flooded area may reduce to only the river channel in dry season.

(local name: mbuta).

The Wasare transect (Figure 7) is characterized by a flat floodplain from homesteads near the floodplain and across Nyando river to the edge of the papyrus wetland. Transect stretches approximately 2300 m from the homesteads near the floodplain and across Nyando River to the edge of the papyrus wetland. In the dry season, shrubs and grasses were observed in the floodplain and include the invasive *Xanthium pungens* and *Mimosa pigra* which the local community considers a nuisance to farming and livestock production. While *M. pigra* is utilised as firewood and browsed by goats, *X. pungens* covers the grassland and compromises the grazing land. The farms are divided into small plots of between 400 to 900 m^2 per person and rice is a predominant crop for most part of the year. Activities change in the wet season when the floodplain, papyrus wetland and river are connected and accessibility becomes difficult. In the dry season, livelihoods activities include sand harvesting from the river bed taking advantage of the low discharge in the river; harvesting of papyrus culms and livestock grazing in the wetland; and crop production (mainly vegetables), brick making and local liquor brewing are economic activities that depend on biomass energy from the wetland particularly dead papyrus and phragmites culms and rhizomes. Fishermen burn the wetland to expose aestivating *Protopterus* sp. and to hunt for the only aquatic antelope, commonly known as Sitatunga (*Tragelaphus spekei*) in dense papyrus area. Burning not only clears the senescing plant biomass but also allows regeneration of wetland vegetation for livestock grazing after flood recession. Human activities in the wet season include rice farming and fishing in the floodplain. Fishing becomes important as the floodplain is filled with fish such as *Synodontis afrofischeri* (okoko rateng'), *Synondontis victoriae* (local name: okoko rachar), *Clarias gariepinus* and *Protopterus aethiopicus.* When the flood water recedes, people prepare farms to grow rice, herd livestock on high grounds along the river levees and harvest papyrus.

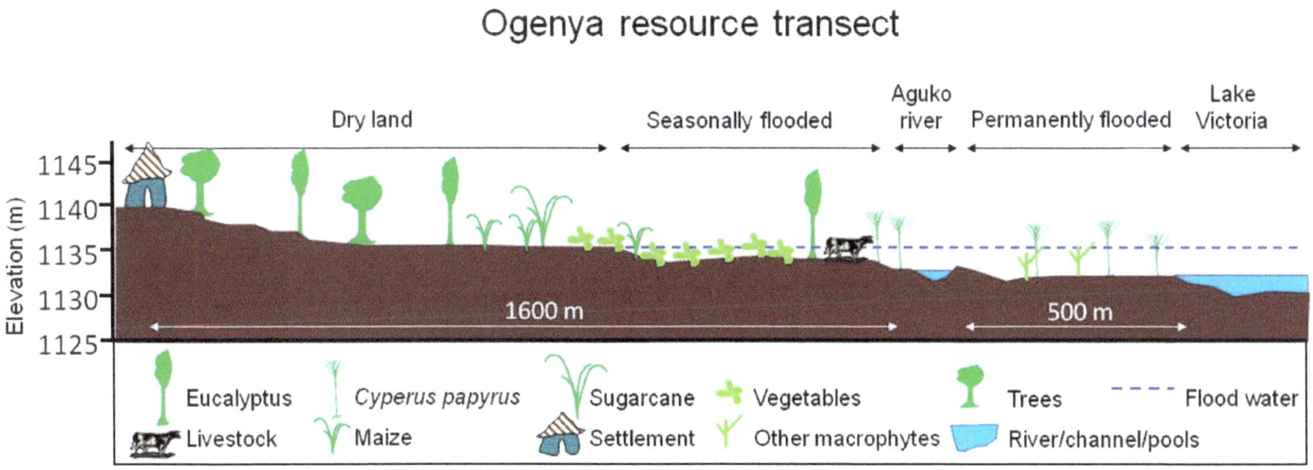

HGMU	Dry land	Seasonally flooded wetland	Permanently flooded wetland
Soil type	Black cotton soil, clay	Clay	Clay, sand, silt
Vegetation	Grass, scattered trees (*Makhamia* sp., *Eucalyptus* sp., *Thevedia peruviana*	*Cyperus papyrus, Cyperus latifolia, Vossia cuspidata, Pennisetum sp., Sesbania sesban, Phragmites sp., Typha sp.*	*Azolla sp., Nymphea sp., floating Cyperus papyrus, Vossia cuspidata, Eichornia crassipes, Trumpheta sp.*
Crops	Maize, beans, kales, cassava, sorghum, sugarcane, green grams, pumpkin	Maize, beans, kales, sweet potato, cassava, cow peas, tomatoes, sugarcane, banana trees, mango trees	Arrow roots, rice
Avg. farm size (ha)	2.0	1.0	0.12
Avg. house-hold size (P/HH)	10	-	-
Permanent water source	Borehole	Shallow well	Lake, Nyaguko river
Domestic animals	Cattle, goats, poultry	Cattle, sheep, poultry	-
Livelihoods activities	Farming, thatching, mat making	Mat making, papyrus harvesting, farming, fishing, trade	Fishing, papyrus and thatch grass harvesting

Figure 6. Resource transect of Ogenya from the beginning of the homestead near wetland to Lake Victoria at Ogenya Point. The lake water influences the permanently flooded zone on a daily basis which influences the livelihoods activities.

DISCUSSION

History of change in Nyando wetland

As far as the residents can remember, Nyando wetland existed before 1960 as a narrow stretch of papyrus vegetation along the lake shore and river banks. The floods of 1962-1963 resulted in extension of the wetland area towards croplands and settled areas. According to Khisa et al., in prep.), the wetland area rose from approximately 6000 ha in 1950 to 10,000 ha in 1973 and dropped to 4000 ha in 2008. Nyando wetland has continued to change as a result of natural and human pressures like changes in rainfall patterns and the demand for land for growing crops (BirdLife International, 2004; Owino and Ryan, 2007). According to the residents, changes in the ecosystem became more pronounced in the 1980s when the community intensified farming, livestock herding and papyrus harvesting. In the past, people depended more on fishing as a source of food and income. However, the community lost income from fisheries due to deterioration of water quality and the spread of water hyacinth in the lake (Juma, 1989; GoK, 2000).

Between 1960 and 2010, Nyando wetland has undergone two key changes in hydrological regime. The first change was the Uhuru rains in the period 1961-1964, coinciding with the commissioning of the Owen Falls dam in Uganda, leading to a 2.3 m rise in lake level (Institute of Hydrology, UK, 1984). With the expansion of the lake boundaries the surrounding wetlands, croplands, grazing areas and villages became permanently inundated. Respondents indicated that their ancestral land in the floodplain was lost and several families migrated to higher areas. To date some members of the local community still believe that their farms are submerged in the permanent wetland and the lake. They recall that rainfall in the basin had increased since the 1961 Uhuru rains, and that the frequency of extreme events; droughts and floods

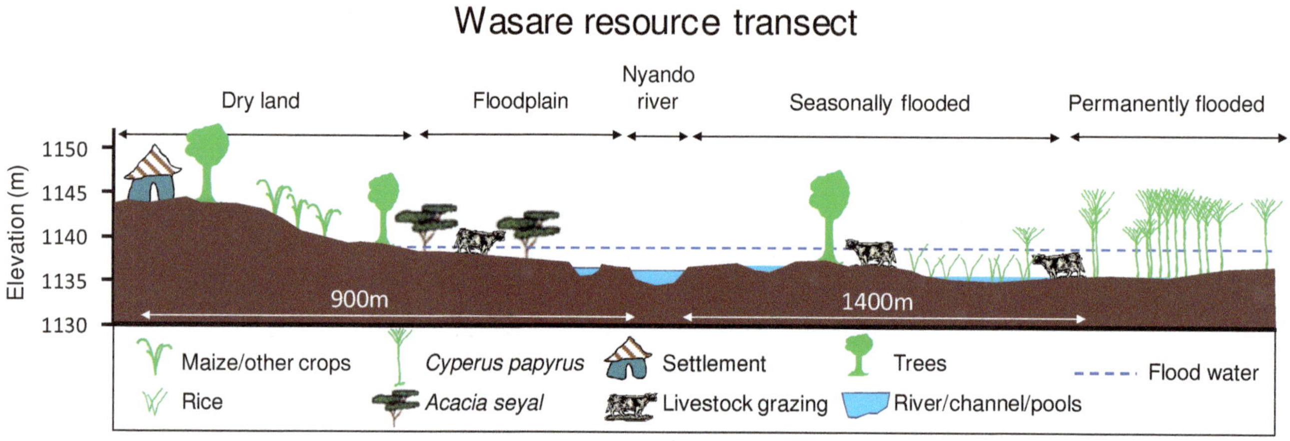

HGMU	Dry land	Floodplain & river channel	Seasonally flooded wetland	Permanently flooded wetland
Soil type	Sandy	Sandy/clay		
Vegetation	*Euphorbia* sp., *Casuarina* sp., sisal, *Eucaplytus* sp.	*Cynodon dactylon, Eucaplytus* sp., *Balaites aegyptica, Acacia* sp., other weeds	*Cyperus latifolia, Datura* sp., *Phragmites* sp., *Cynodon dactylon, Mimosa pigra*	*Cyperus papyrus, Cyperus* sp., *Ipomoea aquatica, Vossia cuspidata*
Crops	Sorghum, maize, groundnuts, green grams, cow peas, cotton, beans	Rice, cow peas, kales	Rice (wet), cowpeas (dry)	-
Avg. farm size (ha)	4	0.2	0.1	-
Avg. house-hold size (P/HH)	13	-	-	-
Permanent water source	Borehole	Borehole, river	Shallow wells, pools, river	-
Domestic animals	Cattle, goats, sheep, poultry, donkeys	Cattle, goats, sheep, donkeys	Cattle, goats, donkeys (dry)	Cattle (dry)
Livelihoods activities	Farming, mat making, trade	Farming, livestock grazing, fishing, sand harvesting, brick making, brewing	Farming, livestock, grazing, firewood collection, bird chasing	Papyrus harvesting, livestock grazing, wild vegetable harvesting

Figure 7. Resource transect of Wasare from homes near Kanyalwal primary school through floodplain and across Nyando River to the wetland with different characteristics and different zones.

have increased and more often they occur in alternation. Residents' recollections of dry and wet years coincide with the published data for the region which shows a shift in the rainfall patterns since 1961/62 (Farmer, 1981). Rainfall was about 8% higher in the 1961-1990 period when compared with the 1930 to 1960 period (Conway, 2002). Annual rainfall totals derived from long-term records of eight stations in the Lake Victoria Basin ranged from 1281 mm in 1949 to 2201 mm in 1961 (Institute of Hydrology, UK, 1984,1985).

The second major change during the last 50 years is the disruption of the regular seasonal flooding pattern of the wetland in the last 10 years, with delays in the onset of rainfall and an increase in duration and frequency of droughts (Ogutu et al., 2007; Awange et al., 2008a). In the early 1960s, hydrology of the wetland was characterized by more or less stable seasonal flooding from Nyando River, which spills into the floodplain over twice per year when the discharge is over 50 $m^3 s^{-1}$ (Muthusi et al., 2005). However, flooding has become both more frequent and erratic, leading to destruction of crops and infrastructure and loss of lives. During this study, we observed a delay in the onset of the long rains with a shift from March to July, and an unusually large flood in December of 2011. Changes in the mean daily flow of rivers discharging into Lake Victoria in the period 1990-2000 has been compared with the period 1950-1960 (Sutcliffe and Parks, 1999) and the results show increase in peak flow by 8, 31 and 38% recorded in Sio, Nzoia and Yala rivers, respectively (Sangale et al., 2005). The change in river flows has been attributed to the decline in vegetation cover in the upper reaches of the rivers by over 50% since 1960s (Melesse et al., 2008).

Apart from river flow, lake level fluctuations cause flooding of the wetland. The areas along the edge of the lake receive daily flooding from waves (local name: "ngeri"), creating a permanent wetland zone in which aquatic plants flourish even when the landward parts of the wetland are dry. Lake Victoria water levels fluctuate seasonally depending on direct precipitation and river flow. Lake water level is controlled mainly by direct precipitation which contributes approximately 85% (Flohn and Burkhardt, 1985; Sutcliffe and Parks, 1999). Between 1896 and 1960, the lake level was stable ranging from 10.2 to 12.0 m at the Jinja gauging station but increased to 13.85-14.46 m in 1961-1964 (Sutcliffe and Parks, 1999; EAC, 2006). Owing to prolonged drought between 1998 and 2005, lake water levels declined by 2.3 m and have not gone back to the 1961-1964 level (EAC, 2006; Swenson and Wahr, 2009). In addition, low lake level was experienced in 2004-2005 which led to loss of wetland area through extensive drawdown and human encroachment (Kiwango and Wolanski, 2008; Obiero et al., 2012). Findings from EAC (2006), Awange et al. (2008b) and Kiwango and Wolanski (2008) show that Lake Victoria water level fluctuation was influenced by over extraction of water at Kiira dam in Uganda. This was an extension project, which was constructed one kilometre downstream from the Owen Falls Dam now called Nalubaale Power Station to supply power to the growing industrial area of Jinja. As a result, the normal water level in the lake decreased as the demand for power increased. However, other factors may have contributed to the dropping of water level which include prolonged drought and demand of water by more than 30 million people living and depending on Lake Victoria.

Population growth influence both hydrology and exploitation indirectly. From a population of 8 million in 1960, Kenya's population is projected to reach 51 million by 2025 (GOK, 2010) which means more land is needed and degradation of the remaining natural forests in the water towers will continue. On the local scale of the wetland, population growth has contributed to a higher need for food production and income generation, and therefore a more intensive exploitation of the wetland. On a wider scale (at the level of the river basin), population growth has led to more intensive land use upstream, leading to deforestation and erosion (affecting both runoff and sediment loss), increases in water abstraction for agriculture and domestic use, and to flood control measures influencing the discharge of Nyando river. Population growth is also related to economic development. When economic growth does not keep pace with population growth, unemployment and poverty are the result. The poverty rate for the Kisumu region in which the Nyando wetland is located (percentage of population earning below US$ 1 per day) was 47.8% (KNBS, 2007). Poverty encourages encroachment and overexploitation of the wetland ecosystem by people with no alternative but to exploit wetland resources (fishing, biomass harvesting and farming) for their survival. The adaptation mechanism of switching between provisioning services (Geheb and Binns, 1997) no longer works as people strive strive to increase their resource exploitation efforts to meet their basic needs. Furthermore, traditional zebu cattle density has steadily increased when compared with dairy cattle which do not need to get to the wetland for herding (Table 5). This has aggravated the problem of degradation in the already stressed ecosystem. Livestock are regarded as savings among the local communities and hence, each individual has the desire to own more cattle as long as there is "free" land (wetland) for grazing.

Status of ecosystem functions and services

Traditionally, the livelihoods of the predominantly Luo community living at the edge of Nyando wetland revolve around provisioning services and depend on three main activities: fishing, farming and livestock herding. The increasing population pressure resulted in increased pressure on the fisheries, overfishing and an increased necessity for farming and other livelihoods activities (Geheb and Binns, 1997). The effect of this increasing pressure has been an increase in the provisioning services. Before 1980, activities of farmers used to be spatially separated in the terrestrial and wetland ecosystems. Depending on the season, livelihoods revolved around fishing in the lake (dry season) and farming or livestock herding in the terrestrial environment (rainy season) (Geheb and Binns, 1997). The changing hydrological regimes led to uncertainty in these provisioning services. Less stable rainfall patterns led to insecurity in dry land agriculture, and stimulated people to make use of the residual moisture in the wetland for crop production.

This study (Table 6) and others (Kipkemboi, 2006; Obiero et al., 2012) show that currently, wetland farming and livestock herding are the most important livelihood activities. Papyrus harvesting is an important provisioning service for earning household income and improving family's wellbeing. The trend of harvesting papyrus increased since the 1990s which coincides with the changes in the hydrological regimes described in the first part of this discussion. The activity becomes the only income earner when both dry and wetland farming have failed as a result of either drought or flooding events. With increasing population pressure, wetland agriculture and papyrus harvesting have become more intensive to the point of becoming commercial in nature. Other vegetation harvesting (*Phragmites* sp., *Vossia cuspidata* and *Cyperus latifolius*) and firewood collection are also important.

Contrary to the maximization of provisioning services, wetland ecosystem functions and cultural services have been declining. Maximization of provisioning services has led to a decline of biodiversity, nutrient and sediment retention and cultural services. While degradation of the ecosystem in terms of species disappearing is perceived by the people, changes in the biogeochemical functions are not immediately visible to them. Individual resource users cannot observe short-term changes in nutrient cycling or soil formation processes. These changes be-

become apparent through changes in water quality or flooding frequency but take place over a relatively long period of time. Yet in the long term, these important factors also affect their livelihoods. An example is Okana wetland in the Nyando floodplain where the community realised that they were poorer after the papyrus wetland had disappeared and consequently embarked on a restoration process (Morrison et al., 2012; Chairman, Okana community wetland group, pers. comm.). There is little data on the impacts of livelihoods activities on the regulating services such as water quality and quantity regulation and on supporting services.

The extent of impact on wetland ecosystem functions depends on the types of livelihood activities and the nature of the disruption. Some activities, such as construction of dikes, digging of channels for irrigation and uprooting of papyrus rhizomes cause structural and irreversible change. Other activities, such as limited burning or vegetation harvesting, is less detrimental for the biogeochemical and hydrological functions because they leave water flow and soil structure intact, and allows the fast-growing papyrus vegetation to recover during the next wet season. There is some evidence that papyrus harvesting may even enhance the nutrient retention function by exporting nutrients from the wetland and increasing the productivity of the plants (van Dam et al., 2007). However, it is likely that most livelihoods activities have a strong impact on the ecological functions of the wetland. Papyrus wetlands serve as a habitat and refuge for fish, birds and wildlife and disturbance, even if non-structural or temporary, can impact negatively on this function. Decrease in wetland area reduces habitat availability for other organisms and also products collected directly from the wetland such as wild vegetables, herbs, fuel and papyrus biomass. These effects have been recorded in other Lake Victoria wetlands like Yala (Thenya, 2006), Sondu Miriu (Gichuki et al., 2001) and Southern Ugandan wetlands (Maclean et al., 2006).

Hydrological and livelihoods gradients

The direct (hydrology and exploitation pressure) and indirect (population growth) drivers of change in Nyando wetland are strongly linked. Exploitation for livelihoods is strongly influenced by hydrological change, both in the long-term history of the wetland and on a seasonal basis. For example, Obiero et al. (2012) suggested that after lake recession in 2005, the contribution of farming to household income increased from 10 to 95%. Farming is done in the seasonal wetland after flood recession, but with more frequent and erratic flooding patterns, the risk of crop loss has been increasing in recent years. In the dry season, the intensity of livelihood activities increases through expansion of farms, burning of vegetation and trampling by cattle in areas with dense papyrus vegetation as access is possible. However, in wet season, livelihoods activities change mainly to fishing and papyrus harvesting. This has been observed in other papyrus wetlands like Naivasha (Harper and Mavuti, 2004) and Yala swamp (Thenya, 2006). Generally, flooding restricts exploitation, while flood recession makes the wetland accessible for livelihoods activities.

The three transects exemplify this strong link between hydrology and livelihoods in Nyando wetland. The three transects chosen for this study represented different combinations of hydrological regimes and livelihoods. Transects were divided into zones related to hydrology (permanent or seasonal inundation) and livelihood activities. The intensity of livelihoods activities depends on hydrology. Understanding the links between hydrology and exploitation pressures in a wetland influenced by wetting and drying is crucial for deriving wise use of wetland resources and wetland management plans. More research would be needed to elucidate the hydrology of the wetland and the impact of human activities upstream, such as water abstraction for irrigation and flood protection schemes, on sediment and nutrient retention. This is important as the growing pressures may interfere with the ecological functions of both the wetland and the lake and reduce services they provide to the local communities.

Conclusions

1. The main direct drivers of change in Nyando wetland were hydrology and livelihoods activities. Irregular rainfall patterns affect the seasonal flooding of the wetland influencing river flows and lake water level fluctuations that are associated with provisioning services. The most important livelihoods activities as perceived by local people were farming, livestock herding and papyrus harvesting. The main indirect driver of change was population growth which led to increased need for exploitation of the wetland.
2. Provisioning services (farming in wetland, herding of livestock and papyrus harvesting) have expanded in Nyando wetland at the expense of the wetland ecosystem (birds, fish, vegetation and soil). However, more research is needed to assess the impact of ecosystem degradation on the regulating and supporting ecosystem services.
3. Hydrology and livelihoods activities are strongly interlinked because flooding prevents people from accessing the wetland. Resource transects identified represent different combinations of hydrological conditions (affected by lake or river flooding or a combination of both) and associated livelihoods activities.

ACKNOWLEGEMENTS

This work was conducted in collaboration with the Victoria Institute for Environment and Development (VIRED International), Kenya. This project was funded by DGIS through UNESCO-IHE under the project "The Ecology of

Livelihoods in East African Papyrus Wetlands" (ECOLIVE). We are thankful to the community members from Singida, Ogenya and Wasare for their contribution. We cannot forget those who took us to the field and helped cross the flooded rivers in their boats. The authors are indebted to Joel Onyango who assisted in the collection of data and bridging the language barrier with the community; and to Serena Nasongo and Patrick Khisa for their collaboration in the field work.

REFERENCES

Asia Forest Network (2002). Participatory Rural Appraisal for Community Forest Management: Tools and Techniques. Asia Forest Network, January 2002. Santa Barbara, California, USA.

Awange JL, Ogallo L, Bae K, Were P, Omondi P, Omute P, Omullo M (2008a) Falling Lake Victoria water levels: is climate a contributing factor? Clim. Change 89: 281-297.

Awange JL, Sharifi MA, Ogonda G, Wickert J, Grafarend EW, Mulo MA (2008b) The falling Lake Victoria water level: GRACE TRIMM and CHAMP satellite analysis of the lake basin. Water Resour. Manage. 22: 775-796.

Bird Life International (2004). Threatened Birds of the World. Bird Life International. Cambridge, U.K.

Boar RR, Harper D, Adams CS (1999) Biomass allocation in *Cyperus papyrus* in a tropical wetland, Lake Naivasha, Kenya. Biotropica 31, 411-421.

Boar RR (2006). Responses of a fringing Cyperus papyrus L. swamp to changes in water level. *Aqua. Bot.* 84, 85-92.

CBS (1999). Population and housing census: counting our people for development Volume I. Central Bureau of Statistics. Ministry of Finance and Planning, Republic of Kenya. January, 2001.

Conway D (2002). Extreme rainfall events and lake level changes in East Africa: recent events and historical precedents. In: The east African Great Lakes: Limnology, Paleolimnology ad Biodiversity Advances in Global Change Research (EO Odada and DO Olago (Eds), Vol. 12, 63-92. Kluwer Dordrecht, The Netherlands.

de Groot RS (1992). Functions of Nature: Evaluation of Nature in Environmental Planning, Management and Decision-making. Wolters Noordhoff BV, Groningen, The Netherlands.

EAC (2006). Special Report on the Declining of water Levels in Lake Victoria. In: K Lubovich (2009). Co-operation and Competition: Managing Transboundary Water Resources in the Lake Victoria Region. East African Community Working Paper No. 5. USAID and FESS.

Farmer G (1981). Regionalisation and study of an alleged change in the rainfall climatology of East Africa. PhD Thesis, University of Sheffield, UK.

Finlayson CM (2012). Forty years of wetland conservation and wise use. Aqu. Cons: Mar. Freshwat. Ecosy. 22:139-143.

Flohn H, Burkhardt T (1985). Nile Runoff of Aswan and Lake Victoria: A case of a discontinuous climate time series Zeitschrift fur Gletscherkunde und Glazialgeologie 21, 125-130.

Geheb K, Binns T (1997). Fishing Farmers or Farming Fishermen? The quest for household income and nutritional security on the Kenyan shores of Lake Victoria. Afri. Aff. 96: 73-93.

Gichuki J, Guebas FD, Mugo J, Rabuor CO, Triest L and Dehairs F (2001). Species inventory and the local uses of the plants and fishes of the Lower Sondu Miriu wetland of Lake Victoria, Kenya. Hydrobiol. 458: 99-106

Gichuki JW, Triest L and Dehairs F (2005). The fate of organic matter in a papyrus (Cyperus papyrus L.) dominated tropical wetland ecosystem in nyanza Gulf (Lake Victoria, Kenya) inferred from ^{13}C and ^{15}N analysis. Isot. Environ. Health Stud. Vol. 41, No. 4. December, 2005, 379-390.

GoK (2000). Kisumu District Development Plan. Government of Kenya. Government Printer, Nairobi, Kenya.

GOK (2010). Population and Housing Census Report 2009. Ministry of State for Planning, National Development and Vision 2030, Government of Kenya Nairobi, Kenya.

Harper DM, Mavuti KM (2004). Lake Naivasha, Kenya: Ecohydrology to guide the management of a tropical protected area. Ecohydrol. and Hydrobiol. 4: 287-305.

Institute of Hydrology (1984). A review of Lake Victoria and the Victoria Nile. Report to Sir Alexander Gibb and Partners. Institute of Hydrology, Wallingford, UK.

Institute of Hydrology (1985). Further review of the hydrology of Lake Victoria. Report to the Overseas Development Administration. Institute of Hydrology, Wellington, UK.

ISRIC (1997). "Impact of soil erosion on maize production in Kenya". International Reference and Information Centre, Wageningen, The Netherlands.

Jackson W, Ingles A (1998) Participatory Techniques for Community Forestry: A field Manual; UK, IUCN Publications Services Unit. Cambridge, UK.

Jaetzold R, Schmidt H (1982). Farm Management Handbook of Kenya: Natural Conditions and Farm Management Information. Ministry of Agriculture Vol. II/A. Kisumu District, West Kenya.

Juma C (1989). Public policy and the Yala Swamp. Resources 1:21-22.

Karicho BM (2010). Regional groundwater flow modelling in the Kano plains, Kenya. MSc. Thesis. UNESCO-IHE. 104 pp.

Kansiime F, Kateyo E, Oryem-Origa H, Mucunguzi P (2007). Nutrient status and retention in pristine and disturbed wetlands in Uganda: management implications. Wetlands Ecology and Management 15: 453-467.

Kelderman P, Kansiime F, Tola MA, and van Dam A (2007). The role of sediments for phosphorus retention in the Kirinya wetland (Uganda). Wetlands Ecol. Manage. 15: 481-488.

KNBS (2009) Kenya Population and Housing Census: Kenya National Bureau of Statistics Vol. 1. Government Printer Nairobi. www.knbs.or.ke. Accessed 11 October 2012.

KNBS (2007). The Kenya Integrated Household Budget Survey 2005/06. Database. Kenya National Bureau of Statistics. , Government Printer, Nairobi.

Khisa PS, Uhlenbrook S, van Dam AA, Wenninger J, van Griensven A, Abira M (in prep.). Eco-hydrological characterization of the Nyando wetland, Lake Victoria, Kenya.

Kibwage JK, Onyango PO, Bakamwesiga H (2008). Local institutions for sustaining wetland resources and community livelihoods in the Lake Victoria basin. Afr. J. Environ. Sci. Technol. 2(5): 97-106.

Kipkemboi J (2006). Fingerponds: Seasonal Integrated Aquaculture in East African Freshwater Wetlands; exploring their potential for wise use strategies. Ph.D Dissertation, UNESCO-IHE Institute for Water Education, Delft, The Netherlands, pp. 73-94.

Kipkemboi J, van Dam AA, Denny P (2006). Biophysical suitability of smallholder integrated aquaculture-agriculture systems (Fingerponds) in East Africa's Lake Victoria freshwater wetlands. Int. J. Ecol. Environ. Sci. 32 (1): 75-83.

Kipkemboi J, van Dam AA, Ikiara MM, Denny P (2007). Integration of smallholder wetland aquaculture-agriculture systems (Fingerponds) into riparian farming systems at the shores of L.Victoria, Kenya: socio-economic and livelihoods. Geogr. J. 173: 257-272.

Kiwango YA, Wolanski E (2008) Papyrus wetlands, nutrients balance, fisheries collapse, food security, and Lake Victoria level decline in 2000-2006. Wetlands Ecol. Manage. (2008) 16:89-96.

Lightfoot C, Gupta MV, Ahmed M (1992). Low external input sustainable aquaculture for Bangladesh -An operational framework. NAGA, the ICLARM Quarterly pp. 9-12.

Loiselle SA, Azza N, Cozar A, Bracchini L, Tognazzi A, Battilo A, Rossi C (2008). Variability in factors causing light attenuation in Lake Victoria. Freshwat. Biol. 53: 535-545.

Maclean IMD, Hassall M, Boar RR, Lake IR (2006). Effects of disturbance and habitat loss on papyrus-dwelling passerines. Biol. Conserv. 131(3): 349-358.

Maltby E (2009). Functional assessment of wetlands: Towards Evaluation of Ecosystem Services. Woodland Publishing Limited (Ed.). CRC Press, New York. pp. 13-15.

MEA (2005). Ecosystems and Human Wellbeing: Wetlands and Water

Synthesis. Millennium Ecosystem Assessment, World Resources Institute, Washington, DC.
Melesse A, McClain M, Abira M, Mutayoba W, Wang XM (2008). Modelling the Impact of land-cover and rainfall regime change scenarios of the flow of Mara River, Kenya ASCE-EWRI. World Environ. Wat. Resour. Congr. "ftp.globalwaters.net/mara.globalwaters.net/EFA/Database/Consultancy/Reports/Melesse et al 2007.pdf Accessed 26th March 2013.
Mitsch WJ, Gosselink JG (2007). Wetlands. 4th Edition. John Wiley & Sons, Inc., New Jersey.
Mnaya B, Asaeda T, Kiwango Y, Ayubu E (2007). Primary production in papyrus (Cyperus papyrus L.) of Rubondo Island, Lake Victoria, Tanzania. Wetlands Ecol. Manage. 15: 269-275.
Morrison EHJ, Odhiambo-K'Oyooh CUK, Harper DM (2012). Managing the natural capital of papyrus within riparian zones of Lake Victoria, Kenya. Hydrobiol. 692: 5-17.
Muthuri FM, Jones MB, Imbamba SK (1989). Primary productivity of papyrus (Cyperus papyrus) in a tropical swamp: Lake Naivasha, Kenya. Biomass 18, 1-14.
Muthusi FM, Gathenya M, Gadain H, Lenga FK (2005). Application of Usgs stream flow model to the Nyando Basin, Western Kenya. Europ. J. Sci. Res. 12: 9-19
Muwanga A, Barifaijo E (2006). Impact of industrial activities on heavy metal loading and their physico-chemical effects on wetlands of Lake Victoria Basin (Uganda). Afri. J. Sci Technol. 7(1): 51-63.
Mwakubo SM, Obare GA (2009). Vulnerability, livelihood assets and institutional dynamics in the management of wetlands in Lake Victoria watershed basin. Wetlands Ecol. Manage. 17(6): 613-626.
Mwaura FB, Widdowson D (1992). Nitrogenase activity in the papyrus swamps of Lake Naivasha, Kenya. Hydrobiologia 232: 23-30.
Nasongo S, Zaal F, Okeyo-Owour JB (in prep) Historical events and land use changes and its impact on biodiversity and livelihoods in the Nyando papyrus wetland, Kenya.
Obiero KO, Raburu PO, Okeyo-Owuor JB, Raburu EA (2012) Community perceptions on the impact of the recession of Lake Victoria Waters on Nyando Wetlands. Sci Res Essays Vol. 7(16), pp. 1647-1661. Academic Journals.
Odada EO, Olago DO, Ochola W (2006) Environment for Development: An Ecosystems Assessment of Lake Victoria Basin, United Nations Environment Program (Eds) (UNEP)/ Pan African START Secretariat (PASS), Nairobi, Kenya.
Ogutu JO, Piepho H-P, Dublin HT, Bhola N, Reid RS (2007) El Niño-Southern Oscillation, rainfall, temperature and normalized difference vegetation index flunctuations in the Mara-Serengeti Ecosystem. Afric. J. Ecol. 46: 132-143.
Ongeri DMK, Lalah JO, Wandiga SO, Schramm KW, Michalke EB (2009) Levels of toxic metals in multisectoral samples from Winam Gulf of Lake Victoria. Bull Environ. Contam. Toxicol. 82: 64-69.
Osumba JJL, Okeyo-Owuor JB, Raburu PO (2010). Effect of harvesting on temporal papyrus (*Cyperus papyrus*) biomass regeneration potential among swamps in Winam Gulf wetlands of lake Victoria Basin, Kenya. *Wetlands* Ecol. Manage. 18: 333-341.
Oyoo-Okoth E, Admiraal W, Osano O, Ngure V, Kraak MHS, Omutange ES (2011). Monitoring exposure to heavy metals among children in Lake Victoria, Kenya: environmental and fish matrix. Ecotoxicol. Environ. Saf. 73: 1797-1803.
Owino AO, Ryan PG (2007). Recent papyrus swamp habitat loss and conservation implications in Western Kenya. Wetlands Ecol. Manage. 15, 1-12.
Rongoei PJK, Kipkemboi J, Kariuki ST, van Dam AA Effects of water level and livelihood activities on plant species composition and diversity in Nyando floodplain wetland, Kenya. In Review.
Sangale F, Okungu J and Opango P (2005). Variation of flow of water from Rivers Nzoia, Yala, and Sio into Lake Victoria. In Mallya GA, Katariga FF, Kang'oha G, Mbwana SB, Katunzi EF, Wambede JT, Azza N, Wakwabi E, Njoka SW, Kusewa M and Busulwa H (eds) Knowledge and Experiences gained from Managing the Lake Victoria Ecosystem. Regional secretariat, Lake Victoria Environmental Management Project (LVEMP): Dar es Salaam: 691-702.
Saunders MJ, Jones MB, Kansiime F (2007) Carbon and water cycles in tropical papyrus wetlands. Wetlands Ecol. Manage. 15: 489-498.
Swallow MB, Sang JK, Nyabenge M, Bundotich DK, Duraiappah AK and Yatich TB (2009). Tradeoffs, synergies and traps among ecosystem services in the Lake Victoria basin of East Africa. Environ. Sci. Policy 12, 504-519.
Swenson S and Wahr J (2009). Monitoring the water balance of Lake Victoria, East Africa, from space. J. Hydrol. 370(2009) 163-176.
Sutcliffe JV and Parks YP (1999). The Hydrology of the Nile. IAHS Special Publication No. 5, 1999: Wallingford, UK.
TEEB (2010). The Economics of Ecosystems and Biodiversity: Mainstreaming the Economics of Nature: A synthesis of the approach, conclusions and recommendations of TEEB. United Nations Environment Program, Nairobi. Available at: http://www.teebweb.org.
Terer T, Triest L, Muthama MA (2012) Effects of harvesting Cyperus papyrus in undisturbed wetland, Lake Naivasha, Kenya. Hydrobiol. 680(1): 135-148.
Thenya T (2006). Analysis of macrophyte biomass productivity, utilization and its impacts on various eco-types of Yala Swamp, Lake Victoria Basin, Kenya. In: P.L.G. VLEK, M. DENICH, C. MARTIUS and C. RODGERS (Eds). Ecol and Develop Series no. 48, 2006.
Van Dam AA, Dardona A, Kelderman P, Kansiime F (2007). A simulation model for nitrogen retention in a papyrus wetland near Lake Victoria, Uganda (EA). Wetlands Ecol. Manage. 15: 469-480
van Dam A, Kipkemboi J, Zaal F, Okeyo-Owuor J (2011). The ecology of livelihoods in East African papyrus wetlands (ECOLIVE). Rev. Environ. Sci. Biotechnol. 10(4): 291-300. doi: 10.1007/s11157-011-9255-6.
Wetlands International (2009). Planting trees to eat fish: Field experiences in wetlands and poverty reduction. Wetlands International, Wageningen, The Netherlands.
World Agro-forestry Centre (2006). Improved Land Management in the Lake Victoria Basin: Final Report of the TransVic Project. World Agro-forestry Centre, Occasion Paper No. 7. World Agroforestry Centre, Nairobi, Kenya.

3

Variations in the mineral composition and heavy metals content of *Moringa oleifera*

Fakankun, O. A., Babayemi, J. O. and Utiaruk, J. J.

Department of Chemical Sciences, Bells University of Technology, Ota, Ogun State, Nigeria.

The parts of *Moringa oleifera* were assessed for mineral composition and some heavy metal contents in this study, which included Ca, Mg, K, Na, Mn, Fe, Zn, Co, Se, Pb and Cd. Parts of the plant were obtained from Badagry in Lagos State, Nigeria. The samples were digested with HNO_3 and analysed for the mineral element content using atomic absorption spectrophotometer. The observed mean concentrations of the mineral elements were 26000, 643, 8210, 2980, 69.9, 169 and 15.3 mg/kg for Ca, Mg, K, Na, Mn, Fe and Zn, respectively. Co and Se, and the heavy metals Pb and Cd were not detected. There seemed to be no significant differences in the overall levels of the mineral elements in the different parts ($P > 0.05$). However, the observations showed that there were strong correlations between the levels of the elements in stem bark and root bark, stem bark and leaf, stem wood and pod, stem wood and seed, root bark and leaf, and pod and seed ($P < 0.01$); and between stem bark and root wood, root wood and root bark, and root wood and leaf ($P < 0.05$).

Key words: *Moringa oleifera*, mineral composition, heavy metals.

INTRODUCTION

Moringa oleifera commonly referred to as "Drumstick tree" belongs to the plant family *Moringaceae*. It is widely cultivated in semiarid, tropical and subtropical areas. The plant may be consumed as vegetable to improve nutriation, taken as medicinal plant to cure ailments and improve health, cultivated as forage for livestock, and used as live fencing.

Uses of various parts of *M. oleifera* plant have been reported. Powder produced from the seed is used as an effective primary coagulant for water treatment (Bhuptawat et al., 2007; Pritchard et al., 2010; Sánchez-Martín, 2010), and also possess the potential to remove cadmium from aqueous system (Sharma et al., 2006). The seed extract has been shown to have ameliorative effect on liver fibrosis in rats (Hamza, 2010), and also contain certain antitumor promoter (Guevara et al., 1999). Crude extracts and essential oil from *M. oleifera* possess anti-fungal activity against certain dermatophytes (Chuang et al., 2007). The leaves possess antioxidant properties (Verma et al., 2009). Some phytochemicals derived from the seeds possess insecticidal properties against mosquito (Prabhu et al., 2011).

This plant contains a considerable amount of various nutrients, and has been suggested as a good supplement for such nutrients as protein, fibre and minerals (Oduro et al., 2008; Jongrungruangchok et al., 2010). Also, including it in diets to supplement daily nutrient needs could help to fight against many diseases as nutraceuticals (Sharma et al., 2012). It is a viable supplement for dietary minerals (Aslam et al., 2005).

Several works have been done, reporting the levels of nutrients, including minerals in *M. oleifera,* with emphasis on variations in the levels as a function of geographical locations where the plant is grown; and the results

Figure 1. Parts of *M. oleifera* (A = leaf, B = seed, C = root, D = pod).

showed that there could be variations in the levels of macro and micro mineral elements in the plant, depending on the soil properties of the geographical locations where it grows (Anjorin et al., 2010). In the geographical location where this study is conducted, the most consumed part is the leaf, alleging that the needed nutrients are concentrated in the leaves. Reports on comparison of the levels of one element and the other in the different parts of the plant are scanty. Hence this study has these as objectives. The knowledge of the levels of various elements in the different parts of the plant may bring about efficient uses of other parts of the plant in addition to the leaves for the formulation as mineral supplement in both human and animal diet. The seeds are used in water purification; the knowledge of mineral composition of the seed will provide information on the elements that are likely to leach out into the water being purified or that may influence the water-purifying property. In addition, the study seeks to find out if relationships exist between the level of one element and the other, to provide an insight into some probable chemical interactions between the elements in the different parts of the plant. With this, one may be able to predict the level and/or behavior of one element by having the knowledge of the other. Others alleged that the parts like the root contain some toxins; but this is not within the objective of this study.

MATERIALS AND METHODS

The leaves (Figure 1A), root (Figure 1C) and stem of *M. oleifera* were obtained from a backyard at Badagry, Lagos state, while the pods (Figure 1D) together with the seeds (Figure 1B) and packaged powder for consumption (mainly produced from leaves) were obtained from University of Agriculture, Abeokuta. The leaves, roots and stem were thoroughly washed with water to minimize contaminations from adhering soil particles. The barks of the stem and the root were carefully peeled differently, while the woody parts were cut into tiny pieces to enhance quick drying. The samples were then air-dried for about four weeks in the laboratory. The dried samples were blended to fine particles, thoroughly cleaning the blender after each blend before the next. The woody parts, as these could not be blended, were cut into tiny particles with stainless scissors. The crucibles and glassware used were washed with detergents, acid solution and then rinsed with distilled water to avoid contaminations and thereby ensured the quality of the results. The samples were labeled SS (stem wood), SB (stem bark), RS (root wood), RB (root back), MP (powder), Lv (leaves), Pd (pod) and Sd (seed).

The analytical method used followed that described by Miroslav and Vladimir (1998) with slight modification. One gram of each sample dried to constant weight at 105°C was weighed in different porcelain crucibles, and then placed in a muffle furnace with the temperature slowly raised over 2 h to reach 550°C. They were left in the furnace at this temperature for 6 h. After cooling, the crucibles with the ash contents were removed from the furnace and allowed for further cooling. Five milliliters of 6 M nitric acid were added to each and carefully swirled to ensure complete dissolution. It was quantitatively filtered into a 100 ml volumetric flask and then made up to mark with distilled water. Blank was prepared in a similar way but omitting the plant sample.

The mineral elements, calcium (Ca), magnesium (Mg), potassium (K), sodium (Na), iron (Fe), zinc (Zn), manganese (Mn), Co and Se, as well as two toxic metals Pb and Cd, were determined using the atomic absorption spectrophotometer (AAS-Buck 205), at the wavelengths of 422.7, 285.2, 766.5, 589, 248.3, 213.9, 279.5, 240.7, 196, 283.3 and 228.9 nm, respectively, at the International Institute of Tropical Agriculture (IITA), Ibadan.

Using SPSS version 17, T-test was conducted on the results to see if there were differences in one set of the result and the other. The data were also analyzed using Pearson correlation analysis to determine if relationships exist. The analysis was done at 95% confidence level.

RESULTS

The observed mean concentrations of the mineral elements in *M. oleifera* plant (Table 1) were 26000, 643, 8210, 2980, 69.9, 169 and 15.3 mg/kg for Ca, Mg, K, Na, Mn, Fe and Zn, respectively. Co and Se, and the heavy metals Pb and Cd were not detected. Considering Figure 2, Ca (94900 mg/kg), Mg (762 mg/kg) and Na (9050 mg/kg) levels were highest in root bark, K (11300 mg/kg) in the seed, Mn (86 mg/kg) and Fe (214 mg/kg) in the leaves, and Zn (18 mg/kg) in powder and stem wood.

Using t-test, no significant difference was observed in the levels of Ca and K ($P = 0.321$), Ca and Na ($P = 0.095$) and Mg and Na ($P = 0.084$) at 95% confidence

Table 1. The distribution or spread of concentrations of the mineral elements and heavy metals in *M. oleifera*.

	Minimum statistic	Maximum statistic	Mean statistic standard error		Standard deviation statistic
Ca	1100	94900	26000	11000	31100
Mg	541	762	643	26	72
K	4750	11300	8210	747	2110
Na	696	9050	2980	1180	3320
Mn	63	86	70	3	7
Fe	154	214	169	7	20
Zn	11	18	15	1	2
*Co	-	-	-	-	-
*Se	-	-	-	-	-
*Pb	-	-	-	-	-
*Cd	-	-	-	-	-

*Not detected.

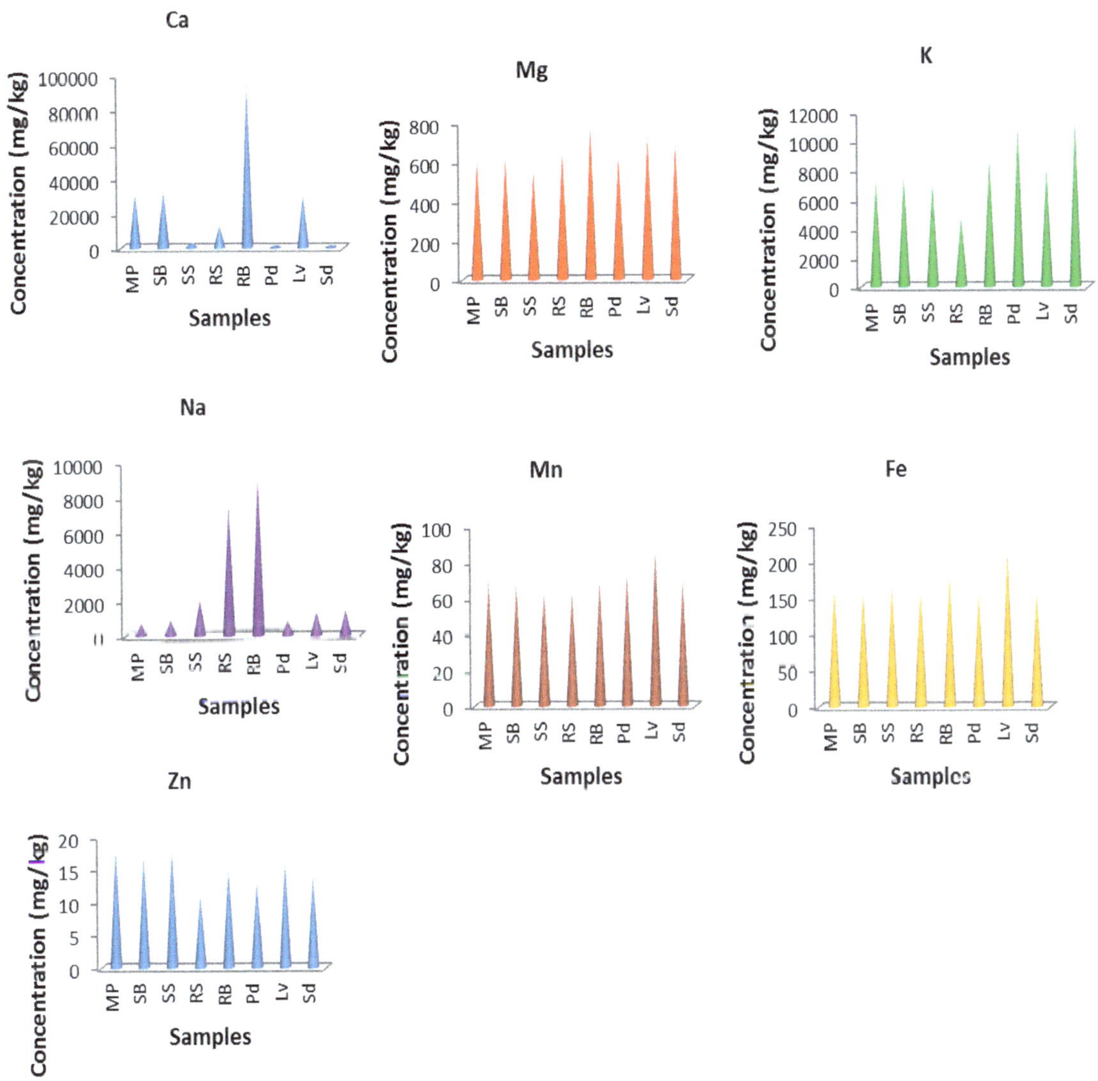

Figure 2. Concentration of mineral elements in MP (powder), SB (stem bark), SS (stem wood), RS (root wood), RB (root back), Pd (pod), Lv (leaves) and Sd (seed) of *M. oleifera*.

Table 2. Differences in the levels of the various mineral elements (paired samples test).

	Paired differences					t	df	Sig. (2-tailed)
	Mean	Std. deviation	Std. error mean	95% confidence interval of the difference				
				Lower	Upper			
Ca - Mg	13600	14900	5280	1110	26100	2.575	7	0.037
Ca - K	6040	16000	5650	-7320	19400	1.069	7	0.321
Ca - Na	11300	16500	5840	-2540	25100	1.929	7	0.095
Ca - Mn	14200	14900	5280	1690	26700	2.685	7	0.031
Ca - Fe	14100	14900	5280	1600	26600	2.667	7	0.032
Ca - Zn	14200	14900	5280	1750	26700	2.695	7	0.031
Mg - K	-7570	2090	740	-9320	-5820	-10.221	7	0.000
Mg - Na	-2340	3290	1160	-5090	411	-2.011	7	0.084
Mg - Mn	573	69	24	515	630	23.607	7	0.000
Mg - Fe	474	63	22	421	527	21.339	7	0.000
Mg - Zn	627	73	26	566	688	24.335	7	0.000
K - Na	5230	4530	1600	1440	9020	3.262	7	0.014
K - Mn	8140	2110	746	6370	9900	10.903	7	0.000
K - Fe	8040	2120	748	6270	9810	10.751	7	0.000
K - Zn	8190	2110	747	6430	9960	10.963	7	0.000
Na - Mn	2910	3330	1180	130	5690	2.475	7	0.043
Na - Fe	2810	3320	1170	34	5590	2.394	7	0.048
Na - Zn	2970	3330	1180	186	5750	2.523	7	0.040
Mn - Fe	-99	15	5	-111	-86	-18.970	7	0.000
Mn - Zn	55	8	3	48	61	20.442	7	0.000
Fe - Zn	153	20	7	137	170	22.271	7	0.000

level (Table 2), while others showed significant differences (at $P = 0.05$). Levels of Mn and Fe showed strong correlation (correlation coefficient of $r = 0.798$ and $P < 0.05$), using Pearson correlation analysis; there was little or no correlation between the levels of the others (Table 3): correlation coefficient between $r = -0.056$ and 0.574 ($0.136 \leq P \leq 0.895$). There seemed to be no significant differences in the overall levels of the mineral elements in the different parts of *M. oleifera* plants ($P > 0.05$) (Table 4). However, the observations showed that there were strong correlations (Table 5) between the levels of the elements in SB and RB, SB and Lv, SS and Pd, SS and Sd, RB and Lv, and Pd and Sd ($P < 0.01$); and between SB and RS, RS and RB and RS and Lv ($P < 0.05$).

DISCUSSION

The level of Ca was the highest (1100 to 94900 mg/kg). The highest level (94900 mg/kg) was observed in RB and the lowest (1100 mg/kg) in the pod. There was no significant difference between the levels of Ca and Na, and Ca and K ($P > 0.05$). There was little or no correlation between the levels of Ca and that of any of the other elements but Ca is a macro nutrient for humans, considerably sought for from other sources such as bones and shells of some crustaceans; if the level is this high, why is the part not commonly consumed? While the pod together with the seed is edible, there is restriction to the extent of human consumption of the root part of *M. oleifera* perhaps because of the information that it has some organic constituents likely to constitute neurotoxic effects (Gupta et al., 1999); however, further study may be needed to unfold the possibility of exploring the high Ca content for possible uses; or some possible food processing methods that may lead to detoxification and thereby making the root part consumable. The most consumed part is the leaf. The level of Ca observed in the leaves was 30300 mg/kg. The levels in literature (Jongrungruangchok et al., 2010) reporting similar works ranged from 15100 to 29510 mg/kg, which are comparable to those recorded in this study. This level of Ca in the leaf of *Moringa* is about four times that in milk (Wikipedia) and six times that observed in *Amaranthus* sp.(a common vegetable in Nigeria) (Sharma et al., 2012), hence as supplement in human diet, has the potential to meet the daily requirement. Ca is an important element in formation of bones and teeth; it is said to prevent osteoporosis (Susan and Lanham-New, 2008). Beneficial effects of Ca exist in the human body up to intake threshold of about 800 mg per day (Boon et al., 2005).

Table 3. Correlations between the concentration of one mineral element and the other in *M. oleifera*.

		Ca	Mg	K	Na	Mn	Fe	Zn
Ca	Pearson Correlation	1	-0.123	-0.439	-0.389	0.340	0.314	0.455
	Sig. (2-tailed)		0.771	0.276	0.340	0.410	0.448	0.257
Mg	Pearson Correlation	-0.123	1	0.293	0.504	0.515	0.574	-0.311
	Sig. (2-tailed)	0.771		0.482	0.203	0.191	0.136	0.453
K	Pearson Correlation	-0.439	0.293	1	-0.358	0.331	-0.073	-0.056
	Sig. (2-tailed)	0.276	0.482		0.384	0.424	0.864	0.895
Na	Pearson Correlation	-0.389	0.504	-0.358	1	-0.315	0.061	-0.479
	Sig. (2-tailed)	0.340	0.203	0.384		0.448	0.886	0.229
Mn	Pearson Correlation	0.340	0.515	0.331	-0.315	1	0.798*	0.072
	Sig. (2-tailed)	0.410	0.191	0.424	0.448		0.017	0.865
Fe	Pearson Correlation	0.314	0.574	-0.073	0.061	0.798*	1	0.227
	Sig. (2-tailed)	0.448	0.136	0.864	0.886	0.017		0.589
Zn	Pearson Correlation	0.455	-0.311	-0.056	-0.479	0.072	0.227	1
	Sig. (2-tailed)	0.257	0.453	0.895	0.229	0.865	0.589	

*Correlation is significant at the 0.05 level (2-tailed).

Table 4. Comparison of the concentration of mineral elements in the different parts of *M. oleifera* (paired samples test).

	Paired difference					t	df	Sig. (2-tailed)
	Mean	Standard deviation	Standard error mean	95% confidence interval of the difference				
				Lower	Upper			
MP - SB	-221	405	153	-596	154	-1.442	6	.199
MP - SS	3940	10900	4130	-6180	14100	.952	6	.378
MP - RS	2130	8060	3040	-5320	9570	.699	6	.511
MP - RB	-10400	23400	8840	-32100	11200	-1.181	6	.282
MP - Pd	3840	11900	4490	-7150	14800	.855	6	.425
MP - Lv	-13	726	274	-684	658	-.049	6	.963
MP - Sd	3680	11900	4520	-7370	14700	.814	6	.447
SB - SS	4160	11300	4280	-6320	14600	.971	6	.369
SB - RS	2350	8420	3180	-5440	10100	.738	6	.488
SB - RB	-10200	23000	8690	-31500	11000	-1.176	6	.284
SB - Pd	4060	12300	4640	-7280	15400	.876	6	.415
SB - Lv	208	1070	405	-782	1200	.513	6	.626
SB - Sd	3900	12300	4660	-7500	15300	.837	6	.435
SS - RS	-1810	4150	1570	-5650	2030	-1.154	6	.292
SS - RB	-14400	34200	12900	-46000	17300	-1.112	6	.309
SS - Pd	-96.4	1780	671	-174	1550	-.144	6	.890
SS - Lv	-3950	10300	3880	-13500	5550	-1.017	6	.348
SS - Sd	-260	1850	699	-1970	1450	-.373	6	.722
RS - RB	-12600	30800	11600	-41000	15900	-1.081	6	.321
RS - Pd	1710	5660	2140	-3520	6950	.802	6	.453
RS - Lv	-2140	7420	2800	-9000	4720	-.764	6	.474
RS - Sd	1550	5660	2140	-3680	6780	.725	6	.496
RB - Pd	14300	35200	13300	-18200	46800	1.075	6	.324
RB - Lv	10400	24000	9080	-11800	32600	1.149	6	.294
RB - Sd	14100	35200	13300	-18400	46700	1.061	6	.329
Pd - Lv	-3850	11200	4240	-14200	6510	-.910	6	.398

Table 4. Contd

Pd - Sd	-164	245	92	-390	62.1	-1.774	6	.126
Lv - Sd	3690	11300	4260	-6730	14100	.867	6	.419

SS (stem wood), SB (stem bark), RS (root wood), RB (root back), MP (powder), Lv (leaves), Pd (pod) and Sd (seed).

Table 5. Correlations between the concentrations of the mineral elements in one part of *M. oleifera* and the other.

		MP	SB	SS	RS	RB	Pd	Lv	Sd
MP	Pearson correlation	1	1.000**	.417	.846*	.986**	.137	.999**	.128
	Sig. (2-tailed)		.000	.352	.016	.000	.770	.000	.784
SB	Pearson correlation	1.000**	1	.417	.848*	.986**	.136	.999**	.127
	Sig. (2-tailed)	.000		.352	.016	.000	.771	.000	.785
SS	Pearson correlation	.417	.417	1	.532	.300	.935**	.449	.941**
	Sig. (2-tailed)	.352	.352		.219	.513	.002	.312	.002
RS	Pearson correlation	.846*	.848*	.532	1	.867*	.196	.858*	.216
	Sig. (2-tailed)	.016	.016	.219		.012	.673	.014	.642
RB	Pearson correlation	.986**	.986**	.300	.867*	1	-.009	.982**	-.013
	Sig. (2-tailed)	.000	.000	.513	.012		.985	.000	.978
Pd	Pearson correlation	.137	.136	.935**	.196	-.009	1	.169	.999**
	Sig. (2-tailed)	.770	.771	.002	.673	.985		.718	.000
Lv	Pearson correlation	.999**	.999**	.449	.858*	.982**	.169	1	.161
	Sig. (2-tailed)	.000	.000	.312	.014	.000	.718		.730
Sd	Pearson correlation	.128	.127	.941**	.216	-.013	.999**	.161	1
	Sig. (2-tailed)	.784	.785	.002	.642	.978	.000	.730	

**Correlation is significant at the 0.01 level (2-tailed); *Correlation is significant at the 0.05 level (2-tailed); SS (stem wood), SB (stem bark), RS (root wood), RB (root back), MP (powder), Lv (leaves), Pd (pod) and Sd (seed).

The concentrations of Mg ranged from 541 to 762 mg/kg, with the highest level in the RB as observed for Ca, and least in the stem wood. Mg and Ca have similar properties (same group in the periodic table of elements), and perhaps these influence the phyto-chemistry in the root of *M. oleifera*, and hence both are found at the highest level in the root. Further study is suggested to establish this. However, from the statistical analysis, there was significant difference in their levels ($P < 0.05$) and there was little or no correlation between their levels ($r < 2.0$ and $P > 0.05$). The level of Mg is not significantly different from that of Na ($P > 0.05$). The level in the most consumed part, the leaf, was 717 mg/kg, next to the level in the RB. These levels in the leaf are comparable to those reported (693 to 725 mg/kg) by Anjorin et al. (2010) for some Nigeria samples analyzed. Mg plays some roles in human health, namely, production and energy transport, contraction and relaxation of muscles, synthesis of protein and function of enzymes.

K levels ranged from 4750 to 11300 mg/kg, with the highest value in the seed and lowest in the root. If there are no anti-nutritional factors or other chemical components that may constitute toxins in the seed of this plant, the seed part may then be recommended as good source of potassium for supplements in diet. Concentration in the leaf was 8010 mg/kg, about three times higher than the level (3070 mg/kg) in *Ocimum gratissimum*, an indigenous green leafy vegetable in Nigeria (Thomas and Oyediran, 2008), and about three times higher than the levels in banana (Wikipedia). Yaméogo et al. (2011) reported K levels of 3086 to 22500 mg/kg in their study of chemical composition of *Moringa* leaf, and the value found in this study falls with this range. K helps in the proper function of the brain as well as nerves, thereby preventing stroke. It plays part in acid-base and water regulation in the blood and tissues. In contrast to Na, it has been reported that a high-potassium diet lowered blood pressure in individuals with raised blood pressure (He and MacGregor, 2008). In addition to its contribution to the electrolytes and function of the nerves in the human body, K has been linked to bone health and osteoporosis prevention according to studies of intake of 2500 mg/day done in United Kingdom, United States and Australia (Susan and Lanham-New, 2008). It may then be inferred from the results of this study that consumption of the leaf of *M. oleifera* could make up for the daily requirement of K in human diet. Observed further by Susan and Lanham-New (2008), the dietary content of pre-agricultural man suggests intake of K up to 7000 mg/day; hence, having the leaves of *M. oleifera* in the diet could bring the modern man to comparable level of diet as the pre-agricultural man.

The highest concentration of Na was observed in root bark. In all the parts studied, the levels ranged from 696 to 9050 mg/kg; and in the leaf, 1350 mg/kg. Observations of Na concentration were about the same for leaves and seeds. Aslam et al. (2005) reported similar levels (1292 to 1837 mg/kg) in M*oringa* leaves and 1032 to 2105 mg/kg in the pods. From this study and the references cited, it may be inferred that the level observed in the leaf is about twenty-three times higher than that in *Colocassia esculenta* (60 mg/kg) and about twelve times higher than the levels in *O. gratissimum* (113 mg/kg), two non-conventional indigenous green leafy vegetables in Nigeria (Thomas and Oyediran, 2008).

The concentrations observed for Mn in this study ranged from 63 to 86 mg/kg. This range is comparable to those reported (47.1 to 113.9 mg/kg) by Foidl et al. (2001). While there were significant differences between the levels of Mn and that of any of the other elements, strong correlation (r = 0.798, $P < 0.05$) existed between the levels of Mn and those of Fe. Mn helps in breaking down of fats, carbohydrates and proteins.

Fe level was highest in the leaves (214 mg/kg) and lowest in the pods (154 mg/kg). The leaves and fruit of *Moringa* are the most consumed, hence potential sources of Fe in human diet. Fe functions in the formation of haemoglobin and myglobin which are carriers of oxygen in the blood and blood vessels. Deficiency of Fe may result in anaemia. In determining the nutritional potential of *M. oleifera*, Oduro et al. (2008) obtained Fe level of 282.9 mg/kg in the leaf. For similar analysis, Jongrungruangchok et al. (2010) reported a range of 203.1 to 376 mg/kg. The level observed in this study is very much comparable to the levels in the latter reference.

Concentrations of Zn observed ranged from 11 to 18 mg/kg. It was 16 mg/kg in the leaf and 14 mg/kg in the seed. The levels are comparable to those recorded by Anjorin et al. (2010): 18 mg/kg in the leaf and 9.6 mg/kg in the seed. Zn enhances the function of immune system and proper functioning of some enzymes such as those involved in cell division, and growth. It has also been said to be involved in the phenomena of taste and smell.

Co, Se, Cd and Pb were not detected because the technique does not have a sufficient detection limit to quantify these elements in these types of samples.

Therefore, the graphite furnace and the generation hydride for selenium may be appropriate in future determinations.

Conclusion

No significant difference was observed in the levels of Ca and K, Ca and Na, and Mg and Na at 95% confidence level, while others showed significant differences. Levels of Mn and Fe showed strong correlation but there was little or no correlation between the levels of the others. There seemed to be no significant differences in the overall levels of the mineral elements in the different parts of *M. oleifera* plants. However, the observations showed that there were strong correlations between the levels of the elements in stem bark and root bark, stem bark and leaf, stem wood and pod, stem wood and seed, root bark and leaf, and pod and seed; and between stem bark and root wood, root wood and root bark, and root wood and leaf.

REFERENCES

Anjorin TS, Ikokoh P, Okolo S (2010). Mineral composition of *Moringa oleifera* leaves, pods and seeds from two regions in Abuja, Nigeria. Int. J. Agric. Biol. 12: 431–434.

Aslam M, Anwar F, Nadeem R, Rashid U, Kazi TG, Nadeem M (2005). Mineral composition of *Moringa oleifera* leaves and pods from different regions of Punjab, Pakistan. Asian J. Plant Sci. 4(4): 417-421.

Bhuptawat H, Folkard GK, Chaudhari S (2007). Innovative physico-chemical treatment of wastewater incorporating *Moringa oleifera* seed coagulant. J. Hazard. Mater. 142: 477–482.

Boon N, Koppes LLJ, Saris WHM, Van Mechelen W (2005). The Relation between Calcium Intake and Body Composition in a Dutch Population. Am. J. Epidemiol. 162:27–32.

Chuang P, Lee C, Chou J, Murugan M, Shieh B, Chen H (2007). Anti-fungal activity of crude extracts and essential oil of *Moringa oleifera* Lam. Bioresour. Technol. 98: 232–236.

Foidl N, Makkar HPS, Becker K (2001). The potential of *Moringa oleifera* for agricultural uses. What Development Potential for *Moringa* Products? Dar Es Salaam. pp. 1-20.

Guevara AP, Vargas C, Sakurai H, Fujiwara Y, Hashimoto K, Maoka T, Kozuka M, Ito Y, Tokuda H, Nishino H (1999). An antitumor promoter from *Moringa oleifera* Lam. Mutation Res. 440: 181–188. PII: S1383-5718_99.00025-X

Gupta M, Mazumder UK, Chakrabarti S (1999). CNS activities of methanolic extract of *Moringa oleifera* root in mice. Fitoterapia 70: 244-250. PII: S 0 3 6 7 - 3 2 6 X _ 9 9 . 0 0 0 2 9 - 5

Hamza AA (2010). Ameliorative effects of Moringa oleifera Lam seed extract on liver fibrosis in rats. Food Chem. Toxicol. 48: 345–355.

He FJ, MacGregor GA (2008). Beneficial effects of potassium on human health. Physiologia Plantarum 133 (4): 725-735.

Jongrungruangchok S, Bunrathep S, Songsak T (2010). Nutrients and minerals content of eleven different samples of *moringa oleifera* cultivated in Thailand. J. Health Res. 24(3): 123-127.

Miroslav R, Vladimir NB (1998). Plant analysis. Practical Environmental Analysis, 379-397.

Oduro I, Ellis WO, Owusu D (2008). Nutritional potential of two leafy vegetables: *Moringa oleifera* and *Ipomoea batatas* leaves. Sci. Res. Essay 3 (2): pp. 057-060.

Prabhu K, Murugan K, Nareshkumar A, Ramasubramanian N, Bragadeeswaran S (2011). Larvicidal and repellent potential of Moringa oleifera against malarial vector, Anopheles stephensi Liston (Insecta: Diptera: Culicidae). Asian Pac. J. Trop. Biomed. 1(2):124-129.

Pritchard M, Craven T, Mkandawire T, Edmondson AS, O'Neill JG (2010). A study of the parameters affecting the effectiveness of Moringa oleifera in drinking water purification. Phys. Chem. Earth 35: 791–797.

Sánchez-Martín J, Ghebremichael K, Beltrán-Heredi J (2010). Comparison of single-step and two-step purified coagulants from Moringa oleifera seed for turbidity and DOC removal. Bioresour. Technol. 101: 6259–6261.

Sharma N, Gupta PC, Rao ChV (2012). Nutrient content, mineral content and antioxidant activity of *Amaranthus viridis* and *Moringa oleifera* leaves. Res. J. Med. Plant 6(3): 253-259.

Sharma P, Kumari P, Srivastava MM, Srivastava S (2006). Removal of cadmium from aqueous system by shelled Moringa oleifera Lam. seed powde. Bioresour. Technol. 97: 299–305.

Susan A, Lanham-New (2008). The balance of bone health: tipping the scales in favour of potassium-rich, bicarbonate-rich foods. J. Nutr. 138: 172S-177S.

Thomas AO, Oyediran OE (2008). Nutritional importance and micronutrient potentials of two non-conventional indigenous green leafy vegetables from Nigeria. Agric. J. 3(5): 362-365.

Verma AR, Vijayakumar M, Mathel CS, Rao CV (2009). In vitro and in vivo antioxidant properties of different fractions of Moringa oleifera leaves. Food Chem. Toxicol. 47: 2196–2201.

Wikipedia (http://en.wikipedia.org/wiki/Dahon_ng_malunggay).

Yaméogo CW, Bengaly MD, Savadogo A, Nikiema PA and Traore SA (2010). Determination of chemical composition and nutritional values of *Moringa oleifera* leaves. Pak. J. Nutr. 10(3): 264-268.

4

Heavy metal concentrations in water, sediment and periwinkle (*Tympanotonus fuscastus*) samples harvested from the Niger Delta region of Nigeria

Otitoju, O.[1] **and Otitoju, G. T. O.**[2]

[1]Faculty of Pure and Applied Sciences, Department of Biochemistry, Federal University, Wukari, Taraba State, Nigeria.
[2]Department of Home Science, Nutrition and Dietetics, University of Nigeria, Nsukka, Enugu State, Nigeria.

Heavy metal pollution of terrestrial and aquatic environments in Niger-delta region of Nigeria is on the increase due to increased urbanization and crude oil exploration. *Tympanotonus fuscastus* are mud dwellers and could bioaccumulate heavy metals; therefore consumption of heavy metal contaminated sea foods like periwinkle may breed lots of health problems. Heavy metal concentration in water, sediment and Periwinkle samples from three locations (Itu-River, Abuloma River and Oron River) in the Niger Delta Region of Nigeria were evaluated using atomic absorption flame photometry. Result showed that cadmium (Cd) concentration was highest in water samples from Abuloma River (0.106 mg/l) while lead (Pb) concentration was highest (0.01mg/l) in the water samples from Itu River. Cd and Pb concentrations (0.127 and 0.08 mg/kg, respectively) in sediment samples were highest in Abuloma River. The concentration of Cd in *T. fuscastus* samples was 0.11 mg/kg in Abuloma River, while concentration was 0.27 mg/kg in Oron River. Copper (Cu) was generally low in the water samples; the highest concentration (0.011 mg/kg) was obtained in water samples from Oron River. Sediment concentration of Cu was high (0.088 mg/kg) in Itu River, while its concentration in the periwinkle samples was 0.54 mg/kg in Abuloma River. The results also showed that Cr, As and Hg were below detectable concentration in tissues, soil and water samples from Itu and Abuloma Rivers, while Hg concentration in Oron River sediment was 64.2 mg/kg.

Key words: Sea foods, heavy metals, pollution, environment, nutrition, bioaccumulation

INTRODUCTION

The Niger Delta region with its creeks and tributaries harbors rich collection of biotopes dominated by vast areas of mangrove swamp forest. However, this region, with its complex ecological form is being subjected to considerable environmental pollutants from industrial activities and crude oil exploration. This has resulted in the release of pollutants (hydrocarbons and heavy metals) capable of contaminating soil and water bodies (Otitoju et al., 2011; Ewa-Oboho, 1994). Heavy metals have been reported to exert negative effect on biological processes in general and may influence the nutritional and biological status of sea foods (Udosen et al., 2001).

Sea-food is a rich source of nutrients; however, its nutritional values may be affected based on the environment in which these organisms live (Nsikak et al., 2007). These important food sources constitute a major part of the diet in the Niger Delta region of Nigeria, where they are traditionally cooked without removing the shells before consumption and these organisms could be contaminated with heavy metals which may be detrimental to the

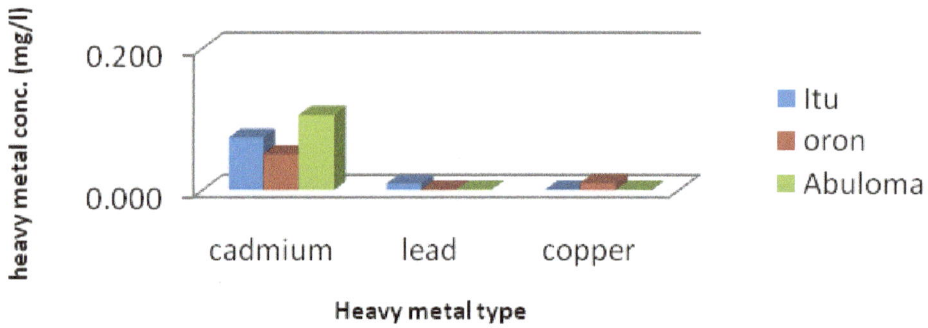

Figure 1. Heavy metal concentration in water samples from Oron, Itu and Abuloma Rivers, Nigeria.

health of children, pregnant women and even the general public.

Large number of people living in the Niger Delta region of Nigeria consumes sea foods harvested from different water bodies laden with history of crude oil spillage and other industrial activities. Heavy metals are high priority pollutants because of their relatively high toxic and persistent nature in the environment. These metals in the form of inorganic compounds from natural and anthropogenic sources continuously enter the aquatic ecosystem where they could pose serious threat to the food chain.

The aim and objectives of this research were to determine the levels of trace metals {lead (Pb), mercury (Hg), chromium (Cr), arsenic (As), cadmium (Cd) and copper (Cu)} in water, sediment and periwinkles harvested from the Niger Delta region of Nigeria.

MATERIALS AND METHODS

Description of the study areas and sampling locations

Smooth specie of periwinkle (*Tympanotonus fuscastus)* commonly found in the Niger Delta region of Nigeria were collected from three different Rivers. This research covered three different locations situated across the Niger Delta region of Nigeria. The studied areas include; Oron River in Idia-oro, Oron Local Government area located in the South-eastern part of Akwa-Ibom state. The second studied area was Itu River in Ikot-Udobia; which is located in the North-western part of Akwa-Ibom state while the third studied area was Abuloma River in Abuloma, Port-Harcourt located in the South axis of Rivers state. The sea food samples were transported in clean polyethylene bags to the Biochemistry Laboratory, University of Uyo were they were washed with deionized water to remove all dirt particles; they were also shelled and kept in an oven at 65°C for 2 days after which they were grounded to powdered form.

Heavy metal analysis

Digestion methods

Nitric acid: Hydrochloric acid digestion- 1.0 g of the powdered sea food sample was digested in 6 ml of HCl and made up to 30 ml with distilled water. Filtration was carried out using an acid wash filter paper (Whatman) and stored in sample bottles. The filtrate was taken to the UNICAM 939 Atomic Absorption Spectrophotometer (AAS) for Hg, As, Cd, Cr, Pb and Cu determination in triplicate.

Statistical analysis

Mean values (±SD) of duplicate experiment were taken for each analysis. Significantly different results were established by one-way ANOVA. The accepted level of significance is $p<0.05$.

RESULTS

Result showed that Cd concentration was highest in water samples from Abuloma River (0.106 mg/kg) while Pb concentration was highest (0.01 mg/kg) in the water samples from Itu River (Figure 1). Cd and Pb concentrations in sediment samples (0.127 and 0.08 mg/kg) were higher in Abuloma River, respectively (Figure 2). The concentration of Cd in *T. fuscastus* samples was 0.11 mg/kg in Abuloma River while concentration was 0.27 mg/kg in Oron River (Figure 3). Cu was generally low in the water samples, the highest concentration (0.011 mg/kg) was obtained in water samples from Oron River. Sediment concentration of Cu was high (0.088 mg/kg) in Itu River while its concentration in the periwinkle samples was 0.54 mg/kg in Abuloma River. The results also showed that Cr, As and Hg were below detectable concentration in Itu and Abuloma Rivers, while Hg concentration in Oron River was 6.42 mg/kg.

DISCUSSION

The aquatic systems are being polluted anthropogenically with chemical pollutants from industrial, domestic and agricultural wastes, which are ultimately absorbed by aquatic animals and plants. Pollution of aquatic ecosystems by heavy metals is an important environmental problem, as heavy metals constitute some of the most dangerous toxicants that can be bioaccumulated in living tissues (Guo et al., 1997; Omoregie et al., 2002). Consumption of periwinkle in the Niger Delta region of Nigeria

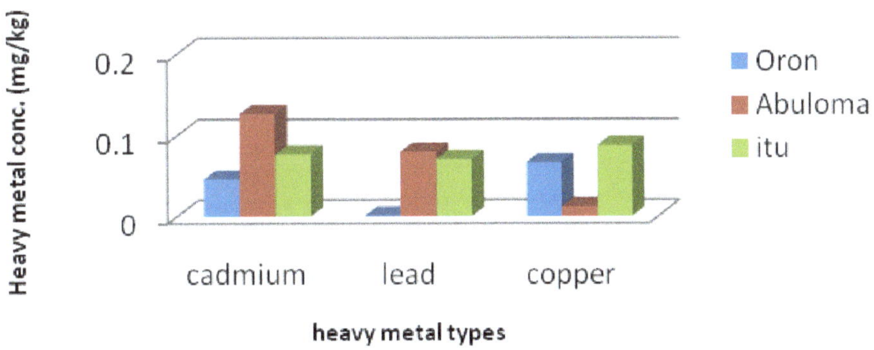

Figure 2. Heavy metal concentration in sediment samples from Oron, Itu and Abuloma Rivers, Nigeria.

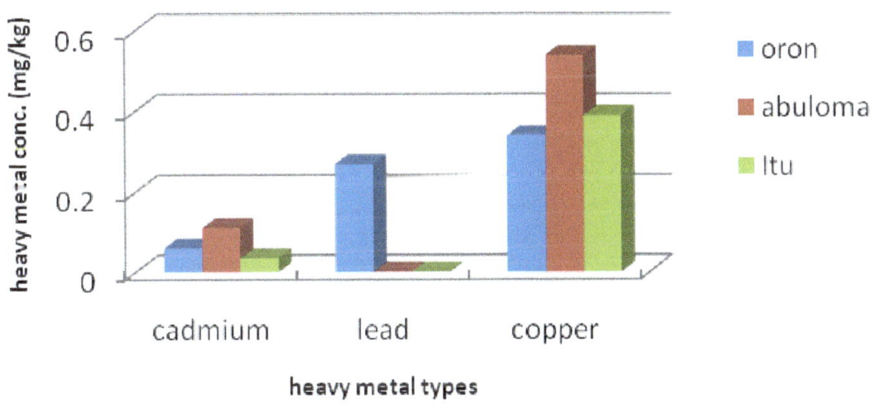

Figure 3. Heavy metal concentration in Sea-Food samples Oron, Itu and Abuloma Rivers, Nigeria.

is considered very common. However, the environments in which these organisms are harvested give reasons to worry.

The results obtained from this research showed that Hg was present in periwinkle samples from Oron River, this result was indeed unexpected, since Oron, is not an industrialized city. Nevertheless, Oron River transverses other locations within and outside Akwa Ibom State, resulting in the presence of Hg in this river and its bioaccumulation in the sea foods. Mercury is a dangerous heavy metal whose toxicity was dated back to the ancient time. The wide use of this liquid metal in medicine, industries and in electrical fittings made the environment especially the aquatic environment vulnerable to its pollution (Nsikak et al., 2007; Gress and Lord, 2002; Wiener, 2002). Similarly, a lot of farming activities where pesticides and other agrochemicals are utilized are ongoing along the shore of this River. The possibility of spillage or and leakage into the soil and consequently into the water bodies by run-off activities is inevitable.

The result also showed that copper (Cu) was present in all the water sediment and tissue samples. Abuloma River had the highest concentration of Cu which ranges between (0.288 and 0.556 mg/kg) and this range generally falls below the World Health Organization (1994) limits of 2.0 mg/kg for food and sea foods. Cadmium was also present in all the samples but the highest concentration of Cd was found in samples obtained from Abuloma River (0.118 mg/kg). This may be as a result of pollutants from industrial and agricultural processes as well as crude oil exploration. Bioaccumulation of this heavy metal most frequently results in kidney damage and can also cause Osteomalacia (Järup, 2003).

Itu River had undetectable concentrations of Pb 0.001 mg/kg, Oron River had low concentrations of Pb (0.027 mg/kg), while Abuloma River had the highest concentration of this metal (0.039 mg/kg). Pb has been reported to be associated with crude-oil exploration, pipe-line transportation, corrosion inhibition as well as many industrial processes (Oguzie and Igwegbe, 2007). High levels of Pb in foods can induce abdominal pains, drowsiness, vomiting, convulsion kidney and reproductive system malfunction (Goldsmith and Hildzard, 1985). The Food and Agriculture Organization and the World Health Organization stipulates the weekly intake of Cd and Pb for adults at 0.42 to 0.49 and 1.5 to 1.7 mg/kg, respectively

(Leski and Rudawska, 2005). From our results, the concentration of Cd and Pb in the samples was lower than the recommended range by FAO and WHO (1999) and Sabin and Wendy (2009). Therefore, these metals may not pose serious health consequence to consumers at short term of exposure but could be in the long run. Similarly, the concentrations of Cr and As were undetec-ted in the sea food samples from the three sampled rivers.

The traditional consumption of periwinkle using 'blowers method' where mouth is used to suck out the unshelled periwinkle may require a change because there could be more risk in consuming sediments from the intestinal tract of these organisms than the shelled periwinkle. Based on the result obtained from this study, it is observed that Cu, Pb, and Cd are present in sea foods but their concentration generally falls below the recommended range by World Health Organization and the Food and Agricultural Organization (FAO). We therefore conclude that heavy metal toxicity may not be an immediate problem in Niger-Delta region of Nigeria but may pose a lot of environmental problems if it is not properly checked. We also suggest that there is a need for regular public health checks on the levels of heavy metals among the communities that border the sampled rivers in order to avoid a possible future toxic exposure. Measures should also be put in place to control the treatment and discharge of effluent into the water bodies.

REFERENCES

Ewa-Oboho IO (1994). Ecological effect of crude oil on a Mangrove Swamp of the Bonny Estuary. M. Phil thesis. Rivers State University of Science and Technology, Port Harcourt. J. Nig. Environ. Soc. (JNES) 1(4): 201-220.

FAO/WHO (1999) .Expert committee on food additives "Summary and Conclusion", in proceedings of the 53rd meeting of joint FAO/WHO expert committee on food additives, Rome, Italy, June.

Goldsmith E, Hildyard N (1985). The earth report. The essential guide to global ecological issues. Price Sterm Sloan Inc, Los Angeles, California, p. 173.

Gress L, Lord J (2002). Return and recycling of used high intensity bulbs for recycling and closed-loop mercury control. In Mercury in Products, Processes, Waste and the Environment: Eliminating, Reducing and Managing Risks from Non-Combustion Sources (ed. Oppelt, E. T.), Proceedings and Summary Report EPA/625/R-00/014.

Guo T, Delaune RD, Patrick Jr WH (1997). The influence of sediment redox chemistry on chemically active forms of arsenic, cadmium, chromium, and zinc in the estuarine sediment. Int. J. Environ. Stud. 23: 305–316.

Järup L (2003). Hazards of heavy metal contamination. Br Med Bull. 68:167-82.

Leski T, Rudawska M (2005). Macro-and Microelement contents in fruiting bodies of wild mushrooms from the Notecka forest in west-central Poland. Food Chem. 92: 499-506.

Nsikak UB, Joseph PE, Akpan BW, David EB (2007). Mercury accumulation in fishes from tropical aquatic ecosystems in the Niger Delta. Nig. Curr. Sci. 92(6): 781-785.

Oguzie FA, Igwegbe AO (2007). Heavy metal concentrations in water and three West African cichlid fishes of Ogba river in Benin City. J. Field Aquat. Studies 3: 41-48.

Omoregie E, Okoronkwo MO, Eziashi AC, Zoakah AI (2002). Metal concentrations in water column, benthic macroinvertebrates and tilapia from Delimi River, Nigeria. J. Aquat. Sci. 17: 55–59.

Otitoju O, Onwurah INE, Otitoju GTO, Ndem JI (2011). Effect of Bonny light crude oil contaminated diet on serum cholesterol and testosterone concentrations on Wistar albino rats. African J. Biotech. 10(53): 10986-10990.

Sabin M, Wendy G (2009). Human health effects of heavy metals. Center for harzadous substance research. Kensas state university www.engg.ksu.edu/CHSR.

Udosen ED, Essien JP, Ubom RM (2001). Bioamendment of petroleum contaminated soil: Effect on oil content, heavy metals and pH of tropical soil. J. Environ. Sci. pp: 92-98.

Wiener JG (2002). Evolution of a contaminant problem: Mercury in fresh water fish. Proceedings and summary report, USEPA/625/R-02/005.

World Health Organisation (WHO) (1994). Assessing human health risk of chemicals, derivation of guidance values for health based exposure limits. Environmental health critical. World Health Organization, Geneva. Vol. 170.

5

Impact analysis of lead, copper and zinc content in selected African indigenous and exotic vegetables from Nairobi markets, Kenya

Mutune, A. N., Makobe, M. A. and Abukutsa-Onyango, M. O. O.

Jomo Kenyatta University of Agriculture and Technology. P.O. Box 62000, 00200. Nairobi, Kenya.

Vegetables consumed in urban sites can be contaminated with harmful metal levels absorbed from planting sources because metals are commonly present in soil. The present study was carried out to evaluate Pb, Cu and Zn concentration in ten of the commonest vegetables in Nairobi markets. Vegetables were collected from 15 markets in urban and peri-urban Nairobi and analyzed for metal content using an Atomic Absorption Spectrophotometer (AAS). The concentrations were in the order Zn > Cu > Pb with values of 15.6 to 120, 0 to 19 and Pb 0 to 1.37 mg/kg, respectively. There were significant differences in markets and within vegetables ($p<0.05$). Sources were not consistent in the levels of metals indicating that they were supplied from different planting sites. Cu and Zn were within permissible consumer limits while Pb in many vegetables exceeded the limit. Continuous control of pollution and evaluation of metal content in vegetables is recommended in monitoring environmental contamination and food safety.

Key words: African indigenous vegetables, heavy metals, pollution, urban and peri - urban markets.

INTRODUCTION

The impact of industrial development on the environment in Kenyan agricultural sites cannot be underestimated. Environmental problems range from water, soil and atmospheric contamination. The pollution has a negative impact on the safety of food produced. Many studies have shown that food crops accumulate trace metals in their tissues when grown on contaminated soil or when watered with contaminated water (Ghosh et al., 2012; Sinha et al., 2010). It is recognized that land used for food production can be contaminated with Pb, Cu and Zn from natural and anthropogenic sources such as metal smelting, irrigation with wastewater, disposal of solid wastes, vehicular exhaust and adjacent industrial activity (Nabulo et al., 2012). Heavy metals are of considerable environmental concern due to their toxicity and cumulative behavior in human tissues (Sinha et al., 2010). Since they are not degradable (Ghosh and Singh, 2005; Salt et al., 1995), they last for long in the environment and thus readily accumulated at toxic levels (Tiwari et al., 2011). Copper and Zinc are essential trace elements (Olowoyo et al., 2012; Audu and Lawal, 2006); however, at high levels, they are known to cause oxidative stress (Sinha et al., 2010). Intake of these metals in low quantities is harmful because they are not easily eliminated from the body and cause cumulative poisoning (Islam et al., 2007). Excess zinc causes copper deficiency, anaemia, neutropenia, pancreatitictis, damage of hepatic parenchyma and vascular shock (Islam et

Table 1. Codes of the vegetables and markets.

Vegetable name	Code	Market	Code	Market	Code
African Nightshade	V1	Githurai 45	M1	Gikomba	M11
Swiss chard	V2	Mutindwa	M2	Kenyatta university KM	M12
African kale	V3	Kangemi	M3	Juja	M13
Spider plant	V4	Kawangware	M4	Ruiru	M14
Amaranth	V5	Dandora	M5	Buruburu	M15
Pumpkin	V6	Lunga Lunga	M6		
Cowpea	V7	Kibera	M7		
Jute mallow	V8	Kayole	M8		
Crotalaria	V9	Kenyatta market	M9		
Kale	V10	Mathare	M10		

al., 2007; Fosmire, 1990). Excess Cu is transported to the liver, and causes liver and kidney damage, anaemia, immunotoxicity, gastrointestinal distress, and genetic disorders such as Wilson's disease and idiopathic copper tocixity (ATSDR, 2004).

Despite the efforts to eliminate lead from sources such as gasoline, paints, batteries and household items, high levels are still reported in soils and vegetables (Nabulo et al., 2012). Harzadous effects of lead depend on dosage but the general effects include increased risk of hypertension, anaemia, reduced lifespan of erythrocytes, impairment of heme synthesis, neurological disorders, impaired cognitive skills in children, in rats, it disrupts normal bone growth and density and causes caries in children (ATSDR, 2007). Consumption of african indigenous vegetables (AIVs) in Kenya is high (Habwe et al., 2008; Shiundu and Oniang'o, 2007; Abukutsa-Onyango, 2005) because they are rich in essential minerals and vitamins, and are medicinal (Shackleton et al., 2009; Smith and Eyzaguirre, 2007; Orech et al., 2005). They are rich in antioxidants, ß-carotene, Fe, Ca and Zn (Smith and Eyzaguirre, 2007). Work on nutrient content of AIVs has gained momentum only in recent years. One study has shown that Amaranth is rich in ß-carotene and vitamin A (Faber et al., 2010), and traditionally, *Cochorus olitorius* is a blood purifier, Amaranth is a tapeworm expeller and *Vigna unguiculata* treats dermatitis and swellings (Ayodele, 2005). Spider plant is used to alleviate migraine, diphtheria, headache, pneumonia, septic ears and stomach ailments (Drissa, 2009). AIVs are easy to grow, many times sprouting on their own in undisturbed sites and are agronomically less-demanding (Abukutsa-Onyango et al., 2005). Vegetables consumed in Nairobi are either supplied from upcountry sources or farmed within the city in all available spaces especially in close proximity to water sources.

Urban and Peri-urban agriculture is an old and common practice in African towns (MOA, 2010; Shackleton et al., 2009). There is need to establish whether there is significant heavy metal accumulation in common AIVs that could poison consumers.

MATERIALS AND METHODS

Collection of samples

A purposive survey and collection was carried out in 15 markets namely, Githurai 45, Kibera, Kangemi, Kawangware, Kenyatta M-arket, Lunga Lunga, Mutindwa, Kayole, Dandora, Mathare, Gikomba, Kenyatta University KM, Juja, Ruiru and Buruburu (Table 1). The choice of sites was purposive taking care to include only markets whose vegetables were likely to have been sourced from contaminated sites. These sites were close to river banks and roadsides where vegetables are often planted. In each market, 3 traders were chosen randomly. The vegetables analysed were African Nightshade (*Solanum villosum*), Swiss chard (*Beta vulgaris*), African kale (*Brassica carinata)*, Spider plant, Amaranth (*Amaranthus)*, Pumpkin (*Curcubita moschata),* Cowpea (*Vigna unguiculata)*, Jute mallow (*Cochorus clitorius)*, Slenderleaf (*Crotalaria*) and Kale (*Brassica oleracea*). These are some of the commonest vegetables consumed in Kenya (Abukutsa-Onyango, 2007; Shackleton et al., 2009). Two kilograms of each vegetable were bought from each trader from all the markets. Not all vegetables; however, were available from the markets. Vegetable samples were composited, one vegetable type from three retailers to make one sample, then taken to the laboratory immediately for analysis. Samples were carried in cooler boxes to avoid undesirable physiological changes. They were dried for three days at 70°C then powdered using a miller.

Digestion of samples for analysis

Samples were digested in two replicates each. Five grams of fine powder was weighed into silica crucibles and charred to remove carbon in a block digester at 200°C until all smoke had gone. Material that is not charred may ignite at high temperatures causing self-combustion (Okalebo et al., 2002). The sample was dry-ashed at 550°C in a muffle furnace for 14 h. To the ash, 15 ml of 6 NHCl was added and the mixture heated for about 10 min at 250°C on a hot plate. This was meant to digest the minerals in the ash. The resulting digest was filtered using Whatmans No. 42 filter paper. The filtrate was topped up to 50 ml using distilled water and stored in plastic bottles in a cold room up to analysis.

Table 2. Pb content of vegetables collected from Nairobi markets.

Market	Vegetable										
	V1	V2	V3	V4	V5	V6	V7	V8	V9	V10	Mean
M1	*	0^{***a}		0.34^{ab}	0.14^{a}	0.224^{a}	0.22^{a}	0.24^{ab}	0.25^{a}	0.27^{ab}	0.21
M2	0.95^{a**}	1.37c	0.43^{a}	0.80^{b}	0.96^{b}		0.36^{a}				0.81
M3	0.65^{a}	0.10^{ab}	0.33^{a}	0.12^{a}	0.67^{ab}					0.18^{ab}	0.34
M4	0.66^{a}	0.27^{ab}		0.23^{ab}	0.38^{ab}	0.087^{a}	0.27^{a}	0.064^{a}	0.26^{a}	0.31^{ab}	0.28
M5	0.76^{a}			0.15^{a}	0.28^{ab}	0.26^{a}	0.22^{a}	0.96^{b}		0.92d	0.51
M6	0.14^{a}			0.07^{a}	0.11^{a}	0.16^{a}	0.27^{a}	0.28^{ab}		0.31^{ab}	0.19
M7	0.33^{a}	0.08^{ab}	0.16^{a}	0.22^{ab}	0.19^{a}	0^{a}	0.24^{a}			0.06^{a}	0.16
M8	0.23^{a}		0.38^{a}	0.41^{ab}	0.48^{ab}	0.06^{a}	0.18^{a}	0.27^{ab}		0.27^{ab}	0.29
M9		0.49^{b}	0.17^{a}	0.41^{ab}	0.71^{ab}	0.37^{a}	0.33^{a}	0.72^{ab}	0.32^{a}	0.03^{a}	0.39
M10	0.33^{a}	0.27^{ab}		0.61^{ab}	0.32^{ab}						0.38
M11	0.42^{a}	0.12^{ab}	0.11^{a}	0.20^{a}	0.22^{ab}	0.25^{a}	0.24^{a}		0.17^{a}	0.70cd	0.27
M12	0.57^{a}	0.53^{b}		0.39^{ab}	0.55^{ab}		0.68^{a}	0.50^{a}		0.49^{b}c	0.53
M13	0.10^{a}			0.3^{ab}	0.13^{a}	0.42^{a}	0.11^{a}				0.21
M14	0.17^{a}			0.08^{a}	0.33^{ab}	0.32^{a}	0.44^{a}			0.27^{ab}	0.27
M15	0.10^{a}		0.15^{a}	0.17^{a}	0.12^{a}	0.22^{a}	0.17^{a}			0.19^{ab}	0.16
Mean	0.42	0.36	0.25	0.30	0.37	0.22	0.29	0.43	0.25	0.33	0.21
SE	0.136	0.07	0.0814	0.156	0.119	0.1^{a}	0.10	0.1246	0.13	0.06	
P	0.008	<.001	0.144	0.003	0.003	0.197	<.001	0.014	0.85	<.001	

*Blanks indicate that the vegetable was missing in a market.
**Alphabetical letters indicate significance of variation within each vegetable among markets.
***Below detection limit.

Heavy metal analysis

Analysis was done using an Atomic Absorption Spectrophotometer (Shimadzu AAS-6200). The AAS was calibrated for all the metals by running standard solutions at concentrations of 0, 1, 1.5, 2 and 2.5 mg/kg. The sensitive wavelength for each metal was specified on the respective cathode lamps.

Data analysis

All data analysis was done using GENSTAT 14. Data obtained from the AAS was analyzed for variance (ANOVA) at $p<0.05$ to determine statistical variation in the concentration of Pb, Cu and Zn in the vegetables. Probabilities less than 0.05 ($p<0.05$) were considered statistically significant. Means were separated using Benferroni test.

RESULTS AND DISCUSSION

African Kale, Crotalaria and Jute mallow were found in fewer markets compared to other vegetables. Tables 2, 3 and 4 show the results of the concentrations of Pb, Cu and Zn, respectively; of the vegetables obtained using the AAS. Differences within vegetables and within markets were significant ($p<0.05$). Metal concentration of this study were in some cases higher while in others were lower than those obtained in other studies. Some vegetables analyzed here were not seen in other recorded studies for comparison purposes. Concentration of lead ranged from 0 (undetectable) to 1.37 mg/kg. The highest Pb content (1.37 mg/kg) was found in Swiss chard from Mutindwa followed by Amaranth (0.96 mg/kg, Mutindwa). Mutindwa had the highest mean Pb content (0.81 mg/kg) followed by Dandora (0.62 mg/kg). Mutindwa market is located along a very busy road while Dandora is surrounded by many sources of urban-planted vegetables and is close to the main Nairobi dumpsite. Othman (2001) doing similar work on various leafy vegetables collected from Dar es Salaam reported Pb concentration of 0 to 7.7 mg/kg in pumpkin and 0.2 to 6.6 mg/kg in cowpea, but in this study, Pb content was lower. There has been a worldwide elimination of Pb in fuel which may account for this discrepancy.

C. olitorius and *Amaranthus hybriduc* from Nigeria had Pb 0.05 to 0.33 mg/kg (Dosumu et al., 2003), while Pb in leafy and non-leafy was 0.1 to 15.8 mg/kg (Salariya et al., 2003), which were much higher than values in these study. Compared to other metals, Pb is low because it is not actively absorbed by plants (Alloway, 1995) and there are no known genes for its accumulation (Ghosh and Singh, 2005). Copper content in vegetables also varied significantly within vegetables and within sites. Values were between 0 and 19.15 mg/kg. The highest copper

Table 3. Cu content of vegetables collected from Nairobi markets.

Market	V1	V2	V3	V4	V5	V6	V7	V8	V9	V10	Mean
M1		6.32^{d}		7.48^{bcd}	13.57^{g}	2.35^{d}	8.67^{f}	6.37^{d}	9.41^{d}	10.88^{bc}	8.36
M2	8.81^{f}	18.00^{f}	8.66^{e}	11.6^{d}	2.24^{bcd}		12.28g				9.01
M3	9.15^{f}	3.31^{b}	7.77^{d}	12.24^{d}	7.78^{f}					13.35^{c}	10.01
M4	7.13^{e}	19.15^{g}		9.11^{cd}	0^{a}	14.88^{a}	0^{a}	0^{a}	0^{a}	3.66^{a}	4.21
M5	8.16^{ef}			9.4^{cd}	6.79^{f}	7.74^{bc}	9.21^{f}	9.81^{e}		5.11^{a}	8.01
M6	2.36^{bc}			4.37^{abc}	2.92^{de}	7.86^{bc}	2.50^{de}	1.91^{e}		1.97^{a}	3.56
M7	1.47^{ab}	1.08^{a}	0.5^{ab}	0.804^{a}	2.84^{de}	1.66^{d}	1.35^{bcd}			0.88^{a}	1.34
M8	4.74^{d}		1.17^{b}	1.754^{a}	2.22^{bcd}	1.91^{d}	0.76^{abc}	6.48^{d}		2.97^{a}	2.50
M9		4.37^{c}	3.84^{c}	4.9^{abc}	0.98^{ab}	6.78^{c}	1.40^{bcd}	3.92^{c}	1.6^{b}	1.22^{a}	3.10
M10	4.93^{d}	1.22^{a}		11.27^{d}	1.26^{abc}						5.73
M11	2.45^{bc}	1.88^{a}	0^{a}	2.07^{ab}	2.38^{bcde}	1.72^{d}	1.88^{cd}		3.37^{c}	5.57^{ab}	2.42
M12	0.75^{a}	11.27^{e}		1.07^{a}	1.24^{abc}		0.00^{a}	3.79^{c}		0.00^{a}	1.45
M13	0.79^{a}			0^{a}	0^{a}	9.13^{b}	3.58^{e}				3.08
M14	3.45^{c}			2.83^{ab}	3.99^{e}	5.88^{c}	0.63^{ab}			1.96^{a}	3.07
M15	1.82^{ab}		4.4^{c}	1.32^{a}	3.09^{de}	7.31^{bc}	0.77^{abc}			3.67^{a}	3.39
Mean	3.96	6.21	3.06	4.75	2.78	6.45	2.12	4.36	2.14	3.72	
SE	0.19	0.15	0.1271	0.896	0.26	0.345	0.19	0.1501	0.209	0.89	
P	<.001	<.001	<.001	<0.001	<.001	<0.001	<.001	<.001	<.001	<.001	

*Blanks indicate that the vegetable was missing in a market.
**Alphabetical letters indicate significance of variation within each vegetable among markets.
***Below detection limit.

Table 4. Zn content of vegetables collected from Nairobi markets.

Market	V1	V2	V3	V4	V5	V6	V7	V8	V9	V10	Mean
M1		51.46^{ef}		46.01^{cd}	47.31^{e}	26.55^{ef}	44.44^{fg}	22.32^{a}	26.82^{a}	54.48^{gh}	39.16
M2	59.29^{g}	110.19^{h}	9.74^{a}	59.33^{f}	81.01^{i}		69.85^{h}				67.14
M3	83.79^{i}	105.3^{h}	77.12^{e}	42.63^{b}	88.78^{j}					76.88^{i}	85.43
M4	21.76^{b}	120.16^{i}		34.56^{a}	37.47^{c}	25.05^{f}	40.99^{ef}	31.82^{c}	32.91^{c}	63.28^{h}	49.63
M5	47.24^{ef}			47.32^{d}	37.66^{c}	29.24^{d}	50.24^{g}	50.24^{e}		33.47^{b}	41.48
M6	42.6^{de}			48^{de}	50.2^{f}	35.24^{c}	37.71^{cde}	42.69^{d}		41.23^{e}	42.51
M7	80.67^{i}	39.19^{bc}	68.55^{d}	44.76^{bc}	57.38^{g}	38.73^{b}	48.03^{g}			52.96^{g}	51.23
M8	45.44^{e}		35.74^{b}	60.75^{g}	26.22^{b}	28.19^{de}	27.89^{ab}	27.02^{b}		24.02^{a}	33.03
M9		27.81^{a}	55.76^{c}	50.6^{f}	60.17^{h}	15.61^{g}	35.96^{cde}	21.25^{a}	31.27^{b}	38.11^{d}	35.93
M10	39.87^{cd}	43.61^{cd}		104.01^{h}	79.71^{i}						63.37
M11	50.24^{f}	31.98^{ab}	77.48	49.63^{e}	51.39^{f}	54.35^{a}	38.96^{def}		39^{d}	50.38^{f}	49.10
M12	24.15^{b}	58.42^{fg}		36.19^{ab}	25.18^{b}		36.95^{cde}	32.3^{c}		35.51^{c}	37.31
M13	15.88^{a}			45.68^{c}	21.31^{a}	33.08^{c}	31.74^{bc}			25.51^{a}	31.03
M14	65.86^{h}			41.16^{b}	45.23^{d}	35.05^{c}	32.65^{bcd}			32.84^{b}	38.18
M15	37.4^{c}		68.41^{d}	48.46^{de}	19.8^{a}	35.14^{c}	22.05^{a}			0.28	32.46
Mean	47.25	66.21	56.11	49.88	48.67	32.91	39.45	33.98	33.92	39.63	
SE	0.0761	0.404	0.422	1.205	0.3067	0.384	1.00	0.205	0.136	<.001	
P	<.001	<.001	<.001	<0.001	<.001	<0.001	<.001	<.001	<.001	11	

*Blanks indicate that the vegetable was missing in a market.
**Alphabetical letters indicate significance of variation within each vegetable among markets.
***Below detection limit.

content was in spider plant from Kawangware, followed by 18 mg/kg from Mutindwa. Kawangware had the highest mean Cu content (10.3 mg/kg) followed by Mutindwa at 9 mg/kg. Othman (2001) reported that copper levels were 0.22 to 1.47 mg/kg in pumpkin and 0.25 to 1.42 mg/kg in cowpea while for this study copper was much higher at 19.3 mg/kg in spider plant. Other studies of leafy vegetables from markets, contaminated sites or vegetables watered with contaminated water recorded up to 2.2 mg/kg (Iyaka, 2007), 0.1 to 1.5 mg/kg (Gogoasai et al., 2005) and 7.8 mg/kg (Yebpella et al., 2010). Besides the variation in conditions, the vegetables were exposed to these differences, may be due to the different capabilities of different vegetables to absorb minerals and ions.

Zinc was also significantly different in vegetables and sites had values of 15.6 to 120.2 mg/kg. The highest level of Zn was found in Swiss chard from Kawangware (120.2 mg/kg). The market with the highest Zn content for all the vegetables was Kangemi (85 mg/kg). Zn was higher in this study (120.2 mg/kg) than Othman (2001) whose highest level was 56.7 mg/kg in pumpkin. Zn in leafy vegetables from contaminated sites or markets in various studies were recorded as 7 mg/kg (Iyaka, 2007), Zn 1.2 to 21 mg/kg (Gogosai et al., 2005), 0.41 to 1.06 mg/kg (Dosumu et al., 2003), 26.7 mg/kg (Yebpella et al., 2010) and 10.5 to 80.92 mg/kg (Salariya et al., 2003) which were below and within range, but all lower than the highest value in this study.

According to Taber (2009), normal Cu content in plants should range from 2 to 20 mg/kg while for Zn it is 25 to 300 mg/kg. The amount found in this study was therefore within acceptable range. Below 2 mg/kg, Cu is deficient; none of the vegetables was deficient. Although, Zn is sometimes deficient in vegetables (Yebpella et al., 2010) from poor soils, none of the vegetables had deficient content. Pb content was 0.3 mg/kg which was above the allowable limit.

Conclusion

Earlier studies of Mutuku (2005) working on kale reported high levels of Pb (300 mg/kg), however, the current study recorded that most sites are within safe limits for consumption. This may be due to efforts to reduce Pb in fuel. Metal contents were higher in some sites compared to others. There is need for relevant authorities to delineate safe planting sites. Although, only Pb in some sites was above the permissible limits, long-term metal exposure by regular consumption of locally grown vegetables poses potential health problems to humans and animals. There is need for more environmental monitoring to ensure Pb sources to water, soil and plants are contained. Urban Agriculture in Kenya is associated with sewage usage for watering vegetables and is perceived illegal, albeit being accepted in MOA (2010). In all the markets; however, some vegetables are from upcountry sources.

REFERENCES

Abukutsa-Onyango M (2007). Seed production and support systems for African leafy vegetables in three communities in Western Kenya. Afr. J. Food Agric. Nutr. Dev. 7:3.

Abukutsa-Onyango MO, Muriithi AN, Ngamau K, Anjichi V, Angong SG, Fricke A, Hau B, Stutzel H (2005). Proceedings of the 3rd Horticultural Workshop on Sustainable Horticultural production in the Tropics. 23rd to 29th Nov. 2003. Maseno University, MSU. Maseno, Kenya.

Alloway BJ (1995). Heavy metals in Soils. 2nd edn. Blackie Academic and Professional. London, England.

ATSDR (2004). Toxicological profile for copper. U.S. Department of Health and Human Services.

ATSDR (2007). Toxicological profile for lead. U.S. Department of Health and Human Services.

Audu AA, Lawal AO (2006). Variation in Metal Contents of Plants in Vegetable Garden Sites in Kano Metropolis. J. Appl. Sci. Environ. Manage. 10 (2):105 - 109.

Ayodele AE (2005). The medicinally important leafy vegetables of south western Nigeria. Available from: http://www.siu.edu/~ebl/leaflets/ayodele.htm. Accessed 8th February 2013.

Dosumu OO, Salami N, Adekola FA (2003). Comparative study of trace element levels in some local vegetable varieties and irrigation waters from different locations. Bull. Chem. Soc. Ethiop.17(1):107-112.

Drissa S (2009). Spider plant: An indigenous species with many uses. AVRDC; The World Vegetable Center.

Faber M, Oelofse A, Van-Jaarsveld PJ, Wenhold FAM, Jansen RWS (2010). African leafy vegetables consumed by households in the Limpopo and KwaZulu-Natal provinces in South Africa. South Afr. J. Clin. Nutr. 23:1.

Fosmire GJ (1990). Zinc Toxicity. Am. Soc. Clin. Nutr. 51(2): 225-227.

Ghosh AK, Bhatt MA, Agrawal HP (2012). Effect of long-term application of treated sewage water on heavy metal accumulation in vegetables grown in Northern India. Environ. Monit. Assess. 184:1025-1036.

Ghosh M, Singh SP (2005). A review of phytoremediation of heavy metals and utilization of its by-products. Appl. Ecol. Environ. Res. 3(1):1-18.

Gogosai I, Jianu I, Rivis A, Parvu D, Gergen I, Jianu C (2005). Heavy metal distribution in soil-water-vegetable system of Banat area. Annals of the Faculty of Engineering Hunedoara 3:1.

Habwe FO, Walingo MK,Abukutsa-Onyango M (2008). Food Processing and preparation Technologies for utilization of African Indigenous Vegetables for nutrition security and wealth creation in Kenya. In: Robertson, G.L. and Lupien J.R. (eds). Using food and technology to improve nutrition and promote national development. International Union of Food Science and Technology.

Islam EU, Yang XE, He ZL, Mahmood Q (2007). Assessing potential dietary toxicity of heavy metals in selected vegetables and food crops. J. Zhejiang Univ. Sci. B. 8(1):1-13.

Iyaka YA (2007). Concentration of Cu and Zn in some fruits and vegetables commonly available in North-Central zone of Nigeria. Electron. J. Environ. Agric. Food Chem. 6(6):2150-2154.

MOA (Ministry of Agriculture) (2010). National urban and Peri-urban agriculture and Livestock policy. May 2010. Ministry of Agriculture, Kenya.

Mutuku J (2005). Lead levels in Nairobi, Kenya. www.unep.org/pcfv. 10th April 2010

Nabulo G, Black CR, Craigon J, Young SJ (2012). Does consumption of leafy vegetables grown in peri-urban agriculture pose a risk to human health? Environ. Pollut. 162:389-398.

Okalebo RJ, Gathua KW, Woomer PL (2002). Laboratory methods of soil and plant analysis. A working manual. TSBF-CIAT ans SACRED

Africa, Nairobi, Kenya.
Olowoyo JO, Okedeyi OO, Mkolo NM, Lion GN, Mdakane ST R (2011). Uptake and translocation of heavy metals by medicinal plants around a waste dump site in Pretoria, South Africa. South Afr. J. Bot. 78:116-121.
Orech FO, Akenga T, Ochora J, Friis H, AagardHansen J (2005). Potential Toxicity of Some Traditional Leafy Vegetables Consumed in Nyang'oma Division, western Kenya. Afr. J. Food Nutr. Sci. 5:1.
Othman OC (2001). Heavy Metals in green vegetables and soils from vegetable gardens in Dar es Salaam, Tanzania. Tanzan. J. Sci. 27:37-48
Salariya AM, Rehman ZU, Ashraf M (2002). Effect of polluted water on accumulation of heavy metals in commonly consumed vegetables. J. Chem. Soc. Pak. 25:3.
Salt DE, Blaylock M, Nanda PBAK, Dushenkov V, Ensley BD, Chet I, Raskin I (1995). Phytoremediation: a novel strategy for the removal of toxic metals from the environment using plants.Biotechnology 13:468-474
Shackleton CM, Pasquini MW, Drescher AW (2009). African indigenous vegetables in urban agriculture. Earthscan. London, England.
Shiundu KM, Oniang'o RK (2007). Marketing African Leafy Vegetables: Challenges and Opportunities in the Kenyan Context. Afr. J. Food Agric. Nutr. Dev. 7:4.
Sinha S, Sinam G, Mishra K, Mallick S. (2010). Metal accumulation, growth, antioxidants and oil yield of *Brassica juncea* L. exposed to different metals. Exoticol. Environ. Safe. 73:1352-1361.
Smith IF, Eyzaguirre P (2007). African leafy vegetables: their role in the World Health Organization's global fruit and vegetables initiative. Afr. J. Food Agric. Nutr. Dev. 7:3.
Taber HG (2009). Plant Analysis Sampling Procedures and Micronutrient Characteristics with Emphasis on Vegetable Crops. Commercial Vegetables, Horticulture. Iowa State University of Science and Technology.
Tiwari KK, Singh NK, Patel MP, Tiwari MR, Rai UN (2011). Metal contamination of soil and translocation in vegetables growing under industrial waste water irrigated agricultural field of Vadodara, Gujarat, India. Ecotoxicol. Environ. Safe.74:1670-1677.
Yebpella GG, Magomya AM, Udiba UU, Gandu I, Amana SM, Ugboaja VC, Usman NL (2010). Assessment of Cd, Cu, Mn and Zn Levels in Soil, Water and vegetable Grown in Irrigated Farm along River Kubani, Zaria, Nigeria. J. Appl. Environ. Biol. Sci. 1(5):84-89.

The cost of poor land use practices in Lake Nakivale Wetland in Isingiro District, Uganda

Bintoora Kamukasa Kakurungu Adonia[1,2]

[1]School of Sciences, Nkumba University, P. O. Box 237, Entebbe, Uganda.
[2]Mt Elgon Conservation Area, P. O. Box 135, Mbale, Uganda.

The land use practices around Lake Nakivale Wetland have over the last five decades changed from dominantly cattle keeping to crop farming, human settlement and urbanisation. This change, has increased food production and security, it has come at a cost to the natural environment in form of land degradation, wetland encroachment, loss of wildlife habitats and crucial wetland resources. Based on a research conducted in Lake Nakivale Wetland in Isingiro District in Uganda, this paper assessed the cost residents incur as wetland resources experience external disturbances from poor land use practices. Collection of data involved a household survey, interviews, focused group discussions and wetland resource ranking. The impact analysis approach was used to estimate the value of the affected wetland resources. This was based on the market price and the concept of willingness to either pay or accept compensation for wetland resources. The findings indicate that cost of poor land use practices on vital wetland resources is enormous, estimated to be worth US $ 2,943,960.3 per annum and is expected to continue to increase in the future, putting the livelihoods of residents in dilemma. Sustainable land use management based on sound land use planning and restoration program is proposed to minimise the negative effects of land use on the wetland.

Key words: Land use practices, wetland resources, livelihoods, sustainable land use management, land use planning, wetland resource value.

INTRODUCTION

Lake Nakivale Wetland (referred to as Lake Nakivali Wetland in the list of Ramsar Sites) is part of the larger Lake Mburo- Lake Nakivale Wetland System (26,834 ha) It was designated as a Wetland of International Importance by the Conference of Parties to the Convention on Wetlands (Ramsar Convention) on 15th September 2006 because it supports globally threatened species of birds, two of the endangered cichlid fish species *(Astatoreochromis* and *Astatotilapia* sp.) and is the only area in Uganda where the Impala (*Aepyceros melampus*) is found (WMD and NU, 2008). The wetland system falls within geographic coordinates of 30°49′ - 31°04′ East and 00°33′ - 00°47′ South. Lake Nakivale is 14 km long, 6 km wide and has a maximum depth of 3.5 m at high water level (NAFIRRI, 2010). Lake Nakivale Wetland intertwines with Lake Kachera and Lake Kijjanibarola Wetlands through River Rwizi Wetland which feeds into Lake Victoria.

The wetland system is not only a site of international importance but plays a great social-economic role at local level as a source of water for domestic use, livestock and wildlife, pasture for the local herds of cattle during droughts, fish and materials for handcrafts and thatching houses. Like any other wetland in Uganda it contributes directly and indirectly to national economy. The wetlands' contribution to the Gross Domestic Product is estimated to be Uganda Shillings 6.5 to 7.0 billion (GOU, 2002). In addition, they provide direct income opportunities to rural communities and indirect benefits in form of environmental goods and services such as purification of water, control of floods and water storage that improves the livelihoods of rural population (Mugisha, 2011; Opio, 2008).

However, in Uganda, conversion of wetlands to other land uses is widespread and increasingly affecting the wetland dependent communities in both urban and rural areas. Poor land use practices around the Wetlands have stimulated human induced environmental problems which have negatively affected the availability and socio-economic value of wetland resources that are crucial to the livelihoods of neighbouring local communities. The lost wetland resources and their socio-economic value indirectly reflect the conservation cost the rural people incur as they search for either scarce wetland resources or substitutes. Consequently, residents walk long distances, and spend more time looking for scarce resources. Before 1994, Lake Nakivale Wetland was well protected and the main human activities that used to take place were cattle keeping, fishing and harvesting of wetland materials for domestic use. However, with time, the land use and farming practices within the landscape systematically changed from cattle keeping to crop farming. The nationals and refugees shifted away from livestock husbandry to crop cultivation. As a result of this shift in land use practice, the wetland started to experience a lot of human pressure in the form of conversion to create more areas for food production, human settlement and urbanisation.

Consequently, crop farming activities and other forms of poor land use practices have had a spill over effects on the wetland. The wetland forests have been converted into crop gardens, leading to soil erosion and subsequently, the siltation of Lake Nakivale. Human settlement on the edge of the wetland coupled with poor human waste disposal has affected the quality of water. Overgrazing of livestock has degraded the shallow parts of the wetland, flood plain and steep slopes of hills around Lake Nakivale Wetland, leading to soil erosion and subsequently, the siltation of water bodies in the area. In addition, the clearance of vegetation cover and conversion of some sections of Lake Nakivale wetland to crop farming has contributed to the scarcity of useful plants that rural people have for centuries depended on as source of medicine and food. By and large, the residents now incur costs in terms of time spent searching for the scarce wetland resources. In practical terms, these costs indirectly represent the socio-economic values of the lost wetland resources.

What is of great concern is that although Lake Nakivale Wetland is serving a multi-ethnic community with people originating from different countries and is of great value to the people of South - Western Uganda due to its ecological, economic and social values, little efforts have been made to study the impact of poor land use practices on the wetland. The few studies with regards to land use which have been carried out (Lambrecht, 1964; Breyer et al., 1997; Place and Otsuka, 1997; Tukahirwa, 2000; Makalle et al., 2008) focus on soil erosion, land use cover, livelihoods and population pressure but not necessarily the wetland resource values lost. Another challenge is that while most of the benefits of wetlands go to the general public, the costs of conservation often fall on the landholders (Musamba et al., 2011). On this basis, wetlands cannot be managed in isolation without taking into account the neighbouring land use (Houlahan et al., 2006). Poor land use practices around Lake Nakivale have resulted into land degradation and other human induced environmental problems such as soil erosion, wetland encroachment, wildlife habitat loss as well as deforestation leading to the scarcity of crucial wetland resources which rural people depend on for survival. Ramsar Convention Secretariat (2007) confirms that degradation of wetlands by poor land use practices have resulted in among others; vegetation destruction and sedimentation. Fennessy and Craft (2011) also attribute the loss of wetland ecosystem services such as the ability to regulate water movement and biogeochemical cycles to the conversion of wetlands into agricultural land.

This paper is a result of a study that was conceived on the realisation that there is urgent need to address threats to the conservation of Lake Nakivale Wetland which emanate from poor land use practices such as poor farming methods, wetland reclamation, human settlement and urbanisation on the fringes of the wetland. Otherwise, this important ecosystem is likely to be lost in the near future. It recognises the fact that although conversion of wetlands to other land uses such as crop production seems to be the cheapest way for local people to meet their basic needs (Musamba et al., 2011), its impact on the wetland resource values is a cost the same people must bear.

Nevertheless, the spill over impacts of conversion of the wetland and surrounding land cover may also be felt at a much larger spatial scale (Marshall and Shortle, 2005). In other words, the cost of poor land use practices are losses in wetland values arising from damages to the wetland resources. Barbier et al. (1997) noted that such damages would amount to the losses in net production benefits and the total cost of this impact in terms of damage to the wetland, are the foregone benefits.

In this regard, however, the affected communities appear not to have fully internalised and appreciated the causes and magnitude of the damage on wetland resources emanating from land use related externalities. The same applies to the decision makers who seem to be constrained by insufficient understanding of the socio-economic values of wetlands as noted by Schuijt (2002). Based on the findings, sustainable land use management

Table 1. Estimated cost of poor land use practice in Lake Nakivale Wetland as perceived by the respondents.

Cost per month (million Ug. Shs)	No. of respondents	Percentage	Cumulative percentage
Less than 1	29	9.2	9.2
1 - 10	201	63.8	73.0
11 - 20	33	10.5	83.5
More than 21	52	16.5	100
Total	315	100	

Table 2. Value of wetland pasture utilized by livestock.

Type of domestic animal	Estimated number	Daily intake of fodder per animal (kg)	Quantity of fodder required (kg)		Value of fodder based on market price (Ug. Shs) '000'	
			Daily	Annual	Daily	Annual
Cows	675	12.5	8,437.5	1,139,062.5	13,500,000	1,822,500
Goats	756	8.0	6048.0	816,480	9,676,800	1,306,368
Total			14.485.5	1,955,542.5	23,176,800	3,128,868

Daily pasture intake rate was adapted from Babu (2002) and value of fodder is based on surrogate market price of livestock feeds in Mbarara Municipality.

Table 3. Value for papyrus reeds.

Use of papyrus	No. of bundles per unit/ house or user	Estimated total bundles	Price of unit/ bundle excluding the cost of harvesting (Ug. Shs.)	Total cost Ug. Shs)
Thatching	65	1,050	2,250	2,262,500
Supporting walls	7	3,500	2,250	7,875,000
Handcraft	4	640	2,250	1,440,000
Total		5,190		11,577,500

The price of bundle of papyrus is based on local market price.

which is hinged on sound land use planning and restoration of already lost important portions of the wetland ecosystem is recommended.

MATERIALS AND METHODS

To assess the cost of poor land use practices on Lake Nakivale Wetland, the impact analysis approach (Barbier et al., 1997; De Groot et al., 2006) was used. This study considers poor land use practices to be external disturbance on the wetland ecosystem that leads to the degradation of resource values. The cost of land use practices is not only reflected in losses of wetland values arising from damage to the wetland system and its resources but time spent by wetland dependent local people while searching scarce resources. Using participatory rural appraisal tools (Conroy, 2002; Abdullah et al., 2012), the most affected wetland resources were identified, prioritized and quantified based on average annual harvest and the local people's preference. The monetary value of affected wetland resources was established based on either market price or indirect substitutes. The wetland resource values were grouped into three categories of direct use values for those resources that local people derive values from direct use, Indirect use values which cover the indirect support and protection and option or existence values (Barbier et al., 1997; Barton, 2002; Millennium Ecosystem Assessment, 2005). The valuation of affected resources took into consideration, the economic concept of willingness to pay (Babu et al., 2002; Ramchandra and Rajinikanth, 2003; Turpie et al., 2010) to attach monetary value on wetland resources that have readily available market price. Where wetland resources or services that do not have market prices such as pasture for livestock were encountered, techniques such as indirect substitute (opportunity cost of using a substitute for the wetland resources or service) and surrogate market prices (the price of related goods and services) were applied. Contingent valuation method (CVM) which involves the asking of individuals how much they are willing to pay to ensure that wetland is preserved was integrated into the survey questionnaire to assess their willingness to pay or accept compensation (Leschine et al., 1997; Subade, 2005) for a change in availability of Lake Nakivale Wetland.

In total, 315 questionnaires were administered to household heads in three randomly selected parishes bordering the wetland. Three out of ten parishes bordering Lake Nakivale Wetland were randomly selected for household survey. The parishes which were selected are Kamuri, Kankingi and Kyabishaho with a total of 1,750

Table 4. Value of water.

Use water	Unit quantity per household/animal (litre)	Total daily quantity (litre)	Total annual quantity (litre)	Value of daily water consumed excluding costs of fetching (Ug. Shillings)	Value of annual water consumption excluding costs of fetching (Ug. Shs)
Domestic use	40	70,000	25,620,000	3,500,000	1,281,000,000
Cows	20	13,500	4,941,000	675,000	247,050,000
Goats	1	756	276,696	37,800	13,834,800
Total		84,256	30,837,696	4,212,800	1,541,884,800

The daily water intake of cows and goats was estimated based on Duguma et al. (2012).

Table 5. Value of fuel wood.

Use of fuel wood	Quantity for a month (bundles)	Annual quantity (bundles)	Monthly value of fuel wood excluding costs of collection (Ug. Shs)	Annual value of fuel wood excluding costs of collection (Ug. Shs)
Cooking and other domestic use	21,000	252,000	63,000,000	756,000,000
Drying fish	864	10,368	4,320,000	51,840,000
Burning of bricks	12,000	144,000	60,000,000	720,000,000
Total	33,864	406,368	127,320,000	1,527,840,000

The value of fuel wood is based on local market price.

Table 6. Summary of the cost of poor land use practices on wetland resources.

Nature of resource	Category	Estimated annual value (Shs)
Fish	Direct use	1,000,771,720
Pasture	Direct use	3,128,868,000
Papyrus reeds/building materials	Direct use	11,577,500
Water	Direct use	1,541,884,000
Firewood	Direct use	1,527,840,000
Herbal medicine	Direct use	38,112,000
Recreation (historical and cultural sites)	Indirect use	6,650,000
		7,359,900,720

households with estimated human population of 10,745 people. The reliability (confidence level) of 95% and the acceptable margin of error of 5% was used to come up with the sample size (number of sampled households). The unit of observation in this study was a household and the unit of analysis was the household head. The questionnaire was supplemented by face to face interviews and focused group discussions in order to minimise biases inherent in CVM. In total, four groups discussion were held as follows: Rukinga Fishing village (n = 10), Kikutsi Fishing Village (n = 6), Rubondo village (n = 8) and Kabazana village (n = 10). For wetland resources, the net monetary value of benefits derived by the wetland neighbouring communities was calculated.

RESULTS

The study results showed that most crucial wetland resources that were being affected by poor land use practices are fish, pasture for domestic animals, papyrus reeds for construction, water for livestock and domestic use, firewood, medicinal plants as well as cultural sites.

Contingent value of Lake Nakivale Wetland resources

The assessment of the residents' willingness to accept somebody from outside their society to buy and own Lake Nakivale Wetland indicated that 94.3% of the respondents (n= 297) would not accept. Only 5.7% of the respondents (n = 18) were willing to accept the idea of selling off the wetland but only at exorbitant price. For instance, nine respondents (2.9%) suggested a price of more than Ug. Shs 20 million (more than US $ 10,000) per acre. Similarly, 2.2% of the respondents proposed a

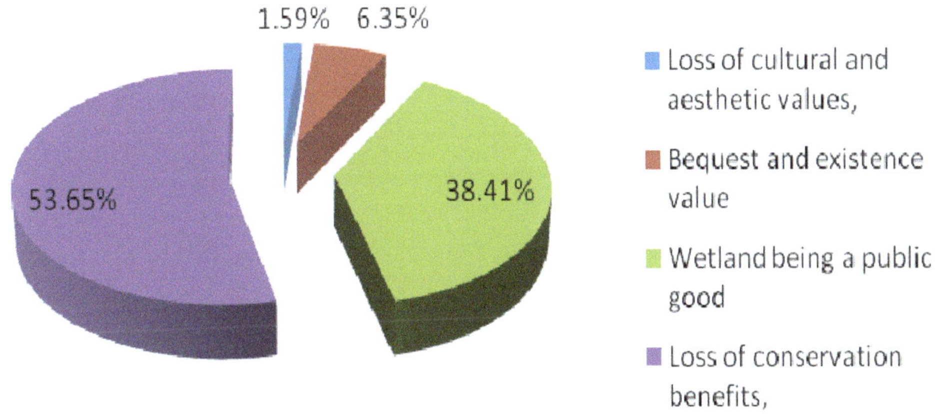

Figure 1. Reasons for not selling Lake Nakivale Wetland.

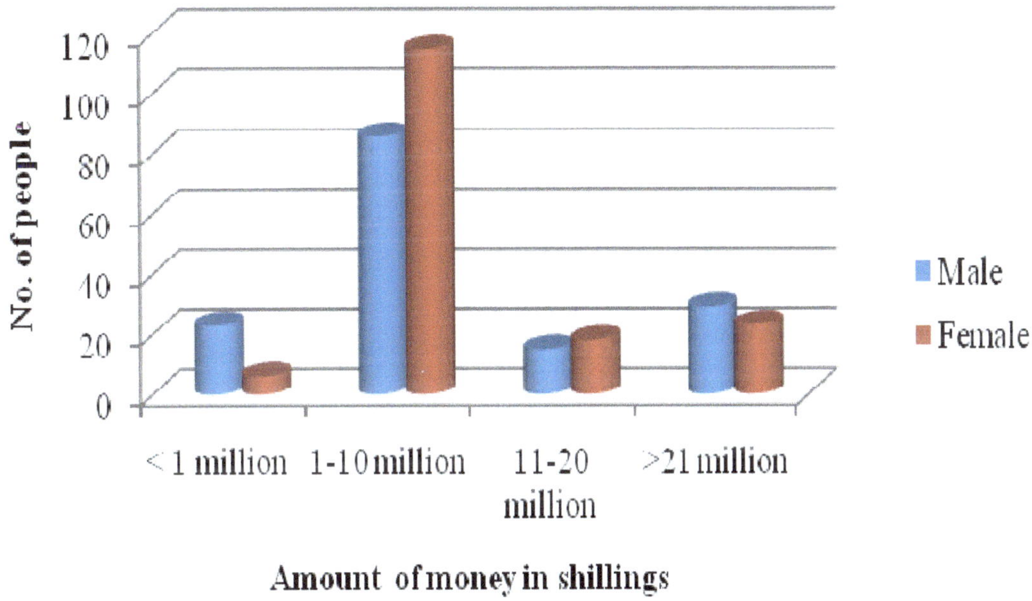

Figure 2. Estimated cost of wetland resources per month according to gender

price ranging from Ug. Shs 2 to 10 million (US $ 500 to 5000) per acre. Only two respondents (0.6%) were willing to accept a price within a range of Ug. Shs 11 to 20 million (US$5500-10,000) per acre.

As to why the majority of respondents (94.3%) were not willing to allow an outsider buy and privately own the wetland, several reasons were given. Most of them (53.7% of the respondents) feared they would lose ecological and economic goods and services the wetland provides. Other 41% of the respondents (n = 121) indicated that Lake Nakivale Wetland was a public good which every member of the society had a right to access and utilise. They considered the wetland to be a public asset that should not be owned by an individual but the whole society. There were 1.6% of the respondents (n = 5) who indicated that selling of the wetland was likely to lead to the loss of cultural values of Lake Nakivale Wetland. Only 6.3% of the respondents (n = 20) attached bequest and existence values to the wetland (Figure 1). They were not sure that a private owner would guarantee the long term existence of the wetland and its future benefits.

As to how much neighbouring communities were likely to lose if the current poor land use practices were not checked, 63.8% of the respondents (n = 201) estimated the loss to be between 1 and 10 million shillings per month (Figure 2). While 10.5% of the respondents (n = 33) indicated that it was in the range of Ug. Shs 11 to 20 million per month, only 9.2% of the respondents (n = 29) suggested a monetary value that was less than one

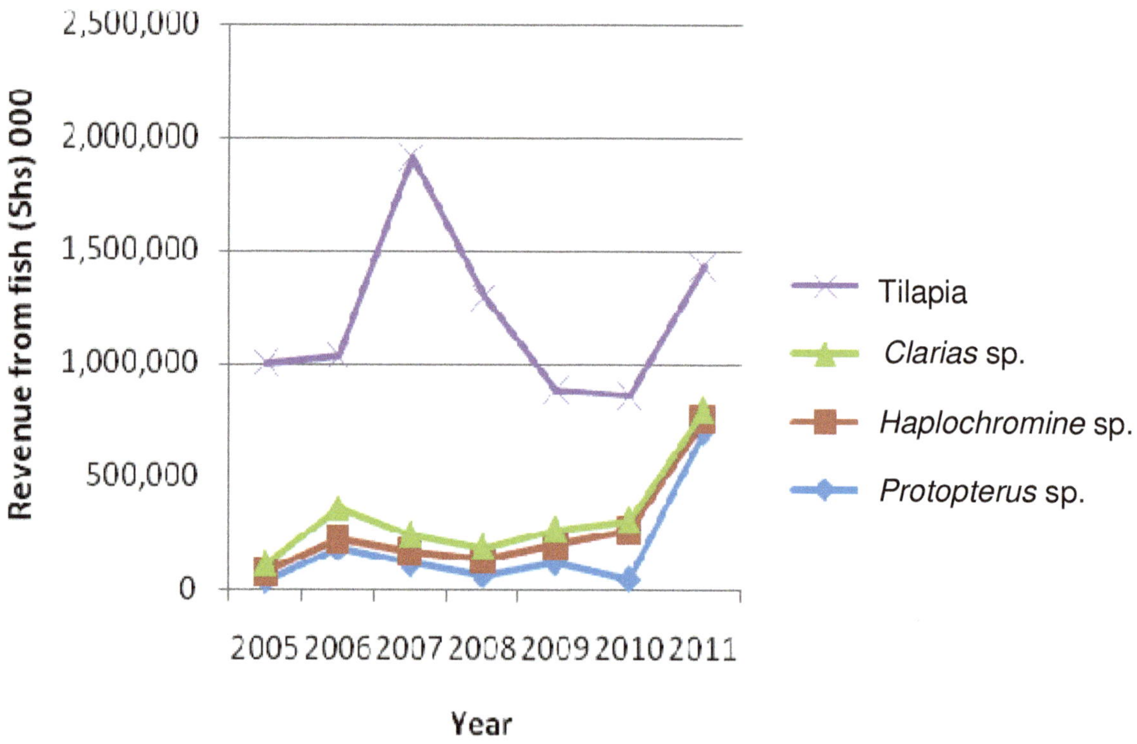

Figure 3. Revenue from fish harvested in Lake Mburo (part of Lake Mburo-Nakivale wetland system) from 2005 to 2011 (Source: Lake Mburo National Park Annual Reports 2005-2011).

million Uganda shillings per month. Using mode as a measure of central tendency of clusters of estimated costs of poor land use practices on the wetland, it was evident that cluster of 1 to 10 million was mentioned most frequently (202 respondents/63.8%)(Figure 2).

When Pearson Chi-square (X^2) for the respondents who considered the magnitude of expected cost/loss of the wetland resources in monetary terms per months in relation to sex was applied, the observed significant level of 0.002 was obtained. This implies that hypothesis that residents' perception on the cost of poor land use practices and gender is independent is rejected. In other words, women were less willing to accept compensation and idea of selling the wetland than men.

Use of market price to establish the cost of losing wetland resources

Fish

The assessment of fish harvest from Lake Mburo which shares the same wetland system with Lake Nakivale shows that neighbouring communities earn a living from fisheries resources. If the current human induced siltation of Lake Nakivale is not checked, residents are likely to lose fisheries resources worth above Ug. Shs 1 billion (US $ 400,000) per annum (Figure 3).

Pasture

The study established that residents were grazing their livestock especially goats and cows in Lake Nakivale Wetland and associated floodplains for a period ranging from 3 to 6 months depending on the intensity of dry seasons and proximity of their homesteads to the wetland. A total of 82 (26%) respondents owned livestock and other domestic animals. Approximately 14.3% of the respondents (n = 45) had cows with average heads of cattle of 15 while 11.4% of the respondents (n = 36) had goats (on average each had 21 goats). Although, all domestic animal owners were found to be relying on Lake Nakivale and associated wetland for watering their animals throughout the year, grazing was seasonal. The mean grazing period X^{-} was 4.5 months (135 days). Based on the average market price of animal feeds of Shs 1,600 per kilo, residents were likely to lose Ug. Shs 3,128,868,000 (US $ 1,251,547) per annum if the current poor land use practices were not checked and the remaining grazing area converted to crop fields.

Papyrus reeds

Local people living close to Lake Nakivale Wetland depended on papyrus reeds for house thatches (Figure 4), make handcrafts and as a source of energy (alternative to fuel wood). On average, 50 to 80 bundles of papyrus were needed to thatch a single house depending

Figure 4. A mud and wattle house thatched with papyrus reeds.

on its size and strength. Even small houses that were thatched with strong polythene covers needed six to eight bundles of papyrus reeds to support mudding of the walls. The market price of each bundle was in the range of Ug Shs 2,000 to 2,500 depending on the proximity of the wetland to the site of the house.

In total, 1050 houses had their roofs thatched with papyrus reeds, 3500 housing units had their walls supported by either papyrus reeds or palm stalks and 40 residents, all of them women, were using on average 160 bundles of papyrus reeds on quarterly basis to make handcrafts such as mats. A total of 5,190 bundles of papyrus reeds valued at Shs11,577,500 (US $4,631) per annum are utilized by the residents in the sampled parishes as building and handcraft materials. This excludes some isolated cases whereby papyrus reeds were being used for house ceiling; fence around homes and makeshift bathrooms.

Water

On average, 30,837,696 L of water from Lake Nakivale Wetland system worth Shs 1,541,884,800 (US $ 616,754) was estimated to be utilized annually for domestic use and watering animals.

Fuel wood

Residents depended on fuel wood harvested from Lake Nakivale Wetland as a source of energy for domestic use, fish drying and brick making. A bundle of fuel wood for cooking was being sold at Ug Shs. 3000, while the one for drying fish and burning bricks was being sold at Shs 5,000. Unlike the bundle of firewood for cooking, the brick burning and fish drying fuel wood bundle was relatively bigger but with fewer stems (pieces of wood). On average, 33,864 bundles of wood worth Shs 127,320,000 (US $ 50,928) were harvested on monthly basis.

Annually, approximately 406,368 bundles worth Shs 1,527,840,000 (US $ 611,136) were collected from the wetland system.

This vital resource was becoming scarce due to conversion of wetland forests into crop fields.

Herbal medicine

Although 62% of the respondents (n = 192) admitted to have used herbal medicine harvested from Lake Nakivale Wetland, not every user was involved in harvesting medicinal plants. There were well known traditional herbalists and healers were specialised in harvesting, processing, prescribing and administering herbal medicine. The common ailments treated by herbs harvested from Lake Mburo Wetland plants are cough, abdomen pains and general body weakness. The market price of a handful herbal medicine after deducting the cost of transport and labour for extraction was on average Ug. Shs 8000 (US $ 3.2). About 70% of the identified traditional herbalists were women who were earning about Shs 3,176,000 (US $ 1,270.4) per person per month by treating the sick residents. The conversion of bush land on the fringes of the wetland system to crop fields affected the herbal plants.

Recreation value

Although only 15.6% of the respondents (n = 49) had used Lake Nakivale and associated wetland system for recreation, the wetland reclamation and deforestation had completely destroyed two important cultural and historical forests of Ishanze and Kabeigarire on the shores of Lake Nakivale. The recreation activities were now restricted in fishing villages (landing sites). Pilgrimages to the cultural and historical sites of Ishanze and Kabaigarire forests where the great Kings of former Ankole Kingdom were laid to rest stopped in early 1990s after the two forested areas were cleared and converted into crop fields

DISCUSSION

The willingness of the neighbouring community members to accept compensation or idea of selling the wetland was very low (5.7%), an indication that residents highly valued Lake Nakivale Wetland. Even those who were willing to accept the idea of selling it proposed exorbitant price per acre. This is attributed to the level of ecological, social and economic benefits the residents get from the wetland and its role in supporting their livelihoods. It also indirectly shows the magnitude of the cost the neighbouring communities were likely to incur when they lose the wetland due to poor land use practices and over exploitation which is estimated to be Ug. Shs 1-10 million (US $ 400 - 4000) per month per person based on contingent approach. As pointed out in the findings, more women attached a higher value on the wetland than men and expressed a much less willingness to accept compensation than men ($X^2 = 0.002 \leq 0.005$). This can be attributed to gender roles as far the harvesting and utilization of the wetland resources are concerned. Women were much more involved in collection of firewood, harvesting handcraft materials, fetching of water and dealing on selling herbal medicine. Therefore, this implies that scarcity or disappearance of wetland resources as a result of poor land practices was likely to affect more women than men. Women would be the ones to bear the cost of poor land use practices since they would be forced to move longer distance and take more time to access essential wetland resources.

The study also shows that neighbouring community's priority list of wetland resources affected by poor land use practices is composed of mainly direct use category of wetland values. This could be attributed to the role of wetland resources such as fish, pasture, water, pasture and fuel wood in sustaining the livelihoods and limited knowledge on the total economic value of wetlands at local level (Schuyt, 2004).

Increased human settlement around the wetland coupled with increased land degradation due to poor land use practices has subsequently increased demand for wetland resources and land for food production. However, other socio-economic factors for instance, low levels of household income (Turpie, 2010) and land tenure (Fraser, 2004) could also be accentuating the destruction of Lake Nakivale Wetland. The majority of the people residing in Nakivale are refugees who have a lot of hopes of returning to their country of origin and therefore perceive the idea of having a semi permanent house made up of iron sheets and bricks as waste of money. On the side of nationals, those who would afford iron sheets and bricks to construct their houses instead of wetland resources were fully aware that land belongs to the government, and they were expecting the government to evict them in future. To them, it was safer, cheaper and justified to continue depending on wetland resources as building materials.

The conversion of wetland and adjacent vegetation cover into agricultural fields has affected the availability of crucial wetland resources such as natural plants of medicinal value and fuel wood; as a result, local people are forced to walk long distances in search of scarce wetland resources. Consequently, residents were paying high price to access the essential resources in addition to environmental problems associated with wetland degradation. From management perspective, the results may therefore suggest that unless sustainable land use management (UN, 2012) that puts into consideration wetland restoration program is adopted, the cost of conservation incurred by neighbouring communities is likely to increase in future.

Conclusion

It is evident from this assessment that cost of poor land use practices on the key resource values and integrity of Lake Nakivale Wetland is so far enormous. The findings indicate that wetland dependant local people are likely to lose vital resources valued at about Shs 7,359,900,720 (US $ 2,943,960.3) per annum if the current situation is not rectified. The effects of poor land use practices are already exhibited in the form of receding Lake Nakivale and associated water levels and scarcity of essential wetland resources such as fish, plants of medicinal values, fire wood, pasture and building materials. The poor and marginalised members of the society, especially the women that either rely on wetland resources or their gender roles, according to the society, are the most affected by the effects of poor land use practices.

REFERENCES

Abdullah MF, Abubakar NR, Sulehan J, Awang AH, Liu OP (2012). Participatory rural appraisal: Analysis of experience in Darmareja Village, Sukabumi District, West Java, Indonesia. Akademika 82(1):15-19.

Babu CR, Kumar P, Prasad L, Agrawal R (2002). Valuation of ecological functions and benefits. A case study of wetland ecosystems along the Yamuna River corridors of Delhi Religion. Indira Gandhi Institute of Development Research, India.

Barbier EB, Acreman M, Knowler D (1997). Economic valuation of wetlands: A guide for policy makers and planners. Ramsar Convention Bureau, Gland, Switzerland.

Barton D (2002). The transferability of benefit transfer: Contingent valuation of water quality improvement in Costa Rica. J. Ecol. Econ. 42:147-167.

Breyer J, Larsen D (1997). Land use cover change in South West Uganda. The case of Katuna and Mparo watersheds in Kabale District. African Highland Initiatives, ICRAF.

Conroy C (2002). Socio-economic methodologies for natural resources research best practice guidelines: Participatory rural appraisal tools used for research into common pool resources. Natural Resources Institute, University of Greenwich, UK.

De Groot RS, Stuip MAM, Finlayson CM, Davidson N (2006). Valuing wetlands: Guidance for valuing the benefits derived from wetland ecosystem services. Ramsar Convention Secretariat, Gland, Switzerland/Convention on Biological Diversity, Montreal, Canada.

Duguma J, Tegegne A, Hegde BP (2012). The effect of location and season on free water intake of livestock under field condition in Ginchi Watershed Area, Ethiopia. World J. Agric. Sci. 8 (1):38-42.

Fennessy S, Craft C (2011). Agricultural conservation practices increase wetland ecosystem services in the Glaciated Interior Plains. J. Ecol. App 21(3):49-64.

Fraser DG (2004). Land tenure and agricultural management: Soil conservation on rented and owned fields in South West British Colombia. Agric. Hum. Values 21:73-79.

GOU (2002). The Wetland Sector Strategic Plan Support Program. Wetlands Inspection Division, Ministry of Water and Environment, Kampala.

Houlahan JE, Keddy PA, Makky K, Findlay CS (2006). Effects of adjacent land use on wetland species richness and community composition. Wet. 26 (1):79-96.

Lambrecht FL (1964). Aspects of evolution and ecology of tsetse flies and *Trypanosomiasis* in Pre-historical Afican Environment. J. Afr. Hist. 5:1-24.

Leschine TM, Wellman KF, Green TH (1997). The economic value of wetlands: Wetlands' role in flood protection in Western Washington. Ecol. Pub. 97-100.

Makalle AMP, Obando J, Bamutaze Y (2008). Effects of land use practices on livelihoods in the transboundary sub-catchments of the Lake Victoria Basin. Afr. J. Environ. Sci. Tech. 2(10):309-317.

Marshall E, Shortle J (2005). Urban development impacts on ecosystem. *In*: Goetz S, Shortle J, Bergstrom J (eds.). Land use problems and conflicts: Causes, consequences and solutions Ch. 7, Routledge Publishing, New York.

Millennium Ecosystem Assessment (2005). Ecosystems and human-beings: Wetlands and water synthesis. World Resource Institute, Washington, D.C.

Mugisha AH (2011). Twenty years of wetland conservation in Uganda: Have Uganda's wetlands become wastelands again? A public talk at Uganda Museum, Kampala to commemorate the World Wetlands Day with a theme: "Wetlands for forests". National Environment Management Authority (NEMA), Kampala.

Musamba BE, Ngaga MY, Boon KE, Giliba AR (2011). Impact of socio-economic activities around Lake Victoria: Land use and land use changes in Musoma Municipality, Tanzania. J. Hum. Ecol. 35(3):143-154.

NAFIRRI (2010). The fisheries resources of Lake Nakivali: Report of rapid assessment of the Lake Nakivali fishery prepared for the Office of the Prime Minister on Lake Nakivali Refugee Settlement. National Fisheries Resources Research Institute, Jinja Uganda.

Opio M (2008). An institutional analysis of the management of the wetland resources: A comparative study of Floahreppur Municipality in South Iceland and Oyam District in Uganda. Land Restoration Training programme, Keldnaholt, Iceland.

Place F, Otsuka K (1997). Population pressure, land tenure and tree resource management in Uganda: Environment and Production Technology Division Discussion Paper No. 24, International Food Policy Research Institute, Washington D.C.

Ramachandra TV, Rajinikanth R (2003). Economic valuation of wetlands. Centre for Ecological Sciences, Indian Institute of Science Technical Report 101, Bangalore, India.

Ramsar Convention Secretariat (2007). National wetlands policies: Developing and implementing national wetland policies. Ramsar handbooks for the wise Use of wetlands 3rd ed. 2, Ramsar Convention Secretariat, Gland, Switzerland.

Schuijt K (2002). Land and water use of wetlands in Africa: Economic values of African wetlands. International Institute for Applied Systems Analysis, Laxenburg, Austria.

Schuyt (2004). Living waters conserving the source of life: The economic values of the world's wetlands. Institute of Environmental Studies, Vrije Universiteit, Amsterdam.

Subade RF (2005). Valuing biodiversity conservation in a World Heritage Site: Citizens' non-use values for Tabbataha Reefs National Marine Park, Philippines. Economy Environment Program for Southeast Asia Research Report No. 2005-RR4, Tanglin, Singapore.

Tukahirwa JMB (2002). Policies and land use change in Uganda: Case study in Ntungamo, Lake Mburo and Sango Bay sites. Land Use Change Impacts and Dynamics (LUCID) Project Working Paper No. 17, International Livestock Research Institute, Nairobi.

Turpie J (2010). Wetland evaluation, vol.3: A tool for the assessment of the livelihood value of wetlands. Water Research Commission, Gezina.

Turpie J, Lannas K, Scovronick N, Louw A (2010). Wetland ecosystem services and their valuation: A review of current understanding and practice. Water Research Commission, Gezina. UN (2012). Sustainable land use for the 21st century. Division of Sustainable Development of the UN, Department of Economic and Social Afffairs, Washington, D.C.

WMD/NU (2008). Implementing the Ramsar Convention in Uganda: A Guide to the management of Ramsar sites in Uganda. Wetland Management Division /Nature Uganda, Kampala.

7

Assessment of oil contamination in the bay of Porto Grande (Cape Verde) using the mullet *Chelon bispinosus*

N. Pinheiro[1], L. Barreira[2], B. Lopes[2] and M. J. Bebianno[2]

[1]Universidade de Cabo Verde, Departamento de Engenharia e Ciências do Mar, CP 163, Mindelo, Cabo Verde.
[2]CIMA, Universidade do Algarve, Campus de Gambelas, 8000 - 117 Faro, Portugal.

Polycyclic aromatic hydrocarbons (PAHs) are a group of persistent organic pollutants, some of which are mutagenic and carcinogenic so PAH concentrations in fish used for human consumption are crucial to assess impact to human health. Total PAH concentrations in muscle and liver of mullets *Chelon bispinosus* from the Bay of Porto Grande (Cape Verde) (four sites in the bay and a control) ranged from 112.7 to 779.5 and 291.5 to 7548.7 ng/g d. w., respectively. Two and three ring PAHs were the most frequent (72.8 to 90.8% in the muscle and 75.9 to 98.3% in the liver), but levels of carcinogenic PAHs (mainly Dibenzo (a,h) antracene) in certain sites (CN and PG) are of concern. Results reflect a chronic PAH pollution in the bay and sources are a mixture of anthropogenic (petrogenic and pyrolytic) and natural sources, making their identification extremely complex. Although, BaP levels were below the threshold established by Cape Verde and the European Union, BaPEs levels in muscle ranged from 0.28 to 3.66 ng/g w. w. and BAPEs and TPAHs exposure for the average adult was 0.02 to 0.26 and 1.6 to 11.2 µg/day, respectively. Further knowledge of PAH concentrations in other species are necessary for a proper environmental risk assessment policy.

Key words: Bay of Porto Grande, Cape Verde, *Chelon bispinosus*, mullets, PAHs, BaPEs, daily intake.

INTRODUCTION

Oil is a complex mixture of compounds with hydrocarbons reaching 98% of the total composition. Among them, Polycyclic Aromatic Hydrocarbons (PAHs) are a group of persistent organic pollutants (POPs) considered priority substances due to their persistence in the aquatic environment. The origin of PAHs results from the incomeplete combustion of fossil fuels, organic materials as well as petroleum. The most common sources of PAHs to the marine environment are natural as well as anthropogenic. Natural sources are restricted to burning of forests and volcanic emissions while anthropogenic sources are mainly from thermal processes and oil spills (Lopes and Andrade, 1996).

PAHs do not easily degrade in the aquatic environment and their persistence increase with the increase of molecular weight. Some of them after metabolic transformation (bioactivation of citochrome P450 enzymes) react with DNA and become potentially mutagenic and carcinogenic to organisms and humans (Neff, 2002). Fish and invertebrates are good indicators to assess POPs contamination of the marine environment and have been widely used to assess the presence and effects of these contaminants worldwide (Reddy et al., 2005).

Cape Verde is an archipelagic state in the Eastern Central Atlantic (14° 50' to 17° 20' N, 22° 40' to 25° 30'

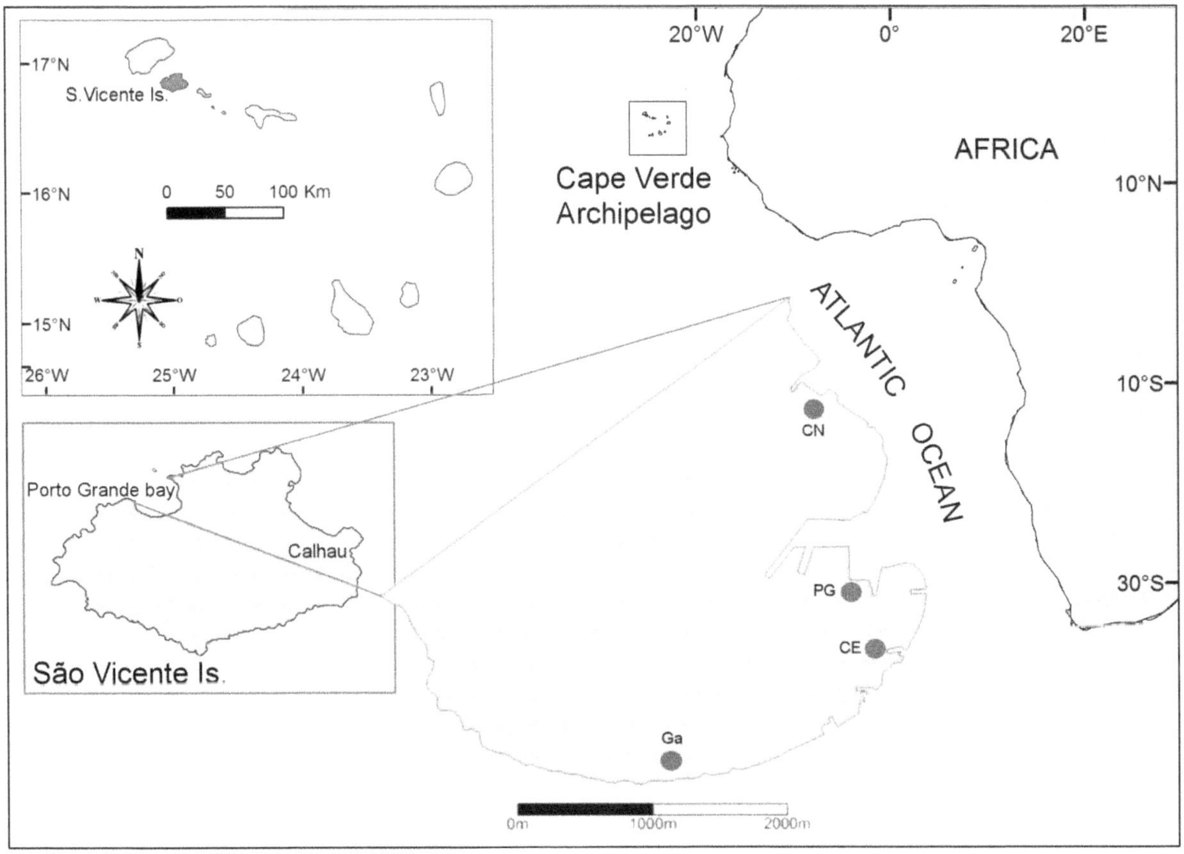

Figure 1. Map showing the geographical position of Cape Verde Archipelago in the eastern southern Atlantic ocean and the area of the island of São Vicente where the mullets *C. bispinosus* (Bowdich, 1825) were collected (shaded island)-Porto Grande bay. Legend: CN - Cabnave, PG - Porto Grande, CE - Cais da Enacol and GA - Galé. The control area [Calhau (C)- East coast] is also shown.

W), 750 km West of Senegal (Menezes et al., 2004). It consists of 10 islands and five islets of volcanic origin with a surface area of around 4,033 km^2 and an Exclusive Economic Zone of about 734,265 km^2 (Bravo de Laguna, 1985, DGMP, 1998) (Figure 1). The maritime traffic in Cape Verde is very intense and the contamination sources in this coastal area results from tanker activities, commercial and touristic vessels (MAAP, 2004). Moreover, the Bay of Porto Grande is at the NW end of Saint Vincent Island (situated West to Northwest of Cape Verde archipelago between latitude and longitude of 16.77 to 16.02° N and 24.86 to 25.09° W) and has a surface area of around 899 ha and a maximum depth of 30 m. The mouth of the bay is roughly 4 km wide and inside, there is a very active port. The currents are weak, the salinity varies between 35.8 and 36.2 psu and the average surface temperature ranges from 24.5 and 26°C (Almeida et al., 2008). Over the years, the bay was the main receptor of point and diffuse contamination sources from industrial activities, as well as from untreated or partially treated domestic effluents from the main urban area (Mindelo city), which lead to the increase of organic and inorganic contamination. Moreover, the intense port activities, the transport and storage of oil from ships, oil loading and unloading through marine terminals, accidental spillage and elimination of oily waste from ships and shipyards are important contamination sources. As PAHs are naturally present in petroleum and its derivatives, such activities constitute an important source of PAH contamination (Almeida et al., 2008). Between 2003 and 2007, several oil spills occurred and even nowadays, the smell of oil is intense and stains of floating oil are visible in the bay.

Therefore, the aim of this study was to assess the impact of PAHs in the bay of Porto Grande in S. Vicente Island (Cape Verde) using mullets *Chelon bispinosus* as a bioindicator. Among the Mugilidae, there are 26 families belonging to 14 genera but along the West Coast of Africa, between Senegal and Angola, only eleven species have been identified. Moreover, *C. bispinosus* (Bowdich, 1825) is the only mugilidae known in Cape Verde Islands (Trape et al., 2012). Therefore, this species was selected, due to their economic importance and food source in this country, to assess the impact of PAHs to human health.

In this prospect, mullets (*C. bispinosus*) were collected from different sites on the bay of Porto Grande and from a reference site (Calhau) and PAH content was analyzed in the muscle and liver. The bay of Porto Grande was selected because it is a marine ecosystem that receives intense pollution from a highly urbanized area and from port activities. This is, to the best of our knowledge, the first PAHs assessment in the bay of Porto Grand and in Cape Verde Archipelago.

MATERIALS AND METHODS

Sample collection and extraction

Fifty mullets *C. bispinosus* were collected in Spring 2009 in Cape Verde Archipelago Saint Vincent Island: ten from each of the four sites at the Bay of Porto Grande namely, from: Cabnave (CN) (16° 53′ 51.64" N; 24° 59′ 40.84" W), Porto Grande (PG) (16° 53′ 11.56" N; 29° 59′ 41.34" W), Cais de Enacol (CE) (16° 52′ 59.79" N; 24° 59′ 31.80" W), Galé (GA) (16° 52′ 39.71" N; 25° 00′ 17.75" W) and from the control site [Calhau, (C)] (16° 50′ 62.06" N, 24° 51′ 51.22" W) (Figure 1) with a gillnet and transported alive to the laboratory. Biometric parameters (weight, size) were measured and the condition factor (CF) and hepatosomatic index (HSI) calculated. Moreover, a portion of the lower back muscle and the liver were removed and individually wrapped in aluminum foil, identified and frozen at -20°C until PAH analysis.

PAH analysis

Muscle (± 16 g) and liver (± 2 g) were thawed and freeze-dried for 48 h. These lyophilized tissues were homogenized with anhydrous Na_2SO_4 and soxhlet extracted with 250 ml of n-hexane: dichloromethane (4:1) for 8 h. The extract was concentrated to about 30 ml in a Rotavapor Büchi R., dried with nitrogen and weighed to determine the percentage of lipids. They were further saponified after re-suspension with NaOH 6% in a water bath at 30°C for 18 h. The organic phase was separated and washed three times with n-hexane: dichloromethane (4:1) concentrated and further re-dissolved in n-hexane and purified on an activated silica/alumina column. The aliphatic fraction was eluted with 50 ml of n-hexane and the aromatic fraction containing the PAHs was eluted with 25 ml of n-hexane/dichloromethane (9:1) and 25 ml of n-hexane/dichloromethane (4:1). After drying under nitrogen, the residue was weighed and kept at -20°C until further analysis. The aromatic fraction was analyzed for PAHs by HPLC equipped with a specific column for PAHs determination (LiChrospher PAH, Merck) and a UV-Vis detector as described in Barreira et al. (2007). The 16 individual EPA (Environmental Protection Agency) recommended PAHs [(naphthalene (N), acenaphthylene (Acfil), acenaphthene (Acflen), fluorene (F), phenanthrene (Phen), anthracene (Ant), fluoranthene (Fu), pyrene (Pyr), benzo (a) anthracene (BaA), chrysene (Chrys), benzo (b) fluoranthene (BbF), benzo (k) fluoranthene (BkF), benzo (a) pyrene (BaP), dibenzo (ah) anthracene (DahA), benzo (ghi) perylene (BPer) and indeno (1,2,3-cd) pyrene (IndP)] were analyzed in the muscle and liver. The individual PAHs were identified by comparison of UV-Vis spectra and quantification performed by comparison of peak areas (254 nm) with an external standard of 16 PAHs (Supelco) by means of calibration curves constructed for each of the analyzed PAHs. Four replicates per sample were injected. The detection limit ranged between 0.01 and 0.24 ng/g w. w. for individual PAHs. Blanks were prepared in the same way as fish tissue samples but PAHs were not detected in the blanks. PAHs analysis was validated using a reference material tissue (SRM 2977) extracted, purified and analysed in the same way as the samples. PAH recovery for the certified material ranged between 73 and 112%. The concentrations of PAHs are expressed as ng/g dry weight.

Several ratios among individual PAHs (Phen/Ant, Flu/Pyr, Ant/Ant + Phen, Flu/Flu + Pyr, BaA/BaA + Chrys and IndP/IndP + BPer) were applied as diagnostic markers to identify the main PAHs sources in mullets from Saint Vincent Island. The limits of these ratios were obtained from the literature (Yunker et al., 2002; Mille et al., 2007; Oros and Ross, 2005; Oros et al., 2006). The formation of oil in which the organic matter maturation processes needs low temperature to originate two or three benzene ring PAHs (such as N, Ant and Phen) are characteristic of petrogenic sources while high temperatures in which 4, 5 or 6 ring PAH compounds are formed [like benzo(a)anthracene, chrysene, benzo(a)pyrene, benzo(g,h,i)perylene or indeno(1,2,3-cd)pyrene] characteristic of pyrolitic inputs. These ratios also allows the distinction between refined products or petroleum combustion (Porte et al., 2001a, b), or derived from biomass combustion (forest fires, fireplaces, incinerators, coal) (Oros and Ross, 2005).

Statistical analysis

Data are presented as mean ± standard deviation. XLSTAT software was used to perform statistical analysis. The experimental data was first tested for normality and homogeneity of variance. As total length, weight, CF and HSI did not meet the assumption of ANOVA (normality and homogeneity of variances), data were log transformed and Tukey test was applied to discriminate differences. To check whether there is any correlation between the variables and if there is any similarity between individual PAH concentrations in different locations, principal component analysis (PCA) was applied. Statistical significance was set at $p < 0.05$.

RESULTS AND DISCUSSION

Biometric data

Total length and weight, condition factor (CF) and hepatosomatic index (HSI) (mean ± S.D.) of the mullets collected from the different sites are in Figures 2 and 3. No significant differences exist between the length and weight of the mullets from the different sites with the exception of those from CN that were larger and heavier (Figure 2A and B). Like for total length and weight, fish from CN has the highest condition factor (Figure 3A) along with that of GA and significantly different from the other three sites whose CF was similar ($p > 0.05$). Moreover, despite the variability in the HSI of the mullets from some of the sites (CE and C), no significant differences exist between the HSI of the mullets from the different sites (Figure 3B) ($p > 0.05$). The percentage of lipids in the muscle and liver of mullets from the different sites are in Figure 4A and B. The lipid content in the liver was significantly higher (around 2-fold) than in the muscle. Moreover, the lipid content in the muscle of the mullets from PG and C had similar high lipid content followed by those from the other sites whose lipid content was also similar (Figure 4A). Regarding the liver, lipid levels of mullets from PG and CE were higher but not significantly

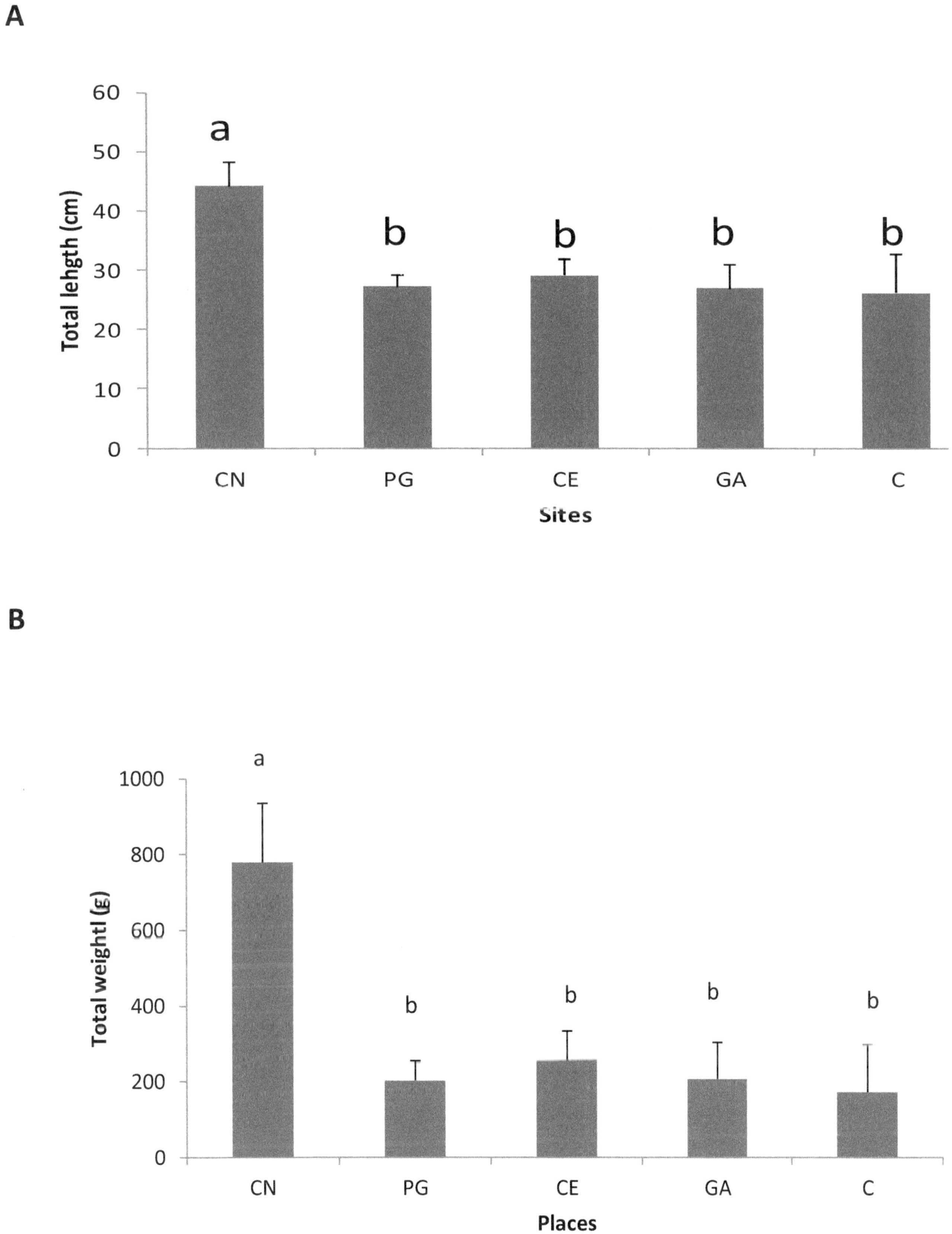

Figure 2. Total (A) length (mean ± S.D.) (cm) and (B) weight (mean ± S.D.) (g) of the mullets *C. bispinosus* from the different sites in Porto Grande Bay and Calhau [different letters indicate significant differences ($p < 0.05$)]. Legend: CN - Cabnave, PG - Porto Grande, CE - Cais da Enacol and GA - Galé. The control area [Calhau (C)- East coast].

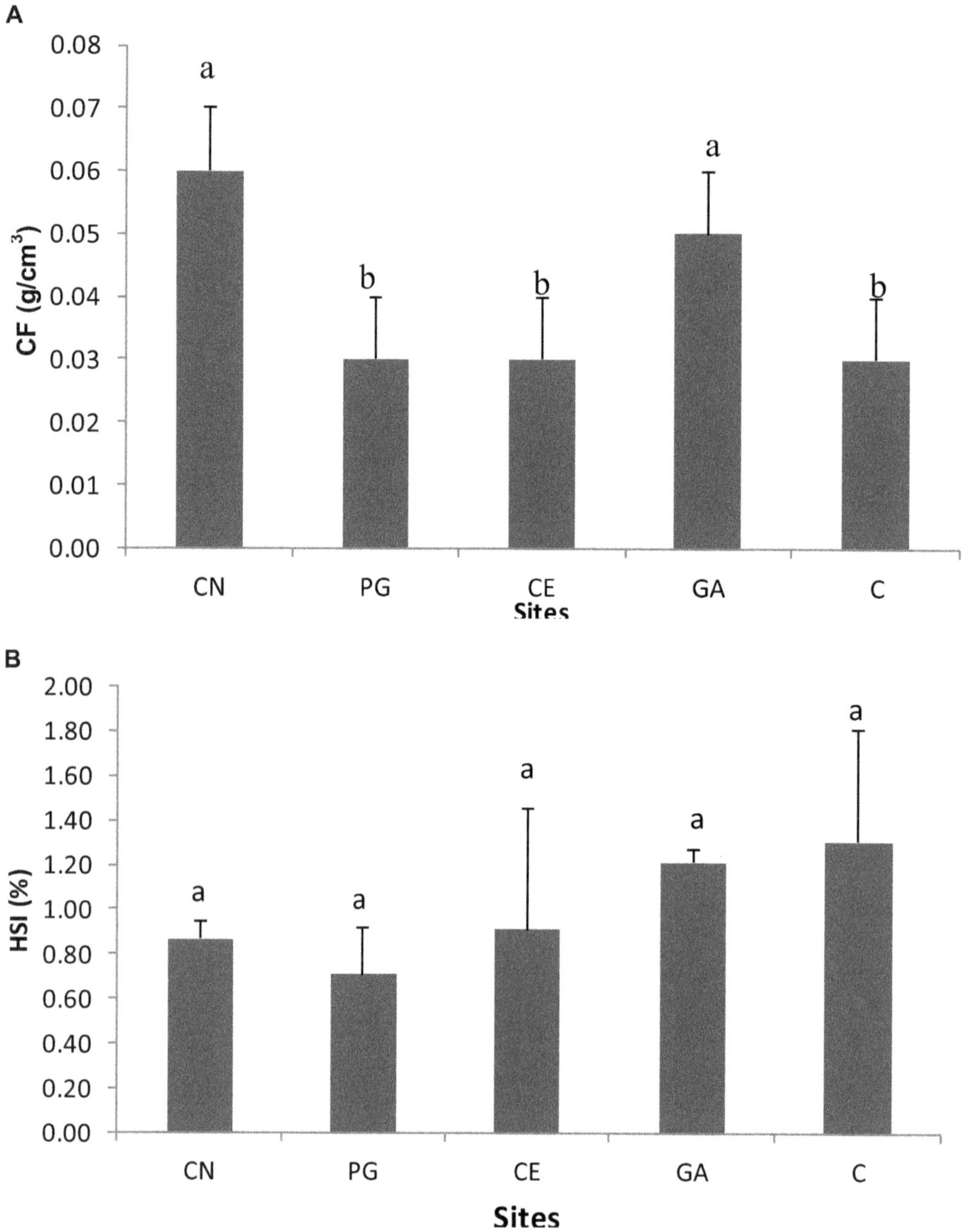

Figure 3. (A) Condition factor (CF) (g/cm^3) and (B) Hepatosomatic index (%) in the mullets *C. bispinosus* from the different sites in Porto Grande Bay and Calhau [different letters indicate significant differences ($p < 0.05$)]. Legend: CN - Cabnave, PG - Porto Grande, CE - Cais da Enacol and GA - Galé. The control area [Calhau (C)- East coast].

different from the other sites (Figure 4B).

PAHs concentrations

The concentration of individual, total and aromatic ring number PAHs in the muscle and liver of mullets from the different sites are in Table 1. Total PAH concentrations (TPAHs) in the muscle and liver of *C. bispinosus* from the different sites are in Figure 5. The results show that PAHs are widespread in the muscle and liver of the mullets from the different sites but levels in the muscle were always lower than in the liver except for mullets from CN (Figure 5). The highest PAH concentrations in the liver compared to muscle are not surprising because the liver is the tissue that usually has the highest lipid content and where the enzymes are able to metabolize PAHs. In the muscle, the highest TPAHs concentrations were in mullets from CN (779.5±17.66 ng/g d. w.) and the lowest in those from CE (112.7±70.67 ng/g d. w.). The

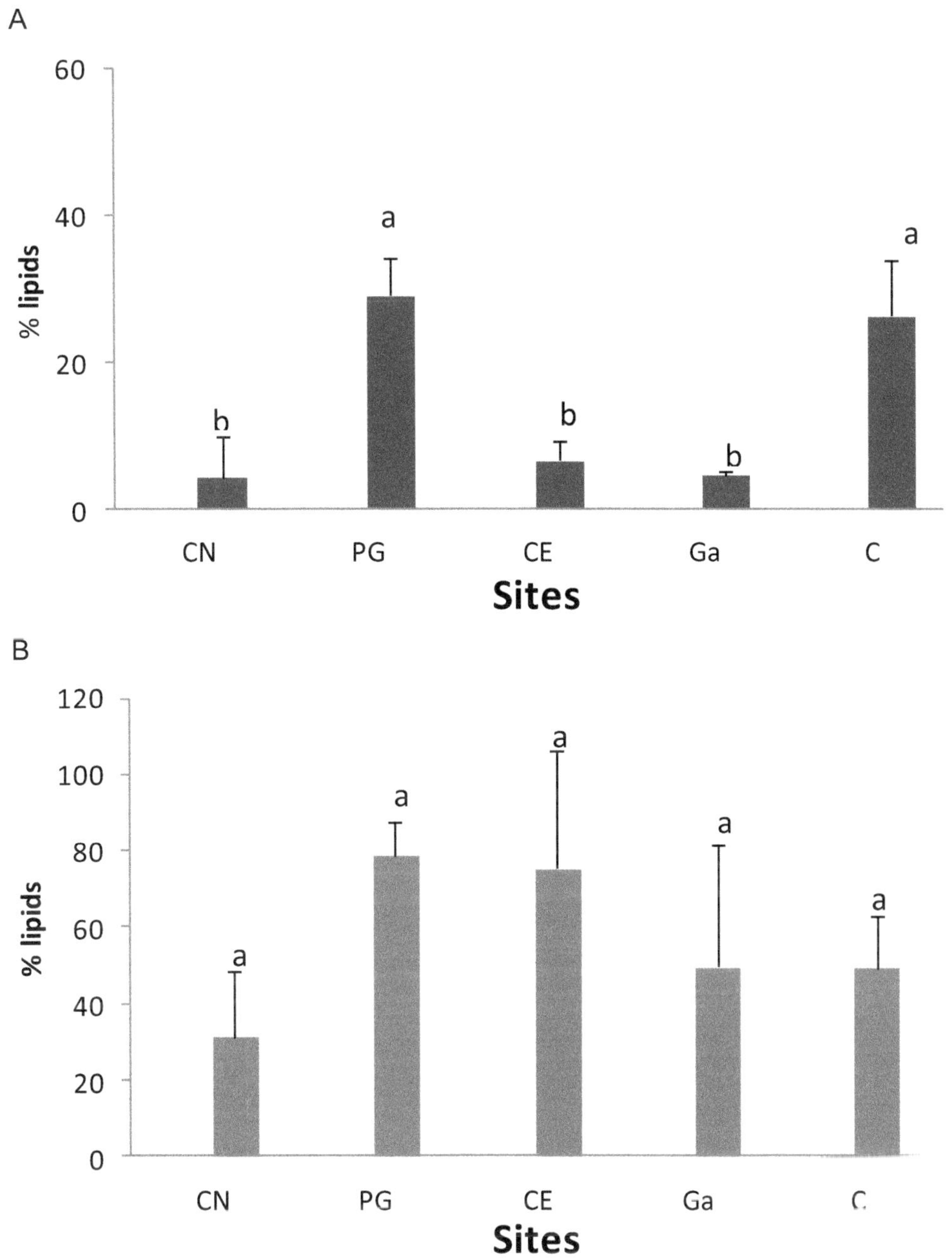

Figure 4. Lipid content (%) in the muscle (A) and liver (B) of mullet *C. bispinosus* from different sites in Porto Grande Bay and Calhau [different letters indicate significant differences ($p < 0.05$)]. Legend. CN - Cabnave, PG - Porto Grande, CE - Cais da Enacol and GA - Galé. The control area [Calhau (C)- East coast].

highest PAH concentrations in the muscle of mullets from CN are directly related to the oil spills that occurred from the shipyard existent in the area. In contrast, PAHs concentrations in the liver were the opposite with the highest TPAHs concentrations in the mullets from CE (7548.7±2.9 ng/g d. w.) and the lowest in those form CN (291.5 ± 30.9 ng/g d. w.). The highest TPAH concentrations in the liver of the mullets from CE reflect the influence of maritime traffic related to oil from small fishing vessels moored in the area and from the impact of urban effluent from the main city area (city of Mindelo).

Total PAH concentrations in the muscle and liver of mullets from the Bay of Porto Grande and Calhau were compared with similar fish species in other parts of the world and the results are in Table 2. TPAHs concentrations in the muscle of the mullets from Porto Grande

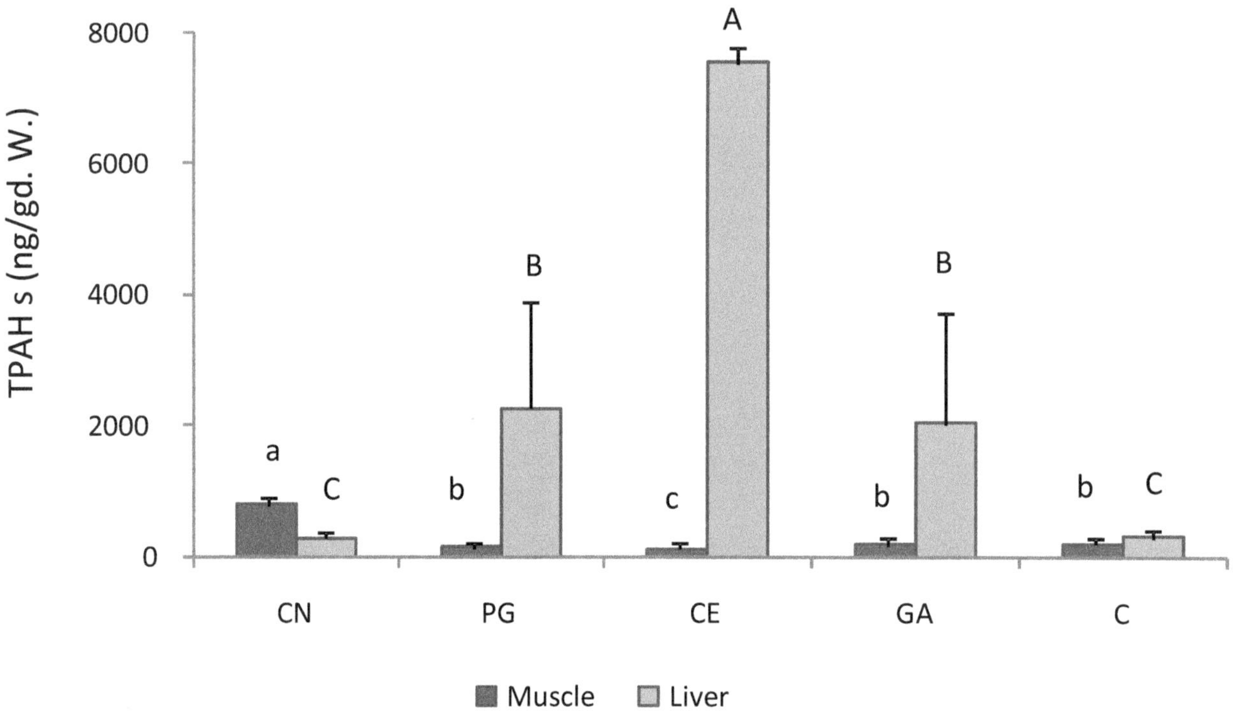

Figure 5. Total PAH concentrations (µg/g d. w) in the muscle and liver of the mullets *C. bispinosus* from the different sites in Porto Grande Bay and Calhau [different letters indicate significant differences (p <0.05)]. Legend: CN - Cabnave, PG - Porto Grande, CE - Cais da Enacol and GA - Galé. The control area [Calhau (C)- East coast].

Bay (113 to 780 ng/g d. w.) are similar to those of the *Mugil* spp. from the eastern Mediterranean Sea and from *Mullus barbatus* from Western Turkey and from *Solea solea* of the Red Sea Coast of Yemen (422 ng/g d. w.) (DouAbul et al., 1979) and higher than *Mugil cephalus* from Brisbane River (Australia); *Mugil lisa* from Guanabara Bay (Brasil) and from other mugil species from Adriatic and Mediterranean Sea. For the liver, the range of TPAHS (291.5 to 7548.7 ng/g d. w.) is also higher than other fish species from the Northeast Atlantic (<LDM to 22 ng/g d. w.) (Hellou et al., 1994) and from the Bay of Archon, France (14.7 to 139 ng/g d. w.) (Baumard et al., 1998), but are within the range of those from the Finnish Archipelago (590 to 2225 ng/g d. w.) (Rainio et al., 1986). Although, PAHs are highly lipophilic, which favors its accumulation in lipid rich tissues, the highest PAH concentrations (CE) in the liver of mullets are not related with the highest lipid content (PG site). Within the muscle, the situation is identical: the mullets from the site with the highest PAH concentrations (CN) are not related with highest fat content.

Individual PAH concentrations in the muscle and liver of mullets from both Porto Grande bay and Calhau are dominated by 2 and 3 ring PAHs with a percentage ranging between 72.8 and 90.8% for the muscle and 75.9 and 98.3% for the liver, respectively (Table 1). In the muscle, the dominant PAH was Acftil at all sites except at CN where the dominant PAH was Acften. In the liver, however, the dominant PAH was Acften (highest at CE) with the exception of the mullets from CN whose dominant PAH was Flu. These results show that organisms accumulated preferentially the lighter PAHs in both tissues (muscle and liver). This accumulation pattern is a characteristic of PAHs contamination from petrogenic origin due mainly from spills of gasoline or diesel at harbors (Martínez-Gómez et al., 2012). Moreover, the percentage of 4 ring PAHs was higher in the muscle (between 2.8 and 18.8%) compared to the liver (1.4 and 7.5%). PAHs with 5 and 6 rings were also present with a similar percentage between the two tissues (ranging from 1.1 and 10.0% and 0.3 and 10.7% in the muscle and liver, respectively) but site dependent (highest in the muscle of mullets from PG and in the liver of mullets from CN) (Table 1). In what concerns the 4 ring PAHs, the dominant PAH on this group was Pyr in both mullet tissues while in the 5 ring group, BaP was present in all mullets from Porto Grande bay with similar levels in the liver of PG, CE and GA. In the muscle, this PAH was only detected in mullets from PG, GA and C. However, the dominant 5 ring PAH was DaA in the muscle and BkF in the liver except from those of GA where DaA was also dominant. Regarding the 6 ring group, Bper was only detected in the liver of the mullets from GA.

To check whether there is any relationship between PAH concentrations, weight, size, CF and HIS and if there is any similarity between the individual PAH con-

Table 1. Average individual and total polycyclic aromatic hydrocarbons (ng / g dry weight) and aromatic ring groups (%) in muscle and liver of fish in five sampling sites.

Site	Tissue									
	CN		PG		CE		GA		C	
PAH compound	**Muscle**	**Liver**	**Muscle**	**Liver**	**Muscle**	**Liver**	**Muscle**	**Liver**	**Muscle**	**Liver**
Naphthalene (N)	98.6	55.1	30.0	161.6	27.9	772.7	45.6	266.2	52.5	n.d
Acenaphthylene (Acftil)	170.6	10.6	61.1	19.1	47.5	1163.5	85.4	223.4	96	n.d
Acenaphthene (Acften)	319.5	61.5	8.3	1300.7	11	5429.3	18.9	1369.1	9.8	186.4
Fluorene (F)	1.6	108.7	1.2	559.5	1.0	43.5	6.6	28.5	1.9	7.7
Phenanthrene (Phen)	3.2	4.8	1.3	47.8	2.9	11.8	0.9	3.7	3.9	1.0
Anthracene (Ant)	0.4	0.3	0.4	1.4	0.3	0.8	0.6	1.8	0.5	n.d.
Fluoranthene (Flu)	5.4	2.7	7.4	30.7	2.3	13	2.2	4.7	2.1	36.9
Pyrene (Pyr)	8.9	8.9	12.0	74.3	12.6	35.3	3.3	17.1	15.3	8.6
Benzo [a] anthracene (BaA)	4.9	5.5	3.5	26.3	5.5	48.4	3.9	11.5	6.7	7.6
Chrysene (Chrys)	1.4	2.2	1.3	38.5	0.8	10.9	0.6	3.1	8.2	0.9
Benzo [b] fluoranthene (BbF)	1.6	7.6	0.4	4.5	0.6	5.8	1.7	2.7	2.2	20.6
Benzo [k] fluoranthene (BkF)	0.2	14.5	0.3	3.8	0.1	10.3	0.1	1.3	0.3	n.d
Benzo [a] pyrene (BaP)	n.d	1.5	1.1	3.2	n.d	3.5	0.1	3.1	0.6	n.d
Dibenzo [a. h] anthracene (DahA)	6.8	7.5	12.2	n.d	n.d	n.d	3.9	97.0	2.4	n.d
Benzo [g. h. i] perylene (Bper)	n.d	n.d	n.d	n.d	n.d	n.d	n.d	6.0	n.d	n.d
Indeno [1.2.3-cd] pyrene (IndP)	n.d	n.d	n.d	n.d	n.d	n.d	n.d	n.d	n.d	n.d
Total PAH (TPAH)	779.5	291.5	140.6	2271.3	112.7	7548.7	173.9	2039.4	202.3	309.7
2+3 rings (%)	76.2	82.7	72.8	92.0	80.6	98.3	90.8	92.8	81.4	75.9
4 rings (%)	2.7	6.6	17.3	7.5	18.8	1.4	5.8	1.8	16.0	17.4
5+6 rings (%)	1.1	10.7	10.0	0.5	0.6	0.3	3.4	5.4	2.7	6.7
Main PAH[a]	Acflen (41)	Flr (37)	Acftil (43)	Acflen (57)	Acftil (42)	Acflen (72)	Acftil (49)	Acflen (67)	Acftil (47)	Acflen (60)

Legend: CN, Cabnave; PG, Porto Grande; CE, Cais da Enacol and GA - Galé. The control area [Calhau (C)- East coast]. TPAH sum of 16 individual PAHs concentration. [a]% of total; n.d. - not detected .

centrations in different sites, PCA was applied (Figure 6A and B). In the muscle (Figure 6A), PC1 and PC2 represent 36.6 and 23.6%, respectively of the total variance of the data (60.1%). CN is isolated in the positive quadrant of the PC1 and PC2, due mainly to the higher concentrations of N, Acftil, Acften and CF, weight and length. PG is separated from the other sites around PC2 with the highest concentrations of BaP, BkF, DahA, Flu and Ant. Mullets from C, CE and GA are the sites with the highest concentration of BaA. In the liver (Figure 6B), the first two components represent 71.9% of the total variance (40.6% for PC1 and 31.3% for PC2, respectively). PC1 separates the sites (CE, GA and PG) with the highest concentrations of Flu, Acftil, Acften, BaA, Pyr, Chrys, Ant, Phen, BaP, N, Bpe and BkF. C and CN have the highest concentrations of BbF and DahA. Mullets from C are directly related with the HSI while those from CN are directly related with CF, weight and length. The lack of a significant relationship between weight, length and individual PAHs, indicate that unlike other classes of persistent organic pollutants, the accumulation of PAHs in mullets are independent of the lipid content. The percentage of lipids is only positively related with Acftil, Acften, Flu, BaA and Pyr but these

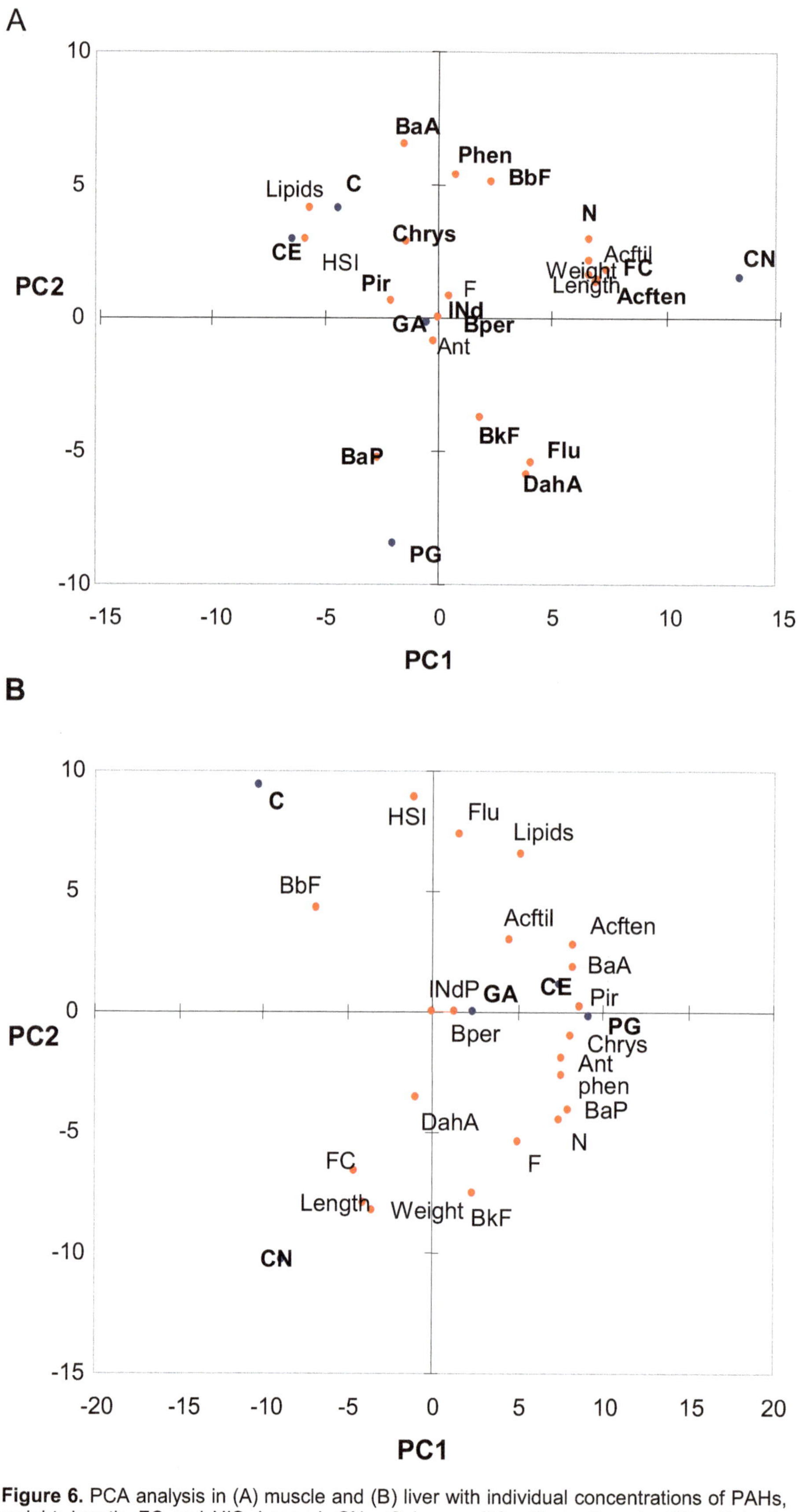

Figure 6. PCA analysis in (A) muscle and (B) liver with individual concentrations of PAHs, weight, length, FC and HIS. Legend: CN - Cabnave, PG - Porto Grande, CE - Cais da Enacol and GA - Galé. The control area [Calhau (C)- East coast].

Table 2. Total PAH concentrations in muscle and liver of fish species from different geographical regions (ng/g d. w.).

Species	Site	Muscle	Liver	Reference
Chelon bispinosis	CN	780	292	Present work
Chelon bispinosis	PG	141	2271	
Chelon bispinosis	CE	113	7549	
Chelon bispinosis	GA	174	2039	
Chelon bispinosis	C	202	310	
Clupea haren	Finnish Archipelago Sea	<0.5 - 33*		Rainio et al. (1986)
Diplodus annularis	Western Turkey	78.7 - 415		Darilmaz and Kucuksezgin (2012)
Gadus morhua	Northwest Atlantic.		<LDM-22	Hellou et al. (1994)
Hippoglossoides platessoides	Northwest Atlantic.	9-15*	14-39*	Hellou et al. (1997)
Liza aurata	Suez Canal	7229		Said and El Agroudy (2006)
Liza dussomeiri	Arabian Gulf	18-36*		DouAboul et al. (1987)
Liza dussomeiri	Arabian Gulf	118		DouAboul et al. (1997)
Lizza dumerillii	Niger Delta	101		Anyakora and Coker (2007)
Lota lota	Finnish Archipelago Sea	26*	445*	Rainio et al. (1986)
Micropogonias furnieri	Guanabara Bay	9.8-52.6		Silva et al. (2007)
Mugil auratus	Barcelona Harbor		57 ±19	Vives and Grimalt (2002)
Mugil cephalus	Bisbane River Estuary, Australia	53.7-195.0		Kayal and Connell (1995)
Mugil lisa	Guanabara Bay	3.7-26.1		Silva et al. (2007)
Mugil, spp.	Eastern Mediterranean	238±39		Barbour et al. (2009)
Mullus barbatus	Mediterranean Sea	14.7-49.6		Baumard et al. (1998)
Mullus barbatus	Mediterranean Sea, Spain	1.91-4.45*		Martínez-Gómez et al. (2012)
Mullus barbatus	Western Turkey	202 -556		Darilmaz and Kucuksezgin (2012)
Mullus barbatus	Adriatic Sea	16.52*		Perugini et al. (2007)
Mullus barbatus	Adriatic Sea	104-147		Della Torre et al. (2010)
Mullus surmuletus	Mediterranean Sea	3.1*		Llobet et al. (2006)
Pleuronectes ferruginea	Northwest Atlantic.	0.6-68*	15-79*	Hellou et al. (1997)
Pseudolithus elongatus	Niger Delta	100.1		Anyakora and Coker (2007)
Serranus scriba	Mediterranean Sea, France		20.4- 139	Baumard et al. (1998)
Solea solea	Red Sea Coast of Yemen	422.1		DouAboul et al. (1997)
Stizostedion lucioperca	Finnish Archipelago Sea	17-30*	118*	Rainio et al. (1986)
*w.w.				

CN, Cabnave; PG, Porto Grande; CE, Cais da Enacol; GA, Galé. The control area [Calhau (C)- East coast].

compounds are not present at higher concentrations in places where there are the highest lipid levels, supporting the information that the PAHs accumulated in mullets tissues are not explained by the lipid content.

In order to identify the PAHs sources in the environment where the mullets were collected, several diagnostic ratios were calculated for the muscle and the liver of *C. bispinosus* (Table 3). The ratio Phen/Ant >10 indicates a petrogenic origin of PAHs while <10 indicates a pyrolytic origin. Pyrolitic PAHs can result from the incomplete combustion (boat traffic, engines) or biomass combustion (coal, wood, fires) that reach the marine environment from road runoff or atmospheric deposition. Moreover, the ratio Flu/Pyr <1 is also related with a petrogenic source while >1 indicates a pyrolytic origin. The Phen/Ant ratio in mullets muscle ranged from 1.5 (GA) and 9.7 (CE) indicating a pyrolytic origin (Phen/Ant <10) of PAHs in this tissue for all sites. However, in the liver, the ratio Phen/Ant was only <10 at GA (2.1) and C (4.0) confirming a pyrolytic origin at those sites while in mullets livers from the other sites, the ratio Phen/Ant was >10 [ranging from 14.8 (CE) to 34.9 (PG)] confirming that at those sites, the origin of PAHs was petrogenic. On the other hand, the ratio Flu/Pyr indicates a petrogenic origin of PAHs in both mullet tissues (Flu/Pyr <1) with the exception of the liver of mullets from C (4.3) that has a pyrolitic origin (Flu/Pyr >1).

Consequently, with the exception of mullets from site C that showed a pyrolitic origin in both tissues, mullets from all the other sites have a mixed origin of pyrolitic and petrogenic sources. Additionally, to better identify the origin of PAHs in fish tissues, other diagnostic ratios

Table 3. Diagnostic ratios of PAHs sources.

Diagnostic ratio	CN		PG		CE		GA		C	
	Muscle	Liver	Muscle	Liver	Muscle	Liver	Muscle	Liver	Muscle	Liver
Fen/Ant	10.90	1.0	0.70	1.0	0.10	13.90	0.40	2.50	0.80	4.60
Flu/Pir	3.0	0.40	0.60	0.70	0.30	0.40	0.40	0.30	0.60	0.0
Ant/Ant + Fen	8.70	0.80	0.20	0.10	0.10	0.10	0.20	0.50	0.90	0.80
Flu/Flu+Pir	3.00	0.30	1.40	0.40	0.40	0.30	0.50	0.20	0.90	0.70
BaA/BaA+Cris	7.0	0.70	0.20	0.20	0.10	0.80	0.20	0.50	0.50	-

CN, Cabnave; PG,Porto Grande; CE, Cais da Enacol; GA, Galé. The control area [Calhau (C)- East coast].

or from combustion of biomass were calculated and the results are in Figure 6. The ratio Ant/Ant + Phen indicate that the origin of PAHs accumulated in the muscle is mainly from petroleum (CE) and petroleum combustion (other sites). In mullet's liver, the main origin is from petroleum (CN, PG and CE) and petroleum combustion (GA) and only those from C have a biomass combustion origin confirming the pyrolytic origin at this site. In the mullets from CN, PG and CE, the ratio Ant/Ant + Phen was <0.10 in both tissues which is consistent with PAHs from oil (or fuel). CN is directly impacted by shipyards and from the effluent of the desalinization plant while PG is located near the harbor and the marina and CE is directly impacted by the oil terminal which can be related with the petroleum origin found.

GA is directly related with the fuel terminal of the oil company present. The ratio Ant/Ant + Fen <0.10 corresponds in most cases to the values of the Flu/Flu + Pyr ratio <0.40 in mullets muscle from CE and in the liver from PG, CE and GA. These results are also in agreement with the ratio Phen/Ant that also shows a predominance of pyrolytic PAHs for both tissues in most of these sites. The Ant/Ant + Phen ratio was >0.10 and the Flu/Flu + Py ratio >0.40 only in the liver of the mullets from C supporting the conclusion that the pyrolytic products are a strong component of PAHs at this site (Figure 7). The trend of the ratio BaA/BaA + Chrys also indicates that the origin of PAHs is primarily from petroleum and petroleum combustion except from the liver of mussels from C confirming the biomass combustion source at this site. Thus, PAHs accumulated in muscle and liver of mullets are mainly from an incomplete combustion of oil and its derivatives (Ant/Ant + Phen; BaA/BaA + Chrys), presenting also petrogenic (oil not burn) and pyrolytic PAHs derived from biomass combustion (Flu/Flu + Pyr).

To assess the health implications of PAHs accumulation in mullets muscle for human consumption, BaP, the most potent carcinogenic of the PAH group after DahA, is used as a toxicological surrogate for all carcinogenic PAHs. Therefore, BaP levels of mullets muscle from the different sites were compared with the guidelines established by the legislation of Cape Verde (Ordinance 24/2009) and of the European Union [Regulation (EC) No 1881/2006 of 19th December 2006] that sets maximum levels for certain contaminants in food stuff and both recommended a maximum allowable BaP concentration in fish muscle of 2 ng/g w. w.. BaP was only detected in mullets from PG, GA and C (at the other sites levels were below detection limit) but levels were well below the standard established by Cape Verde and European Union indicating that from the present results, levels of BaP in *C. bispinosus* muscle are in general safe for human consumption and does not constitute an health risk for Cape Verde consumers.

However, it is also necessary to take into account the daily intake of this species by Cape Verde population. Therefore, based on the approach adopted by EPA to evaluate risk assessment, PAHs were separated into two classes; carcinogenic (BaA, Crys, BbF, BkF, BaP, DahA, and IndP), and non-carcinogenic (N, Aceflen, Acftil, Flu, Phen, Ant, Flu, Pyr and Bper).

The individual carcinogenic potential of each PAH was expressed as BaP equivalents (BaPEs) based on the toxicity equivalent factors (TEFs) defined by Nisbet and LaGoy (1992) for each PAH. TEFs for all 16 PAHs and BaPE (calculated by multiplying the concentration of each individual PAH by the appropriate TEF) and Total BaPEs (the sum of each BaPE from individual PAHs) are in Table 4.

When levels were below the limit of detection, half of the detection limit was used in the calculations. Results indicate that with the exception of fish muscle from CN and PG, higher TPAH concentrations (Table 1) were related to higher total BaPEs. Nevertheless, the highest TPAH levels from CN in muscle mullets had one of the lowest total BaPEs, mainly due to the presence of high levels of 3 ring PAHs (F, Acftil and Acften) that have a TEF of 0.001. Muscle mullets from the other sites had medium levels of PAH (ranging from 140.6 to 202.3 (Table 1) but total BaPEs from PG were the highest. This was mainly due to the highest DaA concentrations (TEF of 1) at this site. With the exception of muscle mullets from CE, BaPEs were well above the levels found for red(Ant/Ant + Phen; BaA/BaA + Chrys and Flu/Flu + Pyr) that distinguish from incomplete combustion of fossil fuels

Table 4. BaPEs (ng/g w. w.) and mean daily intake of BaPEs and TPAHs (µg/day w. w.).

PAH compound	TEFS	BaPEs (ng/g w. w.)					
		CN	PG	CE	GA	C	Mean
Naphthalene (N)	0.001	0.0197	0.0060	0.0056	0.0091	0.0105	0.0102
Acenaphthylene (Acftil)	0.001	0.0341	0.0122	0.0095	0.0171	0.0192	0.0184
Acenaphthene (Acften)	0.001	0.0639	0.0017	0.0022	0.0038	0.0020	0.0147
Fluorene (F)	0.001	0.0003	0.0002	0.0002	0.0013	0.0004	0.0005
Phenanthrene (Phen)	0.001	0.0006	0.0003	0.0006	0.0002	0.0008	0.0005
Anthracene (Ant)	0.01	0.0008	0.0008	0.0006	0.0012	0.0010	0.0009
Fluoranthene (Flu)	0.001	0.0011	0.0015	0.0005	0.0004	0.0004	0.0008
Pyrene (Pyr)	0.001	0.0018	0.0024	0.0025	0.0007	0.0031	0.0021
Benzo [a] anthracene (BaA)	0.1	0.0980	0.0700	0.1100	0.0780	0.1340	0.0980
Chrysene (Chrys)	0.01	0.0028	0.0026	0.0016	0.0012	0.0164	0.0049
Benzo [b] fluoranthene (BbF)	0.1	0.0320	0.0080	0.0120	0.0340	0.0440	0.0260
Benzo [k] fluoranthene (BkF)	0.1	0.0040	0.0060	0.0020	0.0020	0.0060	0.0040
Benzo [a] pyrene (BaP)	1	0.0625	1.1000	0.0625	0.1000	0.6000	0.3850
Dibenzo [a. h] anthracene (DahA)	1	1.3600	2.4400	0.0625	0.7800	0.4800	1.0245
Benzo [g. h. i] perylene (Bper)	0.01	0.0006	0.0006	0.0006	0.0006	0.0006	0.0006
Indeno [1.2.3-cd] pyrene (IndP)	0.1	0.0063	0.0063	0.0063	0.0063	0.0063	0.0006
Total BaPEs (ng/g w.w.)		1.689	3.659	0.279	1.036	1.3250	1.597
Mean daily intake of BaPEs (µg/day w.w.)		0.122	0.263	0.020	0.075	0.095	0.115
Mean daily intake of TPAHs (µg/day w.w.)		11.2	2.0	1.6	2.5	2.9	4.1

CN, Cabnave; PG, Porto Grande; CE, Cais da Enacol; GA, Galé. The control area (Calhau (C)- East coast).

mullet in Italy (0.63 ng/g w. w.) (Perugini et al., 2007) and for the screening value recommended by USEPA for human consumption (0.67 ng/g w. w.) (USEPA, 2000).

The BaPEs of mullets muscle from the different sites were higher (PG and CN) or similar (GA, C) to 1.33, 0.997 and 0.094 ng/g w. w. reported for marine sea food from Kuwait, Italy and Hong Kong, respectively (Alomirah et al., 2009; Perugini et al., 2007; Cheung et al., 2007). Moreover, estimated daily consumption of fish from Cape Verde was 26.5 kg/habitant which is equivalent to 72 g/day (INDP, 2006). This average was similar to that of Kuwait (66.4 g/day) (Alomirah et al., 2009) and lower than the US EPA (142.2 g/day) recommended for subsistence consumers (USEPA, 2002) and for an average person from Hong Kong (164.4 g/day) (Cheung et al., 2007).

The mean exposure to TPAH via *C. bispinosus* muscle for the average Cape Verde consumer is 4.1 µg/day w. w. but those from CN are around 3-fold higher. These levels are higher than those estimated for the average adult Kuwait consumers (1.3 and 0.0013 µg/day for mean daily intake of TPAHs and total BaPEs, respectively). Moreover, the impact of PAHs in the bay of Porto Grande should be attentively followed and to assess the impact of POPs in Cape Verde, further studies are required to ascertain PAH concentrations and BaPE levels in other economically important fish species of the region.

Conclusions

Total concentrations of 16 PAHs accumulated in the mullets *C. bispinosus* ranged from 112.7 to 779.5 ng/g d. w. in muscle and from 291.5 to 7548.7 ng/g d. w. in the liver with mullets from CN and CE having the highest total PAH concentrations in both tissues. These results reflect a chronic PAH contamination in the bay of Porto Grande and show that *C. bispinosus* are useful indicators to detect the presence of PAHs.

The intense maritime traffic of small and large vessels, the transport and storage of petroleum, port operations, leisure and tourism, urban and industrial effluents contribute to the anthropogenic impact of PAHs. There is a mixture of anthropogenic (petrogenic and pyrolytic) and natural sources at this area, making the identification of PAH sources extremely complex due to the diversity of anthropogenic activities and local dynamics. However, the distribution of individual PAHs showed that the two and three ring PAHs which ranged between 72.8 and 90.8% in the muscle and 75.9 and 98.3% in the liver with the predominance of Acften, Acftil and Flu, reflect conta-

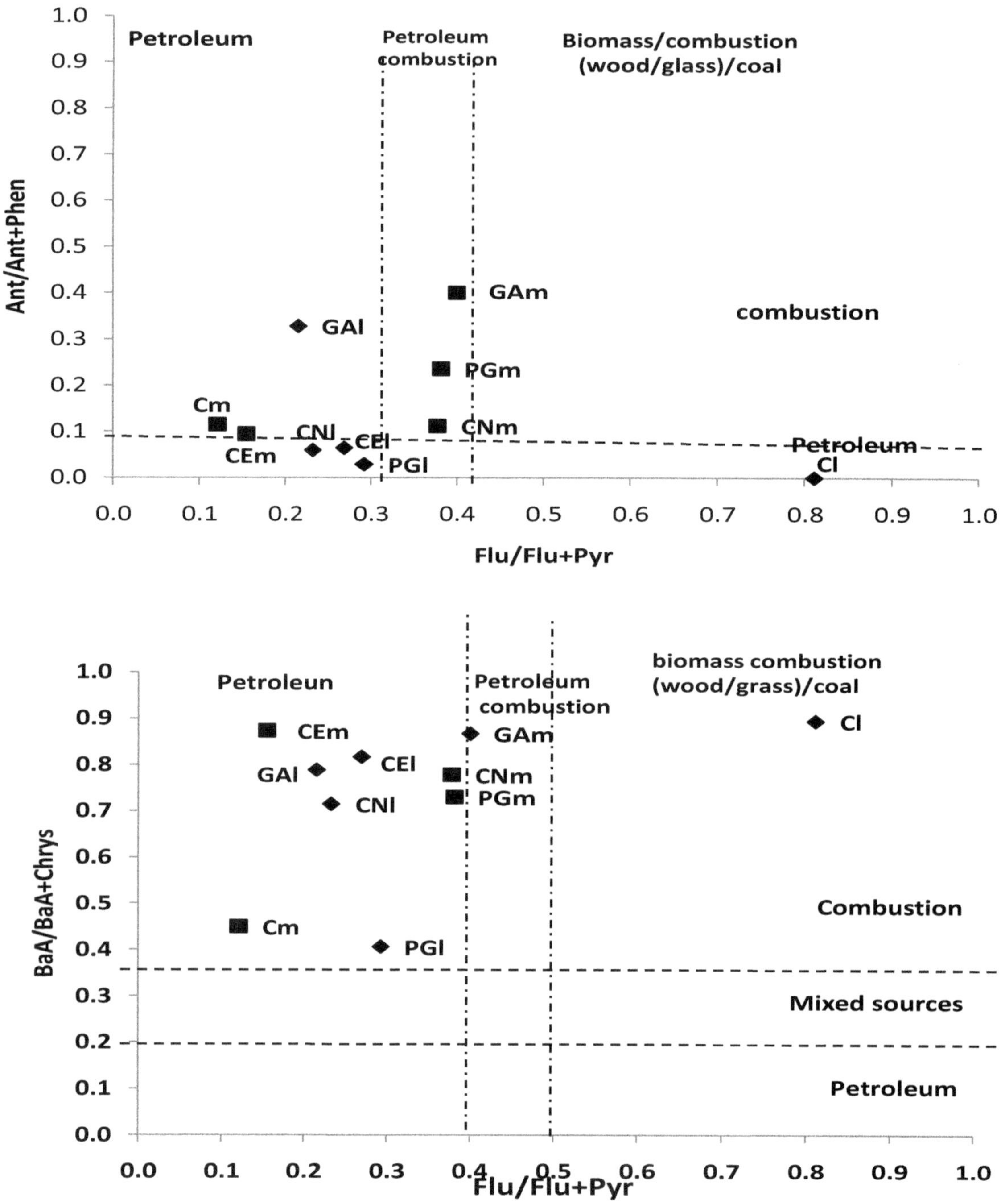

Figure 7. Ant/Ant + Phen, BaA/BaA + Chris versus Flu/Flu + Pyr ratio in muscle (black squares) and liver (black diamonds) of the mullets *C. bispinosus* from the different sites in Porto Grande Bay and Calhau. Legend: CN - Cabnave, PG - Porto Grande, CE - Cais da Enacol and GA - Galé. The control area [Calhau (C)- East coast].

mination by lighter petroleum products but levels of carcinogenic PAHs, especially DaA, are of concern to human health.

Due to the impact of POPs to the environment and human health, it is necessary to improve the knowledge of the different POP concentrations in economically important fish species in Cape Verde for a proper environmental management and risk assessment policy.

The present data is a starting point to assess the impact of PAH levels in this area in the future.

ACKNOWLEDGMENT

The authors would like to thank Câmara Municipal de Portimão for their support to the work of Neusa Pinheiro.

REFERENCES

Almeida FR, Tristão da Cunha Cruz JV (2008). Poluição na Baía do Porto Grande, Mindelo (São Vicente, Cabo Verde): avaliação da situação de referência. In: Ferreira da Silva E Azevedo MR Santos JF Patinha C. Reis A (Coords.). Livro de Resumos, IX Congresso de Geoquímica dos Países de Língua Portuguesa, Praia, Santiago, Cabo Verde. pp. 47.

Alomirah H, Al-Zenki C Husain A, Ahmed N, Al-Rashdan A, Gevao B, Sawaya W (2009). Dietary exposure to polycyclic aromatic hydrocarbons from commercially important seafood of the Arabian Gulf. J. Food Agricult. Environ. 7(1):9- 15.

Anyakora C, Coker H (2007). Assessment of polynuclear aromatic hydrocarbon content in four species of fish in the Niger Delta by gas chromatography/mass spectrometry. Afric. J. Biotechnol. 6(6):737-743.

Barbour EK, Mastori FA, Shaib HA, Abdel Nour AM, Jaber LS (2009). Comparison of the levels of organic contaminants in bile and muscle of *Mugil spp*. following a major oil spill in Eastern Mediterranean. Vet. Ital. 45(3):405-412.

Barreira LA, Mudge SM, Bebianno MJ (2007). Polycyclic aromatic hydrocarbons in clams Ruditapes decussates (Linnaeus, 1758). J. Environ. Monit. 9:187-198.

Baumard P, Budzinski H, Garrigues P, Sorbe J, Burgeot T, Bellocq J. (1998).Concentrations of PAHs (polycyclic aromatic hydrocarbons) in various marine organisms in relation to those in sediments and to trophic level. Mar. Pollut. Bull. 36:951-960.

Baumard P, Budzinski H, Mchin Q, Garrigues P, Burgeot T, Bellocq J (1998). Origin and bioavailability of PAHs in the Mediterranean Sea from mussel and sediment records. Est. Coastal Shelf Sci. 47:77-90.

Bravo de Laguna J. (1985). Plateaux insulaires et Zone Economique Exclusive de la République du Cap Vert. FAO/CVI/82/003/Rapp/Tech/6.

Cheung KC Leung HM Kong KY Wong MH (2007). Residual levels of DDTs and PAHs in freshwater and marine fish from Hong Kong markets and their health risk assessment. Chemosphere 66:460-468.

Darilmaz E, Kucuksezgin F (2012). Distribution of Aliphatic and Aromatic Hydrocarbons in Red Mullet (*Mullus barbatus*) and Annular Sea Bream (*Diplodus annularis*) from the Izmir Bay (Eastern Aegean). Bull. Environ. Contam. Toxicol. 88:283-289.

Della Torre C, Corsi I Nardi F, Perra G, Tomasino MP, Focardi S (2010). Transcriptional and post-transcriptional response of drug-metabolizing enzymes to PAHs contamination in red mullet (*Mullus barbatus*, Linnaeus, 1758): a field study. Mar. Environ. Res.70:95-101.

DGMP (1998). Gestão da Zona Costeira, 1 - Atlas da natureza da costa e da ocupação do litoral. Reconhecimento fotográfico:1-76.

DouAbul AAZ, Heba HMA Fareed KH (1997). Polynuclear aromatic hydrocarbon (PAHs) in fish from the Red Sea Coast of Yemen. Hydrobiol 352:251-262.

DouAbul AAZ, Abaychi J K, Al-Edanee TE, Ghani AA, Al-Saad HT (1987). Polynuclear aromatic hydrocarbons (PAHs) in fish from the Arabian Gulf. Bull. Environ Contam. Toxicol. 38:546-552.

Hellou J, Warren WG (1997). Polycyclic aromatic compounds and saturated hydrocarbons in tissues of flatfish: insight on environmental exposure. Mar. Environ. Res. 43:1-25.

Hellou J, Payne JP, Hamilton C (1994). Polycyclic aromatic compounds in Northwest Atlantic cod (*Gadus moruha*). Environ. Pollut. 4:197-202.

INDP (2006). Boletim Estatístico de Pescas 1991-2006. INDP. Mindelo, 85 p.

Kayal S, Connell DW (1995). Polycyclic aromatic hydrocarbons in biota from the Brisbane River Estuary, Australia. . Est. Coastal Shelf Sci.40:475-493.

Llobet JM Falco GAB Domingo JL (2006). Exposure to polycyclic aromatic hydrocarbons through consumption of edible marine species in Catalonia, Spain. J. Food Protect. 69:2493-2499.

Lopes WA, Andrade JB (1996). Fontes, formação, reactividade e quantificação de hidrocarbonetos policíclicos aromáticos (HPA) na atmosfera. Quím. Nova, 19:497-516.

Martínez-Gómez CB, Fernández J, Benedicto J, Valdés JA, Campillo VM, León AD Vethaak (2012). Health status of red mullets from polluted areas of the Spanish Mediterranean coast, with special reference to Portmán (SE Spain). Mar. Environ. Res. 77:50-59.

Menezes GM, Tariche O, Pinho MR, Duarte PN, Fernandes A, Aboim MA (2004). Annotated list of fishes caught by the R/V Arquipélago off the Cape Verde Arquipélago. Life Mar. Scienc. 21 (A):57-7.

Mille G, Asia L, Guiliano M, Malleret L, Doumenq P (2007). Hydrocarbons in coastal sediments from the Mediterranean Sea (Gulf of Fos area, France). Mar. Pollut. Bull. 54:566-575.

Neff JR (2002). Bioaccumulation in Marine Organisms: Effects of Contaminants from Oil Well Produced Water. Elsevier. The Netherlands. p. 439

Nisbet, ICT LaGoy PK (1992). Toxic equivalent factors (TEFs) for polycyclic aromatic hydrocarbon (PAHs). Regul. Toxicol. Pharmacol. 16:290-300.

Oros DR, Abas MR, Omar NY, Rahman NA, Simoneit BR (2006). Identification and emission factors of molecular tracers in organic aerosols from biomass burning: Part 3. Grasses Appl. Geochem. 21:919-940.

Oros DR, Ross JRM (2005). Polycyclic aromatic hydrocarbons in bivalves from the San Francisco estuary: spatial distributions, temporal trends, and sources (1993-2001). Mar. Environ. Res. 60:466-488.

Perugini M, Visciano P, Giammarino A, Manera MW, Nardo D, Amorena M (2007). Polycyclic aromatic hydrocarbons in marine organisms from the Adriatic Sea, Italy. Chemosphere 66:1904-1910.

Porte C, Biosca X, Solé M, Albaigés J (2001a). The Integrated use of chemical analysis, cytochrome P450 and stress proteins inmussels to assess pollution along the Galician coast (NW Spain). Environ. Pollut. 112:261-268.

Porte C, Solé M, Borgi V, Martinez M, Chamorro J, Torreblanca A, Ortiz M, Orbea A, Soto M, Cajaraville MP (2001b). Chemical biochemical and cellular responses in the digestive gland of the mussel *Mytilus galloprovincialis* from the Spanish Mediterranean coast. Biomarkers 6(5):335-350.

Rainio K, Linko R, R Ruotsila L. 1986. Polycyclic aromatic hydrocarbons in mussel and fish from the Finnish Archipelago Sea. Bull. Environ. Contam. Toxicol. 37:337-343.

Reddy MS, Basha Joshi HV, Ramachandraiah G (2005). Seasonal distribution and contamination levels of total PHCs, PAHs and heavy metals in coastal waters of the Alang-Sosiya ship scrapping yard, Gulf of Cambay, India. Chemosphere 61:1587-1593.

Said TO El, Agroudy NA 2006. Assessment of PAHs in water and fish tissues from Great Bitter and El Temsah lakes, Suez Canal, as chemical markers of pollution sources. Chem. Ecol. 22(2):159-173.

Silva TF, Azevedo DA, Aquino Neto FR (2007). Polycyclic Aromatic Hydrocarbons in fishes and sediments from the Guanabara Bay, Brazil. Environ. Forens. 8(3):257-264.

Trape S, Harrison IJ, Diouf PS, Durand JB (2012). Redescription of *Liza bandialensis* (Teleostei: Mugilidae) with an identification key to mullet species of Eastern Central Atlantic. C. R. Biologies 335:120-128.

USEPA (2000). Guidance for assessing chemical contaminant data for use in fish advisories. Fish sampling and analysis. Vol. 1. 3rd edn. EPA/823-B-00-007USEPA 2000.

USEPA (2002). Polycyclic organic matter. U.S. Environmental Protection Agency, Washington, D.C.20 April 2006.

Vives I, Grimalt JO (2002). Method for integrated analysis of polycyclic aromatic hydrocarbons and organochlorine compounds in fish liver. J Chrom. B 760:247-254.

Yunker MB, Macdonald RW, Vingarzan R, Mitchell RH, Goyette DSS (2002). PAHs in the Fraser River basin: a critical appraisal of PAH ratios as indicators of PAH source and composition. Org. Geochem. 33:489-515.

8

Food items and general condition of *Hyperopisus bebe occidentalis* (Lacepede, 1803) caught in Warri River, Nigeria

Olele, N. F.

Fisheries Department, Delta State University, P.O. Box 1110, Asaba, Nigeria.

A total number of 202 specimens, comprising 75 females and 127 male specimens of *Hyperopisus bebe occidentalis* were caught from Warri River. This gave a 1:1.7 female to male sex ratio. The total length range for male specimens was 189.2 to 355.0 mm and 246.0 to 376.1 mm for females. The body weight range for male specimens was 51.0 to 448.0 g, while that recorded for female specimens was 106.0 to 400.0 g. The correlation coefficient for length-weight relationship revealed significant differences ($P<0.05$) for male ('r' = 0.77) and female ('r' = 0.51) specimens, respectively. The slope of the regression co-efficient 'b' was (2.68) for males and 1.76 for females. Both values were less than 3 implying that the fish increased more in total length than in body weight. The condition index was more favorable for male (4.64) than for female (4.14) specimens based on mean value calculated on fresh and gutted body weights.

Key words: Food items, *Hyperopisus bebe occidentalis*, Warri River, condition factor.

INTRODUCTION

Hyperopisus bebe of the family Mormyridae was formally placed under the genus *Mormyrops* (Malami et al., 2002; Olaosebikan and Raji, 1998). This species is well distributed in swamps, lakes and rivers of most Nigerian fresh water bodies (Ogbeibu and Ezeunara, 2005; Idodo-Umeh, 2003; Malami et al., 2002, 2004; Babatunde and Aminu, 2004); where they are far more abundant than other mormyrids. This species has always been consumed for their oily and tasty flesh. Some researchers (Kauamelan et al., 2002; Nwani, 2004; Malami et al., 2002, 2004) studied the food and feeding habit of this species in different bodies of water. They reported a higher occurrence of plant materials in the guts of juvenile than in the guts of adult specimens examined. Food items of plant and animal origins were also consumed indiscriminately when their preferred food items were scarce.

In Anambra River for instance, this species consumed benthos and allochtonous invertebrates mixed with a limited amount of mud (Malami et al., 2002, 2004; Nwani et al., 2006). The study of food and feeding habits enables farmers to have clear understanding of the fish's dietary requirements and to compound appropriate feeds as supplement in aquaculture (Malami et al., 2004). Sufficient food intake aids optimal growth in fish, resulting in production increases and subsequent economic benefits.

Pius and Benedicta (2002) reported on the benefit of gut content analysis for reducing intra- and inter-specific competition of fish in the ecosystems.

Literature review revealed that information on food and feeding habit of *H. bebe* harvested from the Warri River was lacking.

The present study therefore investigates the recent diet composition and general conditions of this commercially important fish species harvested from Warri River.

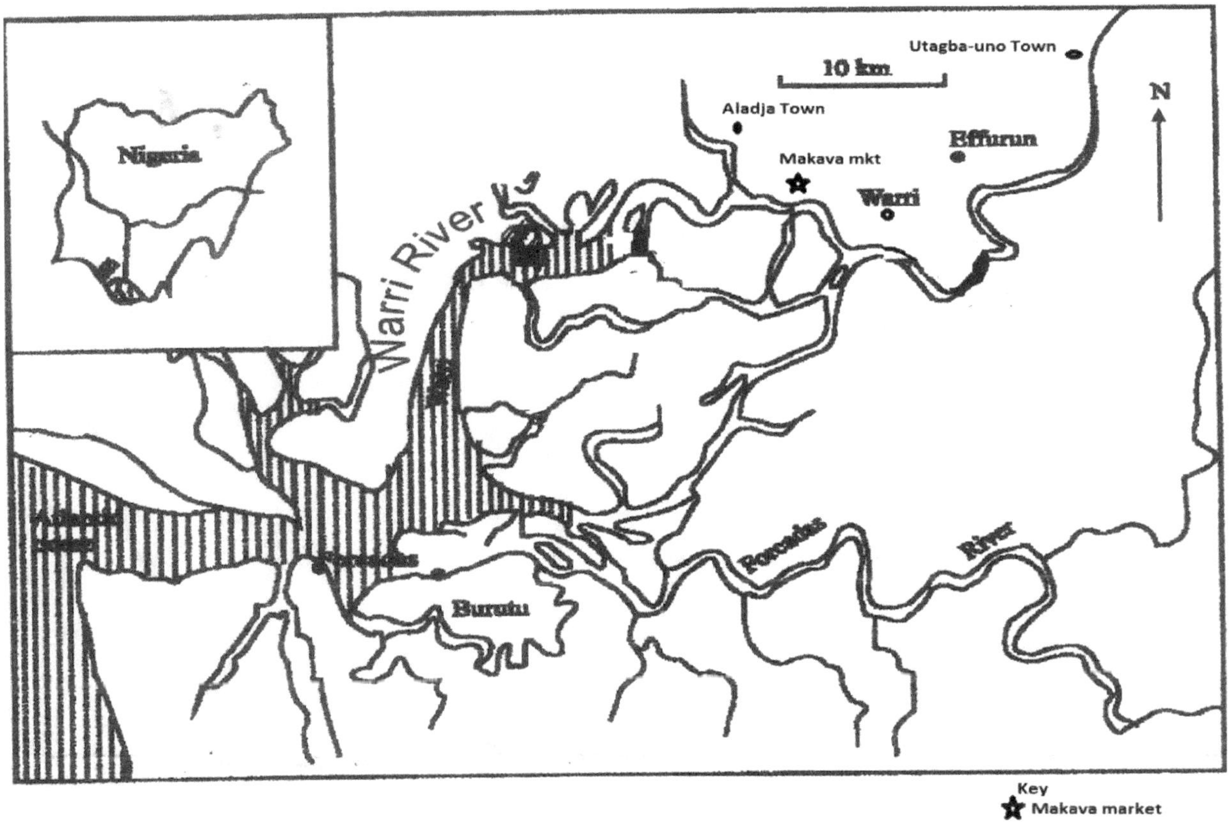

Figure 1. Map of Warri River showing the sample collection station.

MATERIALS AND METHODS

Warri River is one of the most important coastal rivers of the Niger-Delta region. It takes its source about 10 km away from Utagba-Uno which lies North and East of the Equator within latitude 6° 0' N and longitude 5° 2' E (Figure 1). It covers a surface area of 255 km^2 with a length of about 150 km^2 (Olomukoro and Egborge, 2004).

Fish sampling

Fish sampling was conducted on a monthly basis, between April and September, 2010. The gears used were cast nets, traps, fleets of nylon multifilament set gill nets of diverse mesh sizes and other traditional fishing gears, positioned in the morning on each sampling day. Fishes harvested were injected with 5% formalin (in the abdominal region) in order to arrest food digestion during transportation of specimens to the fisheries laboratory located at the Asaba Campus of the Delta State University, where gut content examination was carried out.

Fish identification, length/weight measurements and gut content analysis

Fish specimens were identified to species level with the aid of keys according to Idodo-Umeh (2003) and Nwani (2004) before sorting into male and female sex. Total length and body weight measurements were recorded to the nearest ± 0.01 mm and ± 0.1 g using a measuring board and a triple beam balance (OHAUS 210 Model), respectively. Length weight relationship was calculated with the formula W = aLb according to Nwani et al. (2006) and Erhijowvho (2007). Bodyweight increases more rapidly than total length; hence, the formula was logarithmically transformed for the purpose of data analysis; thus:

$$\text{Log } W = \text{Log } a + b \text{ Log } L$$

Where 'a' was the proportionality constant, 'W' was the weight of fish in grams (g), 'L' was the total length of fish in millimeters (mm) and '*b*' was the allometric growth coefficient.

Guts were opened surgically, and weighed with and without food materials contained in them. Food items in those guts that could not be examined immediately were preserved in sterilized labeled vials containing 1% formalin. Large food items, easily recognized with the naked eyes, were counted, while microscopic ones were teased to disperse their aggregates in accounting chamber. They were examined and counted under a binocular microscope at a magnification of X 10. All recognized food items were identified according to methods described by Kadiri (1987, 2002). Food items were analyzed using three methods of gut content analysis described as follows:

Frequency of occurrence method

The number of guts in which each food item occurred was listed and expressed as a percentage of the total number of guts examined. The proportion of the fish population that fed on a particular food item was estimated according to Odun and Anuta (2001) and Inyang and Nwani (2004).

Table 1. Mean value for total length and body weight of *Hyperopisus bebe* sampled during the study.

Month	Sample size: Male	$\bar{x}$ T_L (mm)	$\bar{x}$ B_W (g)	Sample size: Female	$\bar{x}$ T_L (mm)	$\bar{x}$ B_W (g)	Sex ratio
April	19	279.0	210.0	0	281.0	210.0	1:0
May	25	274.4	163.0	3	280.0	165.0	8.3:1
June	15	313.0	262.0	18	302.7	249.0	1:12
July	20	292.0	187.0	20	312.3	276.0	1:1
August	27	288.7	177.0	13	293.8	176.0	2:1
September	20	285.7	191.0	20	288.5	209.0	1:1
Total	127			75			1:1.7

Note: mm = millimeters, g = gram, $\bar{x}$ T_L = mean total length, $\bar{x}$ B_W = mean body weight.

Numerical method

The total number of each food item from each gut was summed up for all guts and expressed as a percentage of the total number of all food items as expressed by Odun and Anuta (2001) and Inyang and Nwani (2004).

Point method

One hundred points were shared among the food items present in each gut according to their sizes, and expressed in percentage according to Inyang and Nwani (2004).

Condition factor

The condition factor (K), describing the wellbeing of a fish was calculated according to Erhijivwo (2007) thus:

K = 100 W/L3

Where K = condition factor, L = standard length of fish in mm, and W = body weight of fish in grams.

Feeding intensity

Feeding intensity of each specimen was calculated according to Erhijivwo (2007):

$$\text{Feeding intensity} = \frac{\text{Gut weight} \times 100}{\text{Body weight} \times 1}$$

RESULTS

Fish population, length-weight relationship and sex ratio of specimens

Two hundred and two (202) specimens, comprising 127 males and 75 females were examined. Thus, the overall proportion of 1:1.7 female to male sex ratio was calculated (Table 1). Total length ranged between 189 to 379 mm, while body weights varied from 50 to 499 g. The male specimens exhibited both highest and lowest total length frequencies measured (300 to 309 and 180 to 189 mm, respectively, Figure 2). At the weight range of 200 to 249 g, male specimens accounted for the highest population proportion (37 specimens), while at a weight range of 400 to 449 g, only one specimen was encountered. For female specimens, at the highest (350 to 399 g) and lowest (150 to 159 g) body weight ranges, 231 specimen were measured, respectively (Figure 3). Features of *H. bebe* involving body weight and total length measurements were described by the correlation coefficient relationship transformed logarithmically and expressed as log W = aLb, where the 'r' value was 0.77 for males (Figure 4) and 0.51 for females (Figure 5).

Both values were statistically significant at $p<0.05$. The '*b*' value for males (2.68) was higher than that for females (1.76). However, both values indicate negative allometric growth, because they were lower than the value of three.

Condition factor

Our analysis revealed that male specimens (K = 4.64) were in a better condition than the female specimens (K = 4.14) based on mean values calculated for fresh and gutted body weights (Table 2).

Gut fullness, food items analyzed and seasonal variation in feeding habits

Out of the 202 specimens examined for food items in the gut, only four guts were empty, accounting for 1.9%. Twenty-four specimens had full guts accounting for 11.9%, 46 specimen had three quarter full guts (22.8%), 87 specimen had half-full guts (43.1%), and 41 specimens had quarter-full gut content accounting for 20.3%. The three methods used for the analysis of gut contents revealed that the most dominant food item identified were Bacillariophyta. This family contributed 22.8% by numerical method and 23.4% by frequency of

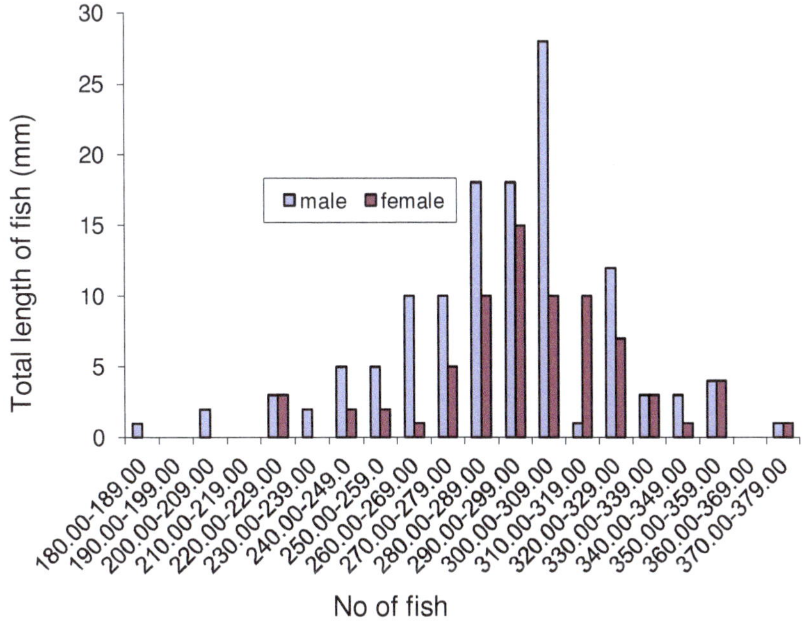

Figure 2. Length frequency distribution of male and female specimens of *H. bebe*.

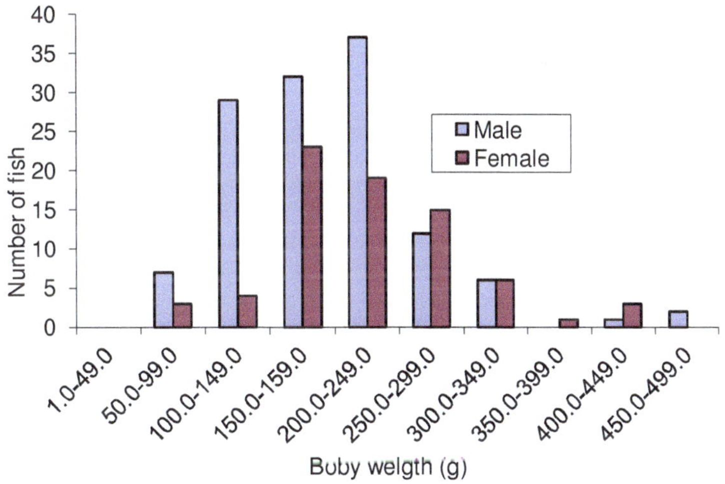

Figure 3. Body weight frequency distribution of male and female *H. bebe*.

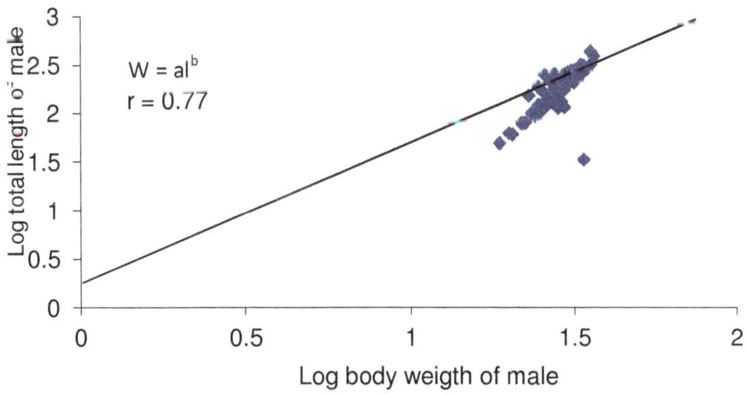

Figure 4. Log relationship between total length and body weight of male *H. bebe*.

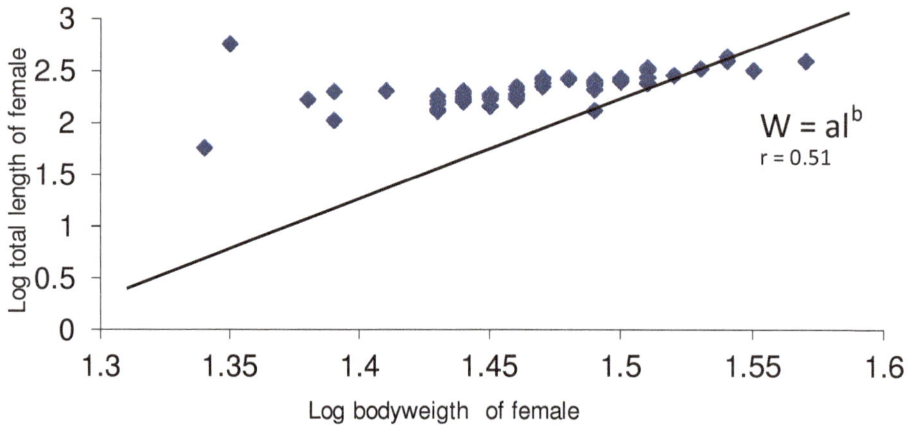

Figure 5. Log relationship between total length and body weight of female *H. bebe.*

Table 2. Mean condition factor for fresh and gutted body weights of *H. bebe.*

Sex	No of specimen	Mean k for fresh body weight	Mean k for gutted body weight
Male	127	0.87	4.64
Female	75	0.68	4.14
Male + Female	202	1.55	8.78

K = Condition factor.

occurrence method to the total diet of the fish analyzed. Dinoflagellates contributed 16.2%, by the numerical method and 20.4% by the point method. Chlorophyta made up 10.4% by numerical method, while insect parts were the least dominant food item contributing 0.7% both by numerical and frequency of occurrence methods (Table 3). No strong seasonality of food items consumed was observed, as the examined fish specimens grazed on those food items which were available throughout the sampling period. However, the diet composition and frequency of food items varied slightly. The fish guts were generally well-filled throughout the study period. The fullest guts were found during the months of April, May and June, when the specimens fed more on green algae, seed grains and Bacillariophyta. Specimens equally fed on dinoflagellates and root hairs within the months of July, August and September. Variations in diet composition were found to be size rather than sex-dependent.

This observation suggests that the plasticity in food abundance and diet composition exhibited by both sexes enables them to exploit various food sources. Thus, they are capable of not only maintaining their distribution in the present habitats, but also to colonize wider habitats.

DISCUSSION

Size frequency distribution

The size class frequency distribution of specimens used for this study reveals differences in body weight, size class and maturity stages within the population of specimens examined. This observation is similar to those of previous studies (Nwani, 2004; Ogheneochuko, 2007; Omorinkoba and Fatuiti, 2009). Omorinkoba and Fatuiti (2009) stated that the size class or body weight of a fish determines the spectrum of food items it will consume. Understanding the relationship between body size/structure and fish food is therefore important for interpreting the fish's diet and feeding mechanisms. Hence, the type of food available in an area influences the distribution, abundance and rate of growth (size) of the fish (Ogheneochuko, 2007).

Sex ratio

Male specimens were caught more than females in April and May. This finding is most likely due to season as female migrate into deeper waters for spawning during this period (Nwani, 2004).

Length weight relationship

There was a significant correlation between body weight and total length of specimens found in this study. Increases in total length resulted in corresponding increases in body weight, as reported (Nwani et al., 2006). According to Olurin and Aderibigbe (2006),

Table 3. Results of gut content analyzed using three methods of analysis of gut content.

Food item	Numerical method		Frequency of occurrence		Point method	
	No	%	No	%	No	%
Bacillariophyta: *Navicula* spp.	182	8.5	103	7.9	120	8.5
Ulothrix spp.	159	7.4	118	9.1	120	8.5
Tabellaria spp.	144	6.7	77	5.9	101	7.1
Coscinodiscus spp.	6	0.2	6	0.5	15	1.0
Total	491	22.8	304	23.4	356	25.1
Chlorophyta: *Closterium* spp.	116	5.4	78	6.0	55	3.9
Cosmarium spp.	63	2.9	42	3.2	30	2.1
Spirogyra spp.	46	2.1	34	2.6	21	1.5
Total	225	10.4	144	11.8	106	7.5
Dinoflagellates: *Gomyaulax* spp.	185	8.6	114	8.8	125	8.8
Ceratium spp.	27	1.3	23	1.8	50	3.5
Merismopedia spp.	136	6.3	88	6.8	115	8.1
Total	348	16.2	225	17.4	290	20.4
Crustaceans: Shrimp parts	10	0.5	9	0.7	21	1.5
Crayfish parts	71	3.3	42	3.2	70	4.9
Total	81	3.8	51	3.9	91	6.4
Chironomid larva	186	8.6	113	8.7	65	4.6
Coleoptera: Whole beetle	100	4.6	71	5.5	70	4.9
Beetle parts	94	4.4	38	2.9	58	4.1
Total	194	9.0	109	6.4	128	9.0
Insects: Whole parts	15	0.7	9	0.7	25	1.8
Root hairs	179	8.3	96	7.4	120	8.5
Sand and stones	146	6.8	51	3.9	44	3.1
Seed grains	141	6.6	68	5.4	98	6.9
Unidentified	145	6.7	117	9.0	95	6.7
Grand total	2,151	100	1,297	100	1,418	100

differences in total length and body weight distributions are dependent of sex and developmental stages of the fish. Aliakbar and Ali (1978) and Kunda et al. (2008) proposed that fluctuations observed in certain length groups might be due to variation in sample size, sex, gonad condition and amount of gut content. Length-weight relationship measurements are important for the estimation of weight where only length data are available, as well as a condition index of fauna (Haimovici and Velasco, 2000). Length-weight relationship has also been shown to be useful in estimating standing stock biomass and densities of various organisms in the aquatic habitat. Specimens used for this study exhibited negative allometric growth having a '*b*' value lower than 3. Olele and Obi (2004) reported positive allometric growth where the '*b*' value for *Citharinus citharus* caught in Onah Lake was higher than 3 (3.1). Other '*b*' values reported were either lower or higher than 3, as been reported by Arawomo (1981) thus (2.8) and (Baijot et al., 1997) (2.5 to 3.5). Differences in '*b*' values may be influenced by sex, maturity stages, seasonality and the time of day the food was consumed by the fish when fullness of gut content is measured.

Fagade and Adebisi (1979) reported that in cichlids, an increase in body weight was associated with an increase in standard length. They reported a '*b*' value of 2.9 and 3.4 for *Tilapia melanotheron* and *T. guinensis*, respectively.

Condition factor

The condition factor calculation for *Hyperopisus* specimens based on fresh and gutted body weights analysis revealed that those for males were higher than those for females. This observation could indicate that the former fed better than the latter at all times. Such conditions may also be responsible for a better survival of male specimens, as was previously reported (Nwani, 2004).

Analysis of food items

Food items identified were both of plant and animal origins, suggesting that the fish acted as an omnivore. This assertion was already proposed by Malami et al. (2004). Much of the variation in diet composition of this species depended on the availability of food items, which has also been previously described (Ogbeibu and Ezeunara, 2005).

Fullness of gut content

The overall results of the gut fullness analyses revealed that 2% of guts were empty, while varying quantities of food items were found in 98% of guts. The observance of higher non empty guts may have resulted from the immediate arrest of food digestion through the injection of formalin in the gut region of the fish before their conveyance to the laboratory for examination. This result is in line with those of Malami et al. (2004) on the same species. The greater number of guts with food is generally attributed to a successful feeding strategy adopted by the specimens (Haroon, 1998; Nwani, 2004), and additionally benefitting from good food abundance during the sampling season. It was observed that more food items were recorded during the rainy and flood seasons. During this period, insects and grains/seeds occured in higher numbers. This observtion is supported by the study of Nwani (2004), who reported on active feeding behavior of *H. bebe* in the River Rima during the early rainy season.

Kouamelan et al. (2002) investigated the food and feeding habits of *Mormyrus rume* in Bia River Cote de' voir, Ghana. They report that young specimens feed more on Chironomid larvae and Bacilliarophyta, while adult specimens feed more on other food items both during the dry and rainy season.

Conclusion

H. bebe belongs to the family of Mormyridae, which is one of the largest groups of fishes in the Nigerian waters. Their ability to feed on a wide range of organisms at different trophic levels (food chain) is the most likely reason for full guts in the majority of analyzed specimens. Thus, they may be described as euryphagous, feeders that consume any food item that could readily come across, particularly when many/different food items are not available. This might also explain their fast growth. Furthermore, the knowledge on the feeding habits of this species resulting from this study should be of benefit in the formulation of supplementary diets necessary for mass production. This will help to meet local demands as well as generate foreign exchange for the country.

REFERENCES

Aliakbar M, Ali ML (1978). Length-weight relationship of *Macrobrachium rosenbergii* (DeMan) of Dakatia River. Bangladesh J. Aquat. 1(1):74-78.

Arawomo GAO (1981). The growth of *Heterotis niloticus* in Opa Reservoir, University of Ife. Proc. of the 2nd Ann. Conf of the Fisheries Society of Nigeria. pp. 34-41

Babatunde DO, Aminu R (2004). Field guide to Nigerian fresh water fishes, second Edition. pp. 61.

Baijot E, Moreau J, Bouda S (1997). Hydrobiologica aspects of fisheries in small reservoirs in Sahel region. Technical Centre for Agricultural and rural co-operation ACP-Eu. pp. 200.

Erhijivwo CA (2007). Food and feeding habits of *Heterotis noliticus* caught in Onah Lake Asaba, Nigeria. B. Sc. Thesis. Department of Animal science and fisheries, Delta State University. Asaba, Delta State, Nigeria.

Fagade SO, Adebisi AA (1979). On the fecundity of *Chrysichthys nigrodigitatus* (Lacepede) in Asejire Dam, Oyo State, Nigeria. Niger. J. Nat. Sci. 1:127-131.

Haimovici M, Velasco G (2000). Length-weight relationship of marine fishes from southern Brazil Naga, International center for living Aquatic Resources management. Quarterly 23:19-23.

Haroon AS (1998). Diet and Feeding ecology of two species of *Barbodes gonionotus* and *Oreochromis niloticus* in rice field in Bangladeh. Naga, the ICLARM Quarterly pp. 13-18.

Idodo-Umeh G (2003). Freshwater fisheries of Northern Nigeria. (Taxonomy, Ecological Notes, Diet and Utilization). ISBN 978-8052-01-0.

Inyang WM, Nwani CD (2004). Food and feeding habits of *Distichodous* spp (Osteichthyes Distichodontidae) in Anambra River. J. Biol. Res. Technol. 4:34-43.

Kadiri MO (1987). Algea and primary productivity studies of Ikpoba Reservoir. Ph.D thesis. University of Benin, Benin City. pp. 298.

Kadiri MO (2002). Periphyton of Ikpoba River in Benin City, Edo State. B.Sc thesis, University of Benin, Benin City. pp. 45.

Kouamelan PE, Juegels GG, Ollevier F (2002). Feeding habits of *Mormyrops anguiloides* (mormyridae) in lacustrine and riverine habitat of a West African basin. Cybium 24 (i) 69-79.

Kunda M, Dewan S, Uddin MJ, Karim M, Kabir S, Uddin MS (2008). Length-weight relationship, and condition factor of *Macrobrachium rosenbergii* in the rice fields. Asian Fish. Sci. 21:451-456.

Malami GZ, Ipinjolu JK, Hassan WA, Magawata I (2004). Feeding adaptations of ten fish species in river Rima, North Western Nigeria. A paper presented at the 2004 Annual conference of Zoological Society of Nigeria held at the Institute of Developmental Research, Ahmadu Bello University, Zaria. pp. 115.

Malami GZ, Ipinjolu JK, Hassan WA, Magawutu F (2002). Food and feeding habits of freshwater mormyrid (*Hyperopisus bebe occidentalis*, Gunther) in River Rimo and Goronyo Dam, Nigeria. FISON conference proceeding (2007). pp. 120 -126.

Nwani CD (2004). Aspects of the biology of mormyrids in Anambra River, Nigeria. Ph.D Thesis University of Nigeria Nsukka. pp. 194.

Nwani CD, Ezenwanji H, Maduka GE, Effiong J, Emmanuel F (2006). Length - weight relationship and condition factor of four mormyrid species of Anambra river. Department of applied Biology, Ebonyi state University PMB 53, Abakaliki, Ebonyi State, Nigeria - Fisheries Hydrobiology Research Unit. Department of Zoology, University of Nigeria, Nsukka.

Odun O, Anuta M (2001). The food and feeding habits of Musansorgi (Boulenger) from Warri River. Niger. J. Aquat. Sci. 21:16-18.

Ogbeibu AE, Ezeunara PU (2005). Studies on the food composition and feeding pattern of fish communities in the Ikpoba River, Southern Nigeria. J. Aquat. Sci. 20 (2):117-129.

Ogheneochuko OS (2007). Food and feeding habits of *Parachanna obscura* in Anwai River. A tributary of River Niger in Asaba. B.Sc. Thesis. Department of Fisheries, Delta State University, Asaba Campus. pp. 10.

Olaosebikan BD, Raji A (1998). Field guide to Nigerian fresh water fisheries. Federal College of Freshwater Fisheries Technology, New Bussa Niger State, Nigeria. pp. 106.

Olele NF, Obi A (2004). Reproductive biology of *Citharinus citharus* in Onah Lake, Asaba. Delta State of Nigeria. Trop. Freshw. Biol. 13:119-129.

Olomukoro JO, Egborge ABM (2004). Hydrobiological Studies on Warri River, Nigeria part 11: Seasonal trend in physiochemical limnology. Trop. Freshw. Biol. 12/13:9-23.

Olurin KB, Aderibigbe OA (2006). Length-weight relationship and condition factor of pond reared juvenile *Oreochromis niloticus*. World J. Zool. 2:82-85.

Omorinkoba WS, Fatuiti EO (2009). Food and feeding habits of *Heterotis niloticus*. Swim up advanced fry from the Kainji Freshwater Reservoir. Department of Fisheries. Niger. J. Fish. Sci. Technol. 1:89-92.

Pius MO, Benedicta OO (2002). Food and feeding inter-relationship. A preliminary indicator to the formulation of the feed of some *Tilapia* fishes. Trop. J. Anim. Sci. 5(1):35-41.

Comparative analysis of municipal solid waste (MSW) composition in three local government areas in Rivers State, Nigeria

Babatunde B. B.[1,2] , Vincent-Akpu I. F.[1], Woke G. N.[1], Atarhinyo E.[1], Aharanwa U. C.[1], Green A. F.[3], Isaac-Joe O.[1]

[1]Department of Animal and Environmental Biology, Faculty of Science, University of Port Harcourt, PMB 5323, Choba, Port Harcourt, Rivers State.
[2]Lancaster Environment Centre, Lancaster University, UK, LA1 4YQ
[3]Department of Applied and Environmental Biology, Rivers State University of Science and Technology, Nkpolu, Port Harcourt, Rivers State.

Rivers State is one of the major oil producing States in Nigeria. Its capital, Port Harcourt and sub-urban areas have witnessed an increased influx of migrants in recent time. Consequently, the consumption of goods and services has also increased leading to generation of unprecedented quantities of municipal solid waste. Previous efforts by relevant authorities to contain this increase in waste generation have proved abortive due largely to among other reasons, inappropriate and unsustainable municipal solid waste (MSW) management systems. The first step in a sustainable MSW management is to understand the types, composition and variation that exist in the waste generated in time and space. Previous reports have been concentrated on the characteristics of waste in Port Harcourt. This present work evaluated solid waste composition in three LGAs in Rivers State with a view of an integrated approach in MSW management in the State. Waste characteristics varied with the different locations studied. Generation rate was 0.45, 0.98 and 1.16 kg/capita/day for Emougha, Obio/Akpor and Port Harcourt, respectively. Most prominent categories were organic waste, paper and nylon. Mean percentage composition was 65, 65.5 and 59% for organic waste, 13, 11 and 6% for paper and 12, 16 and 14% for nylon in Port Harcourt, Obio/Akpor and Emougha LGAs, respectively. It is believed that given the characteristics of these waste streams, a sustainable management strategy could have several benefits including resources recovery and energy generation.

Key words: Municipal solid waste (MSW), characterization, Rivers State, integrated MSW management.

INTRODUCTION

Solid waste management has emerged as one of the greatest challenges facing municipal authorities world wide especially in developing countries. The volume of solid waste been generated has continually increased at faster rate than resources available to contain it. In developing nations, the situation is more critical since their resources are usually meagre and more priority issues like health and education beat municipal solid waste management (MSWM) to the top of the list. The consequence is ill conceived and operated epileptic MSWM systems that leave monuments of solid waste adorning the streets of urban centres in countries like Nigeria posing serious risk to both human and environmental health.

Municipal solid waste (MSW) includes refuse from households, non-hazardous solid waste from industrial, commercial and institutional establishments (including hospitals), market waste, yard waste and street sweepings. MSWM refers to the collection, transfer, treatment, recycling, resources recovery and disposal of solid waste in urban areas. The goals of municipal solid waste management are to promote the quality of the urban environment, generate employment and income, and protect human and environmental health and support the efficiency and productivity of the economy (Ogwueleka, 2009). Igoni et al. (2007) defined MSW as all waste collected by private and public authorities from domestic, commercial and some industrial (non-hazardous) sources. Urban solid waste is a heterogeneous material and its generation rate and composition vary from place to place and from season to season (World Resources Institute, 1996; Gidarakos et al., 2006). The composition and volumes differ between high and low-income locations (UNEP, 2002). According to Ogwueleka (2009), solid waste is wetter, heavier and more corrosive in developing nations, making its management more difficult.

The poor state of solid waste management in urban centres of developing countries is now not only an environmental problem but also a social handicap (Daskalopoulos et al., 1998). Solid waste management in Nigeria is characterized by inefficient collection methods, insufficient coverage of the collection system and improper disposal despite huge budgets that are committed to MSWM. Local authorities in developing countries spend 77-95% of their revenue on collection and balance on disposal but can only achieve 50-70% collection of MSW (Ogwueleka, 2003).

The quantity and generation rate of solid waste in Nigeria have increased at an alarming rate over the years without corresponding efficient and modern technology for managing the waste (Babayemi and Dauda, 2009). The indiscriminate and improper dumping of MSW in developing countries is increasing and is compounded by a cycle of poverty, population explosion, decreased standards of living and poor governance and low level of environmental awareness. Hence, waste generated is illegally disposed of onto any available open space hence the term "Open Dump Site". Solid waste in Nigeria poses many problems including blockage of drainage and channels causing flooding and presenting breeding grounds for mosquitoes and other vectors and pathogens (Igoni et al., 2007). According to Ogwueleka (2009), more than 25 million tonnes of solid waste is generated in Nigeria annually with average rate of generation ranging from 0.44 kg/cap/day in rural areas to 0.66 kg/cap/d in in urban areas.

Rivers state where the present study was undertaken is home to the country's wealth, usually referred to as the treasure base of the nation. It produces crude oil that accounts for 90% of the country's foreign exchange earnings and has a population of 5,198,716 people (NPC, 2006). Port Harcourt, the state capital is the major urban centre and has witnessed high influx of migrants as a result of its rapid urbanization. This has no doubt increased the consumption patterns and consequently waste generation and more pressure on the municipal authorities responsible for solid waste management. Despite efforts made by the State Sanitation Board, refuse still adorn the streets of Port Harcourt and other sub-urban areas in the state and in some cases blocking drainages, roads and channels causing flooding posing danger to public health (Ayotamuno and Gobo, 2004; Igoni et al., 2007). According to Ayotamuno and Gobo (2004), Port Harcourt has lost its glorious name of "Garden City" to Garbage City. Gobo (2002) observed that solid waste management system options that have been carried out without success to solve the problem of refuse disposal in Port Harcourt over the years include incineration, composting, transfer stations and landfilling. Solid waste collection is done by private contractors who pick up waste with trucks from usually overflowing containers stationed for residence on the street sides and disposal is at designated open dump sites in the state where burying, burning and scavenging take place. There are open dumps right in the middle of the urban centres where burning, burying and scavenging usually take place as well. Another set of collectors comprise the cart pushers who collect from residence and dispose to the nearest open dump by the road side.

The first step towards a sustainable solid waste management is to gain understanding of the nature and composition of the waste being generated in time and space. This information is necessary for planning, designing and establishing appropriate and more sustainable collection, transportation and final disposal strategies for the waste. Ogunbiyi (2001) reported that the composition of solid refuse vary from location to location, from household to household and even in the same household, it could vary from week to week, hence solutions should be location specific. Per capita generation of solid waste in Port Harcourt has been reported to range from 0.60-1.25 kg/cap/day and the characteristics of the waste comprise organic matter, plastics, metal, nylon, glass and others (Ibiebele, 1986; Ayotamuno and Gobo, 2004; Igoni et al., 2007; Abah and Ohimain, 2010). But there has been no attempt to compare the waste stream of Port Harcourt metropolis with its sub-urban environs for an integrated catchment management. This paper aimed to present the characteristics of MSW been generated in three local government areas of Rivers State, Nigeria differing in socioeconomic activities for an integrated MSWM scheme. The major consideration was solid waste composition at open dump sites. The categories compared between the three LGA's during the study were organics, paper, plastics, metal, nylon and a category referred to as others comprising wood, diapers, sanitary pads and unidentified materials.

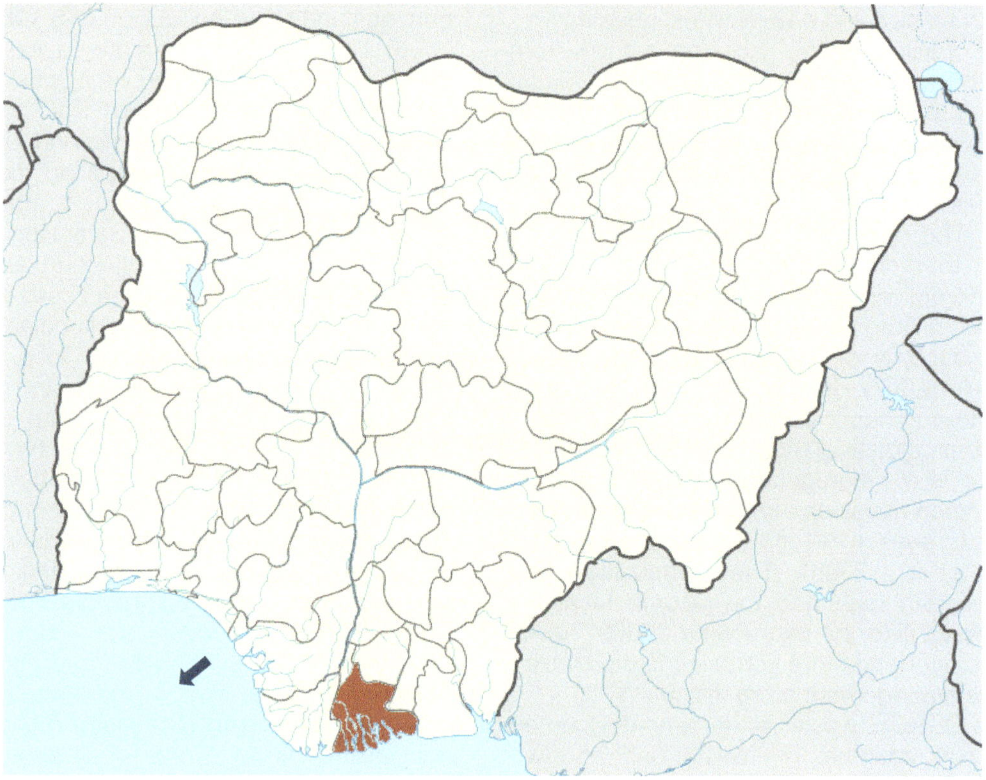

Figure 1. Map of Nigeria showing Rivers State.

MATERIALS AND METHODS

Rivers State where this study was carried out is a maritime state in the southern geopolitical zone of Nigeria (Figure 1) located on 4°45′0″ and 4.75 N and 6°50′0″ and 6°83′3″ E. It has a total population of 5,198,716 (NPC, 2006) comprising 23 local government areas with Port Harcourt, the state capital as one of the Local Government Areas (LGA) (Figure 2). Port Harcourt LGA shares boundary and commerce with the two other LGAs surveyed for solid waste characteristics in this study. The settlements in the three LGAs were chosen according to the category of inhabitants and socioeconomic activities. Port Harcourt LGA is the main urban centre in Rivers State but houses some shanties as well. Obio/Akpor is presumably the richest LGA in Rivers State and houses most of the multinationals. Emuoha LGA is made up of sub-urban and rural settlements. Detail characteristics of the settlements studied under the three LGAs in this study is presented in Table 1.

Sampling and data collection

The determination of the composition of unprocessed municipal solid waste during this study was done according to ASTM D5231 - 92(2008) Standard Test Method which involves the direct sampling of solid waste from specific sources, a labour-intensive manual process of sorting, classifying and weighing all items in each sampling unit and a detailed recording of the data. In each LGA, five distinct communities were selected and five open dumps in each community surveyed. Composite solid waste samples were taken from each dump, mixed and weighed as total weight then placed on a 1 square meter polyethylene material before sorting. The refuse was sorted into seven categories namely; organics, paper, plastic, metal, glass, nylon and others which comprise unidentified waste types. Each category was weighed and its percentage composition determined. Statistical quantities such as the mean, standard deviation and standard error were used to summarise the findings in this study. Standard deviation showed variation in the values of variables from the mean and standard error of the mean (SEM) represents the spread that the mean of a sample of values will have if one keeps taking samples.

RESULTS AND DISCUSSION

The composition of municipal solid waste in the three LGAs studied with mean ± standard error is presented in Tables 2 to 4 and the weighted mean composition is presented in Figure 3. Organic waste composition ranged from 52.08-63.2% in Emuoha LGA, 63.5-66% in Obio/Akpor and 63-69% in Port Harcourt LGA. Figure 3 shows a comparative result of the percentage mean and SEM of waste composition in the three LGAs studied in Rivers State. Organic waste was the highest category in the three LGAs with the least mean value occurring in Emuoha LGA which is a rural area. Paper was highest in Port Harcourt LGA followed by Obio/Akpor and least at Emuoha LGA. Nylon was second highest category overall but recorded the highest mean composition in Obio/Akpor LGA. Metal was highest in Emuoha LGA. Per capital generation was 0.45, 0.98 and 1.16 kg/cap/day for Emuoha, Obio/Akpor and Port Harcourt, respectively. Ayotamuno and Gobo (2004) reported 1.25 kg/cap/day for Port Harcourt while Igoni et al. (2007) reported 1.11

Figure 2. Rivers State showing all the LGAs with the study sites in red circles.

Table 1. The three LGAs studied and characteristics of their settlements.

LGA	Area (sq.km)	Population	Settlement	Settlement type
Port Harcourt	109	541,115	Main township	Planned
			Dioub	NP
			Elekahia/Woji	Semi-planned
			GRA/aba express way	Planed
			Trans Amadi	Industrial layout
Obio/Akpor	260	464,789	Rumuokoro	NP
			Eneka	Industrial layout
			Rumuigbo	NP residential
			Wimpey	NP residential
			Rumuodara	NP residential
Emuoha	831	201,901	Rumuohia	NP rural area
			Isiodu	NP indutrial and residential
			Oduaha	NP rural area
			Mgbutainwo	NP rural area
			Rumuji	NP rural area

*NP = Not planned. Source: NPC (2012).

Table 2: Percentage Composition of Waste categories from five different locations in Port Harcourt LGA.

Category	Elekahia	GRA	Trans Amadi	Diobu	Main township
Organic	65.0±5.5	63.0±.5.1	60±.8.0	69.0±.9.2	66.0±4.9
Paper	14±5.2	16.5±2.3	15.0±3.2	5.6±0.8	13.0±4.1
Plastics	3.3±0.5	4.0±1.2	4.8±1.2	2.2±0.2	3.0±1.4
Metals	1.6±0.1	0.5±0.0	4.0±1.1	1.5±0.1	2.0±0.3
Glass	1.1±0.1	2.5±0.6	2.5±0.3	0.2±0.0	1.0±0.1
Nylon	13.0±2.3	10.5±2.3	10.7±2.8	14.7±4.2	12.0±3.2
Others	2.0±0.5	3.0±0.6	3.0±1.0	6.8±2.2	3.0±0.9

Table 3. Percentage Composition of Waste categories from five different locations in Obio/Akpor LGA.

Category	Rumuokoro	Rumuodara	Eneka	Wimpey	Rumuigbo
Organic	65.8±3.6	63.3±5.8	64.0±4.4	66.0±8.5	66.0±6.9
Paper	12.5±3.1	13.0±2.7	9.3±2.1	10.0±2.8	9.5±2.5
Plastics	1.5±0.1	2.0±0.1	4.0±0.5	2.0±0.5	3.4±1.9
Metals	1.5±0.1	1.2±0.1	2.0±0.1	1.5±0.1	1.3±0.1
Glass	0.2±0.0	0.5±0.0	2.3±0.2	1.5±0.1	0.6±0.0
Nylon	16.3±2.6	17.0±2.4	15.2±3.1	16.0±4.2	16.0±2.1
Others	3.0±0.7	3.1±0.5	3.2±1.0	3.0±0.8	3.2±0.4

Table 4. Percentage composition of waste categories from five different locations in Emuogha LGA.

Category	Rumuohia	Isiodu	Oduaha	Mgbutainwo	Rumuji
organic	63.2±4.5	59.2±6.2	62.0±5.5	59.0±6.3	52.1±8.2
Paper	5.5±0.9	4.3±1.2	4.0±1.1	5.5±0.5	10.4±1.2
Plastic	3.3±0.5	2.4±1.1	5.3±1.5	3.5±1.2	8.3±2.5
Metal	2.6±0.6	15.0±3.1	8.1±3.1	4.2±0.2	4.3±1.1
Glass	3.4±1.1	2.2±0.2	1.2±0.1	2.5±0.2	6.3±1.0
Nylon	15.9±2.8	9.9±2.2	11.2±2.2	18.5±2.2	16.1±2.5
Others	6.1±1.0	7.0±1.8	8.1±2.1	6.8±0.3	2.5±0.2

for Port Harcourt. In 1983, Ibiebele reported 0.19 and 0.22 kg/cap/day in 1984 for the same city. Ogwueleka (2009) has reported similar per capita production of municipal solid waste obtained in this study in other cities in Nigeria. The results of per capital production obtained in the present study also agrees with that reported elsewhere in other parts of the world with similar demographic characteristics as Nigeria, such as India, Bandgladesh and China where per capita production was reported to be0.46,0.49and0.79kg/cap/day,respectively(Chandrappa and Das, 2012). Compilation and comparison of solid waste generation in large cities of various countries showed that waste is generated at an average rate of 0.4-0.6 kg/cap/day in low income countries, as compared to 1.1-5.0 kg/cap/day in high income countries (Chandrappa and Das, 2012). In Japan and the USA, per capita municipal solid waste production was reported to be 1.47 and 2.1 kg/cap/day respectively (Chandrappa and Das, 2012; USEPA, 2008). Earlier in 1990, USA recorded up to 4.5 kg/cap/day before it decreased to 2.1 kg/cap/day (Chandrappa and Das, 2012).

The high percentage of organic waste may be attributed to the consumption of unprocessed food as compared to processed food consumed in more developed countries (Oyelola and Babatunde, 2008). World Bank (1999) reported that generally, low and middle income countries have a high percentage of combustible organic matter in the urban waste stream, ranging from 40-85%. Igoni et al. (2007) reported 69.3% of organic content in MSW in Port Harcourt and opined that the energy content of refuse generated in Port Harcourt gave 1.733 kcal/g which is sufficient for biochemical conversion to yield gas and heat for electricity. This high percentage of organic content in the waste stream in Port Harcourt and its suburbs

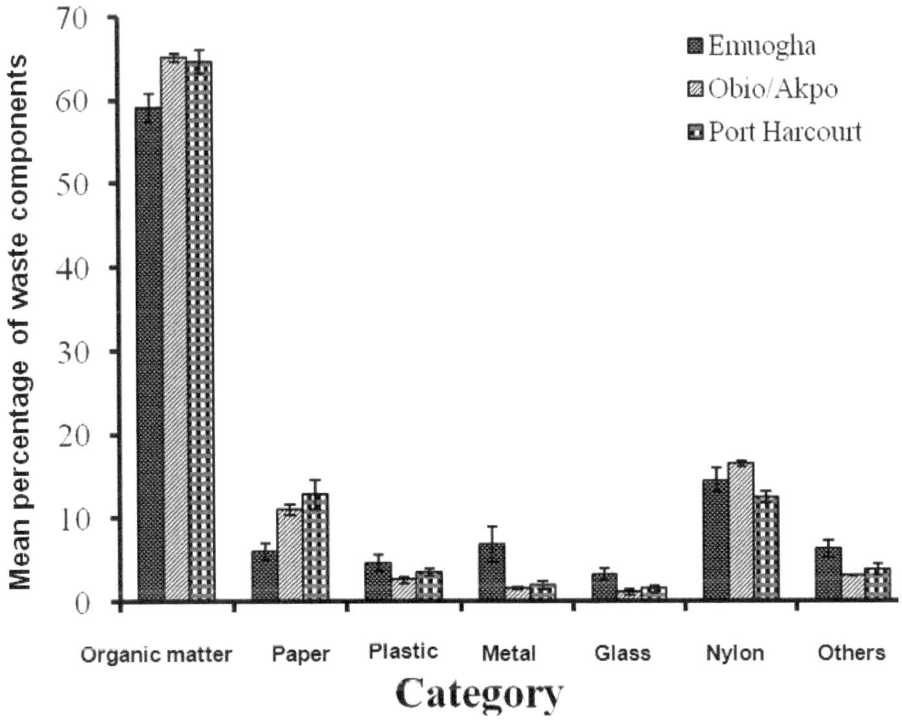

Figure 3. Mean percentage values with SEM of waste composition in the three LGA.

can be harnessed for energy generation. It is estimated that 100 tonnes of municipal refuse with 50-60% organic content can generate 1-1.5 Mega Watt of power depending on the characteristics of the waste (NEED, 2011). Apart from Port Harcourt LGA where paper was the second dominant category (5.6-16.5%), nylon was the second most prominent category in the waste of the other LGAs studied. This was due largely to the presence of nylon sachets used in packaging low cost drinking water popularly referred to as "Pure Water" and sold everywhere especially in the suburbs. Paper's prominence in Port Harcourt could be attributed to the presence of more offices and other paper utilizing outlets. Paper waste is reuseable and recyclable and if adhered to can reduce solid waste significantly in the municipality. According to NEED (2011), one ton of paper recycled from used papers instead of fresh fibres from wood saves 7,000 gallons of water, 17-31 trees, 4,000 kWh electricity and 60 lbs of air pollutants. Metal recorded the highest mean composition of 7% at Isiodu in Emuoha LGA, this must have been due to the presence of a ship building yard at this location. Metals can also be reused or recycled for value added and the safety of the environment. Similar waste characteristics dominated by organics have been reported for countries with similar economic and demographic characteristics with Nigeria such as India, Bangladesh and Ghana (Akinwonmi et al., 2012; Chandrappa and Das, 2012). Whereas in the North such as in the USA, Germany and Japan, the waste stream s are usually dominated by paper, plastics and glass (Chandrappa and Das, 2012; USEPA, 2008).

Port Harcourt LGA recorded the highest organic waste composition at Diobu, a large suburb in Port Harcourt dominated by low income earners and lower organic waste composition was recorded at Trans Amadi and GRA which are industrial layout and high income areas, respectively. Paper recorded the highest composition at the GRA location, followed by Trans Amadi and least at Diobu. There are more offices at GRA and Trans Amadi than at Diobu, however there are several small and medium scale printing press industries at Diobu, so one would have expected high paper composition in the stream of that area. The low paper composition at Diobu could therefore be indicative of high paper reuse or recycle by operators of the printing press industries. Plastic, metal and glass recorded the highest composition at Trans Amadi, this is expected since it is an industrial layout. Nylon and the category others recorded the highest composition at Diobu for reasons of its low income status. Similar waste composition and distribution has been reported by other authors for Port Harcourt city and other cities in Nigeria (Ayotamuno and Gobo, 2004; Igoni et al., 2007; Ogwueleka, 2009).

In Obio/Akpor LGA, organic waste recorded the highest composition at Wimpey and Rumuigbo with 66% each and the least at Rumudara. These settlements are lower income places than Rumuokoro, Eneka and Rumudara. Paper recorded the highest and least composition at Rumuokoro and Eneka which is more of an industrial layout. Plastic, metals, glass and the category others re-

recorded the highest at Eneka. Nylon recorded the highest composition at Rumudara followed by Rumuokoro for reasons stated earlier in this report and Eneka recorded the least nylon composition. In Emougha LGA, organic waste composition was highest at Rumuohia, followed by and Rumuji where the least composition of organic waste was recorded. Rumuohia experiences the highest commercial activities among the five study locations in Emougha LGA. Paper, plastic and glass recorded the highest composition at Rumuji. Rumuji is a community close to the University of Port Harcourt and houses students in the residence. Metal recorded the highest composition at a statistically significant level ($P < 0.05$) at Isiodu due largely to the presence of a ship building company in that location.

The waste stream of these LGAs as observed in the present study show characteristics with high potentials for possible resource recovery and utilization and if properly harnessed, the data presented here can be utilised for a more sustainable MSWM. The results showed that the municipal solid wastes generated at the study locations were heterogeneous and can be sorted at source to ease collection, transportation and ultimate disposal problems allowing the municipalities to adequately recover and utilize the resources abundant in the waste streams. Colour coded disposal bins can be provided by municipal authorities for home sorting at residential areas, market places and offices. An organised collection system can be established by legislation which may include the cart pushers who pick up the sorted waste according to colour codes and deliver to a transfer stations designated for specific waste type under a designated contractor who uses compactor vehicle or convey the waste type to its final destination for recycle, energy generation or resource recovery. This practice has been implemented successfully by the Lagos State Waste Management Authorty (LAWMA) (LAWMA, 2011) and in other parts of the world such as the United States (Painter and Watson, 2008), United kingdom (DEFRA, 2007) and Australia (Moore et al., 1994). The multiplying effect of a sustainable solid waste management cannot be overemphasized as the benefits are indeed many. Greenhouse gases such as methane and CO_2 which are the major gases produced from waste will be substantially reduced if the waste is reduced, reused or recycled (UNEP, 2002).

Conclusion

Information on the characteristics of solid waste is important in evaluating alternative equipment needs, systems, management programs and plans especially with respect to the implementation of disposal, resource and energy recovery options. The waste streams in Rivers State have been shown in the present study to be heterogeneous different in generation rate and characteristics for the three LGAs studied. This means that for a sustainable solid waste management in the state, solutions to the waste problem should be specific to each location, an opinion shared by Igoni et al. (2007). Existing efforts by municipal authorities to manage solid waste in Port Harcourt city may have been insufficient due to lack of comprehensive data on waste characteristics for Port Harcourt and its environs. The importance of understanding the waste streams of Port Harcourt environs cannot be overemphasized as these rural and sub-urban areas are connected to the Port Harcourt main city, thus, waste generated here and not properly managed will end up in the streets of Port Harcourt in time. This will naturally punch a hole in the solid waste management strategy put in place in the main town. It is believed that the results obtained during this study will aid the municipal authorities in taking a holistic and more sustainable step in an attempt to solve the municipal solid waste problem in Port Harcourt and its environs. The benefits are many including a clean and health environment, energy recovery and possibilities of energy generation for economic reasons.

REFERENCES

Abah SO, Ohimain EI (2010). Assessment of Dumpsite Rehabilitation Potential Using the Integrated Risk Based Approach: A case Study of Eneka, Nigeria. J. World Appl Sci. 8:436-442.

Akinwonmi AS, Adzimah SK, Karikari CM (2012). Assessment of solid waste management in tarkwa municipality ghana: time series approach. J. Environ. Earth Sci. 2(10):139

ASTM D5231 – 92 (2008) Standard Test Method for Determination of the Composition of Unprocessed Municipal Solid Waste. ASTM International, West Conshohocken, PA

Ayotamuno JM, Gobo EA (2004). Municipal Solid Waste Management in Port Harcourt, Nigeria: Obstacles and Prospects. Int. J. Mgt. Environ. Qual. 15:389-398.

Babayemi JO, Dauda KT (2009). Evaluation of Solid Waste Generation, Categories and Disposal options in Developing Countries: A Case Study of Nigeria. J. Appl. Sci. Environ. Mgt. 13:83-88.

Chandrappa R, Das DB (2012). Solid waste management principles and practice. Environmental Science and Engineering, DOI: 10.1007/978-3-642-28681-0_2, Springer-Verlag Berlin Heidelberg

Daskalopoulos E, Badr O, Probert SD (1998). An integrated approach to municipal solid waste management. Res. Conser. Recycl. 24:33-50.

Department for Environment, Food & Rural Affairs (Defra) (2007). Incineration of municipal solid waste www.defra.gov.uk

Gidarakos E, Havas G, Ntzamilis, P (2006). Municipal solid waste composition determination supporting the integrated solid waste management system in the island of Crete. Waste Mgt. 26:668-679.

Gobo A (2002). Solid Waste Management Options for the Niger Delta. Proceedings of the 3rd National Workshop on Waste Management and Pollution in the 21st Century, Rivers State Ministry of environment and Natural resources, Port Harcourt.

Ibiebele DD (1986). Rapid Method For Estimating Solid Waste Generation Rate in Developing Countries. Waste Mgt. Res. 4:361-365.

Igoni AH, Abowei MFN, Ayotamuno JM, Eze CL (2007). Effect of total solids concentration of municipal solid waste in anaerobic batch digestion on the biogas produced. J. Food Agricul. Environ. 5:333-337.

Lagos Waste Management Authority (LAWMA) (2011). WASTE MANAGEMENT PROGRESS REPORT FOR THE MONTH OF DECEMBER 2011. www.lagosstate.gov.ng/entities.php%3Fk%3D166

Moore SJ, Kung B, Tu S-Y, Toong P, van den Broek B (1994). Progress Towards the Establishment of a National Waste Database for

Australia, Proc. Second National Solid and Hazardous Waste Convention, Melbourne, May 1994, WMAA, Sydney.

National Energy Education Development USA (NEED) (2011). Museum of Solid Waste. The NEED Project P.O. Box 10101, Manassas, VA 20108

Nigeria Population Commission (NPC) (2012). 2006 Census figures http://www.population.gov.ng

Ogunbiyi A (2001). Local Technology in Solid Waste Mangement in Nigeria. In: Proceedings of the National engineering Conference and General Meeting of the Society of Engineers. 73-9.

Ogwueleka TC (2003). Analysis of Urban Solid Waste in Nsukka, Nigeria. J. Solid Waste Technol. Mgt. 29:239-246.

Ogwueleka TC (2009). Municipal Solid Waste Characteristics and Management in Nigeria. Iranian J. Environ. Health Sci. Eng. 6:173-180.

Oyelola O, Babatunde AI (2008). Characterization of Domestic and Market Solid Waste at Source in Lagos metropolis, Lagos, Nigeria. Afr. J. Environ. Sci. Technol. 3:430-437.

Painter R, Watson VPE (2008). Tennessee Waste Characterization Study. Tennessee State University Department of Civil and Environmental Engineering.

United Nations Environment Programme, UGA (UNEP) (2002). Contribution from Waste to Climate. *http://www.vitalgraphics.net/waste/html_file/42-43_climate_c.*

United States Environmental Protection Agency (USEPA) (2008). Office of Solid Waste (5306P) EPA530-R-08-010 November 2008 www.epa.gov

Wolrd Resources Institute (1996). United Nation Environmental program, United Nations Development Program. The World Bank Resources 1996-1997. The Urban Environment. Oxford: Oxford University Press.

World Bank (1999). What a Waste: Solid Waste Management in Asia. The World Bank Report, Washington D.C. USA.

10

Recovery of pure slaked lime from carbide sludge: Case study of Lagos state, Nigeria

Chukwudebelu, J.A., Igwe, C.C., Taiwo, O.E., and Tojola, O.B.

Federal Institute of Industrial Research Oshodi, Lagos State.

Carbide sludge is the by-product of reaction between calcium carbide and water in the production of acetylene gas for welding purposes. This by-product is discarded as waste due to high content of impurities as a result of the reactants and reaction processes. In this research work an attempt was made at developing an appropriate process technology for the recovery of pure slaked lime from Nigeria's automobile welders' carbide sludge using solubilisation and evaporation process technology. The percentage purity of the slaked lime recovered through the process was 88%. The recovered slaked limes had pH of 11.93, were soluble in glycerol and dilute acid, insoluble in alcohol, and sparingly soluble in water. The optimum percentage yield was 78.2% at a ratio of 1:1000(w/v) of sludge to water held for 24 h at room temperature.

Key words: Carbide, recovery, evaporation, sludge, technology.

INTRODUCTION

Proper management of solid waste is critical to the health and well-being of urban residents (World Bank, 2003). The United Kingdom (UK) Environment Agency classifies waste as either controlled waste or non-controlled waste (Environment Agency, 1999). Controlled waste includes waste generated from households (municipal solid waste), commercial and industrial organizations and from construction and demolition. Non-controlled waste includes waste generated from agriculture, mines and quarries and from dredging operations. At all times, human activities have generated waste in various forms, that is, in gaseous (abattoirs), liquid and solid. These wastes have often been discarded because they were all considered as negative value goods. The more prevalent method of disposal of these wastes have been to first collect them from their source and then burn them in a landfill site or throw them in the surrounding deep erosion gullies (Ogwueleka, 2009). However, the steady increase of landfill site, deposition in the gullies, and waste generally has caused a lot of havoc to the potable water being extracted from downstream and ground water. Currently, the emergence and development of new public environmental consciousness have created a strong negative attitude toward landfill and deposition into gullies. In Nigeria the national and state regulation to protect the environment have increased the cost of developing new landfill and deposition in the gullies, also sitting has become increasingly difficult because the public oppose having such facilities nearby (Ogwueleka, 2009). Solid waste management has become a major concern in developing countries, like Nigeria. The ideal way to improve the situation would be, to reduce the generation of waste. But contrarily, this goes against the people's will to preserve their life style (Otti, 2011).

Solid waste is synonymous with garbage. As long as humans have been living in settled communities, solid waste, or garbage, has been an issue, and modern societies generate far more solid waste than early humans ever did. Daily life in industrialized nations can generate several kilograms of solid waste per consumer, not only directly in the home, but indirectly in factories that manufacture goods purchased by consumers.

Most developing countries, Nigeria, inclusive have solid waste management problems different from those found in industrialized countries in areas of composition, density, political, and economic framework, waste amount, access to waste for collection, awareness and attitude (Ogwueleka, 2009). In Lagos metropolis like most cities in Nigeria and the developing world, several tons of municipal solid waste is left uncollected on the streets each day, clogging drains, creating feeding ground for pests that spread disease and creating a myriad of related health and infrastructural problems (Aliyu, 2010). Nigeria is presently experimenting with the privatization of this sector. The Federal Government has instituted National Integrated Municipal Solid Waste Management Intervention Programme in seven cities of Nigeria. The seven cities are Maiduguri, Kano, Kaduna, Onitsha, Uyo, Ota, and Lagos. Lagos state government established municipal solid waste management policy to encompass private sector participation in waste collection and transfer to designated landfill sites.

The solid waste of interest here is carbide sludge. Calcium carbide is produced industrially in an electric arc furnace from a mixture of lime and coke at approximately 2000°C. This method has not changed since its invention in 1888.

$$CaO + 3\,C \rightarrow CaC_2 + CO \quad \rightarrow (1)$$

Calcium carbide sludge is generated during the process of reaction between calcium carbide and water in the production of acetylene for welding purposes.

The reaction of calcium carbide with water was discovered by Friedrich Wohler in 1962

$$CaC_2 + 2H_2O \rightarrow C_2H_2 + Ca\,(OH)_2. \quad \rightarrow (2)$$

This reaction is the basis for industrial manufacture of acetylene. The sludge has a pH of 12.2 and contains Cu, Pb, Fe, Mn, Ni and Zn ions (Bogner et al., 2002).

No sludge is a waste; all can be used judiciously as per its specification. While carbide sludge may not be considered as toxic, but hazardous material, adequate precaution should be taken to prevent unauthorized discharge.

The Nigerian Automobile Technicians Association (NATA) with average membership of twenty-five thousand panel beaters are scattered all over the country. Each panel beater produces an average of 30 kg of carbide sludge daily. The sludge generated by this activity (industry) has accumulated over the years. The sludge are dumped indiscriminately resulting in environmental problems and health hazard such as adverse effects on the fertility of land, as these wastes are deficient in plant nutrients, drainage problems, contamination of water resources. It also causes air pollution as the dry powder becomes air borne on windy days, wastage of valuable resources of calcium hydroxide (slaked lime) and handling and disposal problem (World Bank, 2003)

Calcium salts currently being marketed derive their calcium from some source which include limestone other than hydrated carbide lime. This is due in large part to the impurities in the hydrated carbide lime, such as metals, slag, minerals and carbon which come from the coal, coke and limestone used in the calcium carbide manufacturing process. These impurities have therefore rendered hydrated carbide lime unsuitable as a feed stock for the manufacture of many purified calcium products (Bunger et al., 1998).

In 2009, 449 metric tons of calcium hydroxide (slaked lime) valued at ₦19,520,000 was imported into the country while in 2010, 364 metric tons valued at ₦20,320,000 was also imported (US, GSM, 2010). If process technology for recovering slaked lime from carbide sludge is developed, this huge amount of money used for importation will be used for other developmental projects. Recovered slaked lime can be collected and reused for many applications such as in water treatment, road stabilization, acid neutralization and carbon dioxide removal. The recovered calcium hydroxide, if treated to a high temperature will decompose into quicklime and water (Halstead and Moore, 1987). Quicklime is a very important compound in our society today due to its many applications, such as in the production of calcium carbide for automobile welders:

$$2CaO \;+\; 4C \;\rightarrow\; 2CaC_2 + O_2 \quad \rightarrow (3)$$

Calcium hydroxide has many and varied uses due to its strong basic properties. The uses include as flocculants, in water and sewage treatment, improvement of acid soils, an ingredient in whitewash, mortar, and plaster, an alkali used as a lye substitute in non-lye hair relaxers, in the liming industry for neutralizing acid, the tanning of hides and skin Calcium hydroxide is used in endodontic (root-canal) dental applications (Mohammadi and Dummer, 2007).

The urgent need to manage this industrial waste in an environmentally friendly manner and also put it to beneficial use is the basis for this research.

MATERIALS AND METHODS

Samples and Chemicals

Carbide sludge samples were collected from twenty automobile welders' shops over a period of 12 months in four mechanic villages located at Ilupeju, Ladipo, Mazamaza and Agoo areas of Lagos mainland (Figure 1).

The chemiclas used in this study include hydrochloric acid, sulphuric acid, calcium hydroxide, glycerol, alcohol, cotton wool, plastic columns, hot plate and distilled water.

Methodology

The collected sludge samples were dried in the oven at 105°C for 1 h and then sieved with a fine mesh sieve of 80 µm. Representative samples were dissolved in the ratio of 1: 100; 1: 200; 1: 300; 1: 400;

Table 1. Physicochemical characteristics of the carbide sludge sample.

Parameter	Value
Moisture content(g)	2.06
Metals:	
Iron	0.9199
Zinc	0.4831
Cadmium	0.6745

Table 2. Determination of optimum percentage yield of slaked lime in a sludge solution for the 4 samples in 100 – 500mls of water.

Sludge(g)	Water (ml)	Ilupeju A (%)	Ladipo B (%)	Mazamaza C (%)	Agoo D (%)
1	100	27.2	27.7	26.0	27.0
1	200	48.2	41.7	39.3	38.9
1	300	55.2	43.3	42.7	44.84
1	400	54.2	49.0	60.7	47.6
1	500	61.7	55.5	58.9	56.4
Average yield		49.24	43.44	45.55	42.94

Table 3. Average yields of different volumes of water with 1g of sludge.

Water (ml)	Yield (g)	Yield (%)
100	0.2699	27.0
200	0.4204	42.0
300	0.4651	46.5
400	0.5288	52.9
500	0.5812	58.1
600	0.6900	69.0
700	0.7434	74.3
800	0.7654	76.5
900	0.7810	78.1
1000	0.7815	78.2
1100	0.7815	78.2
1200	0.7815	78.2
1300	0.7810	78.1
1400	0.7788	77.9
1500	0.7788	77.9

1: 500; 1: 600; 1: 700; 1: 800; 1: 900; 1: 1000 (w/v) sludge to water for 6, 12, 24, 48 and 72 h per batch (Figure 2). This process was repeated three times. At the end of each process the solution was filtered and the filtrate evaporated to get the calcium hydroxide (slaked lime) (Figure 3).

Percentage purity was calculated using molarity formulae

RESULTS AND DISCUSSION

Table 1 shows physicochemical properties of the carbide sludge which include the metal and moisture content. The metal content of the sludge is still within the acceptable limit for discharge into the environment by FME (formally FEPA, 1992) standard.

Table 2 shows the determination of Optimum Percentage yield of Slaked Lime in a sludge solution for the 4 samples in 100 – 500 mls of water.

Table 3 shows the determination of optimum yield of calcium hydroxide using 1 g of sludge in different volumes of water. The highest average yield of 78.2% at a ratio of 1:1000(w/v) of sludge to water held for 24 h at room temperature was obtained. There was some slimy yellowish deposit in the flask when 1300 mls of water was used in the determination. When the slaked lime was scoped, it was found to be less than the previous ones and the same thing occurred with 1400 and 1500 mls. May be part of the slaked lime formed the slimy deposit resulting in the reduction of the quantity recovered. Further research need to be conducted into the occur-rence.

Tables 4 to 8 show the determination of optimum solubilisation time for the 4 samples at different periods at room temperature. The average yield of calcium hydroxide from each sampling location at different periods was determined.

Table 9 shows the average yield of calcium hydroxide at different period of time. From the table it was observed that the optimum yield was obtained when the sludge is solubilised for 24 h at room temperature.

Table 10 shows the physicochemical properties of the recovered calcium hydroxide. Using the calculated molarity of the recovered slaked lime, the percentage purity of the recovered slaked lime is 88%. The pH from the literature is 12.08 while the recovered calcium hydroxide has 11.93 and with this it can be concluded product is highly

Table 4. Determination of optimum percentage yield of slaked lime for the 4 samples in at a 6-hour solubilisation period.

Location	Sludge (g)	Water (ml)	Yield (g)	Yield (%)
A	1	1000	0.2830	28.3
B	1	1000	0.2410	24.1
C	1	1000	0.2560	25.6
D	1	1000	0.4689	46.9

Average yield =0.3872 = 38.72%.

Table 5. Determination of optimum percentage yield of slaked lime for the 4 samples in at a 12-hour solubilisation period.

Location	Sludge (g)	Water (mls)	Yield (g)	Yield (%)
A	1	1000	0.323	32.3
B	1	1000	0.2490	24.9
C	1	1000	0.4188	41.88
D	1	1000	0.5484	54.84

Average yield =0.3848g = 38.48%.

Table 6. Determination of optimum percentage yield of slaked lime for the 4 samples in at a 24-hour solubilisation period.

Location	Sludge (g)	Water (mls)	Yield (g)	Yield (%)
A	1	1000	0.6972	69.7
B	1	1000	0.8830	88.3
C	1	1000	0.9275	92.8
D	1	1000	0.6182	61.8

Average yield =0.7815g = 78.2%

Table 7. Determination of optimum percentage yield of slaked lime for the 4 samples in at a 48-hour solubilisation period.

Location	Sludge (g)	Water (mls)	Yield (g)	Yield (%)
A	1	1000	0.607	60.7
B	1	1000	0.535	53.5
C	1	1000	0.5494	54.9
D	1	1000	0.5031	50.3

Average yield =0.54863g = 54.0%

Table 8. Determination of optimum percentage yield of slaked lime for the 4 samples in at a 72-hour solubilisation period.

Location	Sludge (g)	Water (mls)	Yield (g)	Yield (%)
A	1	1000	0.4324	43.2
B	1	1000	0.2932	29.3
C	1	1000	0.4539	45.4
D	1	1000	0.6003	60.0

Average yield =0.4450 g = 44.5%.

Table 9. Average percentage yield of slaked Lime for different solubilisation periods.

Time (h)	Yield (g)	Yield (%)
6	0.3872	38.72
12	0.3848	38.48
24	0.7815	78.15
48	0.5486	54.86
72	0.4450	44.50

Table 10. Physicochemical characteristics of the recovered slaked lime.

Parameter	Literature	Recovered slaked lime
Density	2.211	2.12
pH	12.08	11.93
Purity (%)		88
Solubility		
Water	Sparingly soluble	Sparingly soluble
Alcohol	Insoluble	Insoluble
Glycerol	Insoluble	Insoluble
Acid	Completely soluble	Completely soluble
Odour	Odourless	Odourless
Appearance	Soft white powder	Soft white powder
Metals	**Sludge**	**Recovered slaked lime**
Iron	0.9199	ND
Zinc	0.4831	0.0454
Cadmium	0.6745	ND

alkaline. The literature has the density of 2.211 g/cm^3 while the recovered slaked line has density of 2.12 g/cm^3. The solubility tests conducted agree with literature. The metal content, though within the limit were reduced more after the recovery.

Bunger et al. (1998) was able to recover slake lime from carbide sludge into solution with subsequent reacting of the solution with carbondioxide to form calcium carbonate. Also Musa et al. (2005) obtained calcium hydroxide from Nigeria limestone and then passed chlorine into the suspension to get calcium hypochlorite. Recovering of slaked lime in solution will limit its usage, because as mentioned above its varied application can only be maximized when it is in powdered form which the present research has actualized.

Production of slaked lime from limestone is a cumbersome process and requires substantial energy to convert the limestone ($CaCO_3$) to lime (CaO) which will then be hydrated. The slaked lime recovered will be of high market value and solves the problem with disposal of low-value hydrated carbide lime from acetylene processes. Calcium hydroxide generally has low solubility in water which is worsened by presence of under/over bur-ning of rocks or impurities that had escaped calcinations during production from limestone (Musa et al., 2005).

The research served dual purpose of preserving the environment as well as making slaked lime available for its many applications. The significant application of slaked lime is as a flocculants, in water and sewage treatment. It forms a fluffy charged solid that aids in the removal of smaller particles from water, resulting in a clearer product. This application is enabled by the low cost and low toxicity of slaked lime. It is also used in fresh water treatment for raising the pH of the water so pipes will not corrode where the base water is acidic because it is self-regulating and does not raise the pH too much. Another large application is in the paper industry, where it is used in the production of sodium hydroxide. This conversion is a component of the Kraft process.

There is no cost analysis but survey was conducted on slaked lime vendors and users in Lagos. Twenty-five (25) copies of questionnaires were distributed to the vendors and users of slaked lime to determine quantity of sales per week, product source, price range and quality require-quirement of their buyers. Twelve (12) questionnaires were retrieved and the analysis of the survey shows that nine of the vendors deal on slaked lime representing 75% of the local vendors, two (2) deals on limbux lime while one (1) did not give any information. The summary of the analysis showed that nine (9) of the vendors obtain their product from China and Great Britain while only two (2) source their own locally with so much difficulty. The vendors sell maximum of seventy-five bags in a week at a price range of between N76, 000 and N122, 500. It was discovered from the survey that slaked lime vendors import their products due to its scarcity locally.

CONCLUSION

The recovered product proved to be calcium hydroxide through the analysis carried out. The 88% purity and 78.1% yield of the recovered slaked lime seems promising for slaked lime recovery from automobile welder's carbide sludge. The dissolution period of 24 h is also not too long to convert the waste to wealth.

The technology needs to be developed so that the problems carbide sludge is causing to the environment will be tackled. There is also need to conduct cost benefit analysis of the process.

REFERENCES

Aliyu BN (2010). An Analysis of Municipal Solid Waste in Kano Metropolis. Niger. J. Hum. Ecol. 31(2): 111-119

Bogner JM, Diaz C, Faaij A (2002). Resources conversion and recycling. Waste management and research series . 20(6): 536-540 .

Bunger JW, Cogs WD, Wiser JW (1998). Process for purifying highly impure calcium hydroxide and for producing high-value precipitated calcium carbonate and other calcium products. United States Patent 5846500.

Environment Agency(1999). Waste Statistics for England and Wales.

Federal Environmental Protection Agency (FEPA) (1992). National Environmental Protection (Pollution Abatement in Industries and facilities Generating wastes) Regulations 1991, Decree No 58 of 1988 as amended by Decree 59 of 1992.

Halstead PE, Moore AE (1987). The thermal dissociation of calcium hydroxide. J. Chem. Soc. 769:3873.

Mohammadi Z, Dummer PM (2007). Properties and applications of calcium hydroxide in endodontics and dental traumatology. Department of Endodontics, Hamedan University of Medical Sciences, Hamedan, Iran Endodontology Research Group, School of Dentistry, Cardiff: University, Cardiff, UK.

Musa IA, Gimba CE, Shide EG(2005). Calcium hypochlorite from Nigeria limestone. Chemclass J. 2 :47-50

Ogwueleka TC (2009) Municipal Solid Waste Characteristics and Management in Nigeria. Iran. J. Environ. Health. Sci. Eng. 6(3): 173-180.

Otti VI (2011). A model for solid waste management in Anambra State, Nigeria. J. Soil Sci. Environ. Manage. 2(2): 39-42.

US Geological Survey publications (2010). Geological survey mineral commodity summaries (lime advance release), 68.

Wohler F (1962). Chemical Engineers' Handbook. Third Edition, McGraw-Hill Book Company. pp. 50-52.

World Bank (2003). Thailand Environmental Monitor 2003. A joint publication of the Pollution Control Department, Royal Thai Government. The World Bank, US Asia Environmental Partnership.

11

Comparison of membrane bioreactor technology and conventional activated sludge system for treating bleached kraft mill effluent

Francis Attiogbe

Department of Chemical Engineering, Kumasi Polytechnic, P. O. Box 854, Kumasi, Ghana.

The purpose of this paper was to review the use of membrane bioreactor technology as an alternative for treating the discharged effluent from a bleached kraft mill by comparing and contrasting membrane bioreactors with conventional activated sludge systems for wastewater treatment. There are many water shortage problems currently in the world, some of which are more serious than others. Public concern over health and the environment, combined with increased requirements for municipalities to reuse wastewater, have created a need for new technologies that can treat wastewater to generate high quality reusable water at low cost. In several of these technologies, membrane technology could make a great contribution since membranes have the ability to produce water of exceptional purity that can be recycled for reuse in a variety of places. This reuse of wastewater is already widely practiced in many countries, which reduces net demand on water supply systems. In industry, in particular the pulp and paper industry, large volumes of water are used with a significant amount of wastewater generated. This effluent needs to be treated prior to final disposal or reuse. The commonly used biological treatment methods of aerated lagoons and activated sludge of bleached kraft mill effluent have been found to be inadequate in achieving the desired level of toxicity removal. There is, therefore, the growing demand for greener/sustainable technologies for reuse/recycling of wastewater and the membrane bioreactors treatment of these effluents has shown some greater potential as it is much cleaner and meet stringent discharge requirements than with other techniques.

Key words: Membrane bioreactor, activated sludge, bleached kraft mill effluent, pulp and paper.

INTRODUCTION

Bleaching effluents from chemical pulp mills are said to be one of the pollution problems of the pulp and paper industry due to the large amounts of chlorinated organics discharged into the environment. Many of these chlorinated organic compounds are toxic, mutagenic, persistent, bioaccumulating and show considerable resistance to biological and chemical degradation (Kringstad and Lindström, 1984). These compounds are produced primarily as a result of complex reactions occurring between the chlorine-containing bleaching agent and the residual lignin (5 to 10%) remains in the wood pulp after the preceding chemical pulping process (O'Connor and Voss,

1992). Pulp and paper industry is a major source of these chlorinated organics worldwide, since pulp bleaching usually involves the use of elemental chlorine or chlorine dioxide for oxidation of the wood pulp (Fulthorpe et al., 1993). According to O'Connor and Voss (1992), about 10% of the chlorine supplied to the pulp in the first stage of bleaching appears in the effluent discharge as organically bound chlorine which has been estimated at 250,000 tonnes/year by Kringstad and Lindström in 1984, while the remaining 90% ends up as chloride ions. The organically bound chlorine is measured as total adsorbable organic halogen (AOX) and kraft mill effluents show that a sizeable fraction of carbon compounds in these effluents is chlorinated (Fulthorpe et al., 1993). It should be noted that AOX discharge has decreased by ~60% since 1984 as Cl_2 has been replaced by ClO_2, ozone, and peroxide in almost all large-scale mills. The AOX is said to be the main source of effluent toxicity of bleached kraft mill effluents. The other environmental impact of the bleached kraft mill effluents is due to suspended solids, organic matter and color.

The systems most commonly employed by the pulp and paper industry in treating their wastewater are biological using stabilization ponds, aerated lagoons and activated sludge processes. These treatments have been successful in lowering the chemical and biological oxygen demands (chemical oxygen demand (COD) and biochemical oxygen demand (BOD), respectively), but their applicability is limited by a great number of problems. These biologically treated effluents still contain significant amounts of colored compounds, microorganisms, recalcitrant organics and a minor amount of biodegradable organics, as well as suspended solids. If water of high organic matter content or biochemical oxygen demand (BOD) value flows into a river, the bacteria in the river will oxidize the organic matter consuming oxygen from the water faster than it dissolves back in the air which has an adverse effect on aquatic life (Attiogbe et al., 2007). Biological treatment does not significantly reduce the inorganic content in the effluent and desalting is sometimes needed before reuse of the effluents in the manufacturing processes (Assalin et al., 2009). Also, these methods are much less efficient in toxicity reduction of the effluent. Given the limitations of the current biological wastewater treatment, there is an increasing interest to develop a more effective treatment approach to reduce the impacts of pulp mill effluents on the environment.

The rapid increase of population and the increased demand for industrial establishments to meet human needs have created problems such as over exploitation of available resources, leading to pollution of the environment. There is a long list of water-related problems worldwide on existing water resources due to increases in human population and activity. Reuse and conservation of water resources has therefore taken a very high priority position. The Global Environment Outlook (GEO) 2000 report of the United Nations indicated that many countries will be experiencing severe water shortages by the year 2025, and this will be especially critical in areas where water from the same inadequate source is required by more than one country. Quantity demands may only be met by re-use and quality demands by advanced treatment in both cases indicating a potential role for membrane technologies. The membrane bioreactor technology has the potential to help industries and municipallities manage their water resources better. It is an innovative wastewater treatment (WWT) technology, based on proven processes of activated sludge biological treatment and membrane separation (Wang and Menon, 2009). The system has been implemented in several full-scale industrial and municipal applications in Europe, North America and Asia. The integration of biological treatment with membrane filtration in MBR produces an excellent effluent quality which is capable of meeting stringent discharge requirements. The possibility of the membrane retaining all bacteria and viruses results in a sterile effluent, eliminating disinfection before discharge or reuse. This provides the opportunity for facilities to recycle/reuse part or all the treated effluent, thereby reducing costs for fresh water and water treatment on one hand and reducing sewer surcharge (for pretreatment facilities) on the other hand (Wang and Menon, 2009).

In this paper, a review comparing and contrasting use of membrane bioreactors and conventional activated sludge system for the treatment of bleached kraft mill effluent (BKME) was done.

CONVENTIONAL ACTIVATED SLUDGE SYSTEM (CAS)

The activated sludge process was developed in England in 1914 by Ardern and Lockett and was so named because it involved the production of an activated mass of microorganisms capable of stabilizing a waste aerobically. There are many version of the original process but fundamentally they are all similar (Metcalf and Eddy Inc., 1995).

The activated sludge process is a biological method of wastewater treatment technique in which a mixture of wastewater and biological sludge (microorganisms) is agitated and aerated. The biological solids are subsequently separated from the treated wastewater and returned to the aeration process as needed (Davis and Cornwell, 1991). The biological wastewater treatment with the activated sludge process is achieved using a

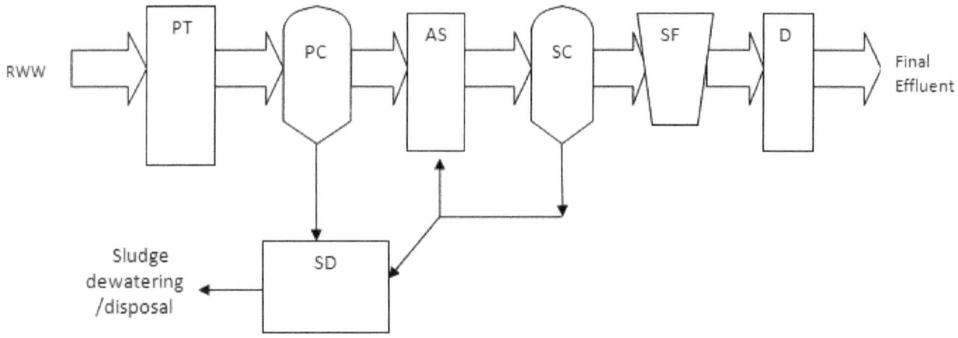

Figure 1. Conventional activated sludge treatment flow diagram. RWW- Raw wastewater; PT- primary treatment; PC- primary clarifier; AS- activated sludge; SC- secondary clarifier; SF- sand filter; D- disinfection; SD- sludge digestion.

flow diagram as shown in Figure 1.

The pretreated wastewater is introduced into a reactor/aeration tank (AS) where an aerobic bacterial culture is maintained in suspension. The wastewater and the microorganisms are thoroughly mixed under conditions that stimulate their growth through the use of the organics as food. The reactor contents are referred to as the mixed liquor. In the reactor, the bacterial culture carries out the conversion in accordance with the stoichiometry shown (Metcalf and Eddy Inc., 1995):

Oxidation and synthesis:

$$\underset{organic\,matter}{COHNS} + O_2 + nutrients \xrightarrow{bacteria} CO_2 + NH_3 + \underset{new\,bacterial\,cells}{C_5H_7NO_2} +$$

other end products

Endogenous respiration:

$$\underset{cells}{C_5H_7NO_2} + 5O_2 \xrightarrow{bacteria} 5CO_2 + 2H_2O + NH_3 + energy$$

In these equations, COHNS represents the organic matter in wastewater. The endogenous respiration reactions results in relatively simple end products and energy, stable organic end products are also formed (Metcalf and Eddy Inc., 1995).

The aerobic environment in the reactor is achieved by the use of diffused or mechanical aeration, which also serves to maintain the mixed liquor in a completely mixed regime. As the microorganisms are mixed, they collide with other microorganisms and stick together to form larger particles called floc. The large flocs that are formed settle more readily than individual cells. These flocs also collide with suspended and colloidal materials (insoluble organic materials), which stick to the flocs and cause the flocs to grow even larger. The microorganisms digest these adsorbed materials, thereby re-opening sites for more materials to stick.

After a specified period of time, the mixture of new cells and old cells is passed into a settling tank/secondary clarifier (SC), where the cells (activated sludge) are separated from the treated wastewater. A portion of the settled cells/sludge is returned/recycle to the aeration tank (AS) to maintain a high population of microbes to permit rapid breakdown of the organics. The volume of sludge returned to the aeration tank is typically 20 to 30% of the wastewater flow. Usually, more activated sludge is produced than is desirable in the process, portion of the return sludge is therefore diverted or wasted to the sludge handling system (SD) for treatment and disposal. The clarified wastewater flows (SC) forward to further treatment or discharge. For further treatment, the effluent is sand filtered (SF) since it could still be high in biological solids and the resultant effluent is disinfected (D) to kill pathogens before final disposal.

In conventional activated sludge systems, the wastewater is typically aerated for six to eight hours in long rectangular basins; about 8 m^3 of air is provided for each m^3 of wastewater treated (Davis and Cornwell, 1991).

Three basic types of organisms important to the operation of an activated sludge system are bacteria, plants and animals. Plants include algae and fungi. The bacteria are the most important and constitute the majority of microorganisms present in activated sludge. Bacteria that require organic compounds for the supply of carbon and energy (heterotrophic bacteria) predominate; whereas bacteria that use inorganic compounds for cell growth (autotrophic bacteria) occur in proportion to concentrations of carbon and nitrogen. In general, bacteria in the activated sludge process include members of

Table 1. Plant efficiencies for activated sludge plants treating papermill wastewater (Thompson et al., 2001).

Activated sludge efficiency (%)		Country	Source
BOD	COD		
73.5-99.2	50.0-92.2	Finland	Saunamäki (1997)
-	90.6	Shotton Papermill, UK	Horan and Chen (1998)
97.9-98.5	72.5-92.4	UK	-

the genera *Pseudomonas, Zoogloea, Achromobacter, Flavobacterium, Nocardia, Bdellovibrio, Mycobacterium*, and the two most common nitrifying bacteria, *Nitrosomonas and Nitrobacter.* Additionally, various filamentous forms of bacteria and fungi, such as *Sphaerotilus, Beggiatoa, Thiothrix, Lecicothrix* and *Geotrichum*, may also be present (Metcalf and Eddy Inc., 1995). Bacteria are primarily responsible for the removal of organic substances from wastewater (Junkins et al., 1983). The algae and fungi in the system play a lesser role than bacteria. Animals include larger microorganisms such as protozoa, crustacians and rotifiers. The animals feed on dispersed bacteria that do not settle well and therefore help polish the quality of the treatment plant effluent (Junkins et al., 1983). Both aerobic and anaerobic bacteria may exist in the activated sludge process.

Activated sludge processes are designed based on the mixed liquor suspended solids (MLSS) and the organic loading of the wastewater, as represented by the BOD or COD. The MLSS represents the quantity of microorganisms involved in the treatment of the organic materials in the aeration basin, while the organic loading determines the requirements for the design of the aeration system. Table 1 (Thompson et al., 2001) shows plant efficiencies for activated sludge plants treating papermill wastewater. These show that very high removal efficiency can be obtained both for BOD and COD removal.

The success of the activated-sludge process is dependent upon establishing a mixed community of microorganisms that will remove and consume organic waste material, that will aggregate and adhere in a process known as bioflocculation, and that will settle in such a manner as to produce a concentrated sludge for recycling. The CAS is efficient in producing clear sparkling treated effluent which is free of odor. The purity of the effluent can be varied as desired depending on the length of time the aeration is carried out. However, the treated effluent from CAS will not be suitable for recycle and reuse, unless tertiary treatment process units are added for further purification.

The operation of CAS requires skilled supervision and constant check on the returned sludge. The process does not work for some industrial wastes and large volume of sludge produced increases difficulties in disposal (Kamala and Kanth Rao, 1988) which are some of the drawbacks of this technology.

MEMBRANE BIOREACTORS (MBR)

The history of membranes applied to treatment of wastewater is relatively new, dating back to the late 1960s (Drioli and Giorno, 2009). The membrane bioreactors can be broadly defined as systems integrating biological degradation of waste products with membrane filtration (Cicek, 2003; Marrot et al., 2004; Wang and Menon, 2009; Drioli and Giorno, 2009). Combining membrane technology with biological reactors for the treatment of wastewaters has led to the development of three generic membrane bioreactors (MBRs): for separation and retention of solids; for bubble-less aeration within the bioreactor, and for extraction of priority organic pollutants from industrial wastewaters (Stephenson et al., 2000). Coupling membranes to biological processes are often used as a replacement of the sedimentation stage, that is, for biomass separation. According to Cicek (2003) the membrane bioreactors (MBRs) have proven quite effective in removing both organic and inorganic contaminants as well as biological entities from wastewater. The bioreactor and membrane module each have a specific function: (i) biological degradation of organic pollution is carried out in the bioreactor by adapted microorganisms; (ii) separation of microorganisms from the treated wastewater is performed by the membrane modules. The membranes constitute a physical barrier for all suspended solids and therefore causes not only recycling of the activated sludge to the bioreactor but also production of permeate free of suspended matter, bacteria and viruses. The use of membranes to separate solids and treated wastewater is the main difference betweens MBRs and conventional activated sludge systems for which the efficiency of the final clarification step depends mainly on the activated sludge settling properties (Marrot et al., 2004).

The evolution and possible application of membranes

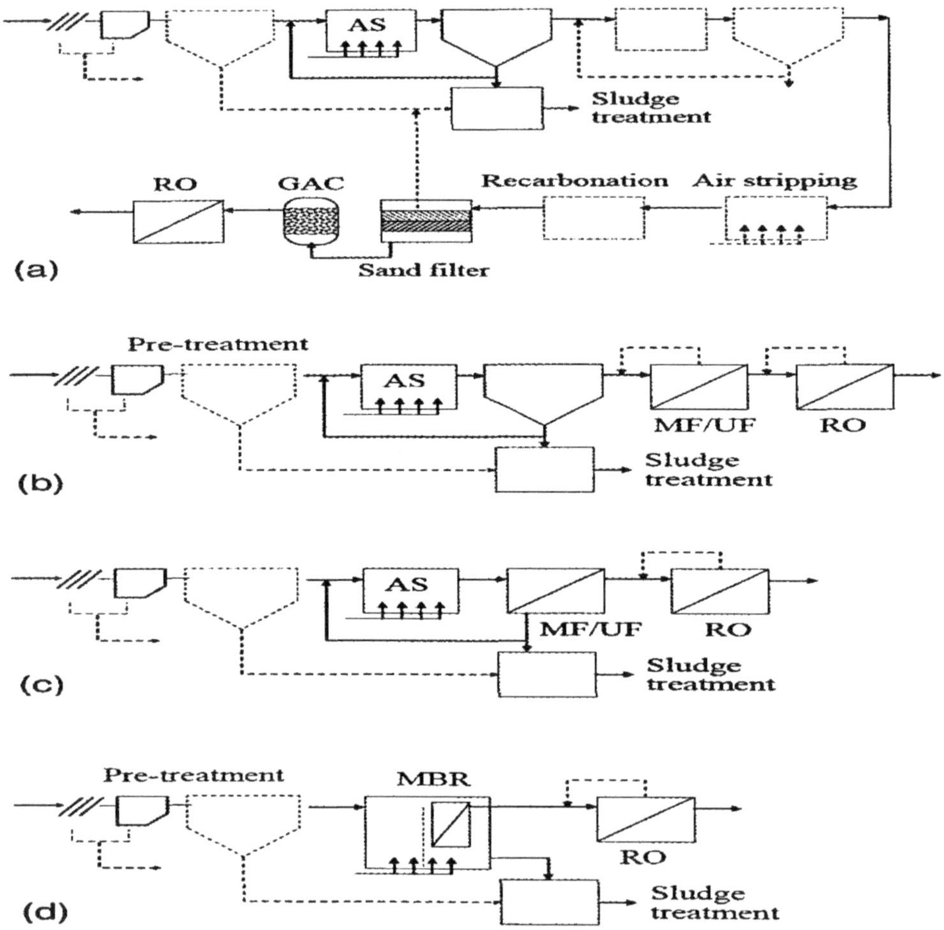

Figure 2. Illustration of evolution and use of membranes in wastewater treatment (Drioli et al., 2009).

in wastewater treatment is illustrated in Figure 2. Treatment of wastewater by membrane technology is an established alternative, particularly in sensitive areas, water scarce regions, and in cases where wastewaters reuse and recycling is required (Drioli and Giorno, 2009). Industries where the membrane bioreactor technology can be implemented include chemical, petrochemical, pharmaceutical, fine chemicals, cosmetics, diary, automotive, pulp and paper, landfill leachate, food, textiles, etc. (Wang and Menon, 2009).

The membrane bioreactor has been demonstrated to achieve higher reductions of bacteriophages as compared to the activated sludge process in the treatment of municipal wastewater (De Luca et al., 2013; Francy et al., 2012; Marti et al., 2011). Bacteriophages were proposed as models of enteric viruses and are considered particularly suitable as viral indicators. In recent years, information on the effectiveness of virus removal by sewage treatment processes has become of major concern, due to the epidemiological significance of viruses as waterborne pathogens (De Luca et al., 2013).

The efficiency of conventional activated sludge process (CASP) in removing pathogenic microorganisms has been investigated in several studies, which have concluded that these treatments may not be sufficient to achieve microbiologically safe effluent to be discharged into natural waters or to be reused. In order to reduce the potential microbiological risk, the secondary effluent is generally subjected to a further tertiary treatment by sand filtration, ultraviolet and ionizing radiation, or, more frequently, by chemical disinfection with chlorine, ozone and peracetic acid. The generation of harmful disinfection by-products such as trihalomethane (THM) and the persistence of disinfection residues are considered adverse environmental effects of chemical disinfection processes. However, the membrane bioreactor is considered an effective, non-hazardous advanced treatment alternative (De Luca et al., 2013).

Membrane materials and classifications

There is a large selection of commercial membranes that can potentially be used in MBR applications. Most polymers can be used to manufacture membranes in principle and there is a wide variety of commercially available polymeric membranes. In water and wastewater applications, most systems are based on a limited set of polymeric materials. The most common membrane materials are: polyvinyl difluoride (PVDF), polyethylsulfone (PES), polyethylene (PE) and polypropylene (PP).

Membranes are generally formed as a flat sheets or tubular/hollow-fiber geometry. Recent developments of manufacturing techniques have led to alternative products such as multibore or multitube in the market. Membranes are commonly given the following classifications:

1. Flat sheet (FS);
2. Hollow fiber (HF);
3. Capillary tubular (CT);
4. multibore or multibular (MT).

The flat-sheet membranes are commonly constructed in a plate-and-frame configuration or as spiral-wound (SW) modules. HF/CT/MT membrane typ; are commonly manufactured into bundles that are installed in housing units or designed to be unconfined in the fluid, that is, immersed units. The plate-and-frame FS and HT/CT membrane modules are the preferred option for MBR applications. The SW membrane modules are not used as the channels within the spiral which are prone to clogging when the feed water has high suspended-solids concentrations. The tubular membrane systems are not common either as they tend to become very expensive due to the low area to volume ratio. Commercial MBR systems today are normally based on immersed FS configurations or HF/CT configurations (Drioli and Giorno, 2009). The pore size of membranes used ranged from 0.01 to 0.4 µm (Cicek, 2003).

There are five main subcategories of membrane processes use for water and wastewater treatment. These are: (i) microfiltration (MF); (ii) ultrafiltration (UF); (iii) nanofiltration (NF); (iv) reverse osmosis (RO) and (v) electrodialysis (ED). Of these, only MF and UF are related to membrane bioreactors.

Microfiltration is a pressure process for the separation of suspended solids in the particle size-range of about 0.08 to 10 µm. The primary function affecting solids separation from water is the size of the suspended solids (SS). The hydraulic pressure (transmembrane pressure) applied in MF is about 1 to 2 bars, or 15 to 20 psig, primarily for overcoming resistance of the filter cake.

Ultrafiltration process is a pressure filtration process for the separation of macromolecular solids in the particle size range of about 0.01 to 0.1 µm. The primary factor affecting solids separation from water relies on the size of macromolecular solids. The hydraulic pressure required by UF for overcoming hydraulic resistance of the polarized macromolecular layer on the membrane surface is about 1 to 7 bars (Wang and Menon, 2009).

Nanofiltration membranes are multiple-layer thin-film composites of polymer consisting of negatively charged chemical groups, and are used for retaining molecular solids such as sugar and certain multivalent salts such as magnesium sulfate, but passing substantial amounts of most monovalent salts such as sodium chloride, at an operating pressure of about 14 bars or 200 psig. Both molecular diffusivity and ionic charge play important roles in the separation process. The sizes of molecular solids and multivalent salts to be rejected by NF are normally in the range of 0.0005 to 0.007 µm.

Reverse osmosis membranes are mainly made of cellulose acetate with the pore sizes of about 5 to 20 Å, and are for rejecting salts as high as 98% and organics as high as 100%, at an operating pressure of about 20 to 50 bars or 300 to 750 psig.

The hydraulic pressure through a pump is used to provide the driving force for permeation, or for overcoming the chemical potential difference between the concentrate and the permeate, expressed in terms of the osmotic pressure. The sizes of molecular solids and salts (multivalent as well as mono-valent) to be rejected by RO are normally in the range of 0.00025 to 0.003 µm (Wang and Menon, 2009).

Electrodialysis uses voltage or current as the driving force to separate ionic solutes. The sizes of ionic solutes to be rejected or separated by ED are normally in the range of 0.00025 to 0.08 µm, depending on the pore size of the ED membranes. Figure 3 illustrates the relationships among the five subcategories of membrane processes:

MBR process configurations

The process configurations of MBR plant depend partly on the type of membrane used (FS or HF/CT) and on the design of the biological treatment. Figure 4 depicts the typical configurations found in MBR treatment schemes. For immersed/submerged membrane designs, the membrane modules are either inserted directly into the biological reactor or placed in a separate reactor constructed to hold the membrane modules only (Figure 4, schemes A and B). In side-stream configuration, the membrane modules are placed outside the biological reactor and can be operated in deadend mode or in cross-

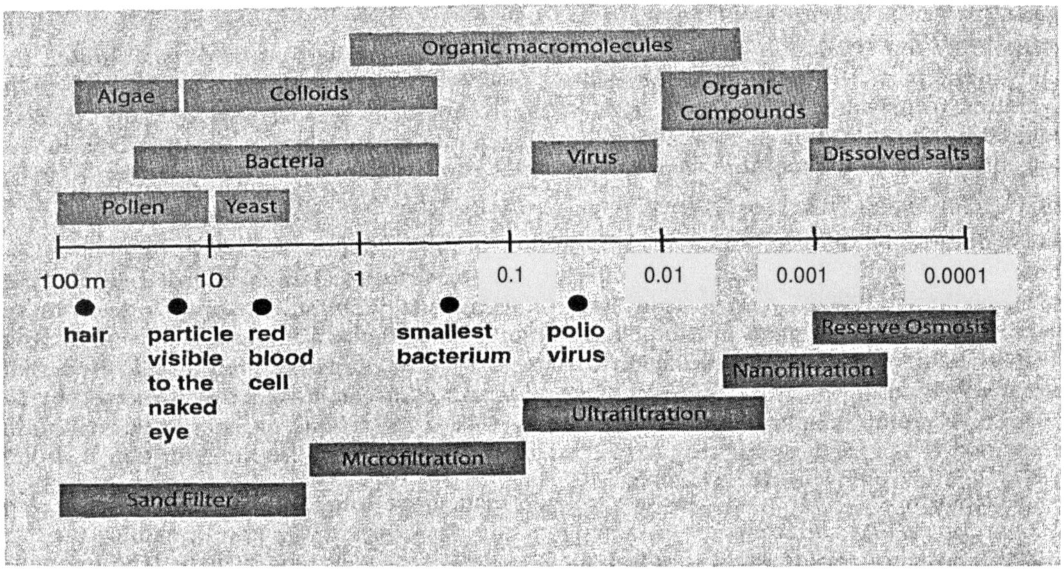

Figure 3. Particle size and separation processes (Wang et al., 2009).

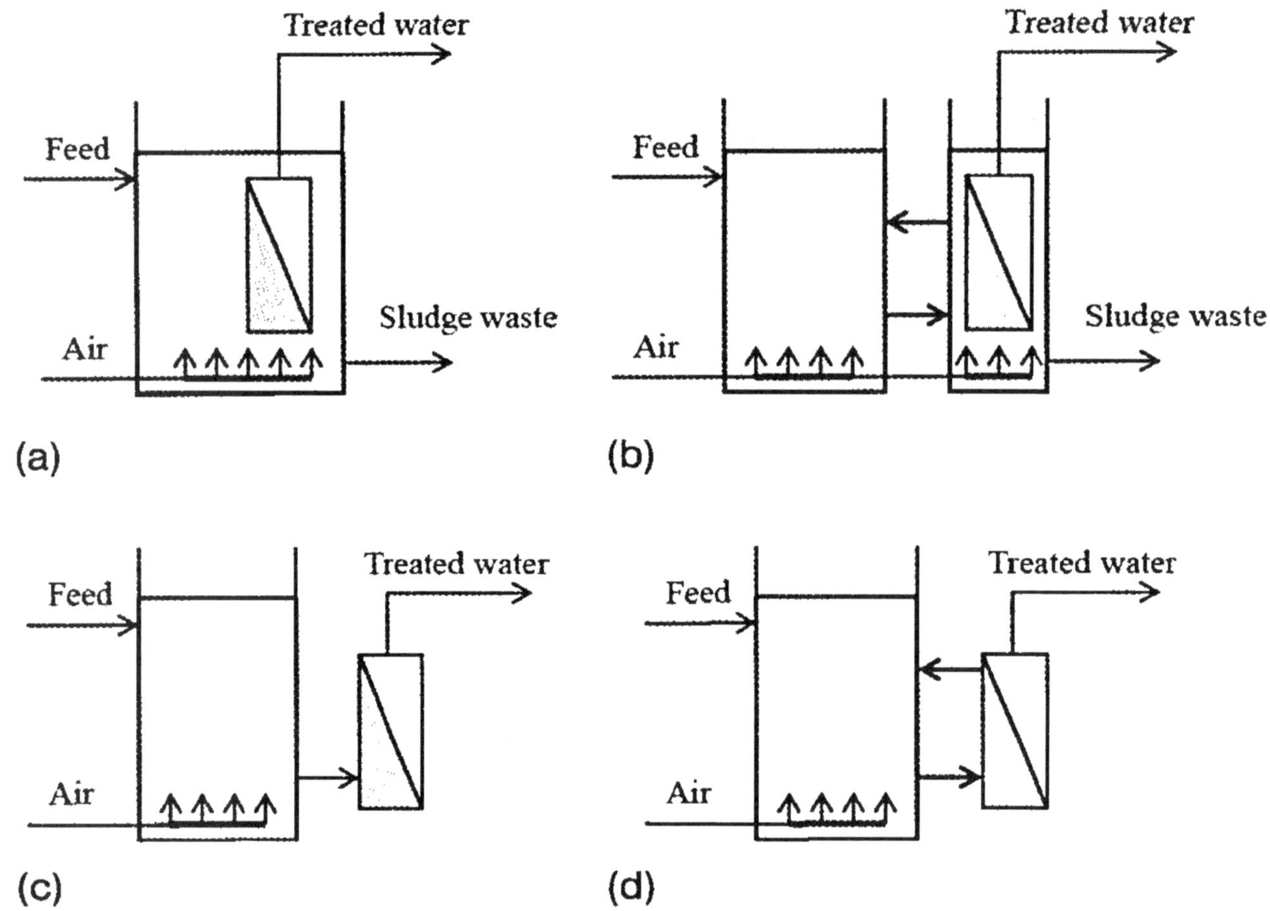

Figure 4. Typical configurations of MBR schemes, immersed vs. side-stream (Drioli et al., 2009).

-flow mode with recycling of the concentrate stream back to the biological reactor (Figure 4, schemes C and D). The treated water (permeate) from the submerged/immersed configuration is extracted from the membrane by vacuum (low pressure) while the permeate from the side-stream configuration is generally produced under pressure. The submerged configuration appears to be more economical based on energy consumption (Marrot et al., 2004) for two main reasons: no recycle pump is needed since aeration generates a tangential liquid flow in the vicinity of the membranes, and the operating conditions are much milder than in an external MBR system because of the lower values of transmembrane pressure (TMP) and tangential velocities. Generally, hollow fiber membranes are used in submerged MBR and tubular membranes are used in external MBR systems.

MBR systems classification

Membrane bioreactor systems are classified into two major groups according to their configuration. The first group is the submerged/immersed MBR, which has a membrane present in the bioreactor itself. The driving force across the membrane is achieved by creating negative pressure on the permeate side of the membrane (Baek and Pagilla, 2006). The submerged MBR has been developed to simplify the system and reduce the power consumption. However, this system operates at a lower transmembrane pressure, and, therefore, a lower permeation flux is achieved. The second group is the sidestream MBR, which recirculates the mixed liquor through a membrane module that exists outside the bioreactor. The driving force for this system is the pressure created by high-cross-flow velocity through the membrane unit. Although the pumping cost of the recirculation of mixed liquors is high, higher effluent fluxes, easier maintenance of the membrane, and less complicated scale up makes this configuration attractive. The MBR systems could also be classified into these two groups: aerobic MBRs and anaerobic MBRs. The aerobic MBR is a combination of a membrane filtration unit and aerobic bioreactor. Most of the aerobic MBRs for municipal-wastewater treatment have been the submerged systems. For industrial wastewater treatment by aerobic MBRs, sidestreamed systems have generally been used. However, the operating costs for the aerobic MBRs are high because of the cost of aeration. Approximately 20 to 50% of the total-process-power requirements are used for aeration in sidestreamed MBRs, and 80 to 100% of the total-process-power consumption is required for aeration in submerged MBRs (Baek and Pagilla, 2006). The aerobic MBR process has been reported to have been used successfully to treat effluents from a range of industrial wastewaters, including cosmetics, pharmaceuticals, metal fabrication, textiles, abattoir, diary, food, beverage, pulp and paper, rendering and chemical manufacture (Stephenson et al., 2000). It has also been used for the treatment of landfill leachate. The anaerobic MBR is a combination of an anaerobic reactor coupled with the membrane unit. The anaerobic MBR has the advantages of aeration-energy savings, possible biogas recovery, and lower sludge production, resulting in competitive capital and operating costs. However, negligible or no ammonia, total nitrogen, or phosphorus removal can be expected from an anaerobic MBR process. Anaerobic MBRs could have potential application in municipal wastewater treatment to remove organic carbon or biochemical oxygen demand (BOD) from the wastewater. Use of an anaerobic process was previously not feasible for BOD removal in municipal wastewater because of the poor settleability of anaerobic sludge in gravity settlers and the potential for odors. In the case of the anaerobic MBR, the bioreactor is a closed unit like an anaerobic digester, and the solid-liquid separation is also a closed unit in the form of membrane filtration unit. Thus, the two drawbacks, which precluded the use of the anaerobic-sludge process for BOD removal from municipal wastewater, could be circumvented by using the anaerobic MBR (Baek and Pagilla, 2006).

Membrane characteristics on MBR performance

Membrane characteristics such as pore size, porosity, surface charge, roughness and hydrophilicity/hydrophobicity, etc., have been said to impact on MBR performance, especially on membrane fouling. Pore size distribution is likely to be one of the parameters affecting membrane performance. A narrow pore size distribution is preferred to control membrane fouling both in MBR process and in conventional membrane separation process.

The membrane materials always show different fouling propensity due to their different pore size, morphology and hydrophobicity. Polyvinylidene fluoride (PVDF) membrane is superior to polyethylene (PE) membrane in terms of prevention of irremovable fouling in MBRs used for the treatment of municipal wastewater (Meng et al., 2009). Regarding MBR processes, the fouling behavior of the membrane used is determined by the affinity between foulants (example, extracellular polymeric substances (EPS)/soluble microbial products (SMP)) and membrane.

Inorganic membranes, such as aluminum, zirconium and titanium oxide, show superior hydraulic, thermal and chemical resistance. A stainless steel membrane was used for MBR, and the result showed that the stainless steel membrane could obtain a higher permeate flux (Meng et al., 2009), and it is a potential alternative for the treatment of high temperature wastewater. According to Meng et al. (2009) in the stainless steel membrane bioreactor, thermophilic bacteria could be cultivated when the MBR was operated at higher temperature. But, these inorganic membranes are not the preferred option for large-scale MBR plants because of their high costs. In addition, inorganic membranes can induce severe inorganic fouling (struvite formation). So, the inorganic membranes might be used only in some special applications such as high temperature wastewater treatment (Meng et al., 2009).

In general, membrane fouling occurs more readily on hydrophobic membranes than on hydrophilic ones because of the hydrophobic interaction between foulants and membranes.

Optimizing MBR operations

When first commercialized, MBR processes were considered to be very expensive systems and only suitable for small-scale plants and for very specific applications. The capital costs are said to have decreased significantly with the advent of several commercially available systems and treatment scheme that are competitive even for large treatment plants. For example, in 1996, ZENON Environmental Inc. was installing municipal ZeeWeed MBR systems with average capacities of 0.2 million gallons per day (MGD). In 2003, just eight years later, MBR plants with capacities >10 MGD were constructed by ZENON (Schneider, 2003). In the infancy of the technology, a major cost was the anticipated membrane-replacement costs, now this item has dropped significantly due to better production of membrane modules as well as an increase in life expectancy gained from operating experiences (Drioli and Giorno, 2009). The energy demands is said to be the largest cost so far today, in particular the need for aeration of the biological process and of the membrane process as depicted in Figure 5

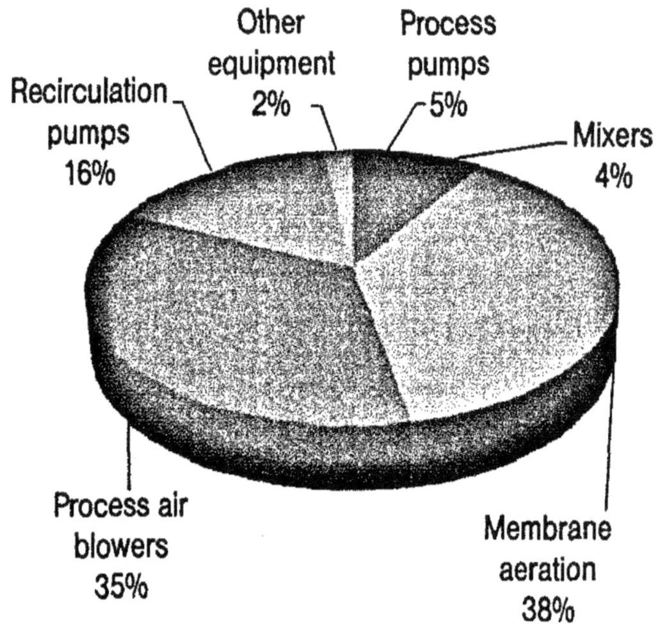

Figure 5. Typical energy demands in operation of a MBR process (Drioli et al., 2009).

COMPARISON OF MEMBRANE BIOREACTORS WITH CONVENTIONAL ACTIVATED SLUDGE SYSTEM

Throughout the world, there are hundreds of MBR systems in operation with many more in the design/construction phase. These range from small to large systems, treating both municipal and industrial wastewater. Small MBR systems are often used for water reuse within commercial developments, such as office complexes or shopping malls, whereas municipalities and industries operate the larger plants.

A major advantage of the MBR system as stated by Drioli and Giorno (2009) is that it can operate at a much higher solids concentration in the bioreactor than that of a CAS-mixed liquor suspended solids (MLSS) concentrations typically in the range 8 to 12 kg/l as compared to CAS that can only work at about 2 to 3 kg/l, because of limitations on settling. This higher sludge concentration is said to permit effective removal, not only of dissolved organic materials but also residual particulate solids. A comparison and assessment of MBR technology verses the conventional activated sludge process generally highlights the following issues:

Improve water quality

1. Meets stringent effluent requirements;
2. Filters out nearly all solids.

Allows wastewater reuse

1. As part of treatment scheme, provides water for potable reuse;
2. Reduces wastewater discharge fees and freshwater costs;
3. Provides water for nonpotable applications where fresh

water is in short supply.

Lowers capital cost

1. Clarifier is not needed;
2. Biological step can be scaled down in volume since bacteria concentration is higher.

Reduces plant space requirements

1. Footprint is up to 50% smaller than conventional plant;
2. Allows for expanded capacity within existing buildings.

Fewer operational problems

Bulking and floating sludge problems are avoided.

In spite of these benefits of the MBR process as compared to CAS, the technology is not without disadvantages. The early years of development of the process was considered to be expensive and this was attributed to high membrane costs, uncertainties of membrane lifetime and anticipated membrane replacement costs. However, as the MBR plants have been in operation for a while and experience has been gained, membrane lifetime appears to be longer than initially thought and replacement costs stipulated in the early 1990s to be 80 to 90% of operation and maintenance (O and M) costs is now estimated to be around 10 to 15% (Drioli and Giorno, 2009). The reduction is due to gain confidence in the technology, better and cheaper production of membrane modules and product development in general. Membrane fouling, an inherent phenomenon in all membrane processes has been mentioned as the main disadvantage of the MBR systems. Strategies and techniques to alleviate fouling coupled with the frequency of membrane cleaning are one main constraints of the process. The high energy demands for aeration, both for the biological process and membrane operation, is currently recognized as another major challenge and drawback of the technology (Drioli and Giorno, 2009).

MBR technology is said to be probably the membrane process that had most success and has the best prospects for the future in wastewater treatment. Trends and developments indicate that this technology is becoming accepted and is rapidly becoming the best available technology (BAT) for many wastewater treatment applications. The cost of an MBR plant for secondary treatment is still higher than that for a CAS plant, but as the numbers of MBR plants increase, and as membrane costs fall, the life cycle cost differential will soon disappear, and the process advantages should lead to rapid installation of the MBR system by the wastewater treatment industry (Drioli and Giorno, 2009).

Table 2. Comparison of MBR and CAS systems (Wang et al., 2009).

Diary application	CAS	MBR
WW flow (m^3/day)	600	600
Influent COD (mg/L)	5,000	5,000
Influent BOD_5 (kg/day)	3,000	3,000
Recycle of treated effluent (m^3/day)	0	400
Aeration volume (m^3)	4,500	600
Total floor space requirement (m^2)	1, 300	260
Effluent COD (mg/L)	90	30
Effluent BOD_5 (mg/L)	30	5
Effluent TSS (mg/L)	30	0

Process comparison

The similarities and dissimilarities of the CAS process system and the MBR process system are shown in Table 2. The data in the table is based on a 5-month study at a diary site in Central France which needs a modern wastewater treatment system to treat its combined wastewater (Wang and Menon, 2009).

From the viewpoint of biochemical engineering, the CAS and MBR process system are similar. The basic process of either CAS or MBR includes the unit processes of: influent feed, biological oxidation, final clarification, treated effluent discharge, return activated sludge (RAS) and excess discharge. Both processes require air supply or oxygen to sustain the biological oxidation and can be operated for the purpose of carbonaceous oxidation, nitrification and denitrification. When compared with conventional activated sludge systems, the MBR offers many attractive advantages:

(i) The traditional secondary clarifier is replaced by a membrane module. This module is more compact and the quality of rejected water is independent on the variations of sludge settling velocity.
(ii) The MBR allows the biomass concentrations to be higher than for traditional treatment plants. Whereas MBR investigations have been reported with biomass concentration of 20 g/l (Marrot et al., 2004; Jefferson et al., 1999) and even as high as 30 g/l (Yamamoto et al., 1989), conventional processes utilize biomass concentrations less than 5 g/l in order to avoid problems inherent to settling of concentrated flocs.

With poor settling flocs avoided, biological degradation is said to be more complete and treatment efficiency is higher. Also a report attributed to LÜbbecke et al. (1995) indicates that, increasing the biomass concentration involves a reduction in the oxygen mass transfer rate depending on the type of wastewater and reactor used. Other advantages of this system are as follows:

(i) The volume of the aeration tank can be reduced since a higher concentration of biomass can be stored in the bioreactor.
(ii) The production of sludge, the disposal of which is often difficult, is decreased by a factor of 2 to 3, resulting in a reduction of the overall operating costs (Marrot et al., 2004).
(iii) The membrane bioreactor is perfectly integrated in the industrial process because the wastewater can directly be treated *in situ*, allowing water reuse and concomitant reduction of the manufacturing costs linked to water consumption.

Unlike the conventional activated sludge system, the membrane bioreactor is characterized by a complete retention of the biomass inside the bioreactor because of the use of membrane separation, which controls and increases the sludge retention time (SRT) independently from the hydraulic retention time (HRT). High SRTs enable one to increase the sludge concentration and the applied organic load, thereby increasing the pollutant degradation. The specific sludge activity during organic matter decomposition and nitrification depends on the SRT. The SRT is a significant operational factor for the biological process (Marrot et al., 2004). Huang et al. (2000) investigated the organic removal performance from a synthetic wastewater treated with a submerged membrane bioreactor, as well as the behavior of soluble microbial products during long-term operation indicating that a satisfactory chemical oxygen demand (COD), total organic carbon (TOC) and biological oxygen demand (BOD) removal efficiencies were achieved, averaging over 90, 94 and 95%, respectively.

The results of a study by Dufresne et al. (1998) was said to be the first ever comparison made of performances between membrane bioreactor (MBR) and conventional activated sludge system treatment of a chemi-thermomechanical pulping (CTMP) effluent indicating that the performances of the MBR were superior for the removals of COD, suspended solids and toxicity. The amount of lignin onto the biosludge in the MBR was also found to be higher as compared to that inside the CAS.

Gao et al. (2004) did a comparison between a submerged membrane bioreactor (SMBR) and a conventional activated sludge system on treating ammonia-bearing inorganic wastewater. The SMBR and the CAS were compared in parallel over a period of 210 days on treating synthetic ammonia-bearing inorganic wastewater under similar conditions. Their result indicated that, the SMBR which contained larger numbers of nitrifiers was more effective and stable than the CAS in treating the synthetic ammonia-bearing inorganic wastewater. Differences were also said to be observed in the microbial community in the two systems. SMPs were reported to tend to accumulate, and then biodegrade in SMBR and the sludge particle size in SMBR were reported to be smaller than that in CAS.

Cicek et al. (1999) studied a system in which the wastewater used contained casein and starch (high molecular weight compounds) and fed to the MBR and CAS at identical conditions except different SRT (20 days for CAS and 30 days for MBR) which showed that approximately 99.0% of COD and 96.9% dissolved organic carbon (DOC) were removed in the MBR as compared to 94.5 and 92.7% in the CAS. The sludge in the MBR system was found to be made up of small flocs of regular size composed of zoogleal bacteria and of a small number of filamentous bacteria. The sludge in CAS system was made up of large flocs and higher amounts of filamentous organisms. Better settleability was observed in the CAS system than the MBR.

Weiss and Reemtsma (2008) carried out a comparative study of the performances of a lab-scale membrane bioreactor system and conventional activated sludge system for polar pollutants removal from municipal wastewater. Their results indicated that for half of the studied compounds, single step MBR treatment was clearly superior to CAS treatment with aerobic and anaerobic stages and provided significantly lower effluent concentrations (22 to 56% lower). They also found that, all the compounds for which no improvement was noted were either well removed by CAS treatment or were hardly degradable in the municipal wastewater. Considering operational conditions on trace pollutant removal by MBR studied by them, no significant effects were found for variation of hydraulic retention time (7 to 14 h) and sludge retention time (26 to 102 days), suggesting that increasing these parameters did not improve removal significantly. They however, observed an improved removal with SRT increasing from 26 to 37 days which was considered as not significant. Results from their work also indicated that MBR is neither superior to well degradable compounds that are already extensively degraded in CAS treatment nor for recalcitrant compounds that are not amenable to biodegradation. For most compounds of intermediate removal in CAS treatment (15to 80%), among them are personal care products, pharmaceuticals and industrial chemicals, the results from their work showed the MBR being superior and reduced the effluent concentration by 20 to 50%.

Mignani et al. (1999) evaluated the economics of using

the membrane processes to treat textile effluents of a factory located in the Northern part of Italy. The effluent after undergoing the conventional biological treatment was found not to meet the quality for discharge into the environment and have to be sent for additional external treatment in order to comply with the discharge standards. In order to avoid the costs of this added process, which represented 59% of the total costs, they installed an ultrafiltration module (standard Flamec-24 assembled modules) following the biological system. This unit was followed, in turn, by a reverse osmosis stage. This new process enabled them to reduce the total treatment costs of 122,000 €/year partly by the reuse of 50,000 m^3/year of water (Marrot et al., 2004). The treated wastewater is recycled and used as cooling water for dyeing machines and/or as water for washing and rinsing. The return on investment of the membrane process (270 000 €) is 2.2 and 2.5 years. This study shows that the equipment cost of a membrane process is not a barrier to its use.

Ciardelli et al. (2000) studied the treatment of effluent of factories that use dyes. The treatment processes studied were activated sludge, sand filtration and ultrafiltration (UF) and reverse osmosis (RO). The study contains a technical and economic analysis of the application of membrane separation technique for the purification of wastewaters targeted at their reuse. The water quality after the membrane processes is much better than that obtained using conventional processes; this treated effluent can be reused at all steps of production, including the most demanding ones concerning water quality. Finally, preliminary analysis of investment and operating costs indicates economic feasibility of the approach by their study.

BLEACHED KRAFT MILL EFFLUENT (BKME) CHARACTERISTICS

BKME is the combined aqueous waste of a major chemical pulp-making process. A mill typically generates waste water from many of its sections, including pulping, chemical recovery, evaporation and condensation, and multistage bleaching operations. Thus, waste waters from acid, alkaline, chlorine oxidant and other chemically diverse processes are sewered together to form the whole-mill discharge. The mixing and ensuing reactions of these streams lead to a final effluent that is highly complex. It consists of simple inorganic salts as well as over 250 identified organic and inorganic compounds of low molecular weight (or mass) with probably many more yet to be identified. The effluent from the production of bleached kraft pulp contains sugars, polysaccharides, organic acids, resin acids, and lignin transformation products and a variety of chlorinated derivatives, 200 of which have been identified (Table 3). In addition, an unusual property of BKME is that many of the organic constituents are of high molecular mass (>1kDa). This material is thought to consist largely of the polar breakdown product(s) of lignin, with lesser amounts of lignin at various degradation stages as well as polysaccharides (Higashi et al., 1992). The high molecular weight constituent of the effluent tended to be resistant to biode-

Table 3. Characteristics of mill effluent (Fulthorpe et al., 1992).

Prior to treatment	
Biological oxygen demand[a]	204 mg/L
pH[a]	6.6
Temperature	32°C
Organically bound chlorine	
During softwood pulp production	38.5 mg/L
During hardwood pulp production	13.9 mg/L
Chlorinated organics[b]	
Total acids	5.0 (mg/L max)
Total phenolics	1.0 (mg/L max)
Total aldehydes, ketones, lactones	1.0 mg/L
High molecular weight lignin derivatives	40.0 (mg/L max)
After Treatment[c]	
Biological oxygen demand	34.3 mg/L
pH	7.2
Total phosphorus	2.0mg/L
Total Kjeldahl nitrogen	1.9 mg/L
Ammonia-ammonium	0.9 mg/L
Chlorinated organics	
Adsorbable organic chloride	25.4 mg/L
Chlorodehydroabietic acid	80 µg/L
Dichlorodehydroabietic acid	100 µg/L
2,4- Dichlorophenol	3.8 µg/L
2,4,5-Trichlorophenol	7.5 µg/L
1,2,3-Trichlorobenzene	0.8 µg/L
Metals	
Aluminum	1.2 µg/L
Copper	4.2 µg/L
Zinc	107.8 µg/L

[a]Twelve-month averages; Liss, unpublished data. [b]Calculated from data in Kringstad and Lindstrom (1984) and the estimate that the mill produces 1000 tonnes of pulp per 100,000 m^3 of water discharged. [c]Average concentrations detected in final effluent in the period January 1 to June 30, 1990; from Ontario Ministry of Environment, 1991.

Table 4. Characteristics of BKME with ECF bleaching sequence (Xavier et al., 2005).

Parameter	Value
pH	3.4±0.17
COD (mg/L)	881.5±24.3
BOD_5 (mg/L)	300.5±9.5
Total phenolic compounds (mg/L)	271.9±14.2
Phytosterols (mg/L)	0.17±0.01
Color (VIS_{440}) (1×1 cm)	0.41±0.01

gradation and has been shown to be the principal source of adsorbable organic halide (AOX), color and chemical oxygen demand present in the effluents discharged (Bullock et al., 1996). The high molecular mass (HMM) material in an untreated effluent was reported to comprise 80% of the AOX but only 20% of the soluble COD. The HMM constituent also were found to be polar, devoid of aromatic structure and had characteristics indicative of lignin breakdown products (Higashi et al., 1992). The majority of effluent toxicity were also attributed to the HMM, chlorinated phenolics and chlorinated lignin derivatives of pulp and paper mill effluents (Afonso et al., 1992). According to them, low-molecular weight chlorinated neutral compounds are major contributors to mutagenicity and bioaccumulation due to their hydrophobicity and ability to penetrate cell membranes.

The pulp and paper industry is reported to be a major source of chlorinated organics worldwide, since pulp bleaching usually involves the oxidation of wood pulp with elemental chlorine or chlorine dioxide (Fulthorpe et al., 1992). However, with the replacement of bleaching sequences using elemental chlorine (Cl_2) by chlorine dioxide (ClO_2), the industry has reduced considerably the formation and discharge of chlorinated organic material into the aquatic environment. The release of chlorinated organic compounds to the environment has an adverse effect because many of these compounds are persistent, bioaccumulable and toxic. The annual estimate of organically bound chlorine discharge from the industry was put at 250,000 tonnes in the mid 1980s (Kringstad and Lindstrom, 1984). That figure has decreased to the 100, 000 tonnes/annum range presently. The organically bound chlorine measured as adsorbable organic halogen (AOX) in kraft mill effluents treated in an aerated stabilization basin show that a sizeable fraction of the carbon compounds in these effluents is chlorinated (Table 3) (Fulthorpe et al., 1992). Chlorinated phenolics are said to account for less than 2% of the organically bound chloride in bleaching effluents but they are greater contributors to effluent toxicity (Afonso et al., 1992).

Table 4 shows the characteristics of bleached kraft mill effluent with elemental chlorine-free (ECF) bleaching sequence.

Treatment methods used for BKME

The activated sludge process and the aerated stabilization basin (aerated lagoons) are the most commonly used methods for the biological treatment of pulp and paper mill effluents (Mueller and Walden, 1976; Ataberk and Gökçay, 1997). Many pulp and paper mills are reported to be relying on the same technologies to reduce toxicity of the effluent to aquatic organisms and the overall levels of chlorinated organics. However, studies have shown that only 10 to 50% of organically bound chlorine is removed in these treatment systems (Fulthorpe et al., 1992).

Aerated lagoons are relatively shallow basins in which wastewater is treated either on a flow-through basis or with solids recycle. Oxygen is supplied by means of surface aerators or by diffused aeration units. The action of the aerators also maintains the solids of the lagoon in suspension. Most of these solids must be removed by settling prior to discharge.

Depending on the degree of mixing, lagoons may be operated as either aerobic or as aerobic-anaerobic systems. In aerobic lagoons, all biological solids are in continual suspension and stabilization of the organics occurs under aerobic conditions. In the case of the aerobic-anaerobic lagoon, a large portion of the solids settle at the bottom of the lagoon. As the solids build up, a portion will undergo anaerobic decomposetion. Therefore, stabilization in this case occurs partly under aerobic conditions and partly under anaerobic conditions. The retention time is a function of the percent removal of BOD and depending on the detention/ retention time, the effluent from an aerated lagoon contains about one-third to one-half the value of the incoming BOD in the form of cell tissue (Metcalf and Eddy Inc., 1995).

The biological treatment is generally effective in reducing the levels of suspended solids and BOD of the effluent prior to discharge but not in eliminating the toxicity and mutagenicity. The chlorinated organic compounds, which are difficult to decompose, remain in the effluent (Afonso et al. 1992).

The AOX removal efficiencies for these plants (CAS and aerated lagoons) are typically around 20 to 60% (Collins and Allen, 1991; Yu and Welander, 1993; Ataberk and Gökçay, 1997; Schnell et al., 2000).

The popularity of the aerated lagoons in the treatment of wastewater is due to its relatively lower cost of installation and also, technically it is one of the simplest forms of engineered biological treatment systems. It is said to be effective in reducing the more readily degradable materials measured as BOD from wastewater as much as 80 to 95% after five to 10 days treatment (Lindström and Mohamed, 1988).

CONCLUSION

A major concern of the pulp and paper industry lies in the fact that even after more than 30 years of consistent efforts, a satisfactory treatment of BKME still remains elusive. This is primarily due to two reasons:

The processes employed in pulping as well as pulp processing (including pulp bleaching) are so diverse that the composition of the resulting wastewaters (in terms of critical components) are very different and no single process or combination of processes that is economical can apply to all;
2. The wastewaters invariably contain considerable quantities of materials that are toxic either to the wastewater treating organisms or to the aquatic species present in the recipient waters or both (Ali and Sreekrishnan, 2001).

MBR for wastewater treatment and reuse are proven systems and are rapidly being accepted by industries. The technology has proven its reliability and efficiency for a variety of industrial plants in North America, Europe and Asia. It applicability for the treatment of pulp and paper effluent has been acknowledged in a few research papers. Membrane bioreactors are effective in treating wastewaters. They offer advantages of compactness over conventional technology, as well as producing a very high quality disinfected effluent. In comparison with the conventional activated sludge system, the MBR systems have better removal efficiency and a potential for water reuse in manufacturing. Membranes bioreactors application is expected to continue to increase in wastewater treatment, with the drivers being the need for compact plant, quality of effluent and value of recycling. The effluent quality for discharge is becoming an issue in many countries now with stringent legislation being put in place or about to be implemented, requiring removal of most of the toxic compounds. These stringent conditions appear to be more achievable with emerging MBR technologies.

REFERENCES

Afonso MD, Geraldes V, Rosa MJ, Norberta de Pinho M (1992). Nanofiltration removal of chlorinated organic compounds from alkaline bleaching effluents in a pulp and paper plant. Wat. Res. 26:1639-1643.

Ali M, Sreekrishnan TR (2001). Aquatic toxicity from pulp and paper mill effluents: a review. Adv. Environ. Res. 5:175-196.

Assalin MR, Almeida E. dos Santos, Nelson Durán N (2009). Combined System of activated Sludge and ozonation for the Treatment of Kraft E1 Effluent. Int. J. Environ. Res. Public Health, 6:1145-1154.

Ataberk S, Gökçay CF (1997). Treatment of chlorinated organics in bleached kraft mill effluents by activated sludge process. Wat. Sci. Tech. 36:147-150.

Attiogbe FK, Glover-Amengor M, Nyadziehe KT (2007). Correlating Biochemical and Chemical Oxygen Demand of Effluents - A Case Study of Selected Industries in Kumasi, Ghana. West African Journal of Applied Ecology 11:110-118.

Baek SH, Pagilla KR (2006). Aerobic and Anaerobic Membrane Bioreactors for Municipal Wastewater Treatment. Water Environ. Res. 78(2):133-140.

Bullock CM, Bicho PA, Saddler JN (1996). The influence of the high and low molecular weight fractions of a bleach kraft mill effluent on the microbial activity of activated sludge. Wat. Res. 30:1095-1102.

Ciardelli G, Corsi L, Marcucci M (2000). Membrane separation for wastewater reuse in the textile industry. Resources Conservation and Recycling 31:189-197.

Cicek M, Franco JP, Suidan MT, Urbain V, Manem J (1999). Characterization and comparison of membrane bioreactor and conventional activated sludge system in the treatment of wastewater containing high-molecular weight compounds. Water Environ. Res. 71(1):64-70.

Cicek N (2003). A review of membrane bioreactors and their potential application in the treatment of agricultural wastewater. Can. Biosyst. Eng. 45:6.37-6.49.

Collins T, Allen DG (1991). A bench-scale aerated lagoon for studying the removal of chlorinated organics. J. Tappi. pp. 231-234.

Davis ML, Cornwell DA (1991). Introduction to Environmental Engineering, 2nd Edition. McGraw-Hill, Inc. USA.

De Luca G, Sacchetti R, Erica Leoni E, Zanetti F (2013). Removal of indicator bacteriophages from municipal wastewater by a full-scale membrane bioreactor and a conventional activated sludge process: Implications to water reuse. Bioresour. Technol. 129:526-531.

Drioli E, Giorno L (2009). Membrane Operations: Innovative Separations and Transformations. Wiley-VCH Verlag GmbH & Co. KGaA, Weinheim, Germany.

Dufresne R, Lavallée HC, Lebrun RE, Lo SN (1998). Etude comparative de l'opération d'un système de bioréacteur a` membranes et d'un système de boues activées conventionnel pour le traitement des effluents papetiers, Pulp & Paper Magazine of Canada 99(2, III). pp. 51-55.

Francy DS, Stelzer EA, Bushon RN, Brady AMG, Williston AG, Riddell KR, Borchardt MA, Spencer SK, Gellner TM (2012). Comparative effectiveness of membrane bioreactors, conventional secondary treatment, and chlorine and UV disinfection to remove microorganisms from municipal wastewaters. Water Research 46:4164-4178.

Fulthorpe RR, Liss SN, Allen DG (1992). Characterization of bacteria isolated from a bleached kraft pulp mill wastewater treatment system. Can. J. Microbiol. 39:13-24.

Gao M, Yang M, Li H, Yang Q, Zhang Y (2004). Comparison between a submerged membrane bioreactor and a conventional activated sludge system on treating ammonia-bearing inorganic wastewater. J. Biotechnol. 100:265-269.

Higashi RM, Cherr GN, Shenker JM, Macdonald JM, Crosby DG (1992). A polar high molecular constituent of bleached kraft mill effluent is toxic to marine organisms. Environ. Sci. Technol. 26:2413-2420.

Horan NJ, Chen W (1998). The treatment of high strength pulp and

Huang X, Liu R, Yi Qian Y (2000). Behavior of soluble microbial products in a membrane bioreactor. Process Biochemistry 36:401-406.

Jefferson B, Laine A, Parsons S, Stephenson T, Judd S (1999). Technologies for domestic wastewater recycling. Urban Water 1:285-paper mill effuent for wastewater re-use 1) The use of modelling to optimise effuent quality from the existing wastewater treatment plant. Environ. Technol. 19:153-162.

292.

Junkins R, Deeny K, Eckhoff T (1983). The Activated Sludge Process: Fundamentals of Operation. Boston: Butterworth Publishers.

Kamala A, Kanth Rao DL (1988). Environmental engineering Water supply sanitary engineering and pollution. Tata McGraw-Hill Publishing Company Ltd, New Delhi.

Kringstad KP, Lindström K (1984). Spent liquors from pulp bleaching. Environ. Sci. Technol. 18(8):236A-246A.

Lindström K, Mohamed M (1988). Selective removal of chlorinated organics from kraft mill total effluents in aerated lagoons. J. Nordic Pulp Paper Res. 3:26-33.

Lübbecke S, Vogelpohl A, Dewjanin W (1995). Wastewater treatment in a biological high performance with high biomass concentration, Water Res. 29(3):793-802.

Marrot B, Barrio-Martinez A, Moulin P, Roche N (2004). Industrial wastewater treatment in membrane bioreactor: A review. Environmental Progress 23(1):59-68.

Marti E, Monclús H, Jofre J, Rodriguez-Roda I, Joaquim Comas J, Balcázar JL (2011). Removal of microbial indicators from municipal wastewater by a membrane bioreactor (MBR). Bioresource Technology, 102:5004-5009.

Meng F, Chae So-Ryong, Drews A, Kraume M, Shin Hang-Sik, Yang F (2009). Recent advances in membrane bioreactors (MBRs): Membrane fouling and membrane material. Water Res. 45:1489-1512.

Metcalf and Eddy Inc (1995). *Wastewater Engineering: Treatment, disposal and reuse,* 3rd edn. (G. Tchobanoglous and F. L. Burton, ed.) Tata McGraw-Hill Publishing Company Ltd, New Delhi.

Mignani M, Nosenzo G, Gualdi A (1999). Innovative ultrafiltration for wastewater reuse. Desalination, 124:287-292.

Mueller JC, Walden CC (1976). Detoxification of bleached kraft mill effluents. J. WPCF, 48(3):502-510.

O'Connor BI, Voss RH (1992). A new perspective (sorption/desorption) on the question of chlorolignin degradation to chlorinated phenolics. Environ. Sci. Technol. 26:556-560.

Schneider C (2003). Membrane Bioreactors (MBR) - Redefining Water Reuse. ZENON Environmental Inc.

Schnell A, Steel P, Melcer H, Hodson PV, Carey JH (2000). Enhanced biological treatment of bleached kraft mill effluents-I. Removal of chlorinated organic compounds and toxicity. Water Res. 32(2):493-500.

Stephenson T, Judd S, Brindle K (2000). Membrane bioreactors for wastewater treatment. IWA Publishing, UK.

Saunamäki R (1997). Activated sludge plants in Finland. Water Sci. Technol. 35:235-242.

Thompson G, Swain J, Kay M, Forster CF (2001). The treatment of pulp and paper mill effluent: a review. Bioresour. Technol. 77:275-286.

*UNEP, GEO-2000, Global Environmental Outlook, 2000.

Wang LK, Menon R (2009). Handbook of environmental engineering: Advanced biological treatment processes. 9:129-156.

Weiss S, Reemtsma T (2008). Membrane bioreactors for municipal wastewater treatment-A viable option to reduce the amount of polar pollutants discharged into surface waters? Water Research 42:3837-3847.

Xavier C, Chamorro S, Vidal G (2005). Chronic effects of kraft mill effluents and endocrine active chemicals on daphnia magna. Bull. Environ. Toxicol. 75:670-676.

Yamamoto K, Hiasa M, Mahmood T, Matsuo T (1989). Direct solid-liquid separation using hollow fiber membrane in an activated sludge aeration tank. Water Sci. Technol. 21(4-5):43-54.

Yu P, Welander T (1993). Comparisons of biological and thermo-alkaline treatment for the removal of chlorinated organic matter from kraft bleach plant effluent. J. Nordic Pulp Paper Res. 8:360-364.

Geographic information system planning and monitoring best practices for West Africa

Akajiaku C. Chukwuocha[1] **and Ngozi AC-Chukwuocha**[2]

[1]Department of Surveying and Geoinformatics, Federal University of Technology, Owerri, Nigeria.
[2]Department of Environmental Technology, Federal University of Technology, Owerri, Nigeria.

Phenomenal increases in the number and sizes of urban settlements across the West African coastal region are leading to massive reclamation of swamps and destruction of natural ecosystems. Poor urbanization policies, inefficient planning and monitoring technologies are evident. The consequences include some of the worst types of environmental hazards. Best urbanization practices require integrated planning approaches that result in environmental conservation. Geographic Information systems (GIS) provide the platform for integration and processing of multi-sector Geosciences data in order to accurately predict results of different planning options. This paper presents the West African urban environmental problems. Using some concluded studies, it illustrates the functionality of integrating data in GIS for efficient planning and monitoring while calling on the governments of West Africa to adopt GIS based planning for best results.

Key words: GIS, Urban Planning, Urban Monitoring, West Africa, Best Practices, Pollution, flooding

INTRODUCTION

As the cities of West Africa are growing in sizes and numbers, the attendant urban problems are increasing. For years now, West African urban areas have recorded devastating environmental problems including such phenomena as flooding, erosion, pollution.

Many researchers have done quite some work on the West African environment and seem to all point to the same environmental planning as a major issue in handling the environmental problems of urban West Africa. Aderogba (2012) studying recent floods and sustainable growth and development of cities and towns in Nigeria concluded, "More attention has to be paid to urban physical planning".

Koenig (2009) writing on the challenges of urban growth in West Africa with case study of Dakar, Senegal writes, "Dakar has made substantive efforts to create forward looking plans that take into account the needs and desires of all major groups. However attempts at participatory planning are not always easy to reconcile with grand strategies proposed by important economic and political actors."

Odufuwa et al. (2012) studying floods in Nigerian Cities concludes, "Impact of floods is more pronounced in low-lying areas due to rapid growth in population, poor governance, decaying infrastructure and lack of proper environmental planning and management."

Karley (2009) writes, "... for the city (Accra Ghana), there is no evidence that unusual rainfall has been occurring recently that could explain the increased occurrences of flooding being experienced. Rather, the cause of the problem is the lack of, drainage facilities to collect the storm water for safe disposal. These could in turn be attributed to the ineffective planning ..."

This paper proposes that planning of the West African urban environments be made more efficient by integrating geographic information systems (GIS). This will enable Urban Planners and policy makers to consider best practices options for the conservation of the urban environment.

The aim of this paper is to propose to West Arican urban planners and policy makers to adopt the best practices methods and technologies offered by the Geographic Information Systems for the sustainable development of the West African environment. The objectives of the paper include: i) to highlight the growing environmental hazards in West Africa due to poor planning of the environment using other research papers; ii) to reference concluded research works in demonstrating that more efficient urban areas result from integrating geosciences data on a GIS platform in different areas of designs and planning for the urban environment; iii) to highlight immediate identifiable areas where GIS may be used in environmental planning and monitoring and iv) to call on policy makers, planners and governments of West Africa to adopt these best practices for the conservation of the West African Urban environment.

URBANIZATION PRACTICES AND PROBLEMS IN URBAN WEST AFRICA

A United Nations report on population growth in the urban centers across the world in the 21st century (Ginkel, 2008), projects that it is expected that by 2030, half of all of Africa's population will live in urban centers. However, the implication of this growth includes the increased pressure on natural resources, and the environment, in order to provide more infrastructures such as housing, water, transport, electricity, schools, etc. The consequences include unprecedented and varied impacts on the environment.

"Since 2007, the flood situation in West Africa is becoming more and more recurrent and the impact on the population and infrastructures is becoming more severe" reports Dieye (2010), for the United Nations Office for the Coordination of Humanitarian Affairs, Regional Office for West and Central Africa (UNOCHA/ROWCA).

The major cities of West Africa lie along the coastal shores of the Atlantic Ocean. Nine of all the 11 countries that line up the shore of the West African coast have their capital cities in the delta swamps by the Atlantic Ocean. It is true that the capital cities of Nigeria and Cote D'Ivoire were later moved away from the coast line to Abuja and Yamoussoukro respectively, the former seats of government and the present business capitals, Lagos, and Abidjan also sit on the coast line. These massive urban settlements are expanding fast. Demand for more land for construction of facilities coupled with citizen poverty across the region is leading to more degrading environmental practices. Ecosystems are being destroyed by efforts that convert the little spaces that can be found anywhere to farm crops for sustenance. Forests are foraged both for fuel wood, and logging to provide timber for new constructions.

Much more seriously is the impact of reclamation of land by both the government and citizens to construct new facilities. A study of the satellite images of the cities will show that the many estuaries, channels, tributaries and mangrove forests that mark the immediate shoreline of the sea are disappearing very fast or have already completely disappeared in these cities. Reclamations are happening in Port Harcourt, Warri, Benin and Lagos Nigeria, Cotonou Benin Republic, Lome Togo, Accra Ghana, Abidjan Cote d'Ivoire, Monrovia Liberia, Freetown Sierra Leone, Conakry Guinea, Bissau Guinea Bissau, Banjul the Gambia, and Dakar Senegal and many other urban centers across the region (Figure 1). The flooding of the urban cities of the West African Coast has now become perennial with its attendant loss of properties and lives.

Several papers such as Atedhor et al. (2010), in "Changing rainfall and anthropogenic-induced flooding: Impacts and adaptation strategies in Benin City, Nigeria", Karley (2009), in "Flooding and physical planning in urban areas in West Africa: situational analysis of Accra Ghana", Integrated Regional Information Networks, IRIN (2009) in "Burkina Faso: Coping with urban flood-displaced", Koenig (2009), in "The Challenges of Urban Growth in West Africa- The Case of Dakar, Senegal", and Aderogba (2012) in "Qualitative Studies of Recent Floods and Sustainable Growth and Development of Cities and Towns in Nigeria" have reported a vast array of the typical urban environmental problems in West African cities. They mostly point to the deficiencies in planning of the urban cities as key in the environmental problems that are about to overwhelm the region.

Federal Ministry of Water Resources and Rural Development Nigeria (undated), writes on a number of urban environmental problems in Nigeria, listing several devastating urban flooding events that have taken place in Nigeria over the years.

On urban resources, the Nigerian Ministry notes: "There is excessive pressure on available urban resources, infrastructure and space, evident in cities such as Lagos, Kano, Port-Harcourt, Aba, Onitsha, Ibadan, Kaduna and of recent Abuja, especially its satellite towns."

The Nigerian Ministry writes further: "Pollution from many industries is a growing problem. In commercial centres like Lagos, Port-Harcourt, Kano, Kaduna, Warri, Aba and Onitsha, coloured hot and heavy metal-laden effluent, especially that from textile, tannery, petrochemicals and paint industries, constitute severe danger to water users downstream."

Federal Ministry of Water Resources and Rural Development [Nigeria] (undated) notes further that, "The following characterise the urban and semi urban centres:

1) The various non biodegradable household petroche-

Figure 1. Map of West Africa.

mical products such as polythene bags, plastic containers, foam packages and tyres; and 2) Million litres of crankcase oil disposed from mechanic workshops, industries, power stations and commercial houses are ischarged carelessly Into drains and ground surfaces at various locations within the watersheds.

The Soil Erosion and Flood Control Department further writes, "In many cities, plants are no longer used for home landscaping. Example, in Abuja many areas earmarked as green belts are being taken over by corner shops, roads and paved concrete surfaces."

Traditional urban plans rely only on the analogue topographic maps which are laid out on drawing boards. When the Computer Aided Design (CAD) is employed, the approach of the planners is fixed on providing the required facilities. Even when they know the need for conserving the environment they are not equipped to generate the information they require for efficient urban planning decisions. The plan calculations and considerations in these cases focus only on what is visible on the drawing board.

Topographic calculations for drainage installations for instance, follow on the general slope aspect. With today's digital geospatial analysis tools, this is quite Inappropriate. Decisions in urban-planning cannot be based only on what is desired to be provided in the city without looking at the effect on the environment. The Geoscientists should come up with plans that will enhance the natural landscape while compromising only what is inevitable in a sustainable way.

The urban practices already noted in the cited paper were observed with visits to the cities and studied high resolution satellite images across West Africa. Some of such practices include bulldozing of large areas of land, sometimes unnecessarily, to site possibly smaller structures.

There are general increases in industrial waste not disposed adequately. Hydrocarbon pollution and other unhealthy activities are the order of the day in these cities. There may be some exception where the government may have shown some interest in preserving the environment, like in Calabar, Nigeria. But the environmental preservation efforts are actually limited.

Governments across West Africa involve in massive constructions to provide needed infrastructure without taking into account in a scientific way the long term implications on the environment. If the environment does not survive, the humans that depend on the environment will not also survive.

NEED FOR EFFECTIVE URBAN PLANNING

Effective urban planning is a very complex issue and cannot be carried out merely to provide only the facilities humans desire to have in place in their settlements and not consider the full implications of the constructions and the altering of the natural habitat. There is an urgent need to adopt a method for regional planning which provides for integration of data from a wide span of Geosciences, Sociology and Economics to monitor and manage the existing cities and plan the next urban cities, predict the future of the cities and hence consciously channel the growth of the urban areas in a sustainable way. Data required here will be in digital form and would include but not exclusive to those from: i) Up to date topographic maps or surveys, ii) current and archived satellite images over the region of interest, iii) Direct field survey captures such as 3-dimensional coordinates of points, iv) aerial photographs, v) Up to date demographic surveys and maps, vi) Soil surveys and maps, vii) Ecosystem surveys and maps, viii) Land use maps of existing urban areas, ix) Data on the major business occupations especially those that have direct impact on the urban environment, x) Up to date data on the natural and artificial drainage and xi) Storm water precipitation data over the years.

The aims here are to: i) integrate planning at regional levels to take care of, for instance, drainage by subsheds of major watersheds using topographic data, land use and storm water data etc; ii) plan the urban areas to sustain the ecosystems and avoid such phenomena as desertification, gully erosion, and flash floods; iii) predict the growth pattern of the cities and provide for the growthin sustainable ways and iv) monitor the growth to ensure that the plan is adhered to strictly and particularly that people do not encroach into reserved areas to use as refuse dump sites or for construction of illegal structures.

Williams (2000) writes on ecosystem planning, "It is clear that a new way of addressing urban problems is needed and that it will have to be more efficiently integrated, more sensitive to ecology and community, more respectful of uncertainties, and more open to citizen involvement than what now prevails. This has led to an ecosystem approach to planning: 'an approach that begins with an ecologically-bounded area, stresses the integration of social, economic, and environmental factors, and seeks to involve all the relevant interests and power-holders in identifying desirable futures, evaluating alternative pathways and implementing preferred solutions'". The article further outlines basic principles that reflect the characteristics of ecosystem planning as follows:

1) Base planning units on natural boundaries, reflecting ecological functions while replacing a politically-oriented hierarchy of units.
2) Design with nature, and respect human activity and its effect on the environment as well as the limits of resource availability and ecosystem resilience.
3) Consider global and cumulative effects, because a much broader and longer perspective must be considered, like attention to off-site, cross-boundary, intergenerational, and cumulative effects.
4) Encourage inter-jurisdictional decision making, and overcome the present fragmentation and isolation with integrated planning and implementation.
5) Ensure consultation and facilitate cooperation and partnering, involving the widest range of stakeholders effectively and openly in the planning process.
6) Institute long term monitoring, feedback, and adaptation of plans, to assess what happens to communities and ecosystems as plan implementation unfolds.
7) Adopt an interdisciplinary approach to information, by greater information gathering (e.g. ecological capacity and functions), more integration of information, and greater cooperation among information providers.
8) Adopt a precautionary but positive approach to development that aims not just to avoid further damage but also to reduce stresses and enhance the integrity of ecosystems and communities.
9) Ensure that land-use planning integrates environmental, social, and economic objectives, but this depends on the planning body having a firm base of established institutional power to foster multi-interest cooperation and implementation.
10) Link ecosystem planning with other aspects of democratic change, social learning, community building, and environmental enlightenment.

The geographic information system (GIS) is a complex data storage, processing and display tool with inestimable capacities to integrate multi-sectored data in a scientific way. It is GIS that is needed to effectively respond to these many urbanization problems in order to generate the information needed for accurate decision making in urban planning and monitoring.

USE OF GEOGRAPHIC INFORMATION SYSTEM (GIS) IN URBAN PLANNING AND MONITORING

GIS and integration of geoscience data

The geographic information system (GIS) is a computer

storage, computing and display system that links inter-related geospatial data through appropriate algorithms, to derive needed information on environmental phenomena and display same in a clearly understandable way. A number of features make the GIS most desirable in handling geospatial matters. The GIS employs the impressive capacities of the computer in storing volumes upon volumes of data, its unrivalled speed in sorting stored data and processing them for needed information. Today, there are in place seamless interfaces for data exchange between the data capture units in the field and the processing computers in the office. The miniaturization of computers with increased computing powers and storage capacities as is in the laptops and palmtops, and the unlimited link between computer systems across the globe through the World Wide Web or locally within the local network, make the system a compulsory necessity if the professional must meet with the demands of high efficiency in today's world. In the field of Environmental Studies, Urban Planning, Engineering, Surveying and Geodesy, the GIS provides the rare opportunity of combining the perceptibility of graphics with the accuracy of digital computations at great speed. Different formats of data storage and exchange, data processing and information display, are combined in achieving and demonstrating results. For example, in considering the hydraulics of storm- water flow in a watershed, the topography of the land spread, the soil types, the land use and the land cover, infiltration rates and the precipitation rates, are all factors that need to be considered. It is obviously difficult if not impossible to independently derive information on each of the factors over a unit of space considered and to integrate these step-wise over the whole land in an analogue manner, in order to decide for example, how to place drainage to mitigate flooding and erosion.

The GIS displays graphical digital elevation models (DEM's) and Digital Terrain Models (DTM's) using elevation and topographic data. The needed land cover or soil type factor in infiltration, and precipitation rate over the land are held in GIS, in form of satellite image data, or map data. The factors influencing storm water runoff flow are integrated by powerful software that drive these complex computations which yield accurate predictions of future possibilities that prove extremely valuable in taking measures to mitigate potential disasters. Additionally, the GIS can handle data over very vast regions beyond what can be laid out on a table at any scale at all. The largest regions can be handled even to the minutest details by the capacity of the computer to organize geospatial data in recognizable spreadsheets and to view information as graphics, coupled with the capacity to zoom in and out on the areas of interest and beyond.

Application of GIS in urban planning

The West African urban centers can no longer be planned just in the traditional form that seeks only to provide facilities needed by humans without accurately determining the consequences of those choices on the environment and allowing alternative options. The use of GIS in urban planning is necessary to provide needed information about the environment of the proposed or existing urban area from a very complex mesh of environmental data. Ginkel (2008), writes "using earth observation and GIS, it will be possible to prepare on time, at much lower cost the optimal direction of urban growth, taking into account the local topography and many other characteristics, to guarantee a sound development of the cities and timely reduction of environmental risks."

Much of the information needed for modern urban planning are products of geospatial data integration and mathematical processing. They are such information that could not be handled independently or purely mathematically without expression in graphical space. For instance in planning a city in a way to preserve habitats, we would most likely be seeking to know what locations and what size of the space hosts a habitat of interest. We may want to know the criteria for choosing the habitat that may be altered for urban use and how the habitat may be altered. All these information are not only displayable on a GIS platform, GIS software are capable to show areas that match prescribed criteria in colour codes. If also it is needed to isolate areas that are up to certain degree of slope, and which are in areas of certain soil types from choice of sites for some facilities, the GIS is instructed by the query tools or scripting instructions. Provided that the background data on the issues in question are integrated by algorithms in the GIS over the area of the proposed or existing urban center, the result of this isolation is displayed on the GIS platform by colour coding. Data required for the slope query is the topographical data. Data required for habitats query is the land cover data. The soil data will provide for the soil type information. Some other very useful data types will include demographic data, rainfall data, land use data. Urban Planners should also be interested in utilities needed to sustain the city. For example what road types are needed for the different areas of the city? Again where will be the most suitable to locate an airport, or university etc considering ecological issues. The GIS also can enable preliminary studies that determine what type of surfacing will be most suitable given for roads given the soil type and rain data.

Availability of GIS data needed for urban planning and monitoring

The GIS data requirements are becoming increasingly available globally. Imagery coverage of the earth at 5 m resolution or higher stands now at nearly 100%. The topography of the earth is fully covered by the Shuttle Radar Terrain Model (SRTM) at 90 m, 30 m and 1 km resolution levels. Other accurate sources of digital elevation data would include all stereo-capable imaging satellites such as SPOT 5. Even though weather conditions change abruptly, GIS has made it possible for weather to be monitored in real time. Weather data have been so monitored that global weather prediction has become a near-totally-accurate science. Satellite weather data are accessible online globally and precipitation rates or volumes are accurately predicted. The Satellite images provide detailed land cover and land use data which are useful in all urban planning exercises all over the world.

The websites for these Satellite data are easily accessible through internet search engines such as Google, Yahoo search, etc. However International Water Management Institute (IWMI) website is an example of sites that warehouse diverse Remote Sensing Satellite data useful for urban planning and monitoring. It also lists a number of web links to important satellite data websites. These include among others:

i) http://www.iwmidsp.org which is a satellite warehouse of sorts.
ii) www.iwmidsp.org for advanced High resolution Radiometric AVHRR satellite data. Cost: Free download. Format: ERMapper.ers
iii) http://free.vgt.vito.bel for SPOT vegetation free download, Format: data binary.
iv) Http://edcimswww.cr.usfs.gov/pub/imswecome/ for MODIS, Terra and Aqua Satellite data, Cost: free. Format: .hdf
v) http://glcf.umiacs.umd.edu/index.shtml for Landsat data of 1970's, 1990's, and 2000's. Cost: Free. Format: GEOTIF or MrSid
vi) http://landsat.org/ for Landsat data. Cost: $50 to $600 per scene. Format GEOTIF
vii) http://glovis.usgs.gov/imgviewer/ or http://edcimswww.cr.usgs.gov/pub/imswelcome for Landsat data of recent and historical times from United States Geological Surveys (USGS). Cost: $450 to $600 per scene. Format: GEOTIF
viii) http://www.spaceimaging.com/ for Landsat (recent and historical) data for commercial sources of Space Imaging: Format:GEOTIF. Costs Approx. $1,000
ix) http://www.spotimage.fr/html/167.php for SPOT HRV data:(Historical and recent) Format:GEOTIF , costs: $2000 for 60km x 60km
x) http://http://glovis.usgs.gov/imgviewer/ or http://edcimswww.cr.usgs.gov/pub/imswelcome/ or http://asterweb.ipl.nasa.gov/ for ASTER data, Format: GEOTIF, cots: approx $50 to $150 for 60km x 60km.
xi) http://eol.usgs.gov/ or http://edcimswww.cr.usgs.gov/pub/imswelcome/ for ALI (next generation landsat data) Format: GEOTIF, Costs: approx $.500 for 100km by 100km.
xii) http://www.nrsa.gov/in/ for IRS LISS data (Historical and recent), Format GEOTIF, Costs: approx. $1,000 for 100 by 100 km.
xiii) http://www.spaceimaging.com/ for IKONOS from Space Imaging, Format: GEOTIF, Cost: 10 by 10km approx. $1500 or more
xiv) http://www.digitalglobe.com/ for quickbird from Space Imaging, Format: GEOTIF, Cost 10 by 10km approx. $1500 or more.
xv) http://www.orbimage.com/index.html Format: GEOTIF, Cost: 10 by 10km approx. $1500 or more.

There are also many other independent sites through which these data can be acquired.

PROPOSALS FOR OPTIMIZING ENVIRONMENTAL CONSERVATION USING THE GIS

Gironas et al. (2007), has proposed a morphological approach in planning urban areas. They note that up to this point, the quantification of changes to the structure of the drainage patterns caused by urbanization has not been developed in detail, and no formal methodologies are used to characterize urban catchments from a morphological and topological point of view. Finally, no special measures in storm water management are being taken in order to imitate natural conditions of the drainage pattern.

In this case, the morphological and topological features of the area of interest should be considered. For instance in planning the storm water drainage, storm water flow from the farthest distances that drain into that area should be considered and taken care of in as natural a way as possible. The urban area planning should not concentrate on the areas of developmental interest only. Every factor that contributes to the functioning of that environment should be considered

The central issues in regional planning are to effectively take care of runoff that may increase as a result of urbanization without causing damage to other habitats and to preserve as much as possible the natural ecosystems.

SOME REPORTED EFFORTS AT GIS BASED FACILITY PLANNING FOR THE WEST AFRICAN URBAN ENVIRONMENT

GIS based urban drainage network design

Chukwuocha (2012) reports on the "GIS based urban drainage network design for Owerri, South East Nigeria". This Ph.D dissertation, reports on drainage design that positioned drainages on the natural flow routes which would effectively drain the watershed subcatchments. Using the GIS, the discrete topographic data of the area was converted into a Digital Elevation Model (DEM). The DEM was then automatically analyzed on the GIS platform to determine the natural drainage routes and the corresponding subcatchments which drained into the determined routes. Needed engineering drainage details such as elevations of the nodes, sizes of conduits that would effectively drain the subcatchments were determined. Simulations were run to ascertain that these drainages would effectively drain the sub catchments.

One big advantage of this GIS based natural drainage route drainage network design ensures efficiency at several levels including the fact that these drainages were located in the best positions that drained the entire sub catchments in a an accountable manner. The method provides needed information in a number of other areas such as the velocity of runoff in each sub catchment which is useful in water-borne diseases studies; the run-off coefficient map of the studied area and so on.

Land use allocation by run - off calculation

When a region is in consideration for urbanization, systematic and scientific approaches should be used in choosing land allocations to the urban facilities. Tang et al. (2005), presents the Runoff MINinimization (ROMIN) model that may be used in sorting between sites for land use allocations. This model compares the soil type of the site in combination with the proposed land use and calculates the resultant runoff.

Comparing different land use options and their runoffs, the minimal runoff land use combination option is adopted. The ROMIN model is quite adaptable for use in a GIS platform. Data and computation of Runoff and decision of the optimal option are automated and in GIS, this can be handled over large regions impossible to be laid out fast in graphical drawing table.

ADDITIONAL POINTS ON GIS BASED URBAN PLANNING

Land use allocation by consideration of existing natural ecosystems

AmphibiaWeb (2010), reports that when amphibian habitat is drained, filled or cut and then converted into parking lots, housing developments or agricultural developments, the natural amphibian habitats are completely destroyed. Habitat alteration occurs when changes made to the environment adversely affect ecosystem function, even if not completely or permanently. An example is when heavy livestock graze in the amphibian environment. These livestock trample on the amphi-bians and other organisms that occur in the aquatic region.

Habitat fragmentation occurs when interlinked habitats are cut apart by land uses introduced between them. Hanski (1999) developed a model for the threshold for metapopulation persistence which predicts that isolated populations are more likely to go extinct in the long run than populations that are slightly connected. With the GIS as a tool we can map the entire region of interest in broad ecosystem groupings. We then consider what optimal areas of each ecosystem grouping must be preserved. The ecosystems must be allowed to link up to more natural unaltered habitats. If habitat destruction must occur, we must employ a gradual process that allows the population to migrate into safer habitats, since the amphibians and of course all other species will have links to other natural habitats.

We should determine what minimum limits of the different habitats that we must not allow urbanization to affect. In GIS, buffering is a toll that enables one to set limits of minimum or maximum distances from a known point or line. Polygonization helps us mark out areas automatically when we have defined the criteria. These tools will be useful in marking out limiting factors of every habitat that we are concerned with. The habitats are also identifiable in satellite images, using certain criteria such as, the predominant plant type, marshiness of environment or otherwise, the topography of the area etc.

Mitigation against urbanization - induced flash floods and gully erosion

GIS-based hydrologic studies and predictions lend so well to the mitigation of flash floods and gully erosions in urban areas.

Lansigan (2007) notes that effective response to these hydrological problems may include:

i) Effective comprehensive land use plan (CLUP) which include the critical areas prone to flooding, landslides and soil erosion.
ii) Regulated land use conversion of lands for urban uses with appropriate provision for adequate drainage system.
iii) Protective management of watershed areas, and clearing of watershed areas and clearing of waterways, rivers and creeks by regular removal of debris that may obstruct smooth flow of water.
iv) Information, education and communication program on flood protection, and response measures at the village-level.

In a paper, Junior et al. (2009) notes the usefulness of the GIS using multi-temporal digital elevation data analysis in studying urbanization-induced gully erosions. These studies are then used in predicting other areas prone to future gullying so that mitigation will be provided. Flood studies must be carried out at regional levels. All areas that drain into the desired urban center or hydrologically connected with it should be considered together. These areas may be vast and can only be handled digitally in a GIS platform.

Integration of vegetative cover in urbanization

The increasing level of urban air pollution has been highlighted. Also, the risk of bare soils due to the scraping of top soil by earth moving machines in urban areas is known to increase the chances of desertification and soil erosion as the soil particles become loose and cake as direct sun heat bakes the particles. Paving of much of the urban areas looks beautiful, but it leads to increase in storm water runoff and runoff velocity and force, hence increasing the potentials for flood disasters and erosions.

It is useful to make policies that encourage the sustenance of green areas in urban areas. One or two average-height trees may be required to be planted in all premises depending on the parcel size. This may mean that the sizes of the parcels allotted for instance for residences will have to be increased, but this should only require policy shift.

Grass lawns and trees planted by the side of paved walkways to increase vegetation life in the cities also help to hold together soil particles and shade them from baking under the sun, while acting as filtering agent of the air pollutants.

Furthermore, the urban planners should incorporate in each layout vegetation areas that must be preserved. The choice of where to locate green areas can also be made in watersheds by setting a given buffer from natural running waters, and marking them out both in the layout plans and on the ground.

GIS BASED URBAN DEVELOPMENT MONITORING

Monitoring urban development is as important as the planning. It is not just enough to plan urban areas, it must be seen that the plans are carefully followed. GIS employing state of the art earth observation and geospatial data capture technology, without doubt are today the most robust and effective environmental monitoring systems in the world. The capacity to detect change in the environment even long before on-site inspectors can see them is particularly very useful in urban development monitoring. The GIS has become an inevitable tool to effectively monitor and accurately predict environmental activities.

In the West African cities, environmental policies are often flouted by desperate developers. The governments obviously do not have the capacity to check every new building sprouting up in every corner. Even when local officials find illegal developments they sometimes get compromised by the developers because the authorities up the line have no means of seeing what is going on at those sites. Building houses across the natural drainage channels or constructions within reserved areas of watersheds are regular occurrences that the governments appear incapable to handle.

Various archived satellite data hold records of the state of a region as the urban area was conceived, the pre-urbanization data. The GIS platform also holds the urban concept in its designed form. As developments continue, more satellite data are captured which record the alterations that are beginning to occur due to the new urban development. These new satellite data records can be matched with the urban plan on a GIS platform to determine compliance with the plan and possibly detect deviations in land use from the plan.

It is also important to monitor the effects of even complying with the new plan. For instance high resolution satellite imagery like IKONOS and Quick bird can reveal flooding where the design had not anticipated it. Unauthorized timbre lumbering can be noticed in forest reserves, or solid waste dumping can be discovered when it has only grown a few meters, and stopped.

Land use/land cover change detection is a key feature of the geographic information system (GIS). When a satellite image or digital area photograph is superimposed over an earlier image, GIS can reveal spots where changes have occurred. The capacity of detecting very small changes is fully dependent on the resolution of the satellite images in use. Every day the resolution of satellite images available in the market are improving. Satellite Imaging systems such as Quickbird, Ikonos Geoeye, WorldView and many others are offering improvements in the already high spatial and temporal f resolution images . For instance new constructions in a reserved area can be discovered in 2 days of the project taking off if the area is monitored by GIS using constantly uploaded WorldView images. Consider the following facts on worldview taken from Mapmart web site:

WorldView-1 Imagery features a high capacity, panchromatic imaging system at half-meter resolution imagery. Operating at an altitude of 496 kilometers WorldView1 has an average revisit time of 1.7 days and is capable of collecting up to 290,000 square miles per day of half-meter imagery. Launched September 18, 2007, WorldView-I has amassed an amazing library of highly-accurate and detailed black and white imagery.

This satellite is also equipped with state-of-the-art geolocation accuracy capabilities and exhibits stunning agility with rapid targeting and efficient in-track stereo collection. This makes for a quick collection of highly-detailed imagery suitable for mapping, feature extraction and orthorectification ground control.

WorldView-1 - 0.5 meter Panchromatic Imagery, Resolution: 0.5 meter Band: Panchromatic only Projection: UTM, Lat/Long, State Plane, Datum: NAD 27, 83, WG 84

Format: Geo TIFF, JPEG, NTIF, Dates: 2007 - presentLAND INFO Worldwide Mapping, LLC (2010) published the following on the WorldView-2 satellite imaging system: WorldView-2, the world's newest high-resolution commercial color imaging satellite, was launched on October 8, 2009 from Vandenburg Air Force Base in California. WorldView-2 is the first high-resolution satellite with 8-Multispectral imaging bands. WorldView-2

Table 1. Technical features of Geoeye and Ikonos.

Satellite feature	Geoeye-1	Ikonos
Resolution	0.50 m	1 m
Blue	450-510 nm	445-516 nm
Green	510-580 nm	505-595 nm
Red	655-690 nm	632-698 nm
Near IR	780-920 nm	757-853 nm
Launch date	06-Sep-08	24-Sep-99
Revisit time	3 days at 40° latitude with elevation > 60°	3 days at 40° latitude with elevation > 60°

will simultaneously collect panchromatic imagery at 0.46m and Multispectral imagery at 1.84m. Due to U.S. Government Licensing, the imagery will be made available commercially as 0.5m imagery. WorldView-2 is capable of collecting up to 975,000 square kilometers of imagery per day (376,000 square miles). The technical features of Geoeye and Ikonos satellite images presented in Table 1 are adapted from Geoeye Inc (2011).

The implication of this is that the West African environment can be monitored to stop illegal reclamations, ecosystem destructions, illegal constructions, and scraping of land surfaces, and illegal lumbering etc. Beyond detecting infrastructural changes, it is possible to detect flooding threats by observing the colour of the water bodies on the satellite images. These colours change with depth of the water body. Integrating the water depth data with weather forecasts is inestimable in predicting flooding.

With 0.5 to 1 m spatial resolution it is possible to detect heaps of refuse being dumped in river banks or other illegal places shortly after they start accumulating. Several other harmful practices to the environment can be monitored using GIS. The long 1m lines of seismic cuttings in the mangrove are visible as grid wire lines. Logging is discovered using the change detection tool of GIS over a time. It is obvious that in the modern West African cities monitoring by GIS is inevitable.

FUTURE WEST AFRICAN URBAN CITIES

Given the improving understanding of the complex problems of urbanization, different approaches are being adopted across the globe to deal with these problems. Debates are also on going on best practices. For instance which is better: to grow a city horizontally or vertically? In horizontal expansion the argument is that lower human population density leads to less human impact on the environment. However some like Ginkel (2008), argue that vertical city growth, where housing in skyscrapers encourage high population density, limit humans from accessing and impacting more environmental space.

The future West African urban cities should be planned with an eye on responding positively to the already noted many urban environmental problems. Training of more personnel on the development and use of the Geographic Information System will be critical in handling the urban planning challenges effectively.

Some of the general points that need to be considered are briefly discussed below:

i) encourage urban cities to be more upland: Given the fact of global warming which is resulting in the overflow of the seas, it is common sense to consider that future cities should be sited on higher grounds than the level of the immediate sea shore where the present West African cities are located. This way the surge of the sea into urban habitations will be mitigated.

ii) Ecosystem planning: Planning of future cities in West Africa should take the factors of ecosystem planning seriously. Such practices such as reclamations, scraping of surfaces, bearing of earth surfaces, paving of entire premises, and disregard for the ecosystems should be discouraged.

iii) Provision of adequate infrastructure: Inadequate infrastructure for the use of the urban community quickly gets worn out and over used due to the pressure of the high population. There is the need to provide adequate infrastructure such as roads, water, sanitation, public transport, housing and electricity and domestic energy in a sustainable way. For the future urban centers of West African urban centers will need to be provided and lack of provision of necessary facilities to serve the urban cities.

iv) Effective drainage systems: The importance of effective drainage systems cannot be overemphasized given the tendencies of the urban cities to flood. Adequate and effective drainage systems should be provided to ensure that storm water runoffs are handled without inducing erosion or flooding. The effective drainage systems design should be carried out taking full consideration of the geomorphology of the area involved and the runoff potentials. For the urban area, the 100- year storm event should be used in calculating runoff.

v) Guided city growth: Urban centre managers must admit from start that the city however planned will outgrow what is provided for initially. Adequate arrangements must be made to manage the growth of the city in such a way that future growths are provided for. It is also to be considered to grow other urban centers so that no one urban area will be under too much growth pressure at any time.

vi) Collaboration among West African governments: West African Urban Planners should group land masses in regions. Normally, regions are quite large, and may incorporate multiple urban settlements, suburbs, rural areas and reserves. For ecosystem planning there is need to integrate some level of general planning by the governments of West Africa. This can be achieved through the Economic Community of West Africa (ECOWAS) or her agency. At this level the aim is to provide policies to ensure that naturally linked up ecosystems are still linked up in the resulting urban plans and that decisions on what should be preserved are taken with the entire region in view.

For optimal use and conservation of the urban environment, planners of urban areas should consider the morphology of the different areas and employ the criteria of environmental homogeneity in grouping areas under regions across a country. This will ensure that measures taken in a region to preserve the environment will have the same positive effect across. It will also provide units for gauging the effect of planning options and decisions. In the developing world, many urban areas are planned in layouts in a strip-wise manner. The detailed planning efforts of providing infrastructures, placing roads, gutters and storm water catchment basins concentrate on the strip of land in consideration. A regional level planning often called master- plan only locates the major roads and railways that will run across the entire region and major features that may serve the region such as military barracks, tertiary healthcare facilities, government centers and airports. Generally, the only scientific criteria considered in choosing the region to locate an urban area, is the extent of land required.

CONCLUSION

Urbanization is the way forward and it will continue as people clog together to offer services to earn their living. Humans cannot however continue to destroy the environment. That should be in the past now. The Geographic Information System offers the 21st century urban areas of West Africa the opportunity to optimize the use and preserve the environment.

Policy makers and urban area monitoring bodies should no longer feel helpless in the face of the rapid growth of urban centers. With GIS, hundreds of squares of kilometers can employ globally attested best practices for satisfactorily planning and monitoring the West African urban environment.

REFERENCES

Aderogba KA (2012). Qualitative Studies of Recent Floods and Sustainable Growth and Development of Cities and Towns in Nigeria. Int. J. Acad. Res. Econ. Manage. Sci. 1(3) ISSN: 2226-3624 http://www.hrmars.com/admin/pics/968.pdf Amphibia Web (2010). Information on amphibian biology and conservation, Berkeley, California. http://amphibiaweb,org/decline/HabFrag.html

Atedhor GO, Odjugo PAO, Uriri AE (2011). Changing Rainfall and Anthropogenic-Induced Flooding: Impacts and Adaptation Strategies in Benin City, Nigeria. J. Geogr. Reg. Plan. 4(1):42-52.

Chukwuocha AC (2012), GIS Based Drainage Network design for Owerri, South East Nigeria", A Ph.D Dissertation, Nnamdi Azikiwe University Awka, Nigeria.

Dieye A (2010), Overview of flood situation in W est Africa in 2010: the role of spatial information management, United Nations Office for the Coordination of Humanitarian Affairs, Regional Office for W est and Central Africa (UNOCHA/ROW CA) Federal Ministry of W ater Resources and Rural Development [Nigeria] (Undated), Sustainable W atershed Management.

Geoeye Inc. (2011). Geoeye and Ikonos Feature comparisonhttp://www.geoeye.com/CorpSite/products-and-services/imagery- sources/Comparison.aspx.

Ginkel HV (2008), Terms, Images and Realities: The future of Urbanization. United Nations University. Hosted at www.merit.unu.edu/archive/docs/.../200905_20090414_vanginkel.pdf

Gironas J, Andrieu H, Roesner L (2007). Alterations to natural catchments due to urbanization, a morphologic approach. http://documents.irevues.inist.fr/bitstream/handle/2042/25294/119_027gironas.pdf?sequence=1.

Hanski I (1999) Metapopulation Ecology. Oxford University Press, Oxford.http://www.un-spider.org/book/5088/4c-challenge-communication-coordination-cooperation-capacity-development

Integrated Regional Information Networks, IRIN, (2009), Burkina Faso: Coping with Urban Flood-Displaced" http://www.irinnews.org/report/87206/burkina-faso-coping-with-urban-flood-displaced.

Karley NK (2009). Flooding and Physical Planning in Urban Areas in W est Africa: Situational Analysis of Accra, Ghana, Theoretical and Empirical Researches in Urban Management Number 4(13) / November 2009, http://um.ase.ro/no13/2.pdf.

Koenig D (2009). The Challenges of Urban Growth in W est Africa: The Case of Dakar, Senegal, Center for African Studies Area Studies, University of Mumbai Vidyanagari, Santacruz (E) Mumbai www.mu.ac.in/arts/social_science/african studies doloreswp .pdf.

LAND INFO W orldwide Mapping, LLC (2010). W orldView-2 High-Resolution Satellite Imagery http://www.landinfo.com/W orldView2.htm

Mapmart (undated), WorldView-1 Imagery, http://www.mapmart.com/Products/SatelliteImagery/W orldView1.aspx?_kk=worldview%20imagery&_kt=4a3df3e7-e137-4887-9874- bf84fadfb8e9&gclid=COyB582ao6oCFQJO4QodrFZwXg

Odufuwa BO, Adedeji OH, Oladesu JO, Bongwa A (2012). Floods of Fury in Nigerian Cities, Journal of Sustainable Development. 5(7) http://www.ccsenet.org/journal/index.php/jsd/article/view/14435.

Tang Z, Engel BA, Lim KJ, Pijanowski BC, Harbor J (2005). Minimizing The Impact of Urbanization on Long Term Runoff (Paper No. 040103). J. Am. Water Resour. Association.https://engineering.purdue.edu/ABE/People/.../minimizing.pdf - United States.

Williams RA (2000). Environmental Planning For Sustainable Urban Development. Proceedings of The Caribbean W ater and W astewater Association 9th Annual Conference and Exhibition at Chaguaramas, Trinidad, October 2000 http://www.bvsde.paho.org/bvsaidis/cwwa9/will.pdf.

Sublethal effects of cadmium, manganese, lead, zinc and iron on the plasma electrolytes regulation of mice, *Mus Musculus*

Osuala, Fidelia Ijeoma[1], Otitoloju, Adebayo Akeem[1] and Igwo-Ezikpe, M. N.[2]

[1]Faculty of Science, Department of Zoology, University of Lagos, P.M.B. 101017, Lagos, Lagos State, Nigeria.
[2]Faculty of Basic Medical sciences, Department of Biochemistry, College of Medicine, University of Lagos, P.M.B. 12003, Lagos, Lagos State, Nigeria.

The toxicological evaluations of cadmium, iron, manganese, lead and zinc were carried out against albino mice model, *Mus musculus.* On the basis of 96 hrLC50 value, cadmium (0.47 mM) was found to be the most toxic followed by zinc (2.40 mM), lead (2.42 mM), iron (4.25 mM) and manganese (5.70 mM) was least toxic. This study also evaluated the sublethal effects of cadmium, manganese, lead, zinc and iron in plasma samples utilising plasma electrolyte parameters as a biomarker using an albino mice model, *M. musculus.* Mice were subjected to sublethal concentrations of the selected heavy metals (1/10th of 96 hrLC_{50}). Blood plasma was collected after 7, 14, 21 and 28 days in long term experiment. Sodium (Na^+), potassium (k^+), chloride (Cl^-) ions bicarbonates (HCO_3^-), calcium (Ca^{2+}) and phosphates ($P0_4^{3-}$). Studies on the effect of heavy metals on plasma electrolytes revealed that Pb caused elevated level of sodium (Na^+), while Cd induced significant ($P < 0.05$) increase in potassium (k^+) and chloride (Cl^-) ions in treated mice groups. In addition, bicarbonates (HCO_3^-) and phosphates ($P0_4^{3-}$) levels increased significantly ($P < 0.05$) in treated mice exposed to Mn and Fe, respectively. The outcome of this study implied that heavy metals have toxic effects and plasma electrolyte is a useful tool for early detection and diagnosis of heavy metals pollution in the mammalian model.

Key words: Electrolyte, blood plasma, heavy metals, *Mus musculus.*

INTRODUCTION

The exploration and exploitation of natural resources using modern technology and the exponential growth of population have inadvertently resulted in the release of varied types and amounts of industrial wastes into the environment. These industrial wastes are complex admixtures of several classes of pollutant such as hydrocarbons and heavy metals (Oyewo and Don-Pedro, 2002). These have contributed immensely to the heavy metal load in the environment (Abdul, 2011; Chin et al., 2012). Heavy metals are metals with densities above 3.5 - 5 g/cm^3 and atomic number greater than 20 (Duffus, 2002). They are toxic at relatively low concentrations and persist in the environment long after the source of emission has been removed. They could therefore be classified as important sources of pollution (DeVagi and Arfiziah, 2009).

Heavy metals also bioaccumulate in one or several compartments across food webs as shown by several scientific observations (Chukwu, 1991; Otitoloju and Don-Pedro, 2002, 2004). Heavy metals are grouped into essential and non essential. The essential elements play important roles as prosthetic groups in enzymes and key metabolic activities in living organisms, for example, iron (Fe), copper (Cu), manganese (Mn), cobalt (Co), vanadium (V), molybdenum (Mo) and zinc (Zn). The non-essential metals, such as arsenic (As), mercury (Hg), cadmium (Cd) and lead (Pb) are not needed in the physiological activities of living organisms hence they are usually toxic at relatively low concentrations (Falusi and Olanipekun, 2007; Raymond and Felix, 2011).

Human activities alter the natural geological and biological redistribution of heavy metals, via mining and other industrial processes which alter the chemical form of heavy metals released into the environment (Adriano, 2001). Presently, the amount of heavy metal exposure is hundred times higher (Howard, 2002) than in the past thus living forms have become a "warehouse" of heavy metals. This has led to a lot of concern about heavy metal pollution especially in industrialized and developing countries such as America, Asia Europe and African. Heavy metals have been found in air, water, land (soil/sediment) and all living organisms, including human samples (Öztürk et al., 2009). Thus living things are inadvertently exposed to different concentrations of heavy metals which accumulate in tissues, as such can be of public health concern. In addition, most of the information concerning the health effects of heavy metals largely stems from studies conducted on populations with relatively high exposure usually to individual metals in industry or in heavily polluted environments (Duruibe et al., 2007). Few studies have addressed the possible effects of chronic environmental exposure especially at short time intervals to these heavy metals. Heavy metals are discharged into the environment, either as constituents of industrial, domestic or agricultural wastes where they come in contact with air, water, soil and living forms.

Dermal absorption (skin), inhalation (lungs) and ingestion (mouth) are the route of entry into living forms which become detrimental to human health when they overwhelm the system. Conditions such as liver damage, cancer and other serious health conditions have been associated with metal exposure (Kan et al., 2011; Sham et al., 2011; Umme et al., 2011). Classical cases of heavy metal toxicities include *Minamata* caused by mercury, *Itai - Itai* caused by cadmium which has been known to contribute to any imaginable illness. For example, calcium when replaced by lead (Pb) in the bones can contribute to weakened bones and osteoporosis. Likewise, zinc (Zn) when displaced by cadmium (Cd) in the arteries cause inflammation and hardening of the arteries. Iron (Fe) that replaces zinc (Zn) and other minerals in the pancreas, adrenals and elsewhere can contribute to impaired blood sugar tolerance and diabetes. The available information on metal toxicity, as it relates to human beings, is derived mostly from health surveys among workers engaged in mining and processing especially in developing countries (Wang et al., 2001). This is a thing of concern especially in terms of biomonitoring where most of these effects become full blown before their detection in the biological system. However, heavy metals exert their toxic effect by generating reactive oxygen species (ROS) such as $O^{\cdot}_{2}{}^{-}$, H_2O_2, and $^{\cdot}OH$ causing oxidative stress. This could cause an imbalance of blood electrolyte, by administration of low initial levels of heavy metals thus illustrates its critical role in heavy metal toxicology and justifies its use in this study as a biomarker of heavy metal toxicity.

Ions of body fluids (electrolytes) have various functions such as contributing a majority of the osmotically active particles and providing buffer systems for the regulation of acid-base balance.

Electrolyte balance of ions (sodium, potassium, or calcium) in the blood is critical in regulating oxygen delivery to fluid imbalance within the cells. Electrolyte imbalance could cause cardiac arrhythmias, and complicate attempted resuscitation and postresuscitation care. Thus, an early warning signal of monitoring underlying electrolyte imbalances can prevent abnormalities such as cardiac arrest. Studies have incorporated electrolyte imbalance as a bioindicator of pollution and it is useful in identifying target organs of toxicity as well as the general health status of animals (Shaista et al., 2010; Chezhian et al., 2011; Mehmet et al., 2011). A lot of research has been done on the effects of heavy metal on electrolytes in fishes (Fernandes et al., 2007; Prem et al., 2008) and in humans especially in relation to disease and drugs activity (Oyewole and Malomo, 2009).

Despite the extensive study of heavy metals all over the world, there is paucity of data on some heavy metals toxicity such as Cd, Fe, Mn, Pb and Zn in relation to plasma electrolyte. Some of these selected heavy metals exhibit varying degrees of toxicity at different exposure periods and dose. There is therefore need to assess the impact of these metals that humans are inadvertedly exposed.

It was thus suggested that plasma ions levels may be employed for quantifying toxic effects of metals. Therefore, measurement of serum/plasma biochemical parameters in response to metal exposures can be especially useful to help identify target organs of toxicity as well as the general health status of animals. This study therefore aims at investigating and determining the effects of the selected heavy metals on plasma electrolyte using mice as a mammalian model.

MATERIALS AND METHODS

A total of seventy two male albino mice (*M. musculus*) of similar sizes (19 - 24 g body weight and 10 - 15 weeks old) which served as test animals were purchased from the animal house in Nigerian Institute of Medical Research (NIMR) Yaba, Lagos, Nigeria. Mice were kept in ventilated plastic cages (20 x 12 x 9 cm) with wood shavings under conventional conditions of natural light-dark cycle in farm house located in NIMR. The room temperature was maintained at 30 ± 2°C (28 - 32°C) and the animals had free access to drinking fluid and a standard rodent laboratory chow purchased from Ladokun feeds Ltd, Ibadan. Acclimatization was done for 14 days during which they were handled with animal care in accordance with the Institute for Laboratory Animal Research (ILAR) guidelines prior to heavy metals treatment. Heavy metals investigated in this work were obtained as metallic salts from Fisons Laboratory Reagents, Analar grades in Nigeria of the following types. Zinc as $Zncl_2.3H_2O$, cadmium as $CdSO_4.8H_20$, lead as $PbSO_4$, manganese as $MnSO_4$ and ferric as $FeSO_4$ we used.

Preparation of test media for acute toxicity test

A predetermined amount of Analar grade salt of each heavy metal compound was weighed (using an Oertling, 30TD top loading balance) and made up to a given volume to obtain a stock solution of known strength. The resulting solution was serially diluted to obtain solutions of required concentrations. The volume of the treated media (toxicant and distilled water) that was administered to the mice varied based on the results of preliminary experiments, to determine the volume of water that can be consumed at once without choking the animal via cannular feeding. In range finding experiments, the mice were exposed to a wide range of concentrations of each test compound to obtain an effective range of activity, initially relying on trial and error techniques.

Differential acute toxicity of heavy metals against mice

Active mice were treated to salts type of the heavy metals by cannular feeding. For series of bioassays, three animals in duplicates were exposed per treatment including untreated control. In these substantive bioassays after range finding preliminary trials, test animals were exposed to graded series of concentrations of each heavy metal compound in milligrams per body weight as follows: Cadmium as $CdSO_4.8H_20$ against mice at 0.28, 0.43, 0.57, 0.71 and 0.85 mM; Lead as $PbSO_4$ against mice at 0.66, 1.98, 3.30, 5.28 and 6.60 mM; Zinc as $Zncl_2.3H_2O$ against mice at 5.29, 6.35, 8.47, 10.58 and 12.70 mM; Manganese as $MnSO_4$ against mice at 1.32, 3.97, 6.62, 10.60 and13.25 mM; Ferric as $FeSO_4$ against mice at 1.64, 3.29, 4.93, 6.58, 8.22 and 9.87 mM

Assessment of quantal response

Mice were taken to be dead if they showed no movement of the body and appendage. Mortality assessments were carried out at definite time interval specified under appropriate bioassays for example 24 h over a 96 h period for the acute toxicity test

Heavy metals administration for chronic toxicity test

Mice were divided into six experimental groups (n = 12 per group) for the sublethal experiment and a control group (n = 3) was also included. They were treated to 1/10 of LC_{50} values of salts of the heavy metals as follows Pb = 0.24 mM, Fe = 0.45 mM, Cd = 0.05 mM, Zn = 0.24 mM and Mn = 0.57 mM (derived from experiments carried out in acute toxicity test) while control group was administered distilled water only. The metallic salts of the corresponding heavy metals of 1000 mg each of the salt was weighed out and diluted in 20 ml of distilled water. Solution of 0.1 ml was administered through cannular method to the mice over the experimental period. The control mice were fed the same volume of distilled water. Prior to sacrificing, 3 mice from each group were separated and fasted for 24 h before dissection. This procedure was maintained every 7 days for analysis under investigation.

Collection and preparation of plasma samples from mice for electrolyte

Mice were sacrificed at the end of each experimental period and their blood was collected into EDTA bottle immediately. Blood was then centrifuged at 4,000 g for 10 min using bench top centrifuge (MSE Minor, England) to remove red blood cells and recover plasma. Plasma samples were separated and collected using dry Pasteur pipette, labelled and stored in the refrigerator at -20°C for analyses. All samples were analysed within 24 h of sample collection.

Measurement of biochemical parameters

Plasma sodium, potassium and bicarbonate, calcium ions were measured by flame photometer as described by Tietz et al. (1994). Chloride ion estimation was measured by precipation titration by the method of Mohr and Volhard as described by Harris (2003). Inorganic phosphate was determined using the direct method of Fiske and Subbarow (1925).

Statistical analysis

Lethal concentration (LC_{50}) was calculated by log/probit regression line method (Finney, 1971). Data was analyzed with oneway analysis of variance (ANOVA). Differences at $P < 0.05$ were considered significant. This was used to compare several treatment means in appropriately designed experiments. Further analysis was carried out only when there is a significant difference at the 5% ($P < 0.05$) level of significance (taken as minimum requirement) based on Duncan multiple range test at 0.05 levels of significance using SPSS 10.0 computer software package (SPSS Inc., Chicago, U.S.A).

RESULTS

Acute toxicity of heavy metals on mice (*Mus musculus*)

On the basis of 96 $hrLC_{50}$ values, cadmium (Cd) with a 96 $hrLC_{50}$ value of 0.47 mM was the most toxic metal tested against *M. musculus* (mice) followed by Zn, Pb, Fe and Mn (96 $hrLC_{50}$ = 5.70 mM) as least toxic in a descending order of toxicity (Table 1) and there was an overlap in 95% confidence limit of 96 $hrLC_{50}$ value. There were similar magnitudes of toxicity for all the tested metals with overlaps in 95% confidence limit of 96 $hrLC_{50}$ value. Zinc (Zn) was also found to be significantly more toxic than Pb and each of the other metallic compounds. Manganese (Mn) was least toxic against the tested animal with no overlaps in 95% confidence limit of 96 $hrLC_{50}$ value. Computed toxicity factors (96 $hrLC_{50}$ ratios) showed that Cd was about 12.0x, 10.0x, 5.3x and 5.0x more toxic than Mn, Fe, Pb and Zn, respectively when

Table 1. Relative acute toxicity of selected heavy metals.

	Time (h)	LC_{50} (95% CL) mM kg^{-1}	LC_{95} (95% CL) mM kg^{-1}	Regression equation	Slope ± SE	DF	TF
Cadmium	24	1.07 (0.54 - 1.19)	7.59 (3.79 - 8.93)	Y= 0.05+ 1.94x	1.94 ± 1.20	3	1.00
	48	0.61 (0.45 - 0.89)	1.82 (1.10 - 3.74)	Y= -2.95+ 3.44x	3.44 ± 1.21	3	1.00
	72	0.54 (0.37 - 1.12)	1.60 (1.01 - 3.55)	Y= -2.89+ 3.49x	3.49 ± 1.19	3	1.00
	96	0.47 (0.37 - 0.56)	0.94 (0.74 - 1.68)	Y= -7.22+ 5.55x	5.55 ± 1.39	3	1.00
Iron	24	8.62 (5.57 - 17.50)	71.90 (21.75 - 148.91)	Y= -0.57+ 1.79x	1.79 ± 0.73	4	8.06
	48	6.13 (4.20 - 10.67)	32.34 (15.29 - 67.64)	Y= -1.76+ 2.28x	2.28 ± 0.74	4	10.05
	72	5.07 (3.58 - 6.96)	19.07 (11.54 - 87.35)	Y= -3.25+ 2.86x	2.86 ± 0.78	4	9.39
	96	4.52 (2.66 - 6.73)	25.46 (12.82 - 102.04)	Y= -1.21+ 2.19x	2.19 ± 0.70	4	9.62
Manganese	24	9.92 (6.51 - 18.46)	66.94 (29.20 - 146.01)	Y= -1.30+ 1.98x	1.98 ± 0.56	4	9.27
	48	6.86 (4.19 - 13.03)	48.50 (22.94 - 105.84)	Y= -0.84+ 1.94x	1.94 ± 0.52	4	11.25
	72	6.06 (3.54 - 9.52)	43.43 (20.97 - 98.19)	Y= -0.69+ 1.92x	1.92 ± 0.52	4	11.22
	96	5.70 (3.25 - 8.88)	41.00 (20.05 - 94.19)	Y= -0.64+ 1.92x	1.92 ± 0.52	4	12.13
Lead	24	11.63 (5.38 - 15.08)	34.78 (20.98 - 43.09)	Y= -7.27+ 3.46x	3.46 ± 2.34	2	10.87
	48	9.77 (3.24 - 14.22)	32.06 (16.62 - 43.79)	Y= -6.06+ 3.19x	3.19 ± 2.25	2	16.02
	72	13.02 (3.60 - 28.91)	37.43 (25.91- 52.11)	Y= -5.03 + 3.09X	3.09 ± 2.25	2	24.11
	96	2.42 (17.93 -43.09)	34.78 (20.98 - 43.09)	Y= -7.27+ 3.46x	3.15± 2.06	2	5.15
Zinc	24	3.83 (2.69 - 5.77)	14.41 (8.16 - 110.39)	Y= -2.90+ 2.86x	2.86 ± 0.86	3	3.58
	48	3.32 (2.69 - 5.77)	20.99 (6.85 - 110.39)	Y= -2.16+ 2.86x	2.81 ± 0.86	3	5.44
	72	2.84 (2.69 - 5.77)	34.14 (14.74 - 110.39)	Y= -2.09+ 2.86x	2.26 ± 0.86	3	5.26
	96	2.40 (1.29- 3.32)	9.41 (5.81 - 49.35)	Y= -2.11+ 2.77x	2.77 ± 0.84	3	5.11

C.L. = 95% confidence limit; S.E = standard error; D.F = degree of freedom.

$$\text{T.F (toxicity factor)} = \frac{\text{24/48/72/96 hrLC}_{50}\text{ value of other metals}}{\text{24/48/72/96 hrLC}_{50}\text{ value of the most toxic (Cd) metals}}$$

tested against mice (Table 1).

Toxicity indices including LC_{50} values derived from 24, 48, or 72 h mortality response data showed similar trends of relative potency as was the case with 96 h data described below (Table 1).

The results (Table 2) showed that elevated levels of plasma electrolytes were induced significantly (P < 0.05) by the test heavy metals in varied levels following the heavy metal administration from 7, 14, 21 and 28 day in comparism with the control.

Cadmium (Cd) induced a consistent increase in Na^+ which was significantly (P < 0.05) different from the control group on the 28th day in the treated group. The lowest and highest level of the Na^+ was recorded on the 7th and 14th (Table 2) in the treated group. Interestingly, K^+ significantly (P < 0.05) increased from the level in control group to that on the 14th day in the treated group. There were significant (P < 0.05) differences in the level of Cl^-, HCO_3^- and PO_4^3 which consistently increased from day 0 to the 28th day.

Manganese (Mn) induced a consistent increase in the sodium (Na^+) which was significantly (P < 0.05) different from the levels in the control to that on the 21st day but decreased on the 28th day while K^+ significantly (P <0.05) increased from day 0 to 21st day but decreased on the 28th day. The HCO_3^- increased consistently on the 28th day while Cl^- increased from day 0 to the 21st day but also decreased on the 28th day. Ca^{2+} significantly (P < 0.05) increased from day 0 to the 14th day but decreased on the 28th day while PO_4^{3-} consistently increased through out the exposure period.

Zinc (Zn) induced a consistent increase in plasma Na^+, K^+which was significantly different (P < 0.05) from 0 - 21st day but decreased on the 28th day but HCO_3^- and Cl^- increased consistently from 0 - 14th day but decreased on the 28th day. Ca^{2+} increased from 0 - 21st day but decreased on the 28th day while PO_4^{3-} consistently and significantly (P < 0.05) increased from 0 - 28th day as in the case with Mn. Lead (Pb) induced a consistent increase in plasma sodium K^+ and (Na^+) which were signi-

Table 2. Plasma electrolyte parameters in mice treated with cadmium, manganese, zinc, lead and iron for 7, 14, 21 and 28 days.

Parameter (mmol / L)	Control	Cadmium (Cd)	Manganese (Mn)	Zinc (Zn)	Lead (Pb)	Iron (Fe)
7 days						
Na^+	23.67±0.27^a	41.33±0.33^c	40.44±0.51^c	36.67±0.33^b	41.00±0.88^c	36.59±0.19^b
K^+	1.17±0.00^a	1.79±0.02^c	1.49±0.02^b	1.99±0.05^d	1.79±0.04^c	1.79±0.02^c
HCO_3^-	4.09±0.17^b	6.68±0.02^e	4.50±0.17^c	5.44±0.19^d	7.67±0.00^f	1.79±0.17^a
Cl^-	26.79±0.15^a	33.52±0.17^d	37.28±0.59^e	29.56±0.19^b	32.78±0.51^c	39.85±0.17^f
Ca^{2+}	0.53±0.00^a	0.60±0.00^b	0.71±0.00^d	0.73±0.01^e	0.69±0.01^c	0.78±0.06^f
PO_4^{3-}	0.42±0.01^a	0.55±0.01^c	0.45±0.00^b	0.46±0.00^b	0.63±0.03^d	0.67±0.00^e
14 days						
Na^+	23.67±0.27^a	74.67±0.29^e	74.00±0.50^e	72.67±0.76^d	71.00±0.87^c	69.33±0.29^b
K^+	1.17±0.00^a	3.05±0.05^b	3.24±0.06^c	3.03±0.03^b	3.05±0.05^b	3.07±0.10^b
HCO_3^-	4.09±0.17^a	8.48±0.08^b	8.58±0.52^b	13.65±0.56^d	8.75±0.22^b	10.33±0.58^c
Cl^-	26.79±0.15^a	52.50±2.50^b	54.67±4.07^{bc}	60.00±6.50^{cd}	62.17±1.04^d	51.17±1.04^b
Ca^{2+}	0.53±0.00^a	1.12±0.01^{bc}	1.18±0.01^{cd}	1.18±0.00^{cd}	1.23±0.10^d	1.07±0.08^b
PO_4^{3-}	0.42±0.01^a	0.80±0.06^b	0.82±0.05^b	0.81±0.06^b	1.25±0.00^c	1.23±0.00^c
21 days						
Na^+	23.67±0.27^a	65.50±0.50^b	75.67±0.29^{de}	72.67±0.29^d	71.50±0.50^c	75.17±0.58^e
K^+	1.17±0.00^a	3.30±0.05^b	3.98±0.06^c	4.27±0.03^d	4.75±0.05^f	4.40±0.05^e
HCO_3^-	4.09±0.17^a	8.50±0.50^b	9.50±0.87^c	11.33±0.29^d	8.43±0.12^b	8.80±0.26^{bc}
Cl^-	26.79±0.15^a	61.67±1.44^d	57.28±0.58^c	59.50±0.50^c	56.50±1.50^c	53.83±0.29^b
Ca^{2+}	0.53±0.00^a	1.13±0.02^c	1.12±0.00^{bc}	1.18±0.02^{cd}	1.23±0.05^d	1.07±01^b
PO_4^{3-}	0.42±0.01^a	0.80±0.00^b	0.72±0.01^b	0.70±0.00^b	1.25±0.01^c	1.23±0.03^c
28 days						
Na^+	23.67±0.27^a	68.83±0.29^{cd}	69.33±5.11^{cd}	68.00±0.50^c	72.33±0.58^d	35.50±0.50^b
K^+	1.17±0.00^a	3.42±0.03^e	2.58±0.08^c	2.70±0.01^d	2.62±0.08^{cd}	2.18±0.03^b
HCO_3^-	4.09±0.17^a	8.83±0.29^b	10.83±0.29^d	9.33±0.29^c	9.45±0.05^c	9.48±0.03^c
Cl^-	26.79±0.15^a	62.33±0.76^d	53.50±0.50^b	52.67±0.29^b	52.83±0.29^b	55.83±0.76^c
Ca^{2+}	0.53±0.00^a	1.15±0.00^d	1.11±0.01^c	0.97±0.00^b	0.99±0.01^b	1.26±0.01^e
PO_4^{3-}	0.42±0.01^a	0.86±0.01^b	0.86±0.04^b	0.86±0.00^b	0.85±0.01^b	0.88±0.03^b

Values are mean ± SD, n = 3. Values with different alphabetical superscripts along a row are significantly different at $P < 0.05$.

significantly ($P < 0.05$) different from that in the control to the 21st and 28th day while HCO_3^- increased consistently from day 0 - 28th^th^. Cl^- increased from 0 - 14th day but decreased on the 28th day while Ca^{2+} and PO_4^3 significantly ($P < 0.05$) increased from day 0 - 21st day but decreased on the 28th day.

Iron (Fe) induced a consistent increase in plasma sodium (Na^+) and K^+ which were significantly ($P < 0.05$) different from the control to the 21st day but decreased on the 28th day. HCO_3^- increased consistently from day 0 - 14th day but decreased ($P < 0.05$) significantly on the 28th day while Cl^- and Ca^{2+}increased significantly ($P < 0.05$) from 0 - 28th day while PO_4^{3-} consistently increased from 0 - 21st day but decreased on the 28th day. At the end of the 28th day, the descending orders of plasma electrolyte with respect to the selected heavy metals were as follows: Sodium (Na^+): Pb > Mn > Cd > Zn > Fe; Potasium (k^+): Cd > Zn > Pb > Mn > Fe; Bicarbonate (HCO_3^-): Mn > Fe > Pb > Zn > Cd; Chloride ion (Cl): Cd > Fe > Mn > Pb > Zn; Calcium (Ca^{2+}): Fe > Cd > Mn > Pb > Zn; Phosphate ($P0_4^{3-}$): Fe > Cd, Mn, Zn > Pb

DISCUSSION

In the single action toxicity studies which showed that Cd was the most toxic of the heavy metals tested followed by Zn, Pb, Fe and Mn, in descending order of toxicity against albino mice, *M. musculus* could be explained based on the fact that the relatively high toxicity of Cd could have been as a result of its high electronegativity and the resultant affinity to sulphydryl (SH) groups which lead to denaturing of many enzyme systems with the SH group. The lowest toxicity which was established as manganese (Mn) could be due to the free inorganic Mn^{2+} which is usually dominant in manganese solutions but which is not lipid soluble hence their transference across membranes would be inhibited. The implications of the low toxicity of Mn if coupled to low excretability in exposed mice is that the mammal could accumulate high concentrations without ill effects or lethal action and thus become a potential danger to the systems in the event of an amplification along the food chain.

The established induction of a significant increase of sodium (Na^+) ions by Pb followed by Mn, Cd, Zn and Fe in descending order is an indication that Pb has an activation action on the monovalent cation transport in the plasma of the mice. Sodium is the major positive ion (cation) in fluid outside the cells. The present finding was supported by Sheikh et al. (2011). They observed increase in Na^+ and Cl^- in Wistar rats exposed to 14 and 28 days treatment with mercuric chloride. The elevation in plasma Na^+ might not be beneficial to exposed living systems for example humans because it has been demonstrated that it induces hypernatrema. This could be used as a biomarker of Pb exposure. The increase in plasma K^+ level (hyperkalemia) by Cd observed in the exposed group might occur due to erythrocyte destruction caused by the Cd exposures which probably lead to the release of K^+. Hyperkalemia has been reported to arise in condition characterized by excess destruction of cells with redistribution of K^+ from the intracellular to the extracellular compartment as in massive haemolysis (Guyton and Hall, 2000). Furthermore, the significant increase in plasma K^+ level may also be attributed to cation exchange of H^+ and K^+ between intracellular and extracellular spaces or erythrocyte swelling. This probably led to hemolysis and contamination of plasma with extracellular K^+ or a reduction in the extracellular spaces and/or an increase in the efflux of plasma K^+ from intracellular compartment. Potassium is the dominant intracellular cation and plasma ionic dilution would favour efflux into extracellular fluid. On the other hand, it is also possible that the alterations observed on the plasma ions may simply result from non-specific stress of the mice.

Generally, iron induction of the higest level of calcium ions (Ca^{2+}) and phosphates ions (PO_4^{3-}) initiated the entrance into the plasma. Ca^{2+} influx activates a number of enzymes, including phospholipases, endonucleases and proteases such as calpain. These enzymes go on to damage cell structures such as components of the cytoskeleton, membrane and DNA. Thus, an increase in Ca^{2+} induced by Fe could be damaging to the cells and therefore should be monitored.

Conclusion

The administration of selected heavy metals over the experimental period has established the fact in this study that Cd, Fe, Mn, Pb and Zn have the ability to induce and synergise significant imbalance in plasma electrolyte in mice in a short period of time and therefore can be used as a biomarker of heavy metal pollution as well as an early warning signal of cell deterioration.

ACKNOWLEDGEMENTS

The authors are thankful to Mr Sunday and Mr Duncan, staff of Lagos University teaching hospital for permission to conduct this research in the central laboratory.

REFERENCES

Abdul RA (2011). Policy of sustainable environment: Malaysian Experience. Eur. J. Sci. Res. 48(3):466-492.

Adriano DC (2001). Trace elements in terrestrial environments: biogeochemistry bioavailability and risk of metals. 2nd edition, Springer - Verlag, New- York. pp. 866.

Chezhian A, Kabilan N, Suresh K, Senthamil S (2011). Influence of different calcium levels and low ph of water on the plasma electrolyte regulation of a fresh water teleost fish cyprinus carpio var. communies, (Linnaeus, 1958). Curr. Res. J. Biol. Sci. 3(2):147-154.

Chin WT, Kok HT, Yit TO, Abdul RM, Sharif HS, Soon HT (2012). Carbon nanotubes applications:solar and fuel cells, hydrogen storage, lithium batteries, supercapacitors, nanocomposites, gas, pathogens, dyes, heavy metals and pesticides. Environ. Chem. Sustain. World 1(1):3-46.

Chukwu LD (1991). Studies of heavy metal contamination of water, sediment and decapods crustaceans from river sasa. Ph.D Thesis University of Lagos. pp.164.

Devagi K, Arfiziah A (2009). Accumulation and Depuration of Lead and Chromium Using Nerita Lineate. World Appl. Sci. J. 6(9):1205-1208.

Duffus JH (2002). Heavy metals - a meaningless term? (Iupac Technical Report). Pure Appl. Chem. 7(5):793-807

Duruibe JO, Ogwuegbu MO, Egwurugwu JN (2007). Heavy metal pollution and human biotoxic effects. Int. J. Phys. Sci. 2(5):112-118.

Falusi BA, Olanipekun EO (2007). Bioconcentration factors of heavy metals in tropical crab (carcinus sp) from river Aponwe, Ado-ekiti, Nigeria. J. Appl. Sci. Environ. Manag. 11(4):51-54.

Fernandes C, Fontainhas-Fernandes A, Monteiro SM, Salgado MA (2007). Changes in plasma electrolytes and gill histopathology in wild liza saliens from the esmoriz-paramos coastal lagoon, portugal. Environ. Contam.Toxicol. 79(3):301-305.

Finney DJ (1971). Probit analysis 3rd edition London: Cambridge University Press. p. 80.

Fiske CH, Subbarow Y (1925). Colorimetric determination of phosphorus. J. Biol. Chem. 66(2):375-378.

Guyton AC, Hall JE (2000). A Textbook of Medical Physiology. 10th Edition. W.B. Saunders Co. Philadelphia. pp. 382-401.

Harris DC (2003). Quantitative chemical analysis (6th Ed.). San Francisco:W.H. Freeman. pp. 142-143.

Howard H (2002). Human health and heavy metal exposure. In:The environment and human health. Michael McCally (Ed). pp 1-14.

Kan U, Koichi K, Keiko O, Shin N, Yumiko S, Yasuhiro K, Akihiro K, Yuko T, Motoshi K, Yasuhisa Y (2011). Toxicological aspects of cadmium and occupational health activities to prevent workplace exposure in japan:A Narrative Review. Toxicol. Ind. Health 27(3):225-233.

Mehmet BE, Sibel A, Kubra K (2011). Alteration in the Hematological and Biochemical parameters and Plasma Ion Concentrations of Common Carp, (*Caprinus carpio* L., 1758) After Short Term Exposure to Sublethal Concentrations of Lead. Kafkas Univ. Vet. Fak. Derg. 20(10):30.

Otitoloju AA, Don-Pedro KN (2002). Bioaccumulation of heavy metals

(Zn, Pb, Cu and Cd) by tympanotonus fuscatus var. radula (l) exposed to sublethal concentrations of the test metal compounds in laboratory bioassay. West Afr. J. Appl. Ecol. 3:17-29.

Otitoloju AA, Don-Pedro KN (2004). Integrated laboratory and field assessments of heavy metals accumulation in edible periwinkle, *Tympanotonus fuscatus var. radula (L).* Ecotoxicol. Environ. Saf. 57(3):354-362.

Oyewo EO, Don-Pedro KN (2002). The toxicity ranking of four heavy metals of industrial sources to six resident animals of a tropical estuarine lagoon. Toxicol. Environ. Chem. 83(1 - 4):87-97.

Oyewole OI, Malomo SO (2009). Toxicological assessment of oral administration of some antisickling agents in rats. Afr. J. Biochem. Res. 3(2):024-028.

Öztürk M, Özözen G, Minareci O, Minareci E (2009). Determination of heavy metals in fish, water and sediment of Avsar Dam Lake in Turkey. Iran J. Environ. Health Sci. 62:73-80.

Prem RA, Amudha J, Sasikala K (2008). Effect of heavy metals, mercury and copper on plasma electrolytes in freshwater fish *Cyprinus Carpio Var. Communis*. Asian J. Water Environ. Pollut. 6(4):107-112.

Raymond AW, Felix EO (2011). Heavy metals in contaminated soils:a review of sources, chemistry, risks and best available strategies for remediation. ISRN Ecol. 2011:1-20.

Shaista N, Saeed AN, Yasar S Armit P (2010). Determination of heavy metals in fresh water fish species of the river ravi, pakistan compared to farmed fish varieties. Environ. Monit. Assess. 167(1-4):461-471.

Sham KW, Tasneem GK, Nida FK, Hassan IA, Sumaira K, Akhtar AC, Abdul QS, Ghulam AK, Syeda N (2011). Case-control study of male cancer patients exposed to arsenic-contaminated drinking water and tobacco smoke with relation to non-exposed cancer patients. Hum. Exp. Toxicol. 30(12):2013-22.

Sheikh TJ, Patel BJ, Joshi DV (2011). Electrolyte Alteration in Plasma And Urine After 28days Repeated Oral Dose Toxicity of Mercuric Chloride in Wistar Rat. J. Appl. Pharma. Sci. 1(10):150-153.

Tietz NW, Pruden EL, Siggaard-Andersen O (1994). In: Tietz textbook of clinical chemistry (Burtis C.A. and Ashwood E.R. Ed.) W.B. Saunders company London. pp. 1395-1406.

Umme SA, Jeremy AS, Amanda C, Greg JE (2011). *In vivo* and *in vitro* Assessment of particulate matter toxicology. Environ. Sci. Eng. 4:427-449.

Wang H, Kimberley MO, Schlegelmilch M (2001). Biosolids derived nitrogen mineralization and transformation in forest soils. J. Environ. Qual. 32:1851-1856.

14

Human health risk characterization of lead pollution in contaminated farmlands of Abare village, Zamfara State, Nigeria

N. Abdu and A.A Yusuf

Departments of Soil Science, Faculty of Agriculture, Institute for Agricultural Research, Ahmadu Bello University, Zaria, P.M.B. 1044, Samaru, Kaduna State, Nigeria.

This study was initiated to assess the level of lead (Pb) contamination in farmlands, crop plants and water sources and the health risk in one of the Pb-contaminated villages of Zamfara State. Soil samples were collected from two depths at intervals of 10, 50, 150, 300, 500 and 1000 m from the last house along each of the four cardinal directions (North, South, East and West). Crop plant samples were also collected at each sampling distance while profile pits were dug at 150 m in each direction. Water samples were collected from all available sources in the villages. Contents of total Pb was determined in all samples following *aqua regia* digestion for soil and plant samples, water samples were digested with HNO_3. Health risk assessment was based on lifetime exposure through ingestion and inhalation of soil and dust based on the USEPA risk exposure models. The average concentrations of Pb in farmlands (515 mg kg^{-1}) and profile pits (365 mg kg^{-1}) were lower than its extent in plant materials (1220 mg kg^{-1}). The average chronic daily intake (CDI) for carcinogenic risk in mg kg^{-1} day^{-1} was 1.3 x 10^5 for Pb. Non carcinogenic CDI in mg kg^{-1} day^{-1} averaged 1.7 x 10^8 in adults. For children, it averaged 9.7 x 10^7 mg kg^{-1} day^{-1}. However, the concentrations in all the samples are far beyond acceptable levels with high repercussion for environment and human health hazard. Suitable intervention measures are required to reduce these high concentrations in order to minimize risks associated with contamination.

Key words: Heavy metals, lead, Zamfara, health risk, soil contamination.

INTRODUCTION

Anthropogenic heavy metal contamination is becoming widespread with ubiquitous nature of heavy metals in the environment. Anthropogenic sources of heavy metal pollution include but not limited to mining, iron smelting, fossil burning and municipal and industrial waste disposal. Illegal mining activities in Zamfara, northern Nigeria was reported to have killed hundreds of people mostly children in 2010 (NNP 2011). This was related to lead (Pb) poisoning or high lead concentration in the blood resulting from widespread contamination in the villages and compounds where gold ore are processed. The gold ore contain Pb as an impurity and hence lead was deposited in the villages during processing. Children were mostly affected due to their smaller bodies and behaviours. Children ingest dust when they put their hands in the ground and eat food. Efforts were made by both governmental and non-governmental organizations within and outside the country to curb the situation. Medicins Sans Frontiers (MSF) in collaboration with the Zamfara State Ministry of Health (ZMoH) administered drugs that can be used to detoxify afflicted patients. Similarly, the Zamfara State Ministry of Environment and Solid Minerals

(ZMoE) requested technical assistance from the Blacksmith Institute (BI) of New York and another US firm, TerraGraphics Environmental Engineering (TG) to secure clean environments in the contaminated villages for children returning from treatments. Contaminated soils were excavated and replaced by clean soil to bring down the lead level. This approach effectively reduced lead toxicity in residential compounds and exterior areas where it was applied (BI, 2011).

Farming activities within the surrounding farmlands in the affected villages is still ongoing which can results to plant accumulation of Pb and other associated heavy metals like cadmium and zinc. Abdu et al. (2011a, b) reported Cd and Zn contamination of vegetables and associated health risk as a result of wastewater irrigation in Kano, northern Nigeria. Sorghum, a widely cultivated crop in the region can accumulate heavy metal in high concentration without showing any symptom of toxicity (Prasad and Freitas, 2003). Extreme of concentrations as reported in this study could lead to heavy metal accumulation by crops and eventual translocation from the shoot to the grain without the crop showing any physiological deficiency. This has a tendency of contaminating and posing a health risk to the ecosystem. Moreover, in the rural northern Nigeria, leaves and stalk of sorghum plant are highly consumed by livestock. The accumulated heavy metal will then be transferred to the animal and eventually contaminating the food chain.

Yusuf et al. (2011) reported contamination of farmlands and various water sources with Pb in the affected villages beyond permissible limits. Contamination could occur due to deposit from dust and underground water movement even from landfills that are not well protected. Even at low level, Pb has no biological or physiological essentiality as a nutrient. It affects the developing brain and causes behavioral problems and intelligence quotient (IQ) deficit. Currently, there is no information on the risk assessment of Pb pollution in the lead-contaminated villages. The objectives of the study were to assess lead status of farmlands and various water sources in one of the lead-contaminated villages of Zamfara State and quantify the potential health risk and hazard from Pb contamination in the affected village.

MATERIALS AND METHODS

A systematic sampling method was used to collect soil, water and plant samples in Abare village in Anka Local Government Areas of Zamfara State. Sampling was done in the direction of each of the four cardinal points (North, West, East and South). In each direction, soil and plant samples were collected at 10, 50, 150, 300, 500 and 1000 m from the last house of the village in that direction. The soil samples were taken at 0 - 15 cm and 15 - 30 cm depth with an auger while the plant samples were collected from both cultivated and non-cultivated plants. In the case of crop plants, samples were taken from the part of high nutrient concentration e.g. flag leaf and ear leaf for sorghum and maize, respectively. All the samples were collected in three replicates. In addition to the surface soil samples, soil profile pits were dug to approximately two meters depth at 150 m in each of the four directions. At least four horizons were delineated in each pit and triplicate soil samples were taken from each horizon. Before analysis, the soil samples were air-dried, ground and passed through 2 mm screen while the plant samples were oven-dried at 65°C to constant weight, ground and sieved through 0.5 mm mesh. Water samples were taken from wells, streams and rivers encountered within and outside the village. One milliliter (1 ml) of concentrated HCl was added to each water sample to suppress microbial activities.

Soil pH was determined in $CaCl_2$ solution at a soil: solution ratio of 1:2.5 using a pH meter, cation exchange capacity was determined by leaching with 1 N ammonium acetate solution (Anderson and Ingram, 1993). Organic carbon was determined by dichromate wet oxidation as described by Nelson and Sommers (1986). Particle size analysis was measured using the hydrometer method (Gee and Bauder, 1986).

Total lead concentrations in the soil, plant and water samples were determined in atomic absorption spectrophotometer (AAS) after hot acid digestion (Lim and Jackson, 1986). Data collected were subjected to suitable statistical analysis such as descriptive statistics and analysis of variance (ANOVA) using Statistical Package for Social Scientists (SPSS, version 17.0).

Risk assessment

Carcinogenic and non-carcinogenic risk for ingestion and inhalation of heavy metals were calculated using the following formulas (USEPA, 2005):

Carcinogenic CDI (mg kg^{-1} day^{-1}) = CS x IF x EF/AT

where IF = (IR x ED/BW) Adult + (IR x ED/BW) Child

Non-carcinogenic CDI (mg kg^{-1} day^{-1}) = CS x IN x EF x ED/BW x AT

CDI (Dust inhalation) for carcinogenic and non-carcinogenic risk were calculated using the relation:

$$\text{CDI (mg m}^{-3}\text{)} = \frac{\text{CS x IF x EF (1/PEF + 1/VF)(ET}_{\text{Outdoor}} + \text{(ET}_{\text{Indoor}}\text{DF}_{\text{Outdoor}}\text{)}}{\text{AT}}$$

Definition of the terms and values for individual parameters are indicated in Table 1.

RESULTS AND DISCUSSION

Assessment of lead concentrations in farmlands in Abare

The average concentrations of Cd, Pb and Zn were 17.5, 1266 and 985 mg kg^{-1}, respectively, in soil. The concentrations of the heavy metals were far beyond minimum threshold values worldwide (300 mg kg^{-1} in EU and UK, 150 mg kg^{-1} in USA and 70 mg kg^{-1} in Canada) (Abdu et al., 2011a). Very high concentrations were observed at the surface soil depth indicating superficial enrichment through anthropogenic mining processing. Sampling depth did not have significant effect on lead (Pb) concentration in all the sampling directions. Similarly, a non-significant effect was found on the interaction between sampling distance and depth in all the directions (Table 2). The concentration of lead between the soil horizons within the profile

Table 1. Risk assessment terms, their definition and values used in the study.

Variable	Value	Variable	Value
CS = Concentration of heavy metal in soil		IR = Ingestion rate (Adult)	100 mg/day
IF = Intake factor	-	IR = Ingestion rate (Child)	200 mg/day
EF = Exposure frequency	350 days/year	BW = Body weight (Adult)	60 kg
AT = Averaging time	365 days/year	BW = Body weight (Child)	15 kg
ED = Exposure duration (Adult)	30 years	IN = Inhalation rate (Adult)	3.6 m^3/hour
ED = Exposure duration (Child)	6 years	IN = Inhalation rate (Child)	2.2 m^3/hour
PEF = Particulate emission factor	1.3 x 10^9 m^3/kg	$ET_{Outdoor}$ = Exposure time	0.68
VF = Volatilization factor	11 m^3/kg	ET_{indoor} = Exposure time	0.07
DF = Dilution factor indoor	0.4		

Volatilization and particulate emission factors are dependent on the climate of the region under study and were calculated based on the mean atmospheric humidity of 9.1 g of water per m^3 of air.

Table 2. Effect of soil depth and sampling distance on lead concentration in Abare, Zamfara, Nigeria.

Sampling	North (mg kg^{-1})	South (mg kg^{-1})	East (mg kg^{-1})	West (mg kg^{-1})
Depth (cm)				
0 – 15	748	392	401	684
15 – 30	626	387	370	508
Mean	687	390	386	596
SED	133	61	44	93
Distance (m)				
10	858^{ab}	398	327^{b}	-
50	684^{ab}	344	549^{a}	987^{a}
150	468^{b}	372	400^{b}	360^{c}
300	521^{b}	265	351^{b}	696^{b}
500	1117^{a}	550	300^{b}	342^{c}
1000	482^{b}	409	384^{b}	-
Mean	688	390	385	596
SED	243	103	69	131
Depth x Distance				
Significance	NS	NS	NS	NS

Means followed by the same letter in a column are not significantly different; NS = not significant at 5% level of significance.

pits was equally found to be at par (Figure 1). However, sampling distance was found to have significant influence on the concen-tration in all the directions except south (Table 2). In the north, soil samples taken at 500 m away from the village had higher concentration of Pb than soil samples taken at 150, 300 and 1000 m. There was no significant difference in Pb concentration of soil samples taken at 500 m as compared to those taken at 10 and 50 m. In the east and west of the village, soil samples taken at 50 m away from the last household had higher concen-tration of Pb than soil samples taken at other distances. In the west, significantly higher concentration was observed at 300 m than 150 and 500 m which were not significantly different from each other. In general, no uniform trend was observed in Pb concentration in all the directions because though hig-her concentrations were found around 50 m distance in the east and west of the village, similar values were observed at much further dis-tance such as 500 m in the north than closer distance like 150 m. This shows lack of uniformity in the distribution of the elements in the soil which could be attributed to soil heterogeneity. Luo et al. (2007) reported that concentra-tions of heavy metals vary remarkably over space due to the heterogeneity of soil itself.

This observed concentration was irrespective of sam-pling depths, distance or direction (Figure 1). The values obtained across the profile pits also fall above these acceptable limits. Although Pb contamination in this village was attributed to lead deposited from gold ore processed, high concentration down the profile pits suggests addi-

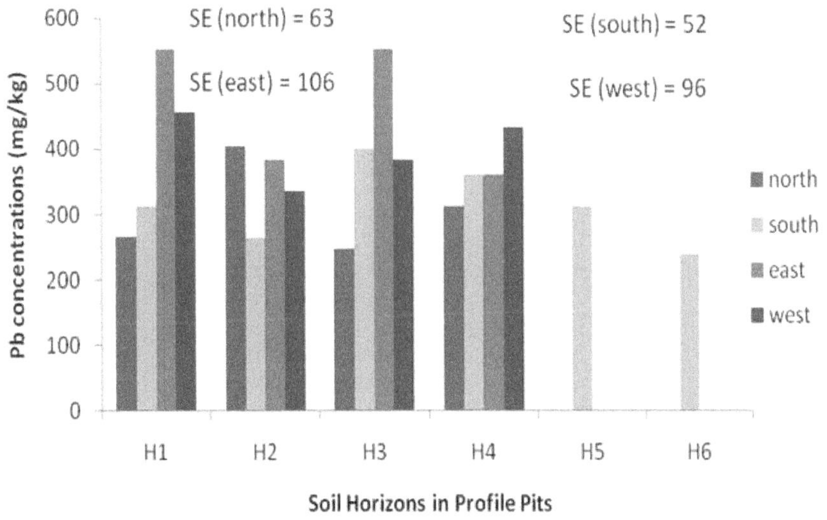

Figure 1. Lead concentration of soil horizons in provfile pits in the four cardinal directions of Abare.

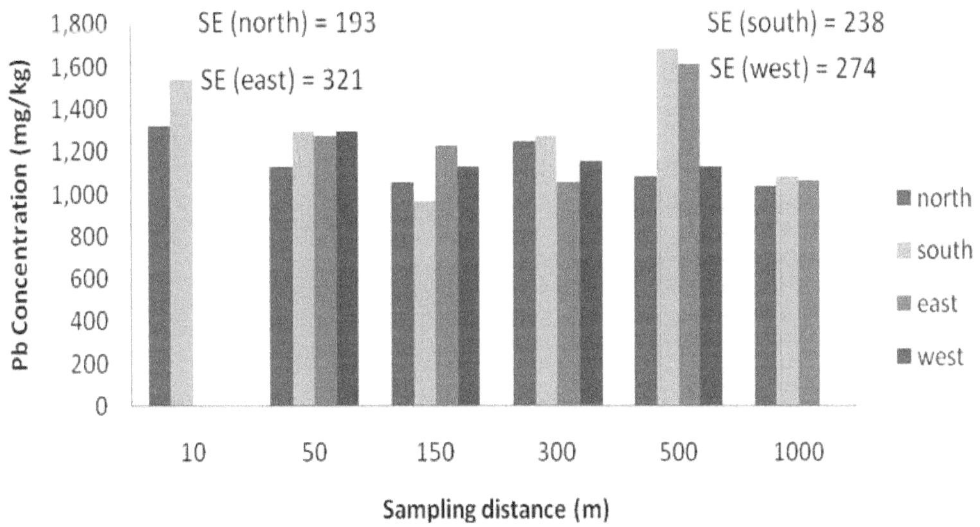

Figure 2. Effect of sampling distance and direction on Pb concentrations in plant samples in Abare.

tional below ground sources of these elements. Parent material could be an important origin of Pb and pedogenesis, a natural input of the Pb into the soils. Phosphate fertilizers are other sources of Pb that may cause pollution in agricultural soils, however, they are not readily available for farmers use in the study area.

Assessment of lead concentrations in plant samples in Abare

Lead determination in plants mainly focused on sorghum and cowpea leaves. These are the dominant crops grown in the study area. Plant samples collected at the various soil sampling points along each direction showed no significant difference in their Pb concentrations (Figure 2). However, Pb concentration in these plant materials was over 3500-fold higher than the recommended thresholds of 50 mg kg^{-1} in edible crops by FAO/WHO (2001). The extremely high levels of Pb in the plant materials could not have been due to uptake from soil solution alone, atmospheric deposition of dust from the contaminated gold ore processing plants within the village must have greatly contributed. Plant sampling was done at the period when contaminated soils were being excavated and replaced by clean soils. This process would generate a lot of dust which will eventually settle on surrounding vegetation. This suggests that the high

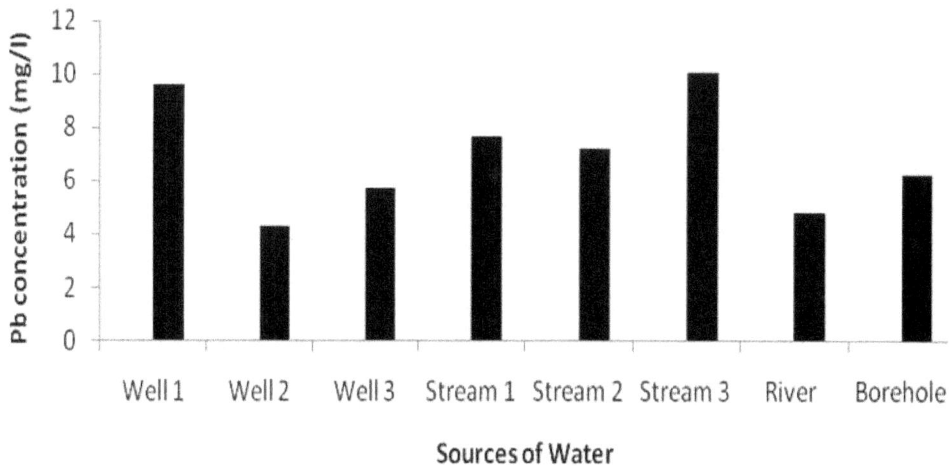

Figure 3. Lead concentration of some sources of water in Abare village.

Pb concentration in the plant materials must have come from dust deposition. High levels of Pb in the soil coupled with human input could result in phytotoxicity at high concentrations and the transfer of these elements to the human and livestock diets from plant uptake. We could not determine the concentrations in the edible part of the crop because sampling was done at the middle of the growing season. The plant parts sampled (leaves and stalk) are essential components of livestock feed especially after harvest and heavy metal accumulation in them could be indirectly used to assess exposure risks.

Assessment of lead concentrations in water samples in Abare

Lead concentration determined in eight sources of water in the village ranged from 4.32 to 10.08 mg/l (Figure 3). The highest concentration was found in the water collected from one of the streams while the lowest was obtained from well water. Generally, stream water had higher concentration of lead than well water. River and borehole water had intermediate concentration of 4.80 and 6.24 mg/l, respectively. The results from all the water sources showed no immediate risk of Pb contamination. Its concentration is not above the threshold limit set by the United States Environmental protection Agency (USEPA, 2004). This limit is in the range of 5 to 10 mg/l.

Risk assessment

Chronic daily intake (CDI) of metals for carcinogenic and non-carcinogenic risks indicated significant danger to the inhabitants of the study area. The average CDI for carcinogenic risk was 1.3×10^5 mg kg^{-1} day^{-1} for Pb. Non carcinogenic CDI were estimated for both adults and children and the average values for adults was 1.7×10^8 mg kg^{-1} day^{-1}. Even for children, the estimated values were very high with an average of 9.7×10^7. Dust inhalation was used as the exposure route of soil considering the exposure time indoor and outdoor. The CDI through this route was calculated to be 1.5×10^6 mg kg^{-1} day^{-1}. These values are far beyond the report of Abdu et al. (2011b) in wastewater irrigated vegetable gardens of Kano, northern Nigeria. They are equally beyond acceptable values worldwide and can thus, be related to high Pb level in blood samples and associated death among inhabitants of this village.

Conclusion

High concentrations of Pb in both soil and plant and the estimated health risk may have posed a health hazard for the environment and the ecosystem as a whole. Based on data description, the primary input of this heavy metal in the soil is the soil parent material. Both human and natural activities influenced concentrations of Pb in the plant material. Processing of contaminated gold ore in this village has been responsible for the severe Pb pollution. Remediation studies are required to tackle Pb contamination in the farmlands.

ACKNOWLEDGEMENTS

The study was sponsored by the Institute for Agricultural Research, Ahmadu Bello University, Zaria. The authors are grateful to Mr. U. O. Bello, Mr. Victor Odigie and Jatau for technical assistance in the field and laboratory and also the people of Abare village for their cooperation.

REFERENCES

Abdu N, Abdulkadir A, Agbenin JO, Buerkert A (2011a). Vertical distribution of heavy metals in wastewater irrigated-vegetable garden soils of three West African cities. Nutr. Cycl. Agroecosyst. 89:387-397.

Abdu N, Agbenin JO, Buerkert A (2011b). Phytoavailability, human risk assessment and transfer characteristics of cadmium and zinc contamination from urban gardens in Kano, Nigeria. J. Sci. Food Agric. 91:2722–2730

Anderson JM, Ingram JSI (1993). Tropical Soil Biology and Fertility. A Handbook of Methods, 2nd Ed. C. A. B. International, Wallingford, United Kingdom.

BI (2011). Blacksmith Institute. UNICEF Progrmme Coperation Agreement. Environmental Remediation – Lead poisoning in Zamfara. Final Report. September 2010 – March 2011. pp. 57.

FAO/WHO (2001). Maximum levels for Lead, Joint FAO/WHO Standards, Codex STAN 230-2001. http//www.codexalimentarius.net/standards_search.asp. (accessed on April 11th, 2011).

Gee GW, Bauder JW (1986). Particle size analysis. In: Methods of soil analysis, Part 1. 2nd ed. Klute A (Ed). Agronomy Monogram. 9. ASA and SSSA, Madison, WI. pp 383–411.

Lim CH, Jackson ML (1986). Dissolution for total elemental analysis. In: Methods of soil analysis: chemical and microbiological properties. Page AL (Ed). pp. 1-12. Soil Sci. Soc. Am. Madison.

Luo W, Wang T, Lu Y, Giesy JP, Shi Y, Zheng Y, Xing Y, Wu G (2007). Landscape ecology of the Guanting reservoir, Beijing, China: Multivariate and geostatistical analyses of metals in soils. Environ. Pollut. 146:567-576.

Nelson DW, Sommers LM (1986). Total Carbon, Organic Carbon and Organic Matter. In Sparks DL (Ed.), Methods of soil Analysis Part 2. Chemical Methods. Am. Soc. Agron. Madison. pp. 961–1010.

NNP (2011). New Nigerian Politics. Lead poison kills 400 infants in Zamfara. March 7th, 2011. http://www.newnigerianpolitics.com/.../lead-poison-kills-400-infants... (accessed on April 8th, 2011).

Prasad MNV, Freitas HMO (2003). Metal hyperaccumulators in plants: Biodiversity prospecting for phytoremediation technology. Electr. J. Biotech. 6:286-321.

USEPA (2004). United States Environmental protection Agency. Estimation of relative Bioavailability of lead in Soil and Soil-Like Materials Using *In Vivo* and *In Vitro* Methods. OSWER-9285.7-77, Washington, DC.

USEPA (2005). United States Environmental protection Agency. Guidelines for carcinogenic risk assessment. EPA/630/P03/001F, Washington, DC.

Yusuf AA, Abdu N, Tanimu B (2011). Assessment of lead contamination of farmlands in Abare village, Zamfara State. In: Hassan WA et al. (Eds). Mobilizing Agricultural Research Towards Attaining Food Security and Industrial Growth in Nigeria. Proceedings of the 45th Annual Conference of the Agricultural Society of Nigeria. pp. 487-491.

Comparative assessment of air pollution tolerance index (APTI) in the industrial (Rourkela) and non industrial area (Aizawl) of India: An eco-management approach

Prabhat Kumar Rai, Lalita L. S. Panda, Biku Moni Chutia and M. Muni Singh

Department of Environmental Science, Mizoram University, Tanhril, Aizawl-796004, Mizoram, India.

Science for phyto-technologies has got immense application in air pollution science. The present study focuses on the determination of air pollution tolerance indices (APTI) from six common road side plant species growing along industrial (Rourkela) and non industrial area (Aizawl), India. The APTI was determined by synthesizing the four different physiological and biochemical parameters, that is, leaf relative water content (RWC), ascorbic acid content (AA), total leaf chlorophyll (TCh) and pH of leaf extract. The plant species selected for the study were *Ficus bengalensis*, *Mangifera indica*, *Bougainvillea spectabilis, Psidium guajava, Hibiscus rosa-sinensis* and *Lantana camara.* Reduction in total chlorophyll content and pH was found in the leaf samples of all selected plants collected from Industrial site (Rourkela) when compared with samples from non industrial site (Aizawl) whereas APTI, ascorbic acid and RWC were found to be higher in the plant samples of Industrial site (Rourkela) as compared to non industrial site (Aizawl). On the basis of APTI, *F. bengalensis* was found to be tolerant (8.64) in industrial site (Rourkela) and *M. indica* (7.95) in non industrial site (Aizawl). Plant species such as *M. indica* and *B. spectabilis*, showing minimum difference in their APTI values may be considered as tolerant for both (industrial and non industrial) sites.

Key words: Air pollution tolerance indices (APTI), biochemical parameters, *Ficus bengalensis*, roadside plants.

INTRODUCTION

Air pollution has become a major problem arising mainly from industrialization and urbanization during the last few decades. Particulate matter is of great concern in relation to their adverse impact on human health and vegetation (Rai, 2013). The particulates and gaseous pollutants, alone and in combination can cause serious setbacks to the overall physiology of plants (Ashenden and Williams, 1980; Mejstrik, 1980; Anda, 1986; Das and Prasad, 2010). Trees experience the greatest exposure and influenced greatly by pollutant concentration due to their perennial habit (Raina and Sharma, 2003; Chauhan, 2010). Regional impact of air pollution on local plant species is one of the major ecological issues. The climate condition, the physico-chemical properties of air pollutants and their residence time in the atmosphere have impact on surrounding plants (Wagh et al., 2006). The most obvious damage occurs in the leaves which include chlorosis, necrosis and epinasty (Prasad and Choudhury, 1992).

The responses of plants to pollutants may provide a

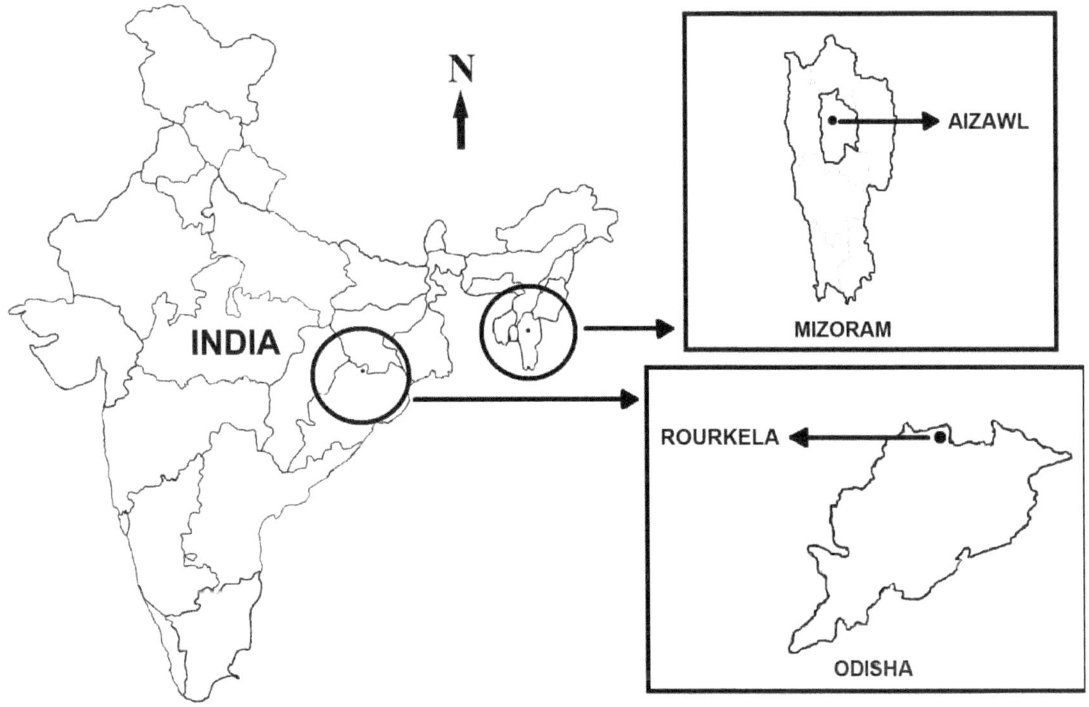

Figure 1. Study area- Aizawl, Mizoram (NE India) and Rourkela, Odhisa, India.

simple method of monitoring air pollutants as well as providing the pollution abatement measures. Plantation of tolerant tree species may have a marked effect on varied aspects of the quality of the urban environment and the cleanliness of life in a city (Bamniya et al., 2011).

Air pollution tolerance index (APTI) is an inherent quality of plants to encounter air pollution stress which is presently of prime concern particularly in industrial and non-industrial areas. Therefore, APTI of the plants needs to be monitored and checked for the predominant species that are present in the polluted and non-polluted areas. In the present study, APTI of common growing roadside plants in industrial and non-industrial site have been investigated.

The present study was conducted in Industrial area (Rourkela) and non-industrial area (Aizawl). The study was carried out on six common roadside plant species growing along Industrial (Rourkela) and non- industrial (Aizawl) sites during winter season (Nov 2011 to Feb 2012).

Rourkela, known as the industrial capital of Odisha is located at 84° 53' E longitude and 22° 12' N latitude in Sundargarh district of Odisha at an elevation of about 219 m above the mean sea level (ASL) (Figure 1). The climate of Rourkela is characterized as tropical monsoon climate, with minimum temperature in December and maximum in May. Rourkela city is famous for its steel industries under steel Authority of India Ltd and is known as steel city. A number of other industries are also present such as cement factory, fertilizer, sponge iron industries, thermal power plants making Rourkela as an industrial site for the present study.

Aizawl (21°58'- 21°85' N and 90°30'- 90°60'E), the capital of the state, Mizoram, is 1132 m ASL and comes under an Indo Burma hotspot region (Rai, 2009) (Figure 1). The altitude in Aizawl district varies from 800 to 1200 m. The climate of the area is typically monsoonic. The annual average rainfall is amounting to ca 2350 mm. The area experiences distinct seasons. The ambient air temperature normally ranges from 20 to 30°C in summer and 11 to 21°C in winter (Laltlanchhuanga, 2006). In Aizawl, there is less traffic intensity and no industries present as compared to Rourkela so it is considered as non industrial site.

The present study deals with bio monitoring potential of six common plant species (*Ficus bengalensis*, *Mangifera indica*, *Bougainvillea spectabilis, Psidium guajava, Hibiscus rosa-sinensis* and *Lantana camara*), along the roadside of industrial and non-industrial site, exposed to air pollution.

METHODOLOGY

The plants selected for the study where those common to both sites. The fresh leaf samples (in three replicates of selected plants) were collected in early morning from height of 1 to 2 m from ground level of both sites. Selected plants were on the edge of the road almost with similar topography or condition and leaf samples were immediately brought to the laboratory in polythene bag, kept in ice box for further analysis of various biochemical parameters such as

Table 1. The biochemical characteristics and the APTI for plants from Industrial site (Rourkela) and Non Industrial site (Aizawl).

Plant species	Rourkela					Aizawl				
	pH	Ascorbic acid	RWC	Chlorophyll	APTI	pH	Ascorbic acid	RWC	Chlorophyll	APTI
Ficus bengalensis	6.42±0.016	0.53±0.03	83.01±2.38	0.1±0.01	8.64	7.44±0.30	0.2 ±0.01	61.45±1.18	0.48±0.02	6.3
Psidium guajava	5.56±0.41	0.47±0.02	86.95±1.68	0.92±0.01	9	6.31±0.07	0.28±0.02	57.68±2.47	1.85±0.03	6
Bougainvillea spectabilis	6.84±0.17	0.2±0.02	75.6±3.84	1.45±0.05	7.72	7.1±0.02	0.1±0.01	68.42±12.4	2.99±0.09	6.94
Mangifera indica	5.88±0.03	0.25±0.02	88.09±1.63	1.1±0.01	8.98	7.07±0.16	0.16±0.02	78.03±4.86	2.41±0.01	7.95
Lantana camara	6.96±0.07	0.28±0.01	78.94±4.30	1.05±0.02	8.11	8.2±0.31	0.19±0.02	57.66±3.76	1.88±0.03	5.95
Hibiscus rosa-sinensis	5.39±0.10	0.45±4.71	90.26±0.63	1.98±0.04	9.35	6.86±0.22	0.24±0.01	72.02±5.85	2.94±0.02	7.43

Statistical analyses were performed with SPSS 11.2 software.

leaf extract pH (Singh and Rao, 1983), relative water content (RWC)RWC (Sen and Bhandari, 1978), total chlorophyll (Arnon, 1949), ascorbic acid (Keller and Schwager, 1977). The APTI was determined by calculating these biochemical parameters, proposed by Singh and Rao (1983). The APTI was calculated using the formula:

$$APTI = \frac{A(T + P) + R}{10}$$

Where, A = Ascorbic acid (mgg^{-1}); T = total chlorophyll ($mggm^{-1}$ - f .w);P = pH of the leaf extract; R = Relative water content of leaf (%).

RESULTS AND DISCUSSION

The biochemical characteristics and the APTI for plants from industrial site (Rourkela) and non industrial site (Aizawl) are shown in Table 1. It is observed that all the plant species collected from industrial site (Rourkela) exhibited a pH towards acidic side, ranging from 5.0 to 7.0 whereas for non industrial site (Aizawl) plant species collected showed neutral to slightly alkaline pH range, that is, 6.0 to 9.0. The acidic nature may be due to the presence of SO_x, NO_x or other acidic pollutants from the industrial emission in the ambient air causing a change in pH of the leaf sap towards acidic (Swami et al., 2004). Low leaf pH extract showed good correlation with sensitivity to air pollution and also reduce photosynthetic process in plants (Yan-Ju and Hui, 2008; Thakar and Mishra, 2010). A pH on higher site improves tolerance against air pollution (Agarwal, 1986; Shannigrahi et al., 2011).

RWC of all the plant species collected from industrial site (Rourkela) were higher as compared to non-industrial site (Aizawl). Highest RWC in industrial site (Rourkela) was seen in *Hibiscus rosasinensis* and lowest in *B. spectabilis* (Table 1). In non-industrial site (Aizawl), the RWC of leaves of the plant species varied from maximum of 78.03% in *Mangifera indica* to a minimum of 57.66% in *Lantana camara.* High water content within a plant body will help to maintain its physiological balance under stress condition such as exposure to air pollution when the transpiration rates are usually high which may lead to desiccation. Therefore, maintenance of RWC by the plant may decide the relative tolerance of plants towards air pollution (Verma, 2003). The higher the RWC in a particular species, the greater is its drought tolerance capacity (Dedio, 1975). Thus, the higher RWC in industrial site sample may be responsible for normal functioning of biological processes in plants (Meerabai et al., 2012).

Ascorbic acid is regarded as an antioxidant found in large amount in all growing plant parts and influence resistance to adverse environmental condition including air pollution (Keller and Schwager, 1977; Lima et al., 2000). The mean concentration of ascorbic acid in leaves of industrial and non-industrial sites is summarised in Table 1. The ascorbic acid content increased in the leaves of plants at industrial site (Rourkela) than those of the non-industrial site (Aizawl). Table 1 shows that in industrial site (Rourkela), the concentration of ascorbic acid ranged from 0.20 to 0.53 mgg^{-1} with *B. spectabilis* and *F. bengalensis* recording the lowest and highest value, respectively, where as in non-industrial site (Aizawl) ascorbic acid concentration was found to be lower with *B. spectabilis* (0.10 $mggm^{-1}$) while *P. guajava* recording highest (0.28 $mggm^{-1}$). Ascorbic acid, a stress reducing factor is generally higher in tolerant plant species. Tripathi and Gautam (2007) reported pollution load dependent increase in ascorbic acid content of all the plant species might be due to the increased rate of production of reactive oxygen species (ROS) during photo-oxidation process. In the present study, concentration of ascorbic acid is high in industrial sample supporting the studies of Meerabai et al. (2012) and Agbaire and Esief (2009).

Chlorophyll content of plant signifies its photosynthetic activity as well as the growth and development of biomass. Chlorophyll content of plant varies from species to species depending upon the age of leaf, pollution level as well as other biotic and abiotic condition (Katiyar and Dubey, 2001). The chlorophyll content was found to be low in the leaf samples collected from industrial site (Rourkela) as compared to the non-industrial site (Aizawl) for all the investigated plant species. The present study revealed that chlorophyll content in all the plants varied with the pollution status of the area. The higher the levels of pollutants, the lower the chlorophyll content as certain pollutants in totality reduce the total chlorophyll content (Allen et al., 1987). Rao and Leblanc (1966) have also reported reduction in chlorophyll content brought by acidic pollutants like SO_2 which causes phaeophytin formation by acidification of chlorophyll. Reduction in chlorophyll content in variety of crop plant due to NO_2, SO_2 and O_3 exposure have also been reported by Agrawal et al. (2003).

The APTI values calculated for each plant species at two sites are presented in Table 1. Table 1 clearly showed an increase in the APTI value of industrial site as compared to the non industrial site. The APTI of plant species from non industrial site ranged from 5.95 to 7.95 and the maximum was observed in case of *M. indica*. The APTI value showed an increase in case of plants from industrial site (Rourkela) ranging from 7.72 to 9.35 with maximum in *Hibiscus rosa-sinensis*. The APTI value estimated using the four biochemical parameters in plant leaves namely RWC, total chlorophyll content, pH and ascorbic acid value can be used as a predictor of air quality. Plants having higher index value are tolerant to air pollution while plants with lower index value show less tolerance (Singh and Rao, 1983). All the plants found in industrial site are indicative of higher pollution exposure as compared to non-industrial site. A plant species known to be sensitive or tolerant in one geographical area may behave differently in another area (Raza et al., 1985).

Dust pollution and chronic concentration of gaseous pollutants may affect the biochemical make up and tolerance capacity of plants to the air pollution. Industrial site (Rourkela) is in the grip of more serious air pollution problem due to heavy vehicular and industrial emission as compared to non industrial site (Aizawl) where there are no industries. It can be concluded that the entire result obtained from this study indicates that different plant respond differently to air pollution hence different indices obtained for both sites suggested that, plants growing in industrial site have higher APTI than those in non-industrial site. *F. bengalensis* (8.64) was found to be tolerant in industrial site (Rourkela) and *M. indica* (7.95) in non-industrial site (Aizawl), whereas, *M. indica* and *B. spectabilis* showed least decrease in their APTI value. They can be considered more tolerant species for both the sites.

Conclusion

Biomonitoring of air pollution and its impact on biochemical parameters is extremely relevant in air pollution science. The study clearly reflects that the tolerance of plants towards air pollution may be site-specific. An overview of the entire result obtained from this study reveals that industrial site (Rourkela) have higher APTI values as compared to non industrial site (Aizawl). Further, *M. indica* and *B. spectabilis* may be suitable for plantation at both the sites.

ACKNOWLEDGEMENTS

PKR wishes to acknowledge the Department of Biotechnology (DBT) vide research and Department of Science and Technology (DST), for providing financial assistance in the form of research project (vide project no. BT/PR-11889/BCE/08/730/2009 and SR/FTP/ES-83/2009, respectively). Thanks are due to Dr. Onkar Nath Tiwari and Dr. Umesh Sharma for their useful discussion and cooperation in this work. LLSP is thankful to DST INSPIRE fellowship for financial assistance.

REFERENCES

Agarwal SK (1986). A new distributional function of foliar phenol concentration in the evaluation of plants for their air pollution tolerance index. Acta Ecologica 8(2):29-36.

Agbaire PO, Esiefarienrhe E (2009). Air pollution tolerance indices (apti) of some plants around Otorogun gas plant in Delta State, Nigeria. J Appl. Sci. Environ. Manage. 13(1):11-14.

Agrawal M, Singh B, Rajput M, Marshall F, Bell JNB (2003). Effect of pollution on peri-urban agriculture: A case study. Environ. Pollut. 126:323-329.

Allen (Jnr) LH, Boote KL, Jones JW et al (1987). Response of vegetation to rising carbon dioxide photosynthesis, biomass and seed yield of soybeans. Global Biogeochemical Cycle 13:1-44.

Anda A (1986). Effects of cement kiln dust on the radiation balance and yields of plants. Environ. Pollut. 40:249-259.

Arnon DI (1949). Coenzyme in isolated chloroplast. Polyphenol oxidase in *Beta vulgaris*. Plant Physiol. 24:1-15.

Ashenden TW, Williams IAD (1980).Growth reduction in Lolium multiflorum Lam and Phleum pratense L as a result of sulphur dioxide and nitrogen dioxide pollution. Environ. Pollut. 21:131-139.

Bamniya BR, Kapoor CS, Kapoor K, Kapasya V (2011). Harmful effect of air pollution on physiological activities of Pongamia pinnata(L.) Pierre. Clean Technol. Environ. Policy 14:115-124.

Chauhan A (2010). Photosynthetic pigment changes in some selected trees induced by automobile exhaust in Dehradun. J. New York Sci. 3(2):45-51.

Das S, Prasad P (2010). Seasonal Variation in Air Pollution Tolerance Indices and Selection of Plant Species for Industrial Areas of Rourkela. Indian J. Environ. Prot. 30(12):978-988.

Dedio W (1975). Water relations in wheat leaves as screening test for drought resistance. Can. J. Plant Sci. 55:369-378.

Katiyar V, Dubey PS (2001). Sulphur dioxide sensitivity on two stage of leaf development in a few tropical tree species. Ind. J. Environ.

Toxicol. 11:78-81.

Keller T, Schwager H (1977). Air pollution and ascorbic acid. Europ. J. Forest Pathol. 7:338-350.

Laltlanchhuanga SK (2006). Studies of the impact of disturbance on secondary productivity of forest ecosystem with special reference to surface, sub-surface litter insect and other non insect groups M.Sc.dissertation. Mizoram University.

Lima JS, Fernandes EB, Fawcett WN (2000). Mangifera indica and Phaseolus vulgaris in the bioindicator of air pollution in Bahia, Brazil. Ecotoxicol. Environ. Saf. 46(3):275-278.

Meerabai G, Venkata Ramana C, Rasheed M (2012). Effect of industrial pollutants on Physiology of Cajanus cajan (L.) - Fabaceae. Int. J. Environ. Sci. 2(4):1889-1894.

Mejstrik V (1980). The influence of low SO_2 concentration on growth reduction of Nicotiana tabacum L.W.Samsum and Cucumis sativus L.cv Unikat. Environ. Pollut. 21:73-76.

Prasad D, Choudhury (1992). Effects of air pollution.In:Misra,S.G.(Ed):Environmental pollution Air, Environmental Pollution and Hazards Series. Venus Publishing House, New Delhi. pp. 58-60.

Rai PK (2009). Comparative Assessment of Soil Properties after Bamboo Flowering and Death in a Tropical Forest of Indo-Burma Hot spot. Ambio: J. Hum. Environ. 38(2):118-120.

Rai PK (2013). Environmental magnetic studies of particulates with special reference to biomagnetic monitoring using roadside plant leaves. Atmos. Environ. 72:113-129.

Raina AK, Sharma A (2003). Effects of vehicular pollution on the leaf micro-morphology, anatomy and chlorophyll contents of Syzygium cumini L. Indian J. Environ. Prot. 23(8):897-902.

Rao DN, Leblanc F (1966). Effects of SO_2 on the lichens and alga with special reference to chlorophyll. Bryologist 69:69-75.

Raza SN, Vijaykumari N, Murthy MS (1985). Air pollution tolerance index of certain plants of hyderabad. In: Symp Biomonitoring State of Environment, New Delhi: Indian National Science Academy pp. 243-245.

Sen DN, Bhandari MC (1978). Ecological and water relation to two Citrullus spp. In: Althawadi, A.M. (Ed.). Indian Arid Zone. Environ. Physiol. Ecol. Plants pp. 203-228.

Shannigrahi AS, Fukushim T, Sharma RC (2004). Anticipated Air Pollution Tolerance of some plant species considered for green belt development in and around an industrial/urban area in India:An overview. Int. J. Environ. Stud. 61(2):125-137.

Singh SK, Rao DN (1983). Evaluation of plants for their tolerance to air pollution. In Proc symp on air pollution control held at IIT Delhi pp. 218-224.

Swami A, Bhatt D, Joshi PC (2004). Effects of automobile pollution on sal (Shorea robusta) and rohini (Mallotus phillipinensis) at Asarori, Dehradun. Himalayan J. Environ. Zool. 18 (1):57 - 61.

Thakar BK, Mishra PC (2010). Dust collection potential and Air pollution Tolerance Index of Tree vegetation around Vedanta Aluminium Limited, Jharsuguda. Int. Q. J. Life Sci. 3:603-612.

Tripathi AK, Gautam M (2007).Biochemical parameters of plants as indicators of air pollution. J. Environ. Biol. 28:127-132.

Verma A (2003). Attenuation of automobile generated air pollution by higher plants. Dissertation, University of Lucknow.

Wagh ND, Shukla PV, Tambe SB, Ingle ST (2006). Biological monitoring of roadside plants exposed to vehicular pollution in Jalgaon city. J. Environ. Biol. 27(2):419-421.

Yan-Ju L, Hui D (2008). Variation in air pollution tolerance index of plant near a steel factory; implication for landscape- plant species selection for industrial areas. Environ. Dev. 1(4):24-30.

Heavy metals found at Umzimvubu River Estuary in the Eastern Cape, South Africa

Songca S.P.[1] , Seteni B.[2], and Oluwafemi O.S.

[1]Faculty of Science, Engineering and Technology, Walter Sisulu University, PO Box 19712, Tecoma, East London, Eastern Cape, South Africa.
[2]Department of Chemistry and Chemical Technology, Walter Sisulu University, Mthatha Campus, Private Bag X1, Mthatha 5117, Eastern Cape, South Africa.

The aim of this research was to detect and measure the concentration of heavy metals found in the water and silt of the Mzimvubu River at Port St Johns in the Eastern Cape. The water from this river estuary is used by the people and animals of that area and is suspected to cause several health conditions. The objective of the research was to undertake a pilot analysis for heavy metals. Atomic Absorption Spectroscopy (AAS) was selected as a tool to determine the concentration of the heavy metals present in the water and sediment samples. Water and sediment samples were collected at the river mouth and 5 kilometers upstream, near the R61 bridge, on either side of the river. The atomic absorption spectrophotometer used to measure the heavy metal concentrations was a Varian Spectra AA 100 equipped with a single slot burner. Measurements were carried out in triplicate. Throughout the study it was evident that there are heavy metals such as lead, cadmium, zinc and nickel in the Umzimvubu River. The bio-accessibility of lead is a concern in view of the fact that it was found in high concentrations. Cadmium contamination was found to be in lower concentrations compared to the South African guidelines. Zinc and nickel were not severe since both were within the WHO and SADWAF guidelines. Therefore, it is concluded that the Umzimvubu River is an intimidation to many living organisms since it contains high concentrations of Lead, significantly higher than the South African guidelines. Severe damage in physical condition could be experienced by the human and animal populations in close proximity to the river. Further and more detailed studies are recommended, including bio-accessibility and bio-accumulation studies.

Key words: Mzimvubu river, heavy metals, atomic absoption spectroscopy.

INTRODUCTION

The Umzimvubu River is no exception to other rivers since it also suffers from pollution by human activities. The river pollution can be caused by three sectors namely; agriculture, households and industries and this can lead to the abundance of heavy metals in river water and sediment. These heavy metals can be life Threatening to plant and animal life.

Heavy metal poisoning can result, for instance, from drinking-water contamination, high ambient air concentrations near emission sources, or intake via the food chain (Njar et al., 2012). Heavy metals can enter a water supply by industrial and consumer waste, or even from acidic rain breaking down soils and releasing heavy metals into streams, lakes, rivers, and groundwater (Dmitri, 1993). Water pollution affects plants and organisms living in these water bodies; and, in almost all cases the effect is damaging not only to individual species and populations, but also to the natural biological communities (Foy et al., 2000). Water pollution occurs when pollutants are discharged directly or indirectly into water bodies without adequate treatment to remove harmful compounds (Lopes et al., 2005).

Among the heavy metals, the most serious effect of pollution is presently associated with lead emission (Olade, 1987). Human exposure to toxic heavy metals has become a major health risk on the continent and is the subject of increasing attention from national and international environmentalists (Yabe et al., 2010). Characteristics of heavy metal pollution in North, West, East and Southern regions of Africa have been described, as have major sources of pollution in the different regions (Yabe et al., 2010). Most of the countries within the West African sub-region are emergent nations which, for a long time, have been grouped among the less developed countries of the world. In recent years, some of these countries, such as Nigeria, Ghana and Ivory Coast have achieved significant strides in their quest for rapid economic growth through industrialization (Mombeshora et al., 1983). Thus, a number of factories, usually situated haphazardly, have mushroomed, and with increased affluence generated by income from the export of natural resources, the problems of urbanization, population explosion and the increased use of automobiles have become very common. At present, relatively little data are available on the extent of environmental pollution because there are no agencies charged with the routine monitoring and protection of the environment in these countries (Nriagu, 1978).

There has been a phenomenal growth in the use of automobiles within the last decade as a result of increased affluence. The level of lead in super grade gasoline in Nigeria is in the range of 0.7 to 0.9 gm/l which is higher than permissible levels in developed countries, where the range 0.4-0.5 gm/l is common. Preliminary studies show that the lead emission in air at ground level in Lagos, Nigeria, is far higher than for cities like London, UK, and New York, USA, but similar to those of Brazil or the Caribbean due to perturbation by roadside traffic. Roadside soils and dust are also getting increasingly contaminated (Olade, 1987). The city of Port Harcourt close to the oil-fields is lit brightly at night by gas being flared in the surrounding areas. No data are available on the extent of metal pollution associated with this practice (Mombeshora et al., 1983).

It is well known that fertilizers may contain up to 400 ppm lead which may constitute a major source of soil pollution, although phosphate lead is not often readily available to plants (Nriagu, 1978). The consumption of fertilizers in South Africa over t he fifty year period from 1955 to 2005 has increased from 800,000 tonnes to 1,500,000 tonnes as shown in Figure 1. The South African fertiliser industry annually supplies about 2,000,000 tonnes of fertiliser products to the local market (van der Linde and Pitse, 2006).

In Lagos no centralized sewage systems existed prior 2012, and the industrial effluents from the factories were usually discharged untreated into streams, lagoons, open drains and other water bodies. This has resulted in high, although not alarming, levels of lead, cadmium and copper in some localized areas. These levels of inorganic pollutants may be reduced to acceptable levels for potable re-usage (Smith et al., 1980). Studies of heavy metal contents of surface waters in Ibadan show that levels of lead may reach 50 µg/L (Mombeshora et al., 1983). In waters of the Lagos lagoon, lead levels exceeding 120 µg/L have been obtained (Olade, 1987). Clearly therefore, it is imperative to conduct environmental impact assessment studies before industries are located, so as to ameliorate environmental pollution from point sources (Vernet, 1991). Acidic mine drainage from disused mine shafts, open pits and dumps have contributed to localized metal pollution in the vicinity of these operations (Olade, 1987).

Municipal composts and farmyard manure are increasingly being applied to croplands. Heavy metal contents of municipal composts are similar to those of sewage sludge and their applications may increase metal contents of farm soils (Nriagu, 1978). South Africa is one of the world's leading producers of precious metals and minerals. The pollution of surface waters by heavy metals and acidic effluents from refineries are a serious problem in the country (van Hille et al., 1999). The majority of mines are located in the most highly industrialized and hence heavily populated areas, where the demand for fresh water for domestic and industrial use is highest (Henzen and Pieterse, 1978).

Heavy metals are used very often in South Africa in the textile industry, especially in the processes of dyeing, finishing, and health care. Many of these metals have unwanted effects on processing and on product quality. They are damaging to the textile material, toxic to man, and hazardous to the environment when released into streams and rivers as wastewater (Binning and Baird, 2004). Scientists need to understand not only the unique features and behaviour of individual heavy metals but also the differences, similarities and interactions of different metals at the ecosystem, systemic and cellular levels if we are to deal with the problems of global heavy metal pollution in a sustainable manner (Vernet, 1991).

In South Africa, the quality of a stream or river is often a good indication of the way of life within the community through which it flows. Healthy streams, wetlands and rivers support a great variety of water life. Rain water and tumbling mountain streams contain high levels of oxygen. Much of the oxygen comes from the atmosphere through rain, tumbling water in fast-flowing streams and

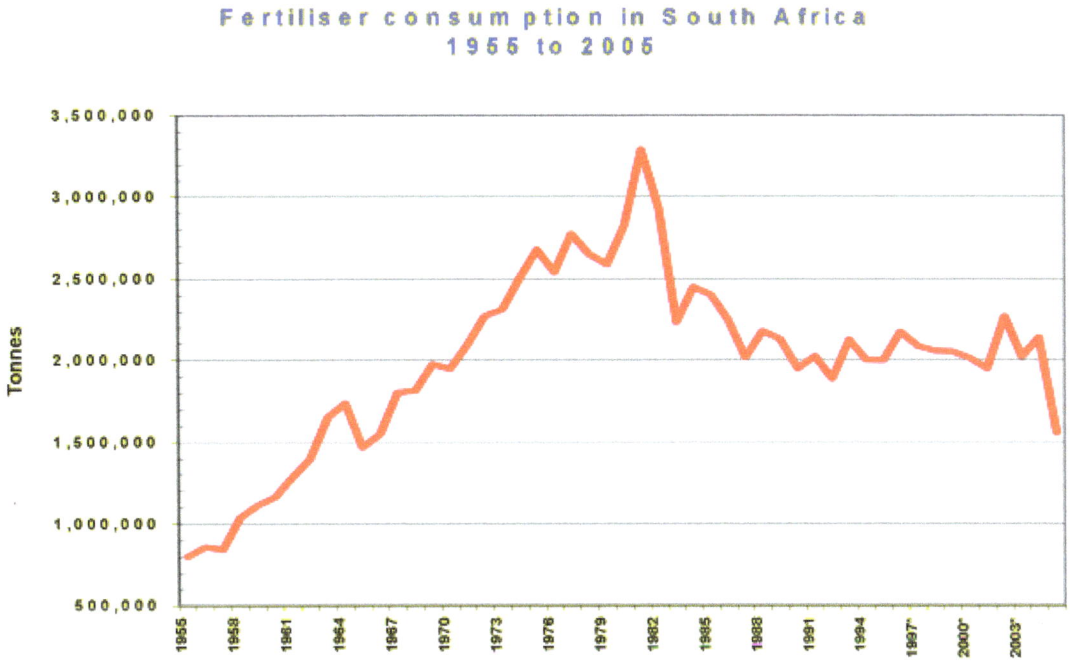

Figure 1. Consumption of fertiliser in South Africa from 1955 to 2005 (van der Linde and Pitse, 2006).

Figure 2. Sampling locations along the Umzimvubu River at the mouth RE and RW and at the bridge BE and BW.

photosynthesis. Water plants, in turn, photosynthesize and provide life-supporting oxygen and other food sources for water organisms that interact in a complex web of life (Eaton, 1994). High quality water is defined as water which is safe, drinkable and appealing to all life on earth (Young, 1992). As the human population increases, there is an increase in pollution and catchment destruction (Olade, 1987). First world countries have been investigated to detect the initial signs of decreasing water quality, to prevent increasing water pollution at an early stage (Robinson, 1996). Organisms in aquatic environment are considered biologically sensitive, due to their ability to respond to changes that occur in water. The biotic integrity of an ecological system is therefore reflected in the health of its fauna (Robinson, 1996).

The Swartkops River catchment contains almost the entire municipal area of Uitenhage and Kwanobuhle, Despatch and Ibhayi and also half of the Port Elizabeth municipal area. It is estimated that approximately one million people live and work presently in the Swartkops River catchment. High-density, low-income housing is developing in the catchment with a concomitant increase in industry, and in the quality and quantity of storm water runoff. These developments will necessitate road, rail, telecommunication and power supply infrastructure over the river and these developments are a source of heavy metals which will lead to physical challenges to humans (Horenz, 1988). Comparison of the concentration of chrome, lead, zinc, titanium, manganese, strontium, copper and tin in the estuary with those obtained in a similar survey made about 20 years ago revealed some remarkable increases. This raises concern over the long-term health of the Swartkops River ecosystem (Binning and Baird, 2001). The distribution of the heavy metals zinc, cadmium, copper, iron, manganese and lead was investigated in seawater directly, and in sediments, using the total digestion method from the East London and Port Elizabeth harbours. Ship repair activities are suspected to be responsible for elevated concentrations reported in this study in the upper reaches of these harbours (Fatoki and Mathabatha, 2001).

The aim of this research was to detect and measure the concentration of the heavy metals Cd, Zn, Ni and Pb in the water and silt of the Mzimvubu River at Port St Johns in the Eastern Cape. The water from this river estuary is used by the people and animals of that area and is suspected to cause several health conditions. To the best of our knowledge, no studies have been reported on the detection and quantification of heavy metals in the Mzimvubu river waters and sediments. However, Moses (2006) makes reference to a study conducted by Van Niekerk (2000), which cites cases of faecal pollution, because of lacking sanitation infrastructure to settlements surrounding the Mzimvubu and other rivers. Furthermore, apart from anthropogenic sources, evidence exists to suggest that heavy metals such as nickel, copper and platinum group metals may be collecting into Mzimvubu river from as far upstream as Mount Ayliff at Intsizwa (Sander and Cawthorn, 1995; Cawthorn and Kruger, 2004).

MATERIALS AND METHODS

A Varian Spectra AA 100 atomic absorption spectrophotometer equipped with a single slot burner was used. Measurements were carried out in triplicate. All glass and propylene ware used were first soaked in dilute HNO_3, thoroughly washed with liquid soap and then rinsed with double distilled de-ionised water and acetone. Thereafter, all the glassware were dried in the oven at 100°C for 24 h, while the plastic ware were left to dry at room temperature and later used for water sample collection. All reagent stock solutions: 1000 mg/mL of the metals cadmium, nickel, lead and zinc used were of analytical reagent grade supplied by Total Laboratory Technology c.c.

Collection of the samples and preparation for testing

The mouth of the Umzimvubu River is situated 31.63° south and 29.55° East, near the Open Africa Route at Port St John in the South East of South Africa in the Eastern Cape province, in an area that was previously known as the Transkei homeland. The river estuary is situated about 96 km from Umthatha along the R61. Water and sediment samples were collected once from the east and west of the river mouth (RE and RW) and then again collected at about 1.5 km from the river mouth near the R61 bridge, east and west side of the river (BE and BW) in August 2010 (Figure 2). The propylene containers were used for sample collection. Before use the propylene containers (500 ml) were first rinsed with the water samples and thereafter immersed about 15 cm below the surface. Water and sediment samples were collected again in January, February and March 2013, from the sampling points used for the August 2010 sample collection.

The samples were transported to WSU Potsdam Campus Laboratory and stored in a refrigerator until analysed. The samples for total metal concentrations were acidified in the lab with grade HNO_3 so that the metals could not be entrapped in other species present in the samples.

Determination of concentrations

Numerous instrumental methods are used to determine heavy metals in natural water and its sediments, mainly with preconcentration, since the sensitivity of direct analysis is often insufficient. But in this research these digestion methods were not employed during the investigation of cadmium, nickel, zinc and lead in water. The reason for this is that the availability of these heavy metals was already observed when the samples were analysed without digestion using the AAS.

Analysis of water

The water samples were taken out of the refrigerator and allowed to warm up to room temperature for about 15 – 20 min before analysis with the AAS. The detection limits for each of the elements were determined using the lowest concentration of each of the elements that gave the least detectable signal using flame AAS for cadmium, nickel, zinc and lead. Three of each of these samples were prepared and analyzed.

Analysis of sediments

Sediment samples were analysed using the total digestion method.

Sediment (0.5 g) was weighed into a 100-ml beaker. Concentrated HNO_3 (5 ml) was added and this was boiled gently for 30 min on a hot-plate without splattering. The beaker was cooled and 2 ml $HClO_4$, 5 ml concentrated HNO_3 and 5 ml HF were added. The mixture was heated to near dryness without splattering. The corners and walls of the beakers were washed with about 5 ml triple-distilled water, and the solution was again heated until dense white fumes developed. The beaker was cooled and 10 ml HNO_3 was added to dissolve the salts. The solution was transferred into a 50 ml standard flask and then diluted with triple-distilled water. The solution was analysed by means of AAS. A blank determination using the same procedure was performed (Fatoki and Mathabatha, 2001).

RESULTS

The average of heavy metal concentrations in µg/l, in the waters of Umzimvubu River are compared to national and international guidelines for drinking water, including the South African guideline for drinking water, the World Health Organisation guideline for drinking water and the World Average of trace elements in unpolluted rivers (Helmer and Meybeck, 1989; Boyle and Schiller, 1987). The results are summarised in Table 1.

In addition, the average of heavy metal concentrations in µg/g, in the sediments of Umzimvubu River is compared to South African Guidelines (Maritz and Swanopoel, 1998). The results are summarised in Table 2. Tables 1 and 2 show the results for four sampling points including the two river mouth sampling points RE and RW and the two sampling points near the bridge BE and BW, the average of the four sampling points as well as the national and international guidelines, SA, WHO, WA and proposed SA.

DISCUSSION

The lead concentrations in water samples from the river shown in Table1 were significantly higher than the values for the other metals except of Zn. The high levels shown by lead may be attributed to the deposition of lead particulates on the road next to the river. Other investigations (Quinche and Zube, 1969; Chow, 1970; Motto, 1970; Fatoki and Hill, 1994) have demonstrated that lead emissions from motor vehicles produce elevated concentrations of the element in roadside water, vegetation and soil. A study by Okonkwo et al. (2003) showed high levels of lead on vegetation and soil as a result of the use of leaded petrol. The same argument is made in this research since gas stations in Port St John's are very close to the river, and farmers near the river have big tanks for the storage of petrol which can contribute to the abundance of lead in river waters through tank leakage. Table 1, BW (29.8 mg/kg) and BE (39.5 mg/kg) shows high concentrations of lead as compared to RW (10.5 mg/kg) and RE (19.3 mg/kg). This is probably because at this location near the bridge it was where the road construction was in progress.

Possible contributors to the high levels of lead include sewage effluent and leaching from the nearby dumping site. Also the agricultural activities around the rivers may have contributed to the observed high levels of lead and cadmium levels, and since these metals can occur as impurities in fertilizers and in metal-based pesticides and compost and manure. With respect to cadmium, contribution from roadside contamination is very unlikely since the contamination of roadside soil and vegetation by cadmium is usually less anomalous than that of lead. The lead mean of sample collection sites exceed the WHO and South African water quality guidelines of 15, as shown in Table 1 (WHO, 1991; SADWAF, 1996). As for cadmium the values shown are well within the guidelines of WHO and South African Department of Water Affairs and Forestry. This comparison is important since the river is used daily by the locals for domestic purposes without further treatment. There is therefore, cause for concern with regards to lead since with respect to human health the effects of lead are the same irrespective of whether it is inhaled or ingested (ATSDR, 1997). Lead can cause irreversible central nervous system damage and decreased intelligence at extremely low doses (ATSDR, 1997). At higher levels of exposure anaemia may result, along with severe kidney damage (ATSDR, 1997). Children are especially susceptible to lead poisoning because they absorb and retain more lead in proportion to their weight than adults (ATSDR, 1997).

Sediment levels of lead can range from 10 to 500 mg/kg. Levels over 500 mg/kg are classified as contaminated, and thus unfit for agricultural or recreational use (South African Guidelines Maritz and Swanopoel, 1998). Therefore, compared to these environmental levels that of the mean of lead which is 662.3 mg/kg in Table 2 is indicative of severe contamination which is very toxic and unfit for human assumption. Figure 3 also shows very high levels of lead in the sediment compared to water samples. The lowest levels were found for cadmium; RW (11.6 mg/kg) and BW (5.6 mg/kg). Background levels of cadmium in the environment are extremely low (Bryan and Langston, 1992). Sediment levels can range from 0.2 to 10 mg/kg, with levels over 3 mg/kg classified as contaminated, and thus unfit for agricultural or recreational use (South African Guidelines, Maritz and Swanopoel, 1998). Therefore, compared to these environmental levels the mean of the Cadmium which is 7.2 mg/kg in Table 2, is indicative of severe contamination which is very toxic and unfit for human assumption. Cadmium has no biological function, and is highly toxic to both animals and plants. The low concentrations of cadmium usually encountered in the environment do not cause acute toxicity, however elevations above background concentrations can have deleterious effects on plant and animal health (Bryan and Langston, 1992; Alloway, 1990). This raises a concern since the river is in constant use by man and animals habituating near the river waters.

The value of the mean concentration of zinc in the sam-

Table 1. Mean of heavy metal concentration (ds/m) in the waters of Umzimvubu River (AAS) compared to international guidelines for drinking water.

RW	RW	RE	BW	BE	Mean
Metal = Cd. SA(5), WHO(3), WA(1)					
August 2010	0.5	0.2	3.0	1.0	1.2
January 2013	0.7	0.3	2.8	1.2	1.3
February 2013	0.5	0.4	2.9	0.9	1.2
March 2013	0.4	0.1	3.1	1.1	1.2
Mean	0.5	0.3	3.0	1.1	1.2
Metal = Zn. SA(2000), WHO(~), WA(200)					
August 2010	29.9	30.0	19.0	12.4	22.8
January 2013	33.0	28.0	19.5	12.6	23.3
February 2013	32.9	29.4	19.1	11.6	23.3
March 2013	29.4	32.o	18.0	12.0	22.9
Mean	31.3	29.9	18.9	12.2	23.1
Metal = Ni. SA(150), WHO(20), WA(0.4)					
August 2010	5.5	7.8	9.2	9.1	7.9
January 2013	5.0	7.7	9.0	9.0	7.7
February 2013	4.8	7.9	9.4	8.8	7.7
March 2013	5.0	7.8	9.5	8.9	7.8
Mean	5.1	7.8	9.3	9.0	7.8
Metal = Pb. SA(10), WHO(10), WA(40)					
August 2010	10.0	20.0	30.0	40.0	25.0
January 2013	12.1	20.2	30.2	37.8	25.1
February 2013	9.9	18.2	33.0	44.0	26.3
March 2013	9.8	18.8	26.1	36.1	22.7
Mean	10.5	19.3	29.8	39.5	24.8

SA = South Africa (Department of Water Affairs and Forestry) guideline for drinking water; WHO = World Health Organisation guideline for drinking water; WA = World Average of trace elements in unpolluted rivers (Helmer and Meybeck, 1989; Boyle and Schiller, 1987); RW = River mouth west side; RE = River mouth east side; BW = West side of the river near the bridge; BE = east side of the river near the bridge.

ples shown in Table 1 is well within the guidelines of South African Department of Water Affairs and Forestry (SADWAF, 1996). This comparison is important since the river is used daily by the locals for domestic purposes without further treatment. The concentrations of zinc at RW and RE are 31.3 and 29.9 µg/l respectively and they are significantly higher than the concentrations of zinc observed at BW and BE which are 18.9 and 12.2 µg/l respectively. These higher concentrations at RW and RE are probably due to the high concentration of low cost houses, roofed with zinc coated corrugated iron sheets. These may have contributed to the higher levels of zinc at RW and RE, since these houses are found in larger numbers near RW and RE. It is therefore possible that zinc may have been leached out from the roofs into the surrounding environment.

Background levels of zinc in sediment are usually quoted as being less than 100 mg/kg (Goncalves, 1990; Bryan and Langston, 1992; Licheng and Kezhun, 1992; ATSDR, 1997). Sediments levels can range from 10 to 750 mg/kg, with levels greater than this being indicative of contamination (Maritz and Swanopoel, 1998). The mean of the zinc concentration for the sediments samples was nearly 61.7 mg/kg as shown in Table 2. The values of RW, RE, BW and BE shown in Table 2 are well within the guidelines of South African Department of Water Affairs and Forestry (SADWAF, 1996). And this data is indicative of the presence of zinc but without any severe contamination. Although zinc is not regarded as being especially toxic, it is sometimes released into the environment in large quantities, and can thus have deleterious effects on certain species at these concentrations. For example, effects on fertilisation and embryonic development have been observed in species of fish and harpacticoid copepods (Ojaveer et al., 1980; Verriopoulos and Hardouvelis, 1988). In terms of human health, zinc exposure via inhalation can cause a specific short-term disease called metal fume fever. Not much is known about the long-term effects of ingesting too much zinc, through food, water or dietary supplements. It is an

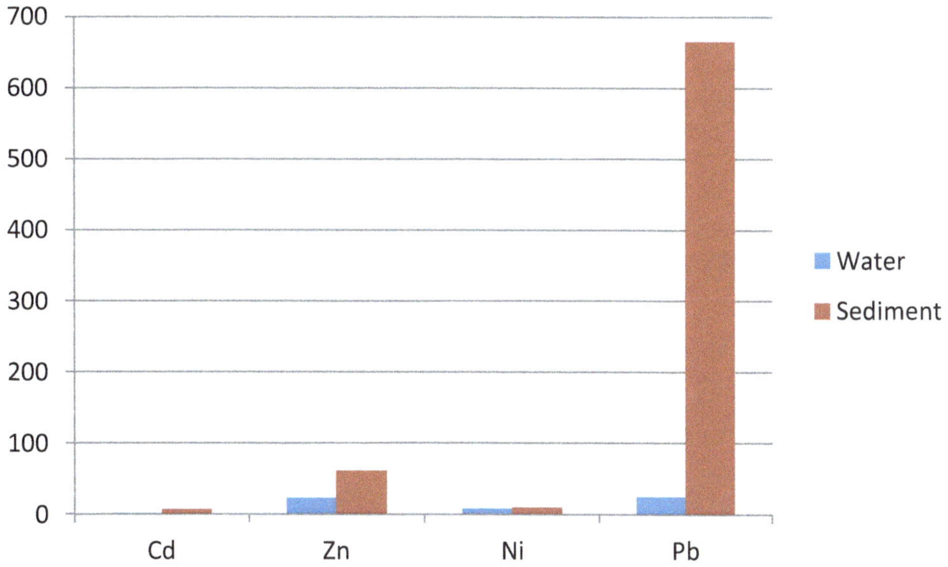

Figure 3. Comparison between sediment and water levels of the heavy metals Cd, Zn, Ni and Pb in Mzimvubu river estuary.

Table 2. Mean of heavy metal concentration in the sediments (mg/kg) of Umzimvubu River (AAS) compared to South African Guidelines.

RW	RW	RE	BW	BE	Mean
Metal = Cd. SA(5), WHO(3), WA(1)					
August 2010	12.1	10.2	5.2	1.3	7.2
January 2013	11.0	10.9	6.0	1.0	7.2
February 2013	13.0	9.2	6.2	1.1	7.4
March 2013	10.3	10.8	4.9	1.7	6.9
Mean	11.6	10.3	5.6	1.3	7.2
Metal = Zn. SA(2000), WHO(~), WA(200)					
August 2010	83.2	61.2	62.1	40.4	61.7
January 2013	77.7	67.0	60.5	42.2	61.9
February 2013	79.2	60.9	61.9	44.0	61.5
March 2013	83.0	59.2	63.7	40.9	61.7
Mean	80.8	62.1	62.1	41.9	61.7
Metal = Ni. SA(150), WHO(20), WA(0.4)					
August 2010	17.8	8.6	10.1	2.1	9.7
January 2013	18.8	9.4	9.8	1.9	10.0
February 2013	20.0	8.5	9.4	2.5	10.1
March 2013	15.8	8.1	10.3	2.0	9.1
Mean	18.1	8.7	9.9	2.1	9.7
Metal = Pb. SA(10), WHO(10), WA(40)					
August 2010	330.2	881.2	741.0	696.7	662.3
January 2013	300.7	890.0	730.1	729.9	662.7
February 2013	340.1	799.3	790.9	669.5	650.0
March 2013	350.2	901.2	778.3	710.2	685.0
Mean	330.3	867.9	760.1	701.6	665.0

Proposed South African Guidelines (Maritz and Swanopoel, 1998); RW = River mouth west side; RE = RIVER mouth east side; BW = West side of the river near the bridge; BE = East side of the river near the bridge.

essential trace element, but ingestion of quantities higher than the recommended levels can have adverse effects on health. The recommended Dietary Allowances for zinc are 15 mg/day for men and 12 mg/day for women. If doses 10 –15 times higher than these recommendations are taken by mouth, even for a short period of time, stomach cramps, nausea and vomiting may occur (ATSDR 1997). Ingesting high levels for several months may cause anaemia, damage to the pancreas, and decreased levels of high-density lipoprotein (HDL) cholesterol (ATSDR, 1997).

The value of the mean concentration of nickel in the samples shown in Table 1 is well within the guidelines of WHO and South African water quality guidelines (WHO, 1991; SADWAF, 1996). The concentrations of RW and RE are 5.1 and 7.8 µg/l respectively and they are significantly lower than those observed in BW and BE which are 9.3 and 9.0 µg/l respectively. This is due to the fact that BW and BE are close to the farms which are suspected to contribute to the availability of Nickel in the river water since these farms use and deposit phosphate fertilizers containing traces of nickel. Another possible contributor may be the anthropogenic sources of nickel including scrap metal waste, notably alloyed metals including stainless steel. Where BW and BE samples collected, it was noted that there were rusted cars beneath the river which can also be a source of the metal in the river.

Background concentrations of nickel are usually quoted as being less than 50 mg/kg (Bryan and Langston, 1992; Solomans and Forstner, 1984). Sediment levels range from 10 to 20 mg/kg, with levels above 50 mg/kg being indicative of contamination (Maritz and Swanopoel, 1998). The concentrations of the sediment samples from RW, RE, BW, and BE are 5.1, 7.8, 9.3 and 9.0 mg/kg, respectively, falling within the recommended standards by WHO and South African water quality guidelines (WHO, 1991; SADWAF, 1996). Hence it is concluded that Nickel in these river waters is not in severe contamination. Nickel is considered an essential trace element at very low concentrations. It does bioaccumulate in aquatic systems, and as such elevations above normal concentrations can result in deleterious effects on aquatic life. The most common adverse health effects of nickel in humans are an allergic reaction and cancer (ATSDR, 1997).

CONCLUSION AND RECOMMENDATIONS

Throughout the research it was evident that there are heavy metals such as lead, cadmium, zinc and nickel in the Umzimvubu River, where the availability of lead is a serious concern since it is in higher concentrations than recommended standards by WHO and South African water quality guidelines. Cadmium was found in low concentrations. Zinc and Nickel were also found but they are not available in severe contaminations since both are within the WHO and SADWAF guidelines. This study was a preliminary pilot, aimed at detecting and determining levels of some of the heavy metals of common concern. It has shown that the concentrations of heavy metals in this river cannot be ignored any more. It is therefore recommended that a monitoring and evaluation study is conducted for a considerable period of time and a substantial distance upstream, in order to fully characterize the heavy metal load. It is further recommended that bioaccumulation studies are also conducted.

ACKNOWLEDGEMENTS

The support of the Walter Sisulu University Research Development Directorate for the research grant is acknowledged.

REFERENCES

Agency for Toxic Substances and Disease Registry (1997). Toxicological Profiles for Cadmium, Zinc, Nicke and Lead: US Department of Health and Human Services, Public Health Service.

Alloway BJ (1990). Heavy metals in soils: Glashow. London. UK: Blackie and Son Ltd ISBN 0-470-21598-4.

Binning K, Baird D (2001). Survey of Heavy Metals: sediments of the Swartkops River Estuary, Port Elizabeth South Africa, Water SA 27(4):461-466.

Boyle EA, Schiller AM (1987). Variability of Dissolved Trace Metals in the Mississippi river, Geochim Cosmochim Acta, 51:3273-3277.

Bryan GW, Langston WJ (1992). Bioavailability, Accumulation and Effects of Heavy Metals in Sediments with Special Reference to United Kingdom Estuaries: A review. Environ Pollut 76:89-131.

Cawthorn RG and Kruger JF (2004). Petrology and Ni–Cu–PGE Potential of the Insizwa Lobe, Mount Ayliff Intrusion, South Africa, Canadian Mineralogy, 42(2): 303-324, doi: 10.2113/gscanmin.42.2.303

Chow JJ (1970). Lead Accumulation: roadside soil and grass. London, Nature, 255:295-296.

Dmitri FM (1993). Mercury Pollution and Cycling in Aquatic Systems: Studies in environmental science. Elsevier B.V. Michigan.

Eaton DL (1994). Toxicology: chemical occupational and environmental medicine. Philadelphia. WB Sounder's Company, pp. 116-117.

Fatoki OS, Hill SJ (1994). Speciation Studies of Tetraalkyllead and Inorganic Pb(II) in Polluted Roadside: vegetation and soil samples. Int. J. Environ Stud. 46:289-301.

Fatoki OS, Mathabatha S (2001). An assessment of heavy metal pollution in the East London and Port Elizabeth harbours, Water SA, 27(2):233-240.

Foy RH, Lennox SD, Smith RV (2000). Agriculture and Environmental Science: assessing the effectiveness of regulatory control on farm pollution using chemical and biological indices of water quality and pollution. Northern- Ireland. Elsevier Science Ltd.

Goncalves EPR (1990). Sediments and Aquatic Mosses as Pollution indicators for heavy metals in the Ave river basin, Portugal, The Science of the Total Environment, 114:7-24.

Helmer R, Meybeck M (1989). The Quality of Water from Pristine Stage to Global Pollution: Palaeography, Palaeoclimatology, Palaeoecology (Global and planetary change section), 75(283):309-319.

Henzen MR, Pieterse MJ (1978). Acidic mine drainage in the Republic of South Africa. Progress in Water Technology, 9:981-1000.

Horenz A (1988). Bioaccumulation of Chromium, Copper and Iron in the Organs and Tissues of Clarias Gariepinus in the Olifants River: Kruger national park. Water SA, 26 (4):569-582.

Licheng Z, Kezhun Z (1992). Background Values: trace elements in the source area of the Yangtze River. The Science of the Total Environment, 125:391-404.

Lopes JF, Dias JM, Cardoso AC, Silva CIV (2005). The water quality of the Ria de Aveiro lagoon, Portugal: From the observations to the implementation of a numerical model. Mar. Environ. Res. 60:594-628.

Maritz NJ, Swanopoel JJ (1998). Chemical Analysis of Dredged Silt from Port Elizabeth Harbour: laboratory buiding, Braamfontein, Johannesburg. Report No CL 41.

Mombeshora C, Osinbanjo O, Ajayi SO (1983). Pollution Studies on Nigerian Rivers: the onset of lead pollution of surface waters in Ibadan, Nigeria, Environment International, 9:81-84.

Moses M (2006). An Investigation into the Negative External Impact of Water Pollution, Public Policy Options and Coping Strategies – with specific reference to the Lotus River Catchment area, Submitted in partial fulfilment of the requirement for the degree of M.Com (Economics), Department of Economics, University of the Western Cape.

Motto HL (1970). Lead in Soil and Plants: its relationship to traffic volume and proximity to highways, USA, Environ. Sci. Technol. 4:231-234.

Njar GN, Iwara AI, Offiong RA, Deekor TD (2012). Assessment of Heavy Metal Status of Boreholes in Calabar South Local Government Area, Cross River State, Nigeria. Ethiop. J. Environ. Stud. Manag. 5(1), 86-91. DOI: http://dx.doi.org/10.4314/ejesm.v5i1.10

Nriagu JO (1978). Lead in Soils, Sediments and Major Rock Types, In: Nriagu JO (Ed): the Biogeochemistry of Lead in the Environment, North-Holland, Elsevier. pp. 15-72.

Ojaveer E (1980). Effects of Copper, Cadmium and Zinc on the Embryonic Development of Baltic Spring Spawning Herring, Marine Research, 247:135-140.

Okonkwo JO, Ndou LW, Marais PJJG (2003). Lead Concentration in Roadside Soils and Vegetation: a measure of motor vehicle pollution in Thohoyandou town. South Africa. Int. J. Environ. Stud. 60:2-3.

Olade MA (1987). Heavy Metal Pollution and the Need for Monitoring: illustrated for developing countries in West Africa, in: Hutchinson TC and Meema KM (Eds) (1987). Lead, Mercury, Cadmium and Arsenic in the Environment, John Wiley and Sons Ltd, Lagos, Nigeria.

Quinche PJ, Zube R (1969). Lead Pollution of Highway Roadside: vegetation. London. Phytopathology, Z.66:259-274.

Robinson J (1996). Evaluation of Health Assessment index with reference to bioaccumulation of metals in Oreochromis Mossambicus and aspects of the morphology of Lernaea cyprinacea, Linnaeus, M.Sc. Thesis. South Africa. Rand Afrikaans University.

Solomans W, Forstner U (1984). Metals in the Hydrocycle. Springer-Verlag, Berlin, Heidelberg, New York, Tokyo. ISBN 3-540-12755-0.

Sander BK, Cawthorn RG (1995). Laminated olivine spinifex textures in the Mount Ayliff intrusion, South Africa, Mineralogy and Petrology, 54(1)1-10, DOI: 10.1007/BF01162753

Smith R, Siebert ML, Hatting WHJ (1980). Removal of inorganic pollutants from waste water during reclamation for potable re-use. Water SA, 6(2):92-95.

South African Department of Water Affairs and Forestry (SADWAF) (1996). Water Quality guidelines for Domestic Use, 2nd Ed, vol. 1.

South African Department of Water Affairs and Forestry (1996). South African Water Quality Guidelines, Volume 7: Aquatic Ecosystems, First Edition.

Van der Linde GJ, Pitse AM (2006). The South African Fertiliser Industry, The Fertilizer Society of South Africa. (Paper presented at the AFA Conference, Cairo, Egypt, February 2006).

Van Hille RP, Boshoff GA, Rose PD, Duncan JR (1999). A continuous process for the biological treatment of heavy metals contaminated acid mine water, Resources Conservation and Recycling, 27:157-167.

Van Niekerk H, (ed) (2000). On the Identification and Prioritisation of Areas in South Africa with a Potentially High Health Risk Due to Faecally Polluted Surface Water. A NMMP First Report, August 2000.

Vernet JP (1991). Heavy metals in the environment. Elsevier Science Publishers, Amsterdam, Netherlands. ISBN0-444-89064-5.

WHO (1991). Guidelines for Drinking Water Quality, Recommendations, vol. 1. World Health Organisation, Geneva.

Yabe J, Ishizuka M, Umemura T (2010). Current levels of heavy metal pollution in Africa. J. Vet. Med. Sci. 72(10):1257-63.

Young JE (1992). Mining the Earth, Worldwatch Paper No. 109, Washington D.C. Worldwatch Institute, p. 21.

17

Assessment of some heavy metals and physicochemical properties in surface soils of municipal open waste dumpsite in Yenagoa, Nigeria

AMOS-TAUTUA, Bamidele Martin W.[1] , ONIGBINDE, Adebayo O.[2] and ERE, Diepreye[1]

[1]Department of Chemical Sciences, Faculty of Science, Niger Delta University, Wilberforce, P.M.B 071, Yenagoa, Bayelsa State, Nigeria.
[2]Chemistry Unit, Basic Sciences Department, Babcock University, Ilishan, Remo, Ogun State, Nigeria.

The current study was designed for the assessment of lead, cadmium and chromium and some physicochemical properties of soils collected from an open dumpsite in Yenagoa, Nigeria. Surface soil samples at two depths (0-10 and 10-20 cm) were randomly collected at the dump field and control site, and were analyzed for physicochemical parameters and contamination by lead, chromium and cadmium using standard analytical methods. The results show that the main dumpsite had higher sand (>80.0%) and lower clay and silt contents than the control site. Soil mean pH varied between 4.89±0.05 in the control and 7.60±0.02 in the dump. Total nitrogen (N) content of the dump soils ranged from 0.06±0.07 to 0.24±0.09% and is slightly higher than that of the control soil. This is reflected in the high value of organic matter (4.71±0.85%) in dump soils. Available P was quite high ranging from 35.00±1.01 to 84.20±1.02 mg/kg. Cation exchange capacity (CEC) varied between 12.98±0.31 and 91.07±0.11 cmol kg^{-1}. ECEC levels were moderate to high ranging from 14.10±0.10 to 91.47±0.11 cmol/kg. All the soil samples had very high base saturation (>90.0%) and exchangeable Ca, Mg, K and Na, far above the critical levels set by FAO for agricultural soil. Average levels of Pb ranged from 14.75±0.05 to 16.14±0.04 mg/kg in the dump and 8.35±0.05 to 8.78±0.07 mg/kg in the control. Mean concentration of Cr in the dump soil varied between 0.05±0.01 and 0.06±0.01 mg/kg, and is slightly higher than the control (0.005±0.01 mg/kg), while Cd was found in trace amounts (<0.0001±0.01 mg/kg). These values are all far below the maximum tolerable levels set by FAO and WHO for agricultural soil. It is suggested that the dumpsite and the control area with their adequate soil nutrients and low levels of metals should eventually be converted to agricultural farmland. No remediation is needed at this time.

Key words: Dump waste soil, heavy metal, soil fertility.

INTRODUCTION

Landfills have long been used as repositories for industries, municipal and commercial wastes. Nigeria, a developing country with non adequate waste disposal or recycling processes is at a risk of metal and organo-metallic contamination of its soil and surface water bodies, which poses health hazard and soil deterioration for agricultural purposes. Yenagoa Township, the capital city of Bayelsa State is located in a humid tropical, wetland area where high rainfall and temperatures favour rapid degradation of organic materials (Ayolagha, 2001).

In Bayelsa State, solid wastes are handled by the Bayelsa State Ministry of Environment. The Ministry has allocated an extent of land along Yenagoa-Tombia road for disposal of solid wastes collected from Yenagoa town and its environs. As Yenagoa is a non-industrialized city, the refuse generated within the city comprise largely of degradable materials from markets, offices, hospitals and households such as garbage, plastics, textiles, stationeries, sludge from sewage, dead animals, ashes, wood, food and farm waste products. However, metallic materials from damaged vehicle parts, electronics, computers, cans, etc, are also disposed in the same way as the other non-metallic materials, thereby constituting a source of metal contamination.

Open dumps are generally unsanitary and constitute malodorous places in which disease-carrying vermin such as rats and flies proliferate (Bellebaum, 2005). Methane and other gases are released into the surrounding air as microorganisms decompose the solid wastes and fires pollute the air with acrid smoke and other numerous volatiles. Liquids that ooze and seep through the solid waste heap ultimately reach the soil, surface water and ground water. Hazardous materials such as heavy metals, pesticides and hydrocarbons that are dissolved in this liquid often contaminate soil and water (Adelekan and Alawode, 2011). Anikwe and Nwobodo (2001) suggested that continuous disposal of municipal waste on soil may lead to increase in heavy metals in the soil and surface water that would be inimical to deep feeding plants. Heavy metals such as arsenic, cadmium, lead, chromium, nickel, cobalt and mercury are of concern primarily because of their ability to harm soil organisms, plants, animals and human beings (Adelekan and Abegunde, 2011). More emphatic are the untreated dumplings that rapidly increase soil toxicity making such large area dumpsites potentially hazardous for agricultural purposes. Yet these workers (Anikwe and Nwobodo, 2001; Adelekan and Alawode, 2011; Adelekan and Abegunde, 2011) also indicate that municipal waste dumpsites bear soils that are sufficiently rich in organic matter that would be acceptable for surface feeder plants. Consequently, Brady (1996) and Helmore and Ralla (1995) reported that open dump fields perform a dual purpose of safe disposal of wastes and simultaneously create improved physical and chemical properties of soils that constitute productive agricultural fields. Other studies have also revealed that dumpsites around two major cities in Nigeria could be effectively utilized for residential and agricultural purposes without risk of heavy metal toxicity (Urunmatsoma and Ikhouria, 2005; Asalawalam and Eke, 2006). Old dump fields therefore can be seen to provide farmers with fertile plots for cultivation of vegetables and other surface feeder crops.

Public concern about environmental pollution has focused attention on the disposal of urban and industrial waste hence this study intends to determine the contents of lead, chromium and cadmium and the fertility status of the soil of the dump field in Yenagoa area in view of interpreting the suitability of its soil for crop production.

MATERIALS AND METHODS

Study area

Yenagoa, the capital city of Bayelsa State, Nigeria lies between latitude 4° 50' to 5° 00' North and longitude 6° 11' to 6° 25' East. The town is located in a humid tropical wetland area with mean annual rainfall of about 2539 mm and an average mean temperature of 26.2°C. The dump field investigated is located opposite HPEB 119 Company along the Yenagoa-Tombia road lying between latitude 4° 58' 59.20"North and longitude 6° 19' 18.20"East (Ayolagha, 2001).

Waste soil sampling

Surface soil samples, after removing the overlying wastes, were collected randomly from the dump field at two depths; 0-10 and 10-20 cm using a Dutch soil auger and a spatula. Another set of soil samples were also collected at the same depths, from uncultivated land, at 100 m away from the dump site, to serve as control. The soils were air dried for three days, ground and sieved through a 2 mm sieve. These were stored in labelled polythene bags and were taken to the laboratory for analysis.

Analyses of some physico-chemical parameters of soil samples

Particle size distribution otherwise known as mechanical analysis was determined by hydrometer method (Bouyoucos, 1962) using sodium hexametaphosphate as dispersant. The texture class was also determined using the 'textured triangular diagram' (Loganathan, 1984).

Soil pH was measured in water suspension (1:2.5) using the glass electrode coupled pH meter. The potential acidity was measured in a 1:10 (w/v) ration of soil to solution of 1 M KCl.

The cation exchange capacity (CEC) was determined by extracting the cations with 1 M ammonium acetate buffered at pH 7. 30 ml of 1 M CH_3COONH_4 was added to 5 g of soil. The suspension was shaken for 2 h and then centrifuged (15 min, 6000 rpm). After centrifugation and filtration, the filtrate was transferred into a 100 ml flask and two other volumes of 30 ml ammonium acetate were added successively after 30 min of agitation and centrifugation. The final filtrates were completed to 100 ml with ammonium acetate solution.

Calcium (Ca) and magnesium (Mg) were determined by EDTA titration while potassium (K) and sodium (Na) were determined by flame photometry. Exchangeable acidity (EA) was determined by titration method (Juo, 1979). The effective cation exchange capacity (ECEC) was calculated as the total exchangeable bases plus exchangeable acidity.

Percentage base saturation (BS%) was calculated as the percentage of the sum of exchangeable bases divided by ECEC. Available phosphorous (Av. P) was extracted with Bray solution 11 and the phosphorous determined by the molybdenum method described by Udo and Ogunwale (1978). The percent organic matter (%OM) was calculated from the percent organic carbon (OC%) measured using Walker-Black (1934) wet oxidation method. Total nitrogen (TN) was determined using the modified Kjeldahl distillation methods (Juo, 1979).

Sample preparation and analyses of heavy metals

One gram of each of the sieved soil samples was digested using the nitric/perchloric acid digestion procedure, as described by Odu et al. (1986). The concentrations of heavy metals, Pb, Cr and Cd

Table 1. Physical properties/particle size distribution and metal concentrations of soil samples

Sample Code	Depth (cm)	Sand (%)	Silt (%)	Clay (%)	Texture	pH H_2O	Pb (mg/kg)	Cr (mg/kg)	Cd (mg/kg)
DS	0-10	87.40 ±0.03	6.12 ±0.02	6.48 ±0.03	Loamy sand	7.09 ±0.02	14.75±0.04	0.05 ±0.01	<0.0001±0.01
	10-20	91.42 ±0.02	3.14 ±0.02	5.46 ±0.02	Sandy	7.60 ±0.01	16.14±0.05	0.06 ±0.01	<0.0001±0.01
CS	0-10	39.41 ±0.01	16.13±0.01	46.47 ±0.04	Clay	5.01 ±0.03	8.78 ±0.07	0.005±0.01	<0.0001±0.01
	10-20	37.40 ±0.03	14.15±0.01	48.45 ±0.03		4.89 ±0.05	8.35 ±0.05	0.005±0.01	<0.0001±0.01

DS = Dumpsite; CS = control site. Data represent mean ± SD of two replicates.

Table 2. Chemical properties of soil samples.

Sample Code	Depth (cm)	pH KCl	pH H_2O	ΔpH	Av. P (mg/kg	OC	OM	TN	C/N	Exchangeable cation Ca	Mg	K	Na	CEC	EA	ECEC	BS (%)
						←	%	→		←			(cmol/kg)			→	
DS	0-10	5.90 ±0.02	7.09 ±0.02	-2.81 ± 0.01	84.20 ±1.02	2.73 ±1.25	4.71 ±0.85	0.24 ±0.09	11.38 ±0.67	72.80 ±0.05	14.00 ±0.12	2.36 ±0.06	1.91 ±0.23	91.07 ±0.11	0.40 ±0.04	91.47 ±0.11	99.56 ±0.03
	10-20	5.86 ±0.01	7.60 ±0.01	-2.26 ± 0.01	68.22 ±0.89	0.60 ±1.07	1.03 ±0.64	0.06 ±0.07	10.00 ±0.57	45.60 ±0.10	12.40 ±0.07	1.13 ±0.10	1.95 ±0.08	61.08 ±0.09	0.96 ±0.06	62.04 ±0.10	98.45 ±0.05
CS	0-10	4.32 ±0.03	5.01 ±0.03	-1.21 ± 0.03	37.12 ±0.85	1.98 ±1.03	3.38 ±0.56	0.15 ±0.86	13.20 ±0.94	10.35 ±0.09	5.72 ±0.11	0.26 ±0.11	0.48 ±0.17	16.81 ±0.12	1.66 ±0.87	18.47 ±0.09	91.01 ±0.02
	10-20	4.26 ±0.02	4.89 ±0.05	-1.37 ±0.04	35.00 ±1.01	1.87 ±1.06	3.22 ±0.55	0.14 ±0.43	13.36 ±0.74	8.00 ±0.08	4.40 ±0.65	0.17 ±0.20	0.41 ±0.09	12.98 ±0.31	1.12 ±0.66	14.10 ±0.10	92.06 ±0.04

DS, Dumpsite; CS, control site; Av. P, available phosphorous; OC, organic carbon; OM, organic matter; TN, total nitrogen; C/N, carbon/nitrogen ratio; CEC, cation exchangeable capacity; EA, exchangeable acid; ECEC, effective cation exchangeable capacity; BS, base saturation. Data represent mean ± SD of two replicates.

were determined using atomic absorption spectrophotometer (UnicamSolaar32 model) following the standard procedures as given in APHA (1995). All analyses were done in duplicates.

RESULTS AND DISCUSSION

Characteristics of the soil samples

The results expressed as mean ± standard deviation of physico-chemical properties of the soils are presented in Tables 1 and 2.

Textural properties

Texture is related to certain physical properties of soil such as plasticity, permeability, ease of tillage, fertility, water holding capacity and overall soil productivity. For instance, for irrigation purposes, loamy and clay textures are classed as soils of high moisture holding capacity while loamy sands and sands have low moisture holding capacity (Brady, 1996).

In this study, the main dumpsite had higher sand and lower clay and silt contents than the control site. This was corroborated by Ideriah et al. (2006) who determined the soil quality around a solid waste dumpsite in Port-Harcourt, Nigeria, which is in the same humid tropical wetland area as Yenagoa.

According to Nyles and Ray (1999), soils with separate high sand and low clay content have high pollutant leaching potentials. Though the soils of the dumpsites predominantly contain high sand fractions (>80.0%) that allows high permeability of water and leacheates, the textural class (loamy sand) may be suitable for sanitary landfills (Loughry, 1973). The soils from the control sites consisted of moderately high clay fractions thus they exhibit plasticity and encourage surface water flooding and pollution. This clayey texture of the control soils also favours low permeability of water and leacheates (Ahn, 1993).

pH measurement

pH affects the mobility of heavy metals in soil. It has been found that soil pH is correlated with the availability of nutrients to the plant (Gray et al., 1998). Consequently, as pH decreases, the solubility of metallic elements in the soil increases and they become more readily available to plants (Oliver et al., 1998; Salam and Helmke, 1998).

Lower pH would favours availability, mobility and redistribution of the metals Pb and Cd in the various fractions due to increased solubility of the ions in acidic environment (Oviasogie and Ndiokwere, 2008). In the present study, soil pH (in H_2O) ranged from 4.89±0.05 in the control to 7.60±0.02 in the dump indicating a moderate acidic to a neutral reaction. The moderately acidic soil from the control site may tend to have an increased micronutrient solubility and mobility as well as increased heavy metal concentration in the soil (Odu et al., 1985). The value of KCl pH bears a strong correlation with aluminium saturation. Aluminium displaced by K^+ on the exchange complex consumes OH^- ions and increases $[H^+]$. The KCl pH values in the soils from the control are less than 5.2 which showed that exchangeable aluminium is present in the soils while the values in the dump soils were greater than 5.2 which are indicative of the presence of non-exchangeable aluminium. This could be attributed to hydrolysis, polymerization and precipitation (USDANRCS, 2004)

ΔpH is the numerical difference between the values of pH measured in KCl and H_2O, which results from the displacement of OH^- ions by Cl^- ions. It also highlights the displacement of the ions H^+ adsorbed on the exchange sites of the adsorbing complex from soil towards the soil solution (Yobouet et al., 2010). Since the ΔpH values for all the soil samples were negative, it means that the colloids have net negative charge which reflects the cation-exchange capacity as seen here which is very high (USDANRCS, 2004).

Organic carbon and soil organic matter

The presence of organic carbon increases the cation exchange capacity of the soil which retains nutrients assimilated by plants. Total organic carbon in the soils under investigation in this study was low to moderate ranging from 0.6±1.07 to 2.73±1.25% as indicated in Table 2. The moderately high amount of organic carbon of the refuse dump soils is suggestive of degradation or presence of degradable and compostable wastes (Munoz et al., 1994).

Soil organic matter (SOM) enhances the usefulness of soils for agricultural purposes. It supplies essential nutrients and has unexcelled capacity to hold water and absorb cations. It also functions as a source of food for soil microbes and thereby helps enhance and control their activities (Brady, 1996). The organic matter in the soil samples varied from 1.03±0.64 to 4.71±0.85%. The dump soils contain high amount of organic matter, about 4.71±0.85% which may be responsible for increase in the soil pH as compared to that of the control soil. This observation was supported by Oyedele et al. (2008) who reported that dump sites had significantly higher pH regime and soil organic matter as compared to the control soil. Ayolagha and Onwugbuta (2001) also demonstrated that high OM (>2.0%) in soils is conducive for heavy metal chelation formation.

The total nitrogen content

The total N content of dump soils ranged between 0.06±0.07 and 0.24±0.09% which is slightly higher than those of the control site which varied between 0.14±0.43 and 0.15±0.86%. The C/N calculated ranged from 10.00±0.57 in dump soil to 13.36±0.74 in control soil. Our result was supported by Brady (1996) who reported that C/N ratio in the topsoil is commonly between 10 and 12 in humid regions. Onwurah (2000) stated that C/N ratio is an index of biomass and they are responsible for the general restoration of microbial flora. Generally, if the ratio is below 20, this is often adequate to satisfy the N requirements of micro flora that decompose the residues (Eja et al., 2003) According to Idoriah et al. (2006), the low values of C/N ratios showed high decomposition and efficient mineralization process of the dump area. The waste dump C/N ratio is between 10.00±0.57 and 11.38±0.67 and it is lower than that of the control soil which is between 13.20±0.94 and 13.36±0.74. Therefore, the waste dump most likely contributed significantly to the very high levels of these soil properties.

Available phosphorous

The dumpsite had higher levels of P ranging from 68.22±0.89 to 84.20±1.02 mg/kg as compared to the control which varied between 35.00±1.01 and 37.12± 0.85 mg/kg; this could be attributed to the presence of high amount of organic matter and plants decomposition

(Ideriah et al., 2006). The high concentration of phosphorous contributes to good growth of plants as was observed. All the soil samples had available P values more than 10 mg/kg considered suitable for crop production (FAO, 1976).

Cation exchange capacity

The cation exchange capacity is the amount of exchangeable cation per unit weight of dry soil that plays an important role in soil fertility. It depends especially on the pH, clay and on the soil organic matter content. Results of this study revealed that soils from the dump had higher values of CEC ranging from 61.08±0.09 to 91.07±0.11 cmol kg^{-1} as compared to that of the control site which varied between 12.98±0.12 and 16.81±0.31 cmol kg^{-1}, although the clay content at the dumpsite was much smaller than that at the control site. So it is possible that a large part of exchangeable bases at the dumpsite must have been existing as a water-soluble form rather than an exchangeable form adsorbed at cation exchange sites. The increase in the TEB and organic matter in the dump field may result in plants taking up nutrients more easily (Aydinalp and Marinova, 2003).

ECEC levels were moderate to high varying from 14.10±0.10 to 91.47±0.11 cmol/kg. However, the ECEC status of dump soils was higher than those of the control and far above 20 cmol/kg regarded as being suitable for crop production (FAO, 1976). All the soil samples analyzed had very high base saturation (>90.0%) and were higher than 60%, the critical limit established for ecological zone (Holland et al., 1989). This also followed the pattern reported by Isirimah et al. (2003) for productive agricultural soil. The exchangeable acidity (EA) was generally low varying between 0.40±0.04 and 1.66±0.87 cmol/kg.

All the soil samples had high Ca values above 4.0 cmol/kg regarded as lower limit for fertile soils (FAO 1976). Exchangeable K varied from 1.13±0.10 to 2.31±0.06 cmol/kg in dump soils; these are higher than the levels in control soils and above 0.2 cmol/kg which is regarded as the critical limit of exchangeable K in soils (Unamba-Opara, 1985). The implication of these is that the soils are rich in nutrients and therefore an indication of good yield potential without any input of fertilizers.

Heavy metal concentrations

The evaluation of dumpsite soils for the concentration levels of toxic elements is essential for healthy crop production, thus this study has endeavoured to determine the levels of Pb, Cr and Cd in the various soil samples.

The distribution of mean concentration± standard deviation of the metals present in the soils is shown in Table 1. The result shows that the metal loads from the refuse dump soils were found to be slightly higher than the control area (100 m from the dumpsite) with the exception of Cd which was found in traces in all the soils. Our results are collaborated by Al-Turki and Helal (2004) and Ren et al. (2005) who reported that lead and cadmium are anthropogenic metals, and without external interference, are normally not abundant in upper layer soils.

Lead

The mean levels of Pb ranged from 14.75±0.04 to16.14±0.05 mg/kg in the soil samples from the dumpsite and 8.35±0.05 to 8.78±0.07 mg/kg in the control. These values were lower than EC (1986) upper limit of 300 mg/kg and the maximum tolerable levels proposed for agricultural soil, 90-400 mg/kg set by WHO (1993) and NEPCA (2010). This is in agreement with the results obtained from similar study by Umoh and Etim (2013) for soils from dumpsites within Ikot-Ekpene in Akwa-Ibom State, Nigeria.

Concentration of lead in the soils from both areas could be as a result of its sources from automobile exhaust fumes as well as dry cell batteries, sewage effluents, runoff of wastes and atmospheric depositions owing to the close proximity of the sites to high vehicular traffic along Yenagoa-Tombia road.

Chromium

The mean concentrations of Cr in the dump soil varied between 0.005±0.01 and 0.200±0.01 mg/kg, which was slightly higher, as compared to the control (0.005±0.01 mg/kg), but are still lower than the critical permissible level which is 50 mg/kg for soil recommended for agriculture by MAFF (1992) and EC (1986). Sources of Cr in the soils could be due to waste consisting of lead-chromium batteries, coloured polythene bags, discarded plastic materials and empty paint containers (Jung et al., 2006).

Cadmium

Cadmium levels in all the soil samples were found in trace amounts (0.0001 ± 0.01 mg/kg) and these values are far lower than the natural limits of 0.01-3.0 mg/kg in soil as given by MAFF (1992) and EC (1986).

The values of the metal concentrations obtained for both sites are all far below the maximum tolerable levels proposed for agricultural soil. This is in agreement with the findings of Asawalam and Eke (2006) and Njoku and Ayoka (2007) who investigated the trace metal concentrations and heavy metal pollutants from dump soils in Owerri, Nigeria.

Even though these heavy metal concentrations fell below the critical permissible concentration level, it seems that their persistence in the soils of the dump site may lead to increased uptake of these heavy metals by plants.

Conclusion

The concentration of the metals, Pb, Cr and Cd are all far below the maximum tolerable levels set by FAO and WHO for agricultural soil. The results of this study have clearly demonstrated that the low levels of heavy metals in the studied soil samples are consistent with the composition of the municipal wastes generated in Yenagoa and its environs. The results also show that the soil is not polluted by various pollutants and not harmful for recreational and agricultural purposes. It is therefore suggested that the dumpsite and the control area with their adequate soil nutrients and low levels of metals should eventually be converted to an agricultural farmland. No remediation is needed at this time.

ACKNOWLEDGEMENTS

The authors are grateful to Mr. Philip of National Root Crop Research Institute, Umuduike and the Staff of Anal Concept Limited, Port-Harcourt, for the analysis and running of the AAS for metal analysis. The assistance received from Mr. Domotimi Gerewei during collection and analysis of samples is acknowledged.

REFERENCES

Adelekan BA, Abegunde KD (2011). Heavy Metals Contamination of Soil and Groundwater at Automobile Mechanic Villages in Ibadan Nigeria. Int. J. Phys. Sci. 6(5):1045-1058.

Adelekan BA, Alawode AO (2011). Contributions of refuse dumps to heavy metal concentrations in soil profile and groundwater in Ibadan, Nigeria. J. Appl. Biosci. 40:7227-2737

Ahn PM (1993). Tropical Soils and Fertilizer Use. Longman Scientific Technical, UK

Al-Turki AI, Helal MID (2004). Mobilization of Pb, Zn, Cu and Cd, in Polluted Soil. Pak. J. Biol. Sci. A. 7:1972-1980.

Anikwe MAN, Nwobodo KCA (2001). Long term effect of municipal waste disposal on soil properties and productivity of sites used for urban agriculture in Abakaliki, Nigeria. Biosour. Technol. 83(3):241-250.

APHA (1995). American Public Health Association. Standard Methods for the Examination of Water and Waste Water, 19th edn. APHA-AWWA-WPCF, Washington, DC. pp. 525-987.

Asawalam DO, Eke CI (2006). Trace metal concentration in soils used for Waste disposal around Owerri, Nigeria. In: Proceedings of the 40th Conference of the Agriculture Society of Nigeria, Michael Okpara University of Agriculture, Umudike, Abia State, Nigeria. pp. 427-430.

Aydinalp C, Marinova S (2003). Distribution and forms of heavy metals in some agricultural soils. Pol. J. Environ. Stud. 12(5):629-630.

Ayolagha GA, Onwugbata GC (2001). Suitability comparison of waste disposal site. 27th Proceedings of the Annual Conference of the Soil Science of Nigeria. Novenber 5-9, 2001, University of Calabar, Nigeria.

Ayolagha GA (2001). Survey and classification of Yenagoa-Meander Belt Soils in the Niger Delta. 27th Proceedings of the Annual Conference of the Soil Science of Nigeria. Novenber 5-9, 200, University of Calabar, Nigeria. pp. 10-16.

Bellebaum J (2005). Between the Herring Gull Larus argentatus and the Bulldozer: Black- headed Gull Larus ridibundus feeding sites on a Refuse Dump. Ornis Fennica 82:166-171. 2005.

Bouyoucos GJ (1962). Hydrometer method improved for making particle size analyses of soils. Agro. J. 54:464.

Brady NC (1996). The Nature and Properties of Soils, 11th ed.; McMillan: New York. p. 621

EC (council of the European Communities) (1986). Directive 86278 EEC on the Protection of the Environment and in Particular of the Soil when Sewage Sludge is used EEC. Brussels

Eja ME, Udo SM, Asikong BE (2003). Bioremediation potential of *Bacillus species* in oil-polluted soil from auto-mechanic workshops in Calabar, Nigeria. Afr. J. Environ. Pollut. Health 2(1):11-18.

FAO (1976). A Framwork for Land Evaluation. FAO Bulletin 32, FAO/UNESCO, France.

Gray CW, McLaren RG, Roberts AH, Condron LM (1998). Sorption and desorption of cadmium from some New Zealand soils: Effect of pH and contact time. Aust. J. Soil Res. 36:199-216.

Helmore K, Ratta A (1995). The Surprising yields of Urban Agriculture. Choices. The Human Dev. Magazines of UNDP. 4(1).

Holland MD, Allen VG, Burton D, Urphy ST (1989). Land Evaluation and Agricultural Recommendations, Cross River National Park, Oban Division prepared by Oversea Development Natural Resources Institute in collaboration with WWF for the Federal Republic of Nigeria and the Cross Rivers State Government

Ideriah TKJ, Omuaru VOT, Adiukwu PA (2006). Soil quality around a waste dumpsite in Port Harcourt, Nigeria. Afr. J. Ecol. 44:388-394.

Isirimah NO, Igwe C, Iwegbue CMA (2003). Important Ions in Soils Environment; In: Introductory Soil Chemistry and Biology for Agric and Biotech. Osia In'l Pub. Lt. Port Hacourt. pp. 34-97.

Juo ASR (1979). Selected Methods for Soil and Plant Analysis. IITA Manual No.1

Loganathan P (1984). Laboratory Manual of Soil and Plant Analysis. Rivers State University of Science and Technology, Port- Harcourt, Nigeria.

Loughry FG (1973). The Use of Soil Science in Sanitary Landfill Selection and Management. Geoderma 10: 131-139.

MAFF (Ministry of Agriculture, Fisheries and Food) and Welch Office Agriculture Department (1992). Code of Good Agriculture Practice for the Protection of Soil. Draft Consultation Document, MAFF, London

Munoz M, Pena L, Halloroms JO (1994). Use of an Industrial by-product as a liming source. J. Agric. Univ. Puerto Rilo 78 (3-4) 73-86.

National Environment Protection Council of Australia (NEPCA) (2010). Limits of heavy metals in soils. Available online at www.newzealand.govt.nz

Njoku PC, Ayoka AO (2007). Evaluation of heavy metal pollutants from soils at municipal solid waste deposit in Owerri, Imo State, Nigeria. J. Chem. Soc. Nig. 32(1):57-60.

Nyles CB, Ray RN (1999). The Nature and Properties of Soils. 12th Ed. United States of America. pp. 743 - 785

Odu CTI, Babalola O, Udo EJ, Ogunkunle AO, Bakare TA, Adeoye GO (1986). Laboratory manual for agronomic studies in soils, plants and microbiology. Dept of Agronomy, Univ. of Ibadan. p. 83.

Odu CTI, Esurosu OF, Nwaboshi IC, Ogunwale JA (1985). Environmental study (Soil and Vegetation) of Nigeria/Agip Oil Company Operation Area. A report submitted to Nigeria Agip Oil Company Limited, Lagos, Nigeria. pp. 102-107.

Oliver DP, Tiller KG, Alston AM, Cozens GD, Merry RH (1998). Effect of soil pH and applied cadmium on cadmium concentration in wheat grain. Aust. J. Soil Res. 36:571-583

Onwurah IN (2000). A Perspective of Industrial Environmental Biotechnology. Snaap Press Ltd. Enugu.

Oviasogie PO, Ndiokwere CL (2008). Fractionation of lead and cadmium in refuse dump soil treated with cassava milling effluent. J. Agric. Environ. 9:10.

Oyedele DJ, Gasu MB, Awotoye OO (2008). Changes in soil properties and plant uptake of heavy metals on selected municipal solid waste dump sites in Ile-Ife, Nigeria. Afr. J. Environ. Sci. Technol. 3(5):107-115.

Ren HM, Wang JD, Zhang XL (2005). Assessment of Soil Lead Exposure in Children in Shenyang, China. Environ. Pollut. 144:327-335

Salam AK, Helmke PA (1998). The pH dependence of free ionic activities and total dissolved concentrations of copper and cadmium in soil solution. Geoderma 83:281-291

Udo EJ, Ogunwale JA (1978). Laboratory Manual for the Analysis of Soils, Plants and water Samples. Dept. of Agronomy University of Ibadan. p. 45.

Umoh SD, Etim EE (2013). Determination of heavy metals content from dumpsites within Ikot Ekpene, Akwa Ibom State, Nigeria. Int. J. Eng. Sci. 2(2):123-129.

Unamba-Opara I (1985). The Potassium Status of the Sandy Soils of Nothern Imo State, Nigeria. Soil Sci. 139(5):437-445.

Urunmatsoma SO, Ikhuoria EU (2005). Effect of Leachates (heavy metal content) from soil waste at "Effruun round about dumpsite" Warri, Nigeria. Chem Tech. J. 1:195-202.

USDANRCS (2004). United State Department of Agriculture and Natural Resources, Conservation Services. Use of Reaction pH in Soil Taxonomy. Available from www.soil.usda.gov/technical/technotes/notes8.html.

Walker-Black IA (1934). An examination of the Degtajareff method for soil organic matter determination and a proposed modification of the chronic acid titration. Soil Sci. 37:29-38.

World Health Organization (WHO) (1993). Standard maxima for metals in Agricultural soils.

Yobouet YA, Adouby K, Trokourey A, Yao B (2010). Cadmium, copper, lead and zinc speciation in contaminated soils. Int. J. Eng. Sci. Technol. 2(5):802-812

18

Batch adsorption of heavy metals (Cu, Pb, Fe, Cr and Cd) from aqueous solutions using coconut husk

I. E. Agbozu and F. O. Emoruwa

Department of Environmental Science, Federal University of Petroleum Resources, P.M.B. 1221, Effurun, Nigeria.

This study was carried out to evaluate the efficiency of metals (Cu, Fe, Pb, Cr and Cd) removal from mixed metal ions solution using coconut husk as adsorbent. The effects of varying contact time, initial metal ion concentration, adsorbent dose and pH on adsorption process of these metals were studied using synthetically prepared wastewater. The percentage removal of metals increased with increasing weight (0.4-1.2 g) in 50 ml of adsorbent dose and the observed trend was: Cr>Cu>Pb>Fe>Cd. The adsorption efficiency increased with increasing initial metal ion concentration (0.3-0.9 mg/l) and the observed trend was: Cr>Cu>Cd>Fe>Pb. Similarly, percentage removal of metal ions increased with increasing pH of the mixed metal ions solution (pH values of 2, 6 and 10). The observed trend of percentage adsorption of metals by varying pH was: Cd>Fe>Cr>Cu>Pb. The effect of contact time on the adsorption efficiency at different time intervals of 20, 40 and 60 min in mixed metal ions solution showed that the removal of tested metals was rapidly achieved during a short interval of 20 min. Generally, the study showed that coconut husk (a waste material) is a viable material for removal of metals from waste water as the percentage adsorbed varies from 95.2-98.8, 91.1-99.3 and 75.0-98.5% for Cd, Cr and Cu, respectively while the percentage removal of Fe and Pb from the waste water varies from 84.9-97.0 and 81.1-98.7%, respectively. Isothermal studies showed that the experimental data are best fitted on Langmuir model.

Key words: Batch adsorption, heavy metals, wastewater, coconut husk.

INTRODUCTION

Increased use of metals and chemicals in process industries has resulted in generation of large quantities of effluent that contain high level of toxic heavy metals and their presence pose environmental-disposal problems due to their non-degradable and persistence nature. Unlike organic pollutants, the majority of which are susceptible to biological degradation, heavy metal ions do not degrade into harmless end products (Gupta et al., 2001). The presence of heavy metal ions is a major concern due to their toxicity to many life forms. Heavy metal contamination exists in aqueous wastes stream of many industries, such as metal plating, mining operations, tanneries, chloralkali, radiator manufacturing, smelting, alloy industries and storage batteries industries, etc. (Goyal and Ahluwalia, 2007; Olayinka et al., 2007, 2009; Kadirvelu et al., 2001). Treatment processes for heavy metal removal from wastewater include: precipitation, membrane filtration, ion exchange, adsorption and coprecipitation/adsorption. The removal of toxic heavy metal contaminants from aqueous waste streams is currently one of the most important environmental issues being investigated (Igwe et al., 2005; Mondal, 2009; Suthipong and Siranee, 2009). Studies on the treatment of effluent bearing heavy metals have revealed adsorption to be a highly effective technique for the removal of heavy metal from waste stream and activated carbon has been widely used as an adsorbent (Chand et al., 1994). Despite its extensive use in the water and wastewater treatment industries, activated carbon remains an expensive material (Abdel-Ghani and El-Chaghaby, 2009).

Adsorption on low cost-adsorbent for removal of toxic metals from wastewater has been investigated extensively. These materials include thioglycolic acid modified oil-palm (Akaninwor et al., 2007), wild cocoyam biomass (Horsfall and Spiff, 2004), brewery biomass (Kim et al., 2005), sodium hydroxide modified Lalang (*Imperata cylindrica*) and leaf powder (Hanafiah et al., 2006). Recently, efforts have been made to use cheap and available agricultural wastes such as coconut shell, orange peel, rice husk, peanut husk and sawdust as adsorbents (Vaishnav et al., 2012) to remove heavy metals from wastewater (Abia and Igwe, 2005).

The use of the coconut shell as a biosorbent material presents strong potential due to its high content of lignin of about 35-45%, and cellulose of about 23-43% (Carrijo et al., 2002). As a result of its low cost, powder of coconut shell- *Cocos nucifera* is an attractive and inexpensive option for the biosorption removal of dissolved metals. Various metal-binding mechanisms are thought to be involved in the biosorption process including ion exchange, surface adsorption, chemisorption, complexation and adsorption-complexation (Pino, 2005; Matheickal et al., 1999).

Coconut shell is a material composed of several constituents, among them lignin acid and cellulose bear various polar functional groups including carboxylic and phenolic acid groups which can be involved in metal binding (Matheickal et al., 1999; Ting et al., 1991). The cellulose and lignin are biopolymers admittedly to be associated to the removal of heavy metals (Gaballah and Kilbertus, 1994; Gaballah et al., 1997; Hunt, 1986). Adsorption on low cost-adsorbent for removal of toxic metals from wastewater has been investigated extensively. These materials include thioglycolic acid modified oil-palm (Akaninwor et al., 2007), wild cocoyam biomass (Horsfall and Spiff, 2004), brewery biomass (Kim et al., 2005), sodium hydroxide modified Lalang (*Imperatacylindrica*) and leaf powder (Hanafiah et al., 2006).

However, these studies did not involve batch adsorption process but rather single adsorption of metal ions from their aqueous solutions, although study carried out by Abdel-Ghani and El-Chaghaby (2009) involved batch adsorption process.

In this study, the use of unmodified coconut husk in the removal of metal ions from aqueous solutions through batch adsorption studies was investigated. Coconut husk which is generally considered as a waste is abundant in Nigeria and has a high sorption capacity due to its high tannin content. This study involved the examination of four variables such as pH of the solution, metal ion concentration, contact time and adsorbent loading on the removal of Cu(II), Fe(II), Cd(III), Cr(III) and Pb(II) ions from aqueous solutions, simultaneously.

MATERIALS AND METHODS

Adsorbent

Coconut *(C. nucifera)* shell/husk were collected from Abusoro at Okitipupa Local Government Area of Ondo State, Nigeria; sun dried for about 2-5 days before being ground into fine particles using the manual grinding machine and sieved with sifter to obtain 120 mm (micrometer mesh) finer dust particles. The finer dust particles were treated with 0.1 M HCl and was later re-introduced into an oven at a temperature of 30°C for 30 min and then preserved in a sample container for future use.

The concentrations of Cu, Cd, Fe, Cr and Pb ions in the adsorbent were determined by placing 5 g of the adsorbent in 50 ml de-ionized water for 50 min in a 50 cm long and 2 cm diameter glass column. Aliquot portions of the eluate from the pre-treatment of the organic waste were carefully decanted into 50 ml plastic bottles and analyzed for the heavy metals using Atomic Absorption Spectrophotometer (AAS) GBC Scientific with oxy-acetylene flame at temperature of about 2500°C.

Adsorption technique

A glass column was fitted with cotton wool and held firmly in a vertical position with the aid of a clamp fixed at one end to a retort stand as illustrated in Figure 1. Atmospheric pressure helped to push the sample through the organic material.

Adsorption experiment was done by measuring 50 ml of the wastewater sample and poured into a 100 ml conical flask. 5 g of the pre-treated fine particle coconut husk was added to the wastewater.

Adsorbates

The solutions of Cu, Cd, Fe, Cr and Pb metal ions were prepared from analytical grade $CuSO_4.5H_2O$, $3Cd(SO_4).8H_2O$, $FeSO_4.7H_2O$, $Cr(NO_3)_3.9H_2O$ and $Pb(NO_3)_2$ respectively from BDH Chemicals Ltd, Pool England. 30 mg/L aqueous solutions (stock solutions) of these salts were prepared with de-ionized water in 250 ml volumetric flask and these stock solutions were diluted with de-ionized water to obtain the working standard solutions.

In each set of experiment, the effect of one factor was evaluated by varying this factor while keeping all other factors constants.

Adsorption experiments

Batch adsorption process was carried out at laboratory room temperature. The different factors affecting adsorption process of

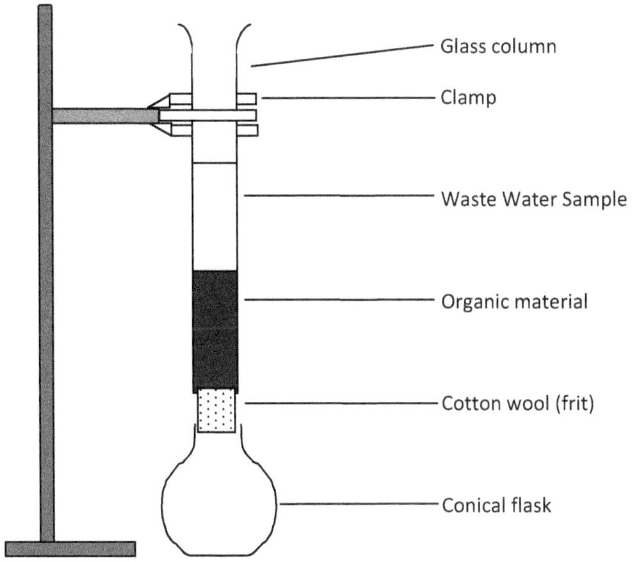

Figure 1. Sketch diagram of experimental set-up.

the metal ions under study (Cu^{2+}, Pb^{2+}, Fe^{2+}, Cr^{3+} and Cd^{2+}) such as contact time, concentration, adsorbent dose and pH have been studied using synthetically prepared wastewater. The sorption capacity qe mg/g and removal efficiency Q were obtained according to the Equations 1 and 2, respectively:

$$qe = \frac{(Co - Ce)V}{W} \quad (1)$$

$$Q = \frac{(Co - Ce) \times 100\%}{Co} \quad (2)$$

Where V is the volume of the solution, W is the amount of adsorbent, C_o and C_e are the initial and concentration of the solution after adsorption in mg/l.

Statistical analysis

The relationship between pairs of metal adsorption at the various variables was tested using Pearson Moment Correlation Coefficient. All statistical analyses were tested using SPSS 21.00 with significance based on 95% confidence level (Ogbeibu, 2005).

RESULTS AND DISCUSSION

Effect of adsorbent dosing

The availability and accessibility of adsorption site is controlled by adsorbent dosage (Rafeah et al., 2009). The effect of mass of adsorbent loading on heavy metal removal using coconut husk was investigated by varying adsorbent loading weight from 0.4 to 1.2 g per 50 ml of mixed metal ions solutions (Appendix 1). The effect of coconut husk weight is graphically presented in Figure 2. It can easily be inferred that the percentage removal of metal ions increased with increasing weight of coconut husk. This is due to the greater availability of the exchangeable sites or surface area at higher dose of the adsorbent. This result is in agreement with previous studies on many other adsorbents (Bin et al., 2001; Ajmal et al., 1998; Abdel-Ghani and El-Chaghaby, 2009). The observed trend of percentage removal of metal ions was: Cr>Cu>Pb>Fe>Cd. Furthermore, the P-values of 0.058 and 0.090 at 95% confidence level ($P>0.05$) shows statistically that there was no significant correlation in adsorption pattern between Cu and Cd and that of Cu and Fe. However, there exist significant correlation in adsorption pattern between Cu and Cr and that of Cu and Pb, since $P<0.05$ at 95% confidence level. Similarly, there exist a significant correlation between the pairs of Cd and Fe ($P<0.05$) and that of Cd and Pb ($P<0.05$) at 95% confidence level. The same trend was observed statistically between the pairs of Fe and Cr ($P<0.05$) and Fe and Pb ($P<0.05$). However, there was no significant correlation statistically between the adsorption pattern of Cr and Pb at 95% confidence level ($P>0.05$).

Effect of concentration on adsorption of the metal ions

The effect of initial metal concentration on the adsorption efficiency of coconut husk is shown in Figure 3. Adsorption experiments were carried out at different initial metal ion concentrations of 0.03, 0.06 and 0.09 mg/l in mixed metal ions solution (Appendix 2). The adsorption efficiency increased with increasing initial metal ion concentration. This result is in accordance with the work of Okieimen and Onyenkpa (2000). It is generally expected that as the concentration of the adsorbate increases the metal ions removed should increase. It is believed that increase in concentration of the adsorbate bring about increase in competition of adsorbate molecule for few available binding sites on the surface of the adsorbent hence increase in the amount of metal ions removed. The observed trend of percentage removal of metal ions was: Cr>Cu>Cd>Fe>Pb. The P-values of 0.041 and 0.018 at 95% confidence level ($P>0.05$) shows statistically that there was significant correlation in adsorption pattern between Cu and Cd and that of Cu and Cr. However, there was no significant correlation in adsorption pattern between Cu and Fe ($P<0.05$). Cd shows statistically that there was significant correlation in the adsorption pattern with the rest metals (Cu, Fe, Cr and Pb) as their $P<0.05$. Fe also shows statistically significant correlation with Pb ($P<0.05$) at 95% confidence level.

Effect of hydrogen ion concentration

The pH adsorption edges of the constant concentration for Cu, Cd, Fe, Cr and Pb for coconut husk are shown in Figure 4. All experiments were carried out in the pH

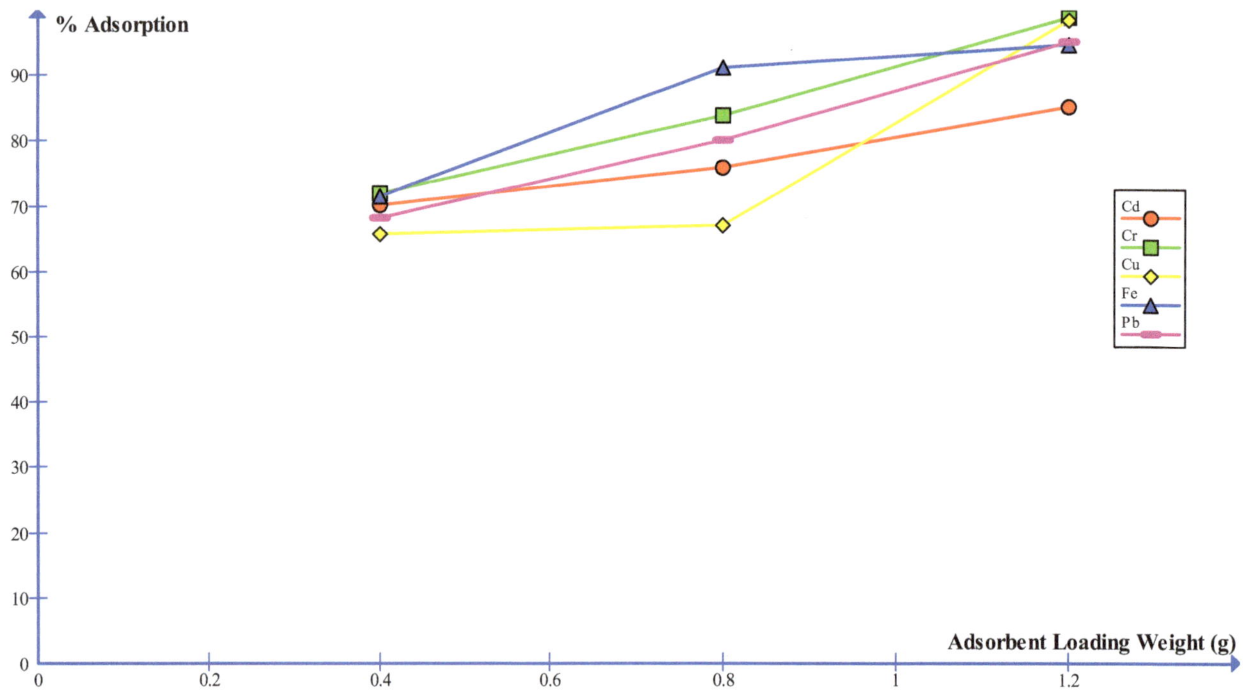

Figure 2. Effect of coconut loading weight on metal ion adsorption in a mixed metal ion solution.

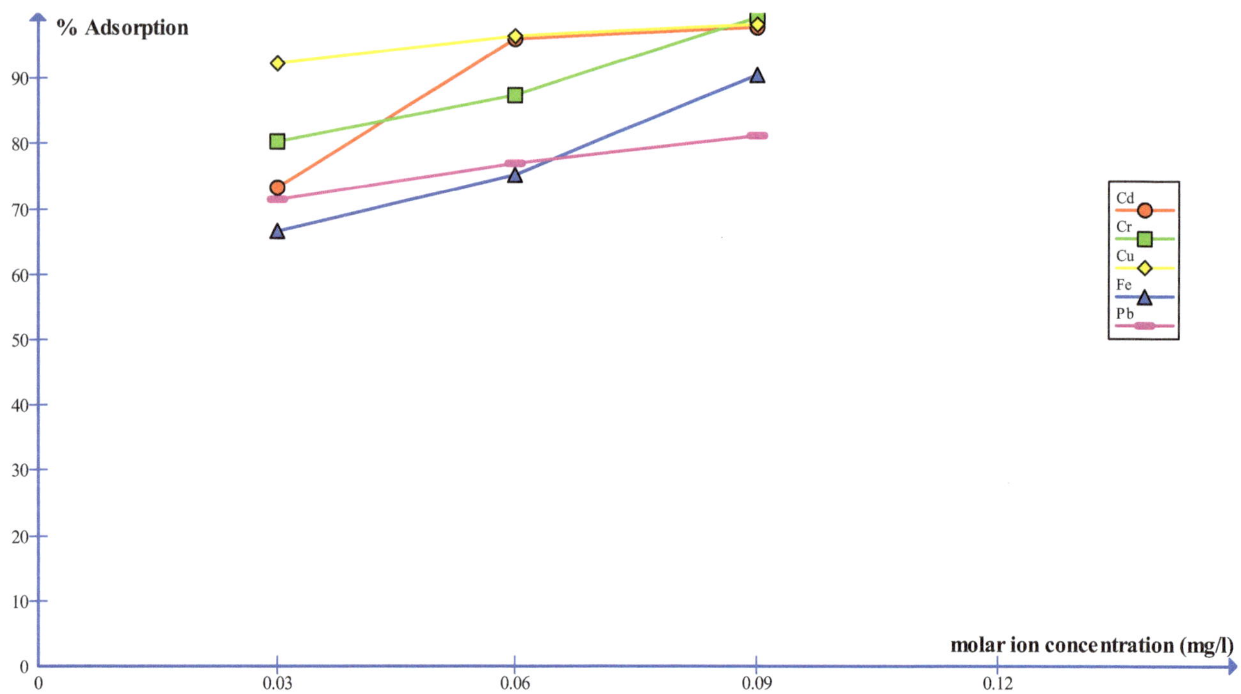

Figure 3. Effect of variation in initial metal ions concentration on adsorption using coconut husk.

values of 2, 6 and 10 (Appendix 3) where chemical precipitation is almost avoided, so that metal removal could be related to the adsorption process (Abdel-Ghani and El-Chaghaby, 2009).

The susceptibility of the system pH changes may be attributed to the nature of the ions in solution and the nature of the adsorbent used. The lower the pH, the more H^+ ions competing with the metal ions for adsorption

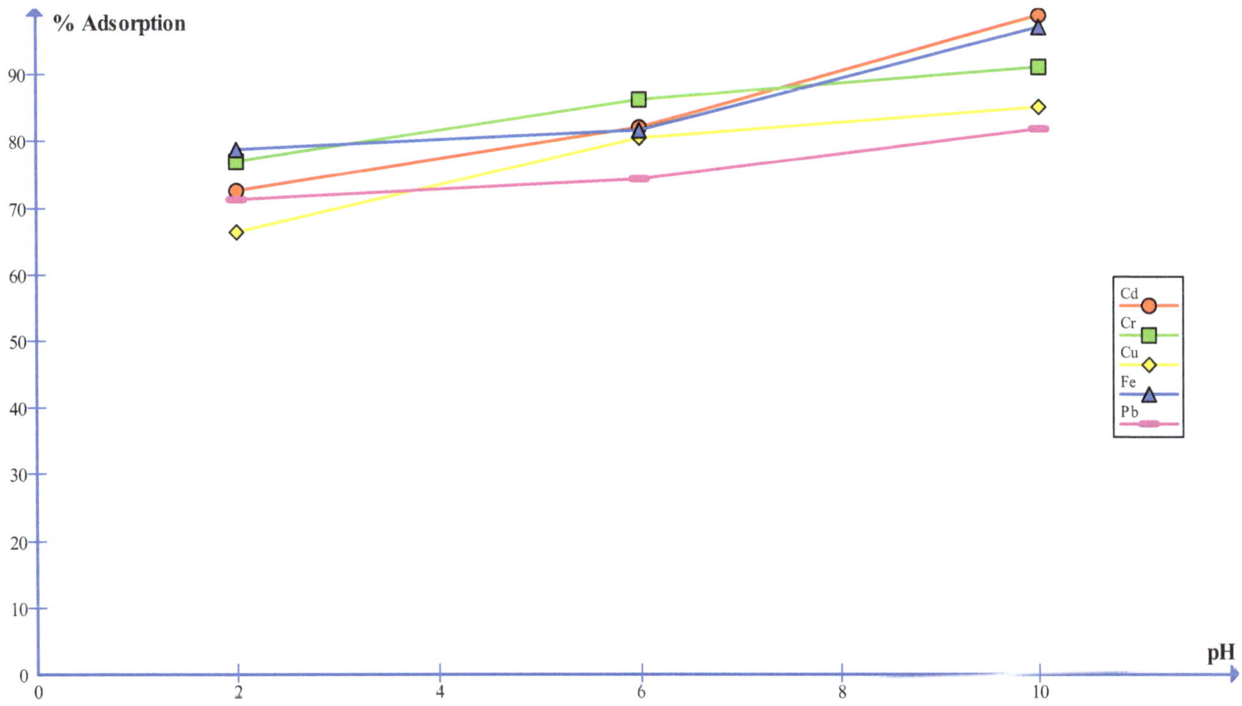

Figure 4. Effect of hydrogen ion concentration (pH) on metal ions adsorption.

sites, thus reducing their adsorption. On the other hand, the higher the pH, the less the H^+ ions competing with metal ions for adsorption sites, thus increasing their adsorption, which explains the obtained results in Figure 4. The observed trend of percentage adsorption of metal ions was: Cd>Fe>Cr>Cu>Pb. Furthermore, the $P<0.05$ at 95% confidence level shows statistically that there was significant correlation in adsorption pattern between Cu and Cd; Cu and Fe; and Cu and Pb, respectively. Similar trend of adsorption pattern statistically was observed between Cd and Cr ($P<0.05$) and Cd and Pb ($P<0.05$).

Effect of contact time

The effect of contact time on the adsorption efficiency is shown in Figure 5. Adsorption experiments were carried out at different time intervals: 20, 40 and 60 min in mixed metal ions (Appendix 4). It was observed that removal of tested metals was rapidly achieved, within a short period of 20 min. Adsorption of Cd and Cr ions attained maximum within 20 min while that of Cu was within 40 min. Adsorption of Fe and Pb increases with increase in contact time. Generally, the observed trend of metal removal was: Pb>Cu>Cr>Fe>Cd. Previous results revealed that removal of all tested metals was rapidly removed within a short period of 30 min (Olayinka et al., 2009). The effect of contact time on adsorption process of metal ions from wastewaters were studied by many authors (Dakiky et al., 2002; Saeed et al., 2005; Abdel-Ghani et al.,2007a; Abdel-Ghani et al., 2007b). The results indicated that the equilibrium time was dependent on the nature of the adsorbent and on metal ions concentration. Furthermore, the $P<0.05$ at 95% confidence level shows statistically that there was significant correlation in adsorption pattern between Cu and the rest metals (Cd, Fe, Cr and Pb, respectively). Similarly, Cd exhibited statistically significant correlation with Fe ($P<0.05$) but there was no significant correlation with Pb ($P>0.05$). However, there exist statistically significant correlation in adsorption pattern between Fe and Cr ($P<0.05$) and Pb ($P<0.05$). The $P>0.05$ shows statistically that there was no significant correlation between the adsorption pattern of Cr and Pb.

Isothermal studies

The analysis of equilibrium data for the adsorption of Cd, Cr, Cu, Fe and Pb on coconut husk was done using the Langmuir and Freundlich isotherm model as shown in Tables 1 and 2, respectively. The extremely high R^2 values provided by the Langmuir isotherm suggest that the data best fitted the Langmuir isotherm given by the equation:

The adsorption capacity of the adsorbent,

$$qe\left(\frac{mg}{g}\right) = \frac{v}{m(Co - Ce)}. \quad (3)$$

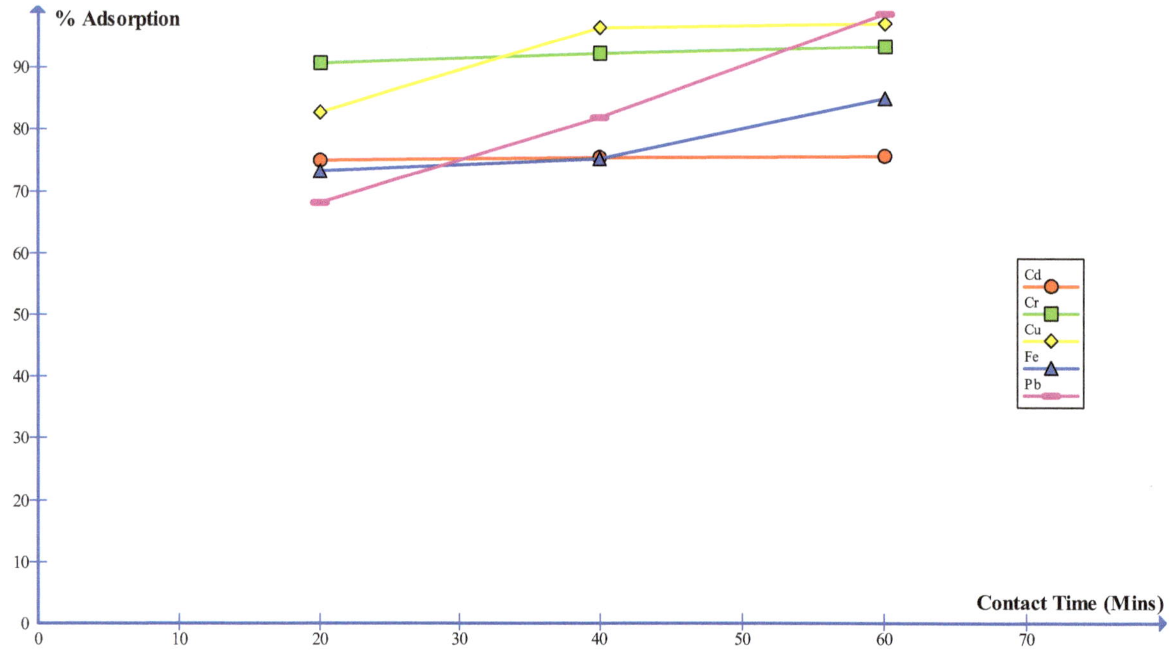

Figure 5. Effect of contact time on metal ions adsorption.

Table 1. Values of Langmuir Isotherm constants for sorption of Cd, Cr, Cu, Fe and Pb metal ions.

Metal ions	q_{max}(mg/g)	K_l (l/mg)	$R_L=1/(1+K_lC_o)$	R^2 values
Cu	0.00010	786.1	0.041	0.965
Cd	0.00010	610.02	0.051	0.992
Fe	0.00009	118.4	0.22	0.919
Cr	0.00031	1355.00	0.024	0.835
Pb	0.00039	38.44	0.46	0.999

Table 2. Values of Freundlich Isotherm constants for sorption of Cd, Cr, Cu, Fe and Pb metal ions.

Metal ions	K_F (mg/l)	$\frac{1}{n}$	N	R^2 values
Cu	4.4E-6	0.810	1.23	0.148
Cd	1.24E-6	1.047	0.96	0.783
Fe	3.7E-7	1.603	0.62	0.346
Cr	2.0E-5	0.585	1.709	0.729
Pb	7.96	2.253	0.44	0.974

Where C_o (mg/l) and C_e (mg/l) are initial and equilibrium concentration of adsorbate solution respectively.

The Langmuir isotherm equation is written as:

$$qe = \frac{qmaxKlCe}{(1 + klCe)} \qquad 4$$

q_{max} (mg/g) is the maximum adsorption capacity upon complete saturation of the adsorbent surface, K_l(dm^3/g) is a constant related to the adsorption/desorption energy.

The equation above can be rearranged to form the Scatchard regression:

$$\frac{qe}{Ce} = qmaxKl - Kl\,qe \qquad (5)$$

A plot of $\frac{q_e}{C_e}$ versus q_e yields a slope $-K_l$ and intercept $q_{max}K_l$

The isotherm constants were determined from the respective plots, and are presented in Table 1. Regression values (R^2) presented in Table 1; indicate that the adsorption data for Cd, Cr, Cu, Fe and Pb metal ion removal fitted well the Langmuir isotherm.

Conclusion

This study was carried out to evaluate the efficiency of metal removal from mixed metal ions solution using coconut husk as adsorbent. Contact time, initial metal ion concentration, adsorbent dose and pH as factors that affect adsorption process of metals were studied using synthetically prepared wastewater. The percentage removal of metals increased with increasing weight of coconut husk and the observed trend of percentage removal of metal ions was: Cr>Cu>Pb>Fe>Cd. The adsorption efficiency increased with increasing initial metal ion concentration and the observed trend of percentage removal of metal ions was: Cr>Cu>Cd>Fe>Pb, while percentage removal of metal ions increased with increasing pH and the observed trend of percentage adsorption of metal ions was: Cd>Fe>Cr>Cu>Pb. The effect of contact time on the adsorption efficiency at different time intervals reveals that the removal of tested metals was rapidly achieved during a short period of 20 min. Generally, the study revealed that coconut husk (a waste material) is a viable material for removal of metals from waste water and therefore could be applied in large scale industrial effluents replete with heavy metals. Isothermal studies showed that the experimental data are best fitted on Langmuir model.

Conflict of Interests

The author(s) have not declared any conflict of interests.

REFERENCES

Abdel-Ghani NT, El-Chaghaby GA (2009). Simultaneous Removal of Chromuim, Copper, Cadmuim, and Lead Ions from Aqueous Solution by Adsorption onto Kaolin. Int. J. Geotechnics Environ. 1(2):161-171.

Abdel-Ghani NT, El-Nashar RM, El-Chaghaby GA (2007b). Removal of Cr(III) and Pb(II) from Solution by Adsorption onto *Casuarina Glauna* Tree Leaves. Electronic J. Environ. Agric. Food Chem. 7(7):3126-3133.

Abdel-Ghani NT, Hefny M, El-Chaghaby Gh. AF (2007a). Removal of Lead from Aqueous Solution using Low-cost Abundantly Available Adsorbents. Int. J. Environ. Sci. Technol. 4(1):67-73

Abia AA, Igwe JC (2005). Sorption Kinetics and Intraparticulate diffusivities of Cd, Pb, and Zn ions on Maize Cob. J. Biotechnol. 4(6): 509-512

Ajmal M, Khan AH, Ahmed-Sh, Ahmad-A (1998). Role of Sawdust in the Removal of Copper (II) from Industrial Wastes. Water Res. J. 32(10): 3085-3091.

Akaninwor JO, Wage MO, Iba IU (2007). Removal of Iron, Zinc and Magnesium from polluted waste water samples using thioglycolic acid modified oil-palm. Afr. J. Biochem. Res. 1(2): 011 013.

Bin Y, Zhang Y, Shukla A, Shyam SS, Dorris KL (2001). The Removal of Heavy Metals from Aqueous Solutions by Sawdust Adsorption-Removal of Lead and comparison of its Adsorption with Copper. J. Hazard. Mater. 84(1):83-94

Carrijo OA, Liz RS, Makishima N (2002). Fiber of green coconut shell as agricultural substratum. Braz. Hortic. 20:533–535 (in Portuguese)

Chand S, Aggarwal VK, Kumar P (1994). Removal of Hexavalent Chromium from the Wastewater by Adsorption. Indian J. Environ. Health 36(3):151-158 pp.

Dakiky M, Khamis M, Manassra A, Mer'eb M (2002). Selective Adsorption of Chromium (VI) in Industrial Wastewater using Low-cost Abundantly Available Adsorbents. Adv. Environ. Res. 6(4):533-540.

Gaballah I, Gey D, Allain E, Kilbertus G, Thauront J (1997). Recovery of copper through decontamination of synthetic solutions using modified barks. Metallurgical and Materials Transactions B 28B,13-23.

Gaballah I, Kilbertus G (1994). Elimination of As, Hg and Zn from synthetic solutions and industrial effluents using modified barks. In: Misra, M. (Ed.), Separation Process: Heavy Metals, Ions and Minerals. The Minerals, Metals & Materials Society.

Goyal D, Ahluwalia S. S (2007). Microbial and plant derived biomass for removal of heavy metals from wastewater. Biores. Technol. 98:2243-2257.

Gupta VK, Gupta M, Sharma S (2001). Process development for the removal of lead and chromium from aqueous solution using red mud - an aluminum industry waste. Water Res. 35(5):1125-1134.

Hanafiah MAK, Ibrahim SC, Yahaya MZA (2006). Equilibrium adsorption study of lead ions onto sodium hydroxide modified Lalang (*Imperatacylindrica*) leaf powder. J. Appl. Sci. Res. 2:1169-1174.

Horsfall M (Jnr), Ayebaemi IS (2004). Studies on the effect of pH on the sorption of Pb^{2+} and Cd^{2+} ions from aqueous solutions by *Caladiumbicolor* (Wild cocoyam) biomass. Environ. Biotechnol. 7(3):1-11.

Hunt S (1986). Diversity of biopolymer structure and its potential for ion binding applications. In: Eccles, H., Hunt, S. (Eds.), Immobilization oflons by Bio-sorption. Ellis Horwood Limited Publishers.

Igwe JC, Ogunewe DN, Abia AA (2005). Competitive adsorption of Zn (II), Cd (II) and Pb(II) ions from aqueous and non-aqueous solution by maize cob and husk. Afr. J. Biotechnol. 4(10): 1113-1116.

Kadirvelu K, Thamaraiselvi K, Namasivayam C (2001). Removal of Heavy Metal from Industrial Wastewaters by Adsorption onto Activated Carbon Prepared from an Agricultural Solid Waste. Bioresour. Technol. 76: 63-65.

Kim TY, Park SK, Cho SY, Kim HB, Kang Y, Kim SD, Kim SJ (2005). Adsorption of heavy metals by brewery biomass. Korean J. Chem. Eng. 22(1): 91-98.

Matheickal JT, Yu Q, Woodburn GM (1999). Biosorption of cadmium (II) from aqueous solutions by pre-treated biomass of marinealga DurvillAeapotatorum. Water Res. 33:335-342.

Mondal MK (2009). Removal of Pb(II) ions from aqueous solution using activated tea waste: Adsorption on a fixed-bed column. J. Environ. Manage. 90: 3266-3271.

Ogbeibu AE (2005). Biostatics- A Practical Approach to Research and Data Handling. Mindex Publishing Company Limited, Nigeria. pp. 139-141.

Okieimen FE, Onyenkpa VU (2000). Binding of Cadmium, Copper, Lead and Nickel ions with Melon (Citrullus vulgaris) seed husk. Biol. Waste 29: 11-16.

Olayinka KO, Alo BI, Adu T (2007). Sorption of heavy metals from electroplating effluents by low-cost adsorbents II: Use of waste tea, Coconut shell and Coconut husk. J. Appl. Sci. 7(16): 2307-2313.

Olayinka KO, Oyedeji OA, Oyeyiola OA (2009). Removal of chromium and nickel ions from aqueous solution by adsorption on modified coconut husk. Afr. J. Environ. Sol. Technol. 3(10). 286-293.

Pino GH (2005). Biosorption of heavy metals using powder of greencoconut shell. Master's Degree Dissertation. Catholic University of Riode Janeiro, Brazil (in Portuguese).

Rafeah W, Zainab N, Veronica U (2009). Removal of Mercury, Lead and Copper from aqueous solution by activated carbon of Palm Oil Empty Fruit Bunch. World Appl. Sci. J. 5;84-91,

Saeed A, Igbal M, Akhtar MW (2005). Removal and Recovery of Lead(II) form Single and Multimetal (Cd, Cu, Ni, Zn) Solutions by Crop Milling Waste (Black Gram Husk). J. Hazard. Mater. 117(1):65-73.

Suthipong S, Siranee S (2009). Utilization of pulp and paper industrial wastes to remove heavy metals from metal finishing wastewater. J. Environ. Manage. 90: 3283-3289.

Ting YP, Prince IG, Lawson F (1991). Uptake of cadmium and zinc by the alga Chlorella vulgaris: II. Multi-ion situation. Biotechnol. Bioeng. 37:445–455.

Vaishnav V, Daga K, Chandra S, Lal M (2012). Adsorption Studies of Zn (II) ions from Wastewater using Calotropis procera as an Adsorbent. Res. J. Recent Sci. 1:160-165.

Appendix 1. Percentage removal (%) of metal ion from synthetic wastewater by varying adsorption weight.

Metals	Cu	Cd	Fe	Cr	Pb
0.4 g	65.80	70.03	71.47	71.83	68.10
0.8 g	67.03	75.97	91.13	83.77	80.03
1.2 g	98.53	85.17	94.70	98.93	95.20

Appendix 2. Percentage removal (%) of metal ion from synthetic wastewater by varying molar concentration.

Metals	Cu	Cd	Fe	Cr	Pb
0.03 M	92.27	73.27	66.63	80.37	71.40
0.06 M	96.47	96.07	75.23	87.43	77.03
0.09 M	98.33	97.83	90.47	99.30	81.10

Appendix 3. Percentage removal (%) of metal ion from synthetic wastewater by varying pH.

Metals	Cu	Cd	Fe	Cr	Pb
pH 2	66.40	72.53	78.70	77.03	71.17
pH 6	80.47	82.03	84.93	86.37	74.43
pH 10	85.13	98.83	97.03	91.10	81.90

Appendix 4. Percentage removal (%) of metal ion from synthetic wastewater by varying contact time.

Metals	Cu	Cd	Fe	Cr	Pb
20 Min	82.77	74.90	73.30	90.73	68.17
40 Min	96.57	75.50	75.30	92.23	81.90
60 Min	97.10	75.60	84.93	93.27	98.67

Effect of heat build-up on carbon emissions in chimato compost piles

Angstone Noel J. Thembachako MLANGENI[1] , Shaibu KANANJI[2] and Sosten S. CHIOTHA[3]

[1]NR Department, Natural Resources College, Box 143, Lilongwe, Malawi.
[2]Agriculture Department, Natural Resources College, Box 143, Lilongwe, Malawi.
[3]Biology Department, Chancellor College, University of Malawi

A study was conducted to determine impacts of heat build-up of chimato compost piles TD0, TD20, TD40, TD50, TD60, TD80 and TD100, made by blending maize stalks with 0, 20, 40, 50, 60, 80 and 100% *Tithonia diversifolia*, respectively, on carbon losses and emissions during composting. Compost piles temperatures that determined heat built-up were obtained from previous studies. Organic carbon and total carbon of chimato composts were determined using Kjeldahl method. Relatively, greater carbon reductions were observed in compost piles TD0, TD20, and TD40 (that experienced prolonged high thermophilic temperatures (temperature >60°C)) than compost piles TD50 and TD60 (that experienced a short-lived thermophilic temperatures (45≤60°C)). The prolonged high temperatures increased kinetic energy of chemical species CO_2 and CH_4 that became more volatile and probably escaped from the compost piles in large quantities resulting in significant carbon emissions. Relatively, short-lived heat built-up in TD50 and TD60 resulted in minimal carbon reduction, hence minimal carbon emissions. Therefore, chimato composts TD50 and TD60 significantly reduce compost pile carbon emissions (p=0.05, α=0.001) and should be recommended to mitigate effects of climate change.

Key words: *Tithonia diversifolia*, heat built-up, carbon emissions, carbon reduction, prolonged thermophilic temperatures.

INTRODUCTION

Agriculture sector plays vital role in sustaining growth and reducing poverty in many countries in Africa including Malawi. However, depletion of soil fertility as well as organic matter has been reported to continuously and negatively affect sustainable agricultural production in Malawi (Elias et al., 2013). Application of composts in agricultural filed has been identified as one of the potential initiative to replenish soil fertility and organic matter. It is reported that application of well matured and stable composts into agricultural fields significantly increases soil carbon content by diverting inorganic atmospheric carbon compounds such as dioxide (CO_2) and methane (CH_4) into soil organic carbon compounds (Biala, 2011; Gill et al., 2012; Biddlestone and Gray, 1987). Maturity and stability of compost is partly dependent on type of feedstock that influence compost pile moisture content, aeration and heat build up. Well matured and stable composts gradually decompose under aerobic conditions (Biddlestone and Gray, 1987; Christensen and Peacock, 1998).Compost piles that mature into well matured and stable composts usually meet certain minimum blending composition of nitrogen rich ingredients and carbon rich ingredients (WSU, 2010; Mlangeni et al., 2013). Such requirements make ingredients to be digested very fast by microbes and the compost pile stays in high thermophilic stage for relatively short time (Mlangeni, 2013).

On the contrary, immature and unstable compost is usually produced when fewer quantities of nitrogen rich have been blended with disproportional large quantities of carbon rich feedstock. Such ingredient composition limits microbial activities. Microbes use limited available nitrogen to slowly digest ingredients in prolonged high thermophilic temperatures. The high temperatures ensure availability of microorganisms that multiply rapidly at specific temperature range (WSU, 2010). Therefore, compost piles that stay in the high thermophilic phases for a long time are symptomatic to prolonged exothermic oxidation reactions that built-up large amount of heat (WSU, 2010). Large generation of prolonged heat is attributed to high active microbial activities taking place in the compost piles due to quantities of high carbon content cellulosic materials (lignin) whose oxidation and digestion generate large amounts of heat (Agromisa, **1990**). The compost piles stay longer in the thermophilic stage because of greater quantities of hard to digest materials in the feedstock that prolongs maturity period. Thus, the compost piles' prolonged stay in the thermophilic stage may have a negative impact on kinetic energy of chemical species available in the compost pile, which may also be symptomatic to volatilization and consequently to loss of such chemical species.

Smallholder farmers in Malawi are encouraged to make and use chimato compost in their fields. However, such farmers use low quality ingredients such as grass and maize stalks. Use of nitrogen rich ingredients such as *Tithonia diversifolia* biomass is recommended (Mlangeni et al., 2013). Temperature progression of chimato compost piles made using 80 and 100% of *T. diversifolia* was reported to rapidly increase to high thermophilic peak temperature (54.5 and 55°C respectively) before rapidly dropping down to mesophilic phase (35.0°C) (Mlangeni, 2013). Temperature progression of chimato compost piles made using 60 and 50% of *T. diversifolia* was reported to experience prolonged stay in the high thermophilic phase peak temperature (65.0 and 70.0°C), respectively, symptomatic to well matured compost (Biddlestone and Gray, 1987; Biddlestone and Gray, 1987; Darlington, 2010). The blending is reported to generate a balanced heat build-up conducive to initiate a self-accelerated process of decomposing the feedstock (Mlangeni et al., 2013). Chimato compost piles' heat build-up determines both nitrogen and carbon content of the chimato composts. Too high temperatures deactivate mesophilic microbes whereas a longer duration of high temperatures accelerates losses of methane (CH4), carbon dioxide (CO_2) and nitrogen loss as ammonia NH_3 and nitrogen oxide through volatilization (Carr, 1998; CAW, 2012). Heat built-up in compost pile is dependent on quality of ingredients used in constructing the compost piles (Mlangeni, 2013), whereas quantity of CO_2, methane and other nitrides diverted is dependent on type, stability and quality of composts. High quality ingredients such as *T. diversifolia* enhances rapid and active microbial activity that break down chemical bonds of compounds and transforms chemical energy (Biddlestone and Gray, 1987; WSU, 2010) contained in the bonds to heat, which raises the temperature of the compost pile (Mlangeni, 2013). Blending maize stalks with *T. diversifolia* of greater than 40% ensures availability of both carbon and nitrogen in sufficient amounts to enhance optimal active and rapid microbial activities. It is suggested that optimal heat build-up may have significant impacts on residual compost nitrogen and carbon. This study was conducted to determine impacts of compost piles heat build-up of chimato compost produced by blending maize stalks with different quantities of *T. diversifolia* on carbon losses and carbon emissions during composting.

MATERIALS AND METHODS

Study site

The study was carried out at Natural Resource College farm (130 85' S 330 38, E) in Lilongwe, Malawi. Natural Resource College farm is about 13 km south west of Lilongwe town and 3.5 km off Lilongwe-Mchinji road. It lies on an altitude of 1146 m above sea level. Natural Resource College experiences mean annual temperature of 20°C and mean annual relative humidity of 68%. It receives an annual mean rainfall of 892 mm of which 85% falls between November and March (DARETS, 2002).

Experimental design and treatments

Six standard chimato composting treatments Td0, Td20, Td40, Td50, Td60, Td80 and Td100 were made. In each case, *T. diversifolia* biomass was blended with maize stalks in the ratio (TD/MS) 0:100, 20:80, 40:60, 50:50, 60:0, 80:20 and 100:0 respectively by mass. Tender green *T. diversifolia* shrubs of about 8 weeks old (counted from the date the *T. diversifolia* shrubs were slashed for regeneration) were used. Both maize stocks and the *T. diversifolia* materials were cut into pieces ranging from 5.0 to 10.0 cm. The small sizes were chosen to increase surface area onto which microbes would act on, to enhance efficient diffusion of air throughout the entire pile and to enhance active and rapid aerobic decomposition (Nalivata, 2008; Michel et al, 2010). Each compost pile was adequately watered to optimize compost pile moisture content. Initial diameter of each compost pile was 1.5 m wide, 1.5 m high and conical shape. The piles were then plastered on the outside with mud as shown in Figure 1. Air vents (Figure 1c) were made at the bottom and middle of the piles to allow air circulate through the biomass to enhance aerobic degradation (Nalivata, 2008).

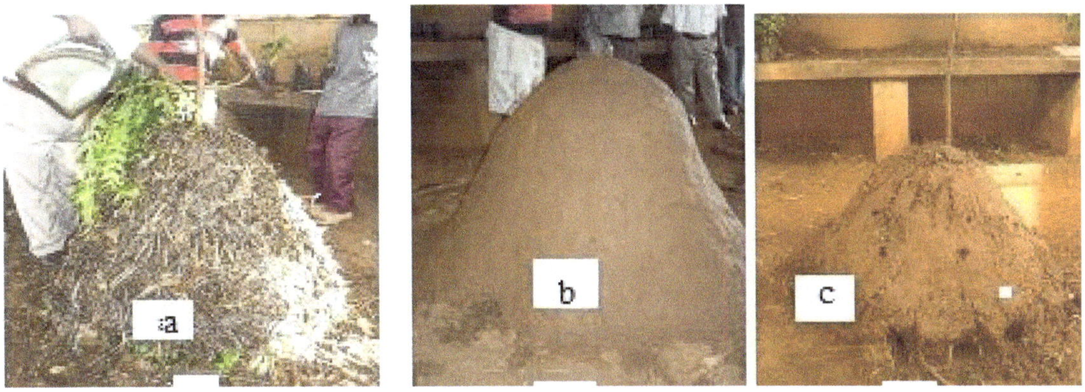

Figure 1. a) Un-plastered compost pile; **b)** freshly plastered conical piles; **c)** matured chimato compost piles.

Analysis of chimato composts

Compost samples were purposely collected from different locations of the matured chimato compost pile. Samples from each treatment were homogeneously mixed to get a composite representative sample. The samples were analyzed for total Kjeldahl carbon and organic carbon using Kjeldahl method (Jeffery et al., 1989).

Data analysis

Data was analyzed using SPSS 17.5 version. Level of variation of organic carbon and total Kjeldahl carbon were determined using analysis of various (ANOVA).

RESULTS AND DISCUSSION

Effect of prolonged stay in high thermophilic phase on carbon losses

Study results indicates that amounts of carbon losses in chimato compost TD0 and TD20 were significantly different from carbon losses of chimato composts TD50, TD60, TD80 and TD100 ($p=0.018$, $\alpha=0.05$; Wilcoxon). Carbon losses in TD0 was not significantly different from carbon losses in treatment TD20 and TD40 ($p=0.207$; $\alpha=0.05$; Wilcoxon). However, greatest carbon losses were observed in chimato compost TD0 followed by chimato compost TD20 then chimato compost TD40 (Figure 2). Carbon losses in TD0, TD20 and TD40 highly correlated with chimato compost piles' prolonged stay in the high thermophilic stage, implying that prolonged stay in the high thermophilic stage (prolonged heat built-up) was responsible for the observed carbon losses (Carr, 1998; CAW, 2012). The heat increased kinetic energy of the chemical species which probably became more volatile and escaped the compost pile. High temperatures in compost piles are symptomatic to rapid active microbial activities likely to generate large heat (WERL, 2005; WSU, 2010). Thus, the observed carbon losses could be attributed to prolonged exothermic decomposition reaction of carbon atoms that induced the observed high-prolonged temperatures (heat built-up) and that released significant quantities of carbon dioxide (Biala, 2011; Biddlestone and Gray, 1987).

Since prolonged stay in the thermophilic stages suggests prolonged active microbial activities, production of excessive methane and carbon dioxides in compost piles TD0 and TD20 was inevitable. TD0 and TD20 relatively contained large quantities of large quantities of hard to digest carbon containing polymers, such carbon atoms were digested for a longer time thereby releasing enormous heat which raised the temperature of the compost piles. In this process, microbes used and recycled limited quantities of nitrogen used as source of building materials of microbe structure and carbon atoms as source of energy (Biala, 2011; Gill et al., 2012).

Secondly, prolonged stay in high thermophilic phase is symptomatic to prolonged generation of large amounts of heat (WERL, 2005; Mlangeni, 2013) likely to increase kinetic energy of chemical species in the compost piles. The prolonged stay in the thermophilic phase cause chemical species such as carbon dioxide (CO_2), methane (CH_4), ammonia (NH_3) and nitrogen oxides (NO_x) acquire kinetic energy that significantly increases chemical species' random motions in the compost piles. The rapid random motion would either trigger faster rates of reaction with other chemical species in the compost pile or accelerates volatilization from the compost piles. Methane (CH_4) species, which has about 300 times impact on climate change (EPA, 2008; Bernal et al., 1998), could readily react with oxygen under aerobic conditions to produce carbon dioxide, which is of less impact thereby significantly reducing carbon emission of more dangerous greenhouse gas. In addition, methanogenesis would likely occur under conditions of high temperatures as well as low supply of ammonia (Biala, 2011; Gill et al., 2012). Ammonia would be readily available in compost piles with greater quantities of nitrogenous feedstock. Volatilization and escape of chemical species especially carbon dioxide (CO_2) and

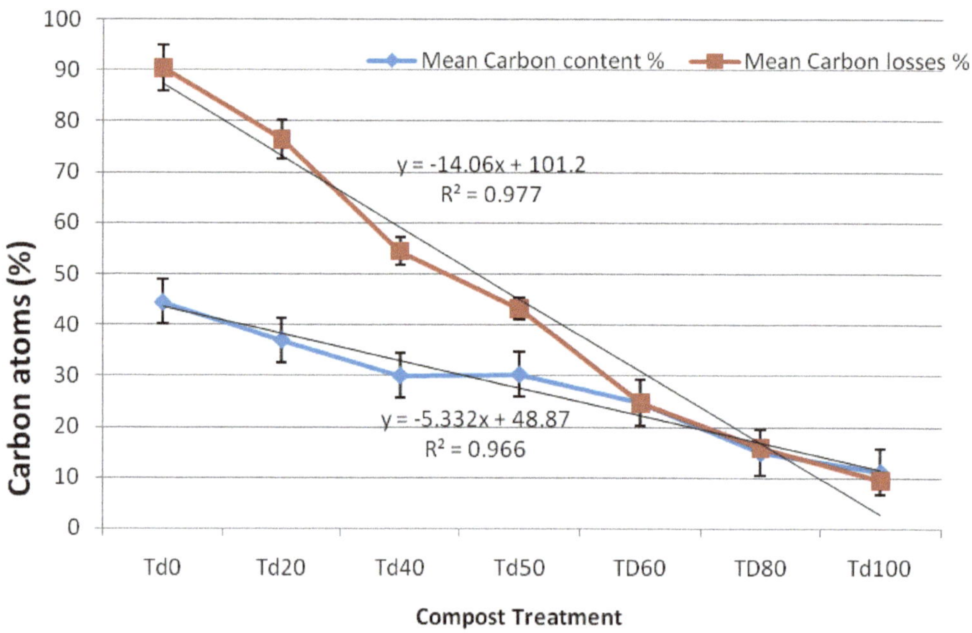

Figure 2. Carbon retainance and losses in various compost treatments.

methane (CH_4) could be enhanced by cracks of compost piles that quickly develops due to high heat built up of compost piles (WERL, 2005; Gill et al., 2012; Biddlestone and Gray, 1987). Thus, carbon dioxide (CO_2) and methane (CH_4) escaped from the piles at fast rate greatly increasing carbon losses of the resultant chimato composts. Therefore, the significant carbon reduction observed in TD0 could be attributable to described carbon losses which would increase carbon emissions. Secondly, microbes had probably plenty nitrogen atoms used to digest available little carbon atoms, which was accomplished with greater speed and significantly short-lived heat build-up of compost piles.

Further analysis has shown that quantity of *T. diversifolia* is negatively correlated with carbon losses in the compost pile up to 60% of *T. diversifolia* content in the compost piles. A linear equation model was developed as follows:

$$Y = -17\text{-}46x + 106.63 \quad (1)$$

Where, y=quantities of *T. diversifolia* and x = carbon lost in the compost piles.

A linear model revealed that *T. diversifolia* is inversely related to carbon losses suggesting that as quantities of *T. diversifolia* increase amounts of carbon lost decreases. Carbon is lost asbeyond 60% of *T. diversifolia* content in the compost piles, carbon losses increased. Carbon could be lost into the surrounding environment as carbonates leachate hence not contributing to carbon emission.

Effect of shorter stay in lower thermophilic stage on carbon emission

As shown in Figure 2, lowest carbon losses were observed in compost TD60 followed by TD50 whose compost piles experienced moderate stay in the high thermophilic stage. The observation implies that compost piles' moderate stay in the high thermophilic phase was due to a balanced blending proportion of carboceous and nitrogenous feedstock. A balanced blending proportion of carboceous and nitrogenous provided an optimum microbial activities that generated an optimum heat build-up in compost piles (Mlangeni and Chiotha, 2013); EPA, 2008; Bernal et al., 1998) that greatly limited production of excessive methane and carbon dioxides. Td50 and Td60 did not experience prolonged heat build-up that forced microbes to reuse the available nitrogen to build up their structures thereby causing significant nitrogen losses nor did they experience short-lived heat build-up which was wet and induced favorable conditions for anaerobic decomposition and significant nitrates losses.

Effect of longer stay in mesophilic stage on carbon emission

As shown in Table 1, TD100 and TD80 experienced a short lived stay in the thermophilic phase and relatively longer stay in the mesophilic phase. The short lived stay in thermophilic phase and longer stay in mesophilic phase resulted from rapid active microbial activities that quickly degraded the ingredients in relatively short time. Thus, less heat was generated in the compost piles

Table 1. Effect of *T. diversifolia* on carbon and nitrogen losses.

Treatment	Initial N (%)	Percentage of N lost (%)	Total C (%)	C lost	Std Error
Td0	1.2	56.67	44.5	90.5	5.0
Td20	1.4	53.57	37.0	76.5	5.0
TD25	1.57	56.05	29.4	67.5	5.0
Td40	1.7	51.76	30.1	44.5	3.5
Td50	2.3	47.83	30.3	33.3	2.0
Td60	2.5	44.80	24.7	24.8	2.5

implying minimal occurrence of decomposition of carboceous materials. The feedstock provided little quantities of hard to digest macromolecules such as cellulose and polyphenols which were easily digested thereby releasing less heat than those with lower composition *T. diversifolia* in the feedstock. Since ammonia volatilization occurs in compost piles with high nitrogen content, Td80 and Td100 were certain to experience reductions of both total Kjeldahl nitrogen and nitrate-N. Nitrates were leached out of the compost piles as leachate or as slurry (Manahan, 2008) collected under the piles. Hence, the short-lived heat build-up in Td80 and Td100 is associated with decreased recorded amounts of total Kjeldahl nitrogen and nitrate-N while the less prolonged were recorded in Td80 and Td100 (Mlangeni, 2013).

Conclusion and recommendations

The results have shown that compost piles that experienced moderate heat build-up experienced minimum carbon losses. Either the heat build-up or the microbial activities or both were responsible for the low carbon losses. However, balanced blending proportion of carboceous and nitrogenous materials in feedstock of compost piles TD50 and TD0 played a significant role in limiting rapid microbial activities which subsequently limited generation of heat in the compost piles. Subsequently, the limited generated heat also limited kinetic energy of chemical species such as NH_4, CO_2 and CH_4 and became more volatile and escaped from the compost piles in large quantities hence the observed greater carbon reduction which increased carbon emission. Making and using chimato composts produced by blending *T. diversifolia* biomass with maize in the ratio of 50:50≤Td/MS≤60:40 greatly reduce compost pile carbon emissions and should be recommended to mitigate effects of climate change.

Conflict of Interests

The author(s) have not declared any conflict of interests.

ACKNOWLEDGEMENTS

I express my profound gratitude to Flanders International Co-operational Agency (FICA) and Leadership for Development (LEAD), southern and eastern Africa for their support.

REFERENCES

Agromisa. (1990). Preparation and use of compost. Agrodox series 8. Wageningen. AGROMISA.

Bernal M, Sanchez-Mondero M, Paredes C, Roig A (1998). Carbon mineralization from organic wastes at different composting stages during their incubation with soil. Agric. Ecosyst. Environ. 69(3): 175–189.

Biala J (2011). The benefits of using compost for mitigating climate change. Department of Environment, Climate Change and Water NSW. DECCW 2011/0171. Available on: www.environment.nsw.gov.au

Biddlestone AJ, Gray KR (1987). Production of Organic Fertilizers by Composting, In. Moriarty, DJ, Pullin, RSV (Eds), Detritus and Microbial Ecology in aquaculture: ICLARM Conference proceeding. Philippines: International Centre for living aquatic resources management. pp.151-180.

Californian against Waste (CAW) (2012). Composting: A Greenhouse Gas Mitigation Measure. Available at: http://www.cawrecycles.org/issues/ghg/compost

Carr L (1998). Composting fundamentals. Maryland (USA): Department of Biological Resources Engineering, University of Maryland.

Christensen, P, Peacock, B (1998). Manure as a Fertilizer. California: University of California, Cooperative Extension -Tulare County.

Darlington W (2010). Compost: A guide for evaluating and using compost material as Soil amendments. Soil and Plant Laboratory, Inc. 714(282):8777-8788.

Department of Agriculture Research Extension and Technical Services (DARETS) (2002). About Chitedze. Accessed on 22/07/2010 from: http://www.agriresearch.gov.mw./research/chitedze

Elias A, Nohmi M, Yasunobu K, Ishida A (2013). Effect of Agricultural Extension Program on Smallholders' Farm Productivity: Evidence from Three Peasant Associations in the Highlands of Ethiopia. J. Agric. Sci. 5:8.

EPA (2008). Municipal Solid Waste Generation, Recycling, and Disposal in the United States: Facts and Figures for 2007. Washington, DC: U.S. Environmental Protection Agency, Office of Solid Waste and Emergency Response. Available at: http://www.epa.gov/epawaste/nonhaz/municipal

Jeffery GH, Bassett J, Mendham J, Denney RC (1989). Vogel's textbook of quantitative chemical Analysis (5th Ed.). New York: Longman.

Manahan SE (2000). Fundamentals of environmental chemistry. (2nd Edition). New York: Taylor and Francis Group.

Michel FC, Heimlich JE, Hoitink HA (2010). Composting at home. (Horticultural and Crop Sciences Composting Series). Ohio: Ohio State University Extension.

Mlangeni A (2013). Effect of *Tithonia diversifolia* on compost pile heat built-up and physico-chemical quality parameters of chimato Compost. Environ. Nat. Resour. Res. 3(3):63-71.

Mlangeni ANJT, Chiotha SS (2013). Potential of *T. diversifolia* chimato composts in enhancing soil carbon sequestration. Environment and Nat. Resour. Res. 3(4):58-67.

Mlangeni ANJT, Sajidu S, Chiotha SS (2013). Total Kjeldahl-N, nitrate-N, carbon/nitrogen ratio and pH improvements in chimato composts using *T. diversifolia*. J. Agric. Sci. 5(10):1-7.

Nalivata PC (2008). Evaluation of factors affecting quality of compost made by smallholder farmers in Malawi. Unpublished doctoral thesis. Cranfield University (UK). National Soil Resources Institute.

Woods Ends Research Laboratory (WERL) (2005). Interpreting Waste and Compost Tests. J. Woods End Res. Lab. 2:1-10.

Washington State University (WSU) (2010). Fundamentals of composting: Why compost, material and methods to ensure quality compost. (Whatcom Extension Research Report). Washington: Washington State University.

Prospects and challenges of vermiculture practices in southwest Nigeria

Aladesida, A. A.[1] , Owa, S. O.[2], Dedeke, G. A.[1] and Adewoyin O. A.[3]

[1]Department of Biological Sciences, Federal University of Agriculture, Abeokuta, Ogun State, Nigeria.
[2]Department of Biological Sciences, Landmark University, Omu-Aran, Kwara State, Nigeria.
[3]Department of Zoology, University of Ilorin, Kwara State, Nigeria.

The prospect of vermiculture in the south-western coast of Nigeria was studied. One hundred and two respondents were interviewed in the three coastal towns of Badagry, Epe and Igbokoda. The results show that 94% of respondents made use of earthworms as fish bait, 37% buy their worms and 57% collected by themselves. The respondents, who buy, however, noted that their suppliers do not breed the worms but search for them from marshy areas. The price value placed on the worms ranged between 0.35 and ₦3.10 per kg; while the most quoted prices were 0.35 and ₦0.80 (16.7% each). The average quoted price was ₦0.70 per kg. Forty-six percent (46%) of respondents were willing to buy earthworms if supplied to them; 49% were willing to serve as sales agents if contacted and 66.7% opined that earthworm was their choicest fishing bait. The results indicate positive expectation for vermiculture as a business venture. Vermiculture should, however, put into consideration the production of *Alma millsoni* and *Libyodrilus violaceus*, the species quoted as the most preferred and effective for fishing. Vermiculture will go a long way in solving the problem of earthworm scarcity among these people, reduce the stress put on natural populations earthworms, thus conserving the worms and also reduce damage done to the soil environment in the course of excavating for worms.

Key words: Vermiculture, earthworm, prospects.

INTRODUCTION

Earthworms, today, are being considered world over in the degradation of waste and as an alternative to inorganic fertilizers. There role in improving soil fertility is well known (Edwards and Lofty, 1977; Owa et al., 2003; Owa et al., 2004a, 2004b). Earthworms have also been suggested by Deolalikar and Mitra (1996) as good substitute or feed supplement for fishmeal in aqua culture.

Deolalikar and Mitra (1996) have used vermicompost prepared from paper mill solid waste for fertilizing aquacultural tanks and found an increase in net primary productivity from 32.08 to 220.83 MgC/m/h. Vermicompost application also facilitates better growth of rohu fish (*Labeo rohita*) when compared with other commercially available organic manures (Deolalikar and Mitra, 1997). There is an increasing demand for protein-rich raw materials in fish and other animal feed industry. Fishmeal is the main protein component of fish feed.

Table 1. Species of earthworms used for fishing.

Earthworm species used	Frequency	Percent (%)
Alma millsoni only	33	37.5
A. millsoni and Libyodrilus violaceus	34	38.6
A. millsoni, L. violaceus and *Eudrilus eugeniae*	1	1.1
L. violaceus only	13	14.8
Unspecified choice	7	8.0
Total	**88**	**100.0**

Earthworm is generally used as bait by the anglers. However, large-scale vermiculture has the potential of supplying earthworm meal as a substitute of fishmeal. The earthworm meal contains all the essential amino acids required in fish feed (Dedeke et al., 2010). The methionine and lysine availability are higher than that of fishmeal.

Vermicomposting and vermiculture technology is applicable to the rural as well as the urban society. It not only helps in commercial aqua-farming but also acts as a convenient source of earthworm for growing ornamental fishes in aquarium. Thus, Vermicomposting can be included as a component of sustainable life-style. Application of vermiculture and vermicomposting in aquaculture is eco-friendly and bio-ethically acceptable (Deolalikar and Mitra, 1997).

Vermiculture, which involves the breeding of earthworms under controlled conditions, has been carried out on species in temperate regions and a few studies have been done in India. Here in Africa, Nigeria in particular, the prospects of vermiculture has not been duly explored, particularly as regarding indigenous species. The present study was therefore carried out to establish a baseline for future works on the prospects of vermiculture in Nigeria, using the coastal areas of south-western states as case study. The research also put into consideration the economic value of the earthworm, which is much needed in the study of the economics of the belowground biodiversity.

METHODOLOGY

The study was carried out in the coastal towns of Badagry, Epe and Igbokoda, southwest Nigeria. This area is characterized by heavy rainfall spanning more than six months of the year and reaching 1200 mm in the months of June and July. Temperature here ranges between 29 and 32°C.

One hundred and two respondents were interviewed, using a structured questionnaire having both multiple choice and open-ended questions. The questionnaire was divided into three (3) sections, with section A addressing the socio-economic characteristics of the respondents. Sections B and C addressed the issues related to knowledge of earthworm types and uses and the culturing and sourcing of earthworms, respectively.

The questionnaires were administered to fishermen along the coastal line of the sampling locations. These areas are well known fishing spots in the study area and are mainly visited by fishermen who are involved in commercial fishing, thus their need for and use of earthworms is believed should be regular.

The data obtained was analysed for frequencies and simple percentage of the earthworm species used by respondents, cost of purchasing earthworms and alternative baits, aside from earthworm, used mostly for fishing by the respondents.

RESULTS

The questionnaires were administered to 46, 25 and 31 individuals at Igbokoda, Epe and Badagry respectively. Almost all the respondents practiced fishing either as full-time or as part-time occupation. Ninety-four percent (94%) of them make use of earthworms as fish bait and the most commonly used species were *Alma millsoni* and *Libyodrilus violaceus* (Table 1). Most of the respondents collected earthworms by self-sourcing (60%), while some 37% both buy and collect earthworms directly. Most collect earthworms on their own because suppliers were scarce and there might be the need to cover long distances in search of suppliers. However, 3% of the respondents do not collect worms themselves but go in search of suppliers. These along with those who both buy and collect noted that they travel long distances in search of suppliers. For instance, some of the respondents in Epe (one of the sampling location) claimed that they go as far as Ibadan, a city which is about 250 km away, in search of suppliers. The respondents, however, noted that none of their suppliers bred earthworms.

Some respondents (46%) showed willingness to buy earthworms if suppliers were available, while some 49% indicated readiness to serve as earthworm sales agent if contacted.

The price range for one kilogram (1 kg) of earthworms was put at between 0.35 and ₦3.10 by the respondents (Table 2). The highest prices quoted were 0.35 and ₦0.80 (16.7% of respondents for each price). The average price was ₦0.70/kg. The most frequent rate of purchase of earthworms by the respondents was twice per week (Table 3). Seventy-three per cent (73%) of respondents picked earthworm as their choicest fishing bait (Table 4) while other baits used for fishing include fish organs, insect larvae, crayfish, small snail, bath soap, etc (Table 5).

Table 2. Prices at which respondents valued 1 kg of earthworms.

Price (₦)	Frequency	Percent (%)
0.01 to 0.35	23	29.7
0.36 to 0.70	17	21.8
0.71 to 1.40	35	44.9
1.41 to 2.11	1	1.3
Above 2.11	2	2.6
Total	**78**	**100.3**

Table 3. How frequently respondents buy earthworms.

Time of purchase	Frequency	Percent (%)
Daily	7	18.9
Once a week	11	29.7
Twice a week	12	32.4
Every fortnight	2	5.4
Once a month	1	2.7
Thrice a week	4	10.8
Total	**37**	**100**

Table 4. Choicest bait.

Choicest bait	Frequency	Percent (%)
Anything available	19	20.7
Earthworms	68	73.9
Fries	1	1.1
Crab	1	1.1
Crayfish	3	3.3
Total	**92**	**100.0**

DISCUSSION

The results show that there is the need to develop the breeding of limicolous earthworm species, particularly *A. millsoni* and *L. violaceus*. There are no literatures indicating any efforts so far on breeding these species of worms. Most commonly used earthworm for culturing is *Eisenia fetida* that has wide range of temperature tolerance, has very high reproductive potential and is less sensitive to density pressure (Watanabe and Tsukamoto, 1976; Tsukamoto and Watanabe, 1977; Hartenstein et al., 1979; Hartenstein, 1981; Reinecke and Kriel, 1981). *Eudrtilus eugeniae,* which is commonly called 'African night crawler', has also been reported as an efficient earthworm to maintain as culture (Kale, 1994). Other species that have been successfully cultured are *Perionyx excavatus* and *P. sansibaricus* (Kale, 1998, 2002). However, these species, except *E. eugeniae* are not found in the West African sub-region, where the present study was carried out. The possibility, though of adapting such species to the tropical climate of this region has not been tested.

Since vermiculture is not expected to serve the fishing industry alone, there might also be the need to ascertain the usability of these species in other areas of need of vermiculture such as waste degradation and vermicomposting. There is, however, the possibility that the reason why these species are widely used in fishing in this area is their availability within the marshy terrains of these communities. Introducing other species of earthworms may not easily attract patronage as locals do not give in to change easily.

Those who travel long distances in search of suppliers claimed the reason is the cumbersome process involved in digging for the worms, while some who collected themselves complained of the cost of transportation if they have to go in search of suppliers. Vermiculture would go a long way in solving these problems, particularly if the vermiculture farmer either supplies the anglers at the harbour or set up the tanks near the fishing harbour. The vermiculture farm also have the benefit of helping to check the problem of habitat destruction done during the process of excavating for earthworms, by both anglers and their earthworm suppliers. This will also go a long way in helping to conserve the natural populations of earthworm, hence the productivity of the soil environment as relating to fertility and water movement.

Conclusion

The science of vermiculture in Nigeria has not been explored because little is known about the biology of our indigenous species of earthworms, particularly regarding reproductive timing and hatching rates. The present research serves as a baseline for intending earthworm breeders. Vermiculture has a viable prospect among the people of south-western Nigeria.

Vermiculture, however, still needs to be explored here by scientists, for information on the biology of the local species, vermicomposting potentials, and the protein content of the individual species need to be known to guide in specifying species to use in feed formation. Scientist equally needs to explore the fields of biotechnology and genetics as a way of developing better varieties of worms taking into consideration:

1. Age at maturity;
2. Number of cocoons produced; and
3. Size of worms.

Conflict of Interests

The author(s) have not declared any conflict of interests.

Table 5. Alternative baits to earthworms.

Bait	Frequency	Percentage (%)
Crayfish	3	3.1
Insect larvae, and adults	6	6.1
Small fishes and fish organs	11	11.2
Amphibian larvae	1	1.0
Bivalve and small fishes	1	1.0
Insect larvae and eba	1	1.0
Amphibian and insect larvae	1	1.0
Insects, small snail and crabs	1	1.0
Snail and small fish	3	3.1
Insects, crabs and amphibian larvae	1	1.0
Small fishes, snails and amphibian larvae	1	1.0
Fish, insects and eba	1	1.0
Insects, eba, amphibian larvae and diatoms	3	3.1
Fish, insects and snails	3	3.1
Fish and eba	2	2.0
Insects, snails and diatoms	11	11.2
Fish and insects	12	12.2
Insects and diatoms	4	4.1
Fish, insect and amphibian larvae	2	2.0
Fish, insect, eba and amphibian larvae	1	1.0
Fish, and amphibian larvae	1	1.0
Fish, insect, eba and snail	1	1.0
Fish, insect, bivalve, amphibian larvae, diatoms and snail	1	1.0
Fish, insect, snail and amphibian larvae	1	1.0
Insects and crabs	1	1.0
Fish and crayfish	6	6.1
Fish, crayfish and crab	3	3.1
Fish and crab	1	1.0
Fish, insect, crayfish, toilet soap and crab	1	1.0
Crayfish and snail	1	1.0
Fish, bread, eba and toilet soap	1	1.0
Fish, eba and toilet soap	1	1.0
Fish, insect, and crayfish	1	1.0
Fish, crayfish and snail	4	4.1
Fish, eba and crayfish	3	3.1
Fish, insect, eba, crab and crayfish	1	1.0
Insect, eba, crab and crayfish	1	1.0
Total	98	100

REFERENCES

Dedeke GA, Owa SO, Olurin KB (2010). Amino acid profile of four earthworm species from Nigeria. Agric. Biol. J. N. Am. 1(2):97-102.

Deolalikar AV, Mitra A (1996). Vermicompost of paper mill solid waste: a potential low cost source of organic manure for primary production in aquaculture. Aquaculture (in press).

Deolalikar AV, Mitra A (1997). Application of paper mill solid waste vermicompost as organic manure in Rohu (Labeo rohita Hamilton) culture—A comparative study with other commercial organic manure. In: Azariah, J., et al., Eds., Proc. Int. Bioethics Workshop: Biomanagement of Biogeoresources, University of Madras, Chennai.

Edwards CA, Lofty JR (1977). Biology of Earthworms (2nd edition), London. pp. 132-197.

Hartenstein R (1981). Use of *Eisenia fetida* in organic recycling based on laboratory experiments. In: workshop on the role of earthworms in the stabilization of organic residues. M. Appelhof-Compiler. Vol. I. Proc. Beech Leaf Press. Kalamazoo, USA. pp.155-165.

Hartenstein R, Neuhauser EF, Kaplan DL (1979). Reproductive potential of the earthworm *Eisenia fetida*. Oecologia (Berl.) 43: 329-340.

Kale RD (1994). Promotion of vermicomposting for production of organic fertilizer. Report of the Adhoc scheme sponsored by ICAR, New Delhi.Sanction no. ICAR No.-F.No. 13(3)91-SW&DFdt. 26. 3.

1991.
Kale RD (1998). Earthworm; Cinderella of Organic Farming. Publ. Prism Books Pvt. India. p. 88.
Kale RD (2002). Vermicomposting technology in India: An answer to shortages in nutrient supplies. In: Earthworms in the processing and utilization of organic wastes (Ed. C. A. Edwards). Publ. J. G. Press, PA. Chapter 22.
Owa SO, Dedeke GA, Morafa SOA, Yeye JA (2003). Abundance of earthworms in Nigeria ecological zones: implications for sustaining fertilizer-free soil fertility. Afr. Zool. 38(2):235-244.
Owa SO, Moreyibi OH, Dedeke GA, Olojo FO, Fashunwon OO (2004a). Earthworm-created microenvironments around roots of lowland rice. J. Sci. Eng. Technol. 11(1):5261-5270.
Owa SO, Moreyibi OH, Morafa SOA, Dedeke GA (2004b). Contributions of earthworms to soil temperature and its physicochemical implications on crops. J. Sci. Eng. Technol. 11(1):5343-5350.
Reinecke AJ, Kriel JR (1981). Influence of temperature on the reproduction of the earthworm *Eisenia fetida* (Oligochaeta). S. Afr. J. Zool. 16: 96-100.
Tsukamoto J, Watanabe H (1977). Influence of temperature on hatching and growth of *Eisenia foetida* (Oligochaeta, Lumbricidae). Pedobiologia 17:338-342.
Watanabe H, Tsukamoto J (1976). Seasonal change in size and stage structure of Lumbricid *Eisenia foetida* population in field compost and its practical application as the decomposer of organic waste matter. Rev. Ecol. Biol. Sol. 13: 141-146.

21

An appraisal of the handler awareness of electrical and electronic equipment toxicity in Nigeria

Onyenekenwa Cyprian Eneh

Institute for Development Studies, Enugu Campus, University of Nigeria, Nsukka, Enugu, Nigeria.

Some developing countries in Africa are waking up to the broadband race by providing public access to information communications technologies (ICTs) to instigate social and economic development and to narrow the digital divide. One of such countries is Nigeria, which ranked 60th in 2006 in e-readiness report by the Economist Intelligence Unit. Consequently, the market, marketers and repairers of electrical electronic equipment (EEE) are growing geometrically in Nigeria. Incorporated in EEE are about 46 separate chemical substances or elements, some of which are hazardous. Their emission from EEE translates to toxicity through inhalation of the polluted air, especially by marketers and repairers. Becoming aware of this danger is key to exercising the necessary caution in handling EEE. This study assessed the awareness of marketers and repairers of EEE toxicity and adoption of safety measures against the same. Questionnaires were administered on 80 purposively selected marketers in Alaba International Market, Lagos and on 20 repairers in C-to-C Plaza, Enugu. Data were analysed using average mean score technique. Results show that all participants were completely ignorant of the toxicity associated with EEE and did not associate it with various symptoms of ill-health.The study recommends education programmes for marketers and repairers of EEE and developing policies on how to handle EEE.

Key words: Hazardous chemical components, electrical electronic equipment, toxicity awareness, handlers, safety measures.

INTRODUCTION

Some developing countries in Africa are adopting the information communications technologies (ICTs) for the needed leverage in their development trajectory in the belief that it will contribute to social and economic development and in narrowing the digital divide. Little wonder, the 2006 global e-readiness rankings positioned South Africa 35th, Egypt 55th, Nigeria 60th and Algeria 63rd. This is an indication that Africa is waking up to the broadband race (Gomez et al., 2009; Mutula, 2005; Nkamnebe, 2010).

Also, a World Bank publication reports that "the digital divide" between the rich and poor nations is narrowing fast; telecommunications services to poor countries were growing at an explosive rate and the digital divide was rapidly closing. "People in the developing world were getting more access, especially to cell phone communications, far faster than they got access to new technologies in the past" (Mutula, 2005).

In the information age, EEE drive development and have consequently become ubiquitous even in developing countries. Therefore, their markets, marketers and repairers are growing geometrically. Researches have shown that about forty-six (46) separate chemical substances or elements are incorporated in EEE. Some are toxic by inhalation of the air polluted by gaseous emissions of these chemical components from EEE (Slade, 2006; Murali, 2009). For instance, the emission of lead from EEE constitutes health hazards.

An earlier study revealed a perfect positive correlation between the log dose of crude lead-extract from waste-EEE materials and % mortality of the rats to which the crude lead-extract from waste-EEE had been exposed by oral administration. The degree of confidence that crude lead-extract had lethal effect on rats was high, as well as the reliability. The lethal dose for rat was higher than the reported lowest lethal dose for man and the lowest lethal dose for guinea pig and dog.

Since the EEE marketers and repairers spend greater part of their time in the presence of EEE, the lethal concentrations of the various chemical components of EEE are easily attained (by inhalation). Osibanjo and Ogundiran (2010), submit that risk is the sum of hazard and exposure. In this case, emissions of various chemical components of EEE constitute the toxicity. The exposure or regular inhalation is high beacuse the EEE marketers and repairers keep close with the equipment most day. Thus, the risk is high. This is a serious cause for worry over the toxicity of EEE. More so, as indoor air pollution and urban air quality, of which emission is a factor, are listed as two of the world's worst pollution problems in 2008 World's Worst Polluted Places report (Blacksmith Institute, 2008).

Awareness of this danger is key to being informed, enlightened, knowledgeable, mindful of and alert or alive to the need for exercising the necessary caution in handing EEE. The present study was aimed at minimising the environmental health effects of the chemical components of EEE on marketers and repairers in Nigeria. The specific objectives of the study were to (i) assess the awareness of EEE toxicity by marketers and repairers, (ii) assess the adoption of safety measures in handling EEE by marketers and repairers, and (iii) recommend measures to ameliorate the environmental health effects of EEE on marketers and repairers. The study assumed that EEE makrters and repairers (i) were not aware of the toxicity of EEE, and (ii) were not adopting safety measures for handling EEE.

The study will greatly benefit EEE marketers and repairers, as it offers them a thought regarding EEE toxicity and adopting safety measures for handling the equipment. The findings will guide government and non-governmental organisations on the need to create awareness on safety practices among EEE marketers and repairers. Policy makers will find the results of the study useful in formulating policies for ameliorating the environmental health effects of EEE on marketers and repairers.

METHODOLOGY

Key informant technique was used to elicit information from purposively selected fifty (50) EEE marketers in Alaba International Market, which is famous for EEE trading business, and twenty (20) repairers in C-To-C Business Plaza, Nkpokiti Street, Enugu, Nigeria, where significant EEE repairing activities take place at commercial scale. All trader respondents had lasted 20 years upwards in the business, while repairers had lasted ten (10) years upwards on the job, to qualify for participation. Questionnaire copies were administered to the key informants to elicit information from them.

Part A of the questionnaire elicited demographic information on participants, while the Part B elicited information on the awareness of the participants of the toxicity of EEE and their adoption of safety measures for handling them. Answer options to each question were rated on a five-point lycert scale: strongly agree (5), agree (4), not sure (3), disagree (2) and strongly disagree (1).

The Average Mean Score test technique was used to analyse the data using the formula (Osuala, 2007):

$$\text{Calculated value (CV)} = \frac{\sum Fx}{\sum F}$$

Where, F is frequency and x is scale point. The CV was compared with the decision value (DV) of 3 (neutrality value). If CV was greater than DV, then the answer was regarded as being in the affirmative, otherwise it was in the negative. Similarly, null hypothesis was rejected if CV was greater than DV, and accepted if otherwise.

RESULTS AND DISCUSSION

Table 1 contains demographic information on participants. No woman owned EEE marketing shop or repairs workshop; but, their spouses did. All 100 participants were male. Therefore, the businesses were not gender balanced, but that of males.

All participants were literate, having obtained the first School Leaving Certificate (FSLC) (7%), attained secondary education (21%), obtained the Ordinary/Higher National Diploma (OND/HND) (45%), bagged a degree (25%) and/or obtained a higher degree (2%). Therefore, the businesses were for the literate class of the society.

The mode age bracket of participants was 18 to 25 (28%), followed by 26 to 35 (24%), 36 to 45 (23%), 46 to 55 (15%), 56 to 65 (8%) and above 65 years (3%). Thus, the businesses were for the youth (18^+ years) and productive age (25 to 45 years) in the society. Although, 37% of the participants were single, 56% were married, and 7% either separated or divorced, showing that most of the participants were married. This implies that victims of EEE toxicity were mostly married people and the brunt of its ugly effects was borne by families. These ugly effects include loss of breadwinners, ophaning children,

Table 1. Personal data of participants.

Subject		% Score
Sex	Female	0
	Male	100
Highest educational level	FSLC	7
	Secondary education	21
	OND/HND	45
	Degree	25
	Above degree	2
Age bracket	18-25	28
	26-35	24
	36-45	23
	46-55	15
	56-65	8
	Above 65	3
Marital status?	Single	37
	Married	56
	Separated/divorced	7
No.of years on the business involving EEE	11 - 20	68
	21 - 30	21
	31 - 40	7
	41-50	4
	50 upwards	-

Source: Field work, 2012.

child labour, street children, widowing women, widowhood practices, among others (Eneh and Nkamnebe, 2011).

Sixty-eight per cent of participants had been in the businesses for between 11 and 20 years, 21% for between 21 and 30 years, 7% for between 31 and 40 years, 4% for between 41 and 50 years, and none of them for above 50 years. Thus, people drop from the businesses as years go by, and hardly continued till old age. Those of them who drop into joblessness and idleness may get into addictions, thereby creating some problems in the society.

Table 2 shows the data on the awareness of the marketers and repairers of the toxicity of EEE and their adopton of safety measures. Forty-nine (49) or 49% of participants strongly disagreed that EEE emit hazardous gases that they inhale, 29 disagreed, and 21 were not sure. None of them (0%) agreed and none (0%) strongly agreed. The CV (1.70) was less than DV (3.0), showing that the answer was in the negative. That is, all participants were completely ignorant of the fact that EEE emit toxic gases. And, this ignorance is suicidal.

Forty-eight (48) or 48% of participants strongly disagreed that inhaling the emissions from EEE caused them some health problems, 33 disagreed, and 19 were not sure. None of them (0%) agreed and none (0%) strongly agreed. The CV (1.71) was less than DV (3.0), showing that the answer was in the negative. Therefore, all participants were completely ignorant of the fact that their inhalation of hazardous gases emitted from EEE caused them some health problems. In their ignorance, more havock would likely result because they would not care to take precaution.

Thirty-eight (38) or 38% of participants strongly disagreed that inhaling emissions from EEE had anything to do with some symptoms of ill-health, 31 disagreed, 28 were not sure, 3 agreed, and none (0%) strongly agreed. The CV (1.76) was less than DV (3.0), showing that the answer was in the negative. That is, most (97%) of the participants did not associate any symptoms of ill-health with inhalation of emissions from EEE. This ignorance would only occasion more harm.

Fifty-nine (59) or 59% of participants strongly disagreed that some EEE marketers and repairers have had to quit the trade because of EEE toxicity and its effects on them, 39 disagreed, and 2 were not sure. None (0%) agreed and none strongly agreed. The CV (1.43) was less than DV (3.0), showing that the answer was in the negative.

Table 2. Awareness of health effects and compliance with manufacturer's guides.

Parameter	Lycert scale (x)					∑F	∑Fx	CV	Remark	Decision
EEE gas emission	**1**	**2**	**3**	**4**	**5**	100	170	1.70	CV < DV.	Accept null
Frequency (F)	49	29	21	0	0				Answer is neg.	hypothesis (i)
Fx	49	58	63	0	0					
EEE toxicity						100	171	1.71	CV < DV.	Accept null
Frequency									Answer is neg.	hypothesis (i)
Fx	48	33	19	0	0					
	48	66	57	0	0					
Symptoms knowledge						100	176	1.76	CV < DV.	Accept null
Frequency									Answer is neg.	hypothesis (i)
Fx	38	31	28	3	0					
	38	62	64	12	0					
People quiting trade						100	143	1.43	CV < DV.	Not apllicable
Frequency	59	39	2	0	0				Answer is neg.	
Fx	59	78	6	0	0					
Need to quit trade						100	144	1.44	DV > CV.	Not applicable
Frequency	61	34	5	0	0				Answer is	
Fx	61	68	15	0	0				affirmative	
Quiting trade before old age						100	168	1.68	CV < DV.	Not applicable
Freuency	48	38	12	2	0				Answer is neg.	
Fx	48	76	36	8	0					
Need for precaution						100	155	1.53	CV < DV.	Accept null
Frequency	63	26	5	4	2				Answer is neg.	hypothesis (ii)
Fx	63	52	15	16	10					
Knowledge of caution						100	156	1.56	CV < DV.	Accept null
Frequency	60	31	4	3	2				Answer is neg.	hypothesis (ii)
Fx	60	62	12	12	10					

Source: Field work, 2012.

That is, although all participants were aware of some of their colleagues quiting the trade, they were certain that none of such cases had to do with EEE toxic effects on them. But, this ignorance would not stop the harm.

Sixty-one (61) or 61% of participants strongly disagreed that they could be required to quit the EEE trade on account of EEE toxic effects on them, 34 agreed, and 5 were not sure. None (0%) agreed and none strongly agreed. The CV (1.44) was less than DV (3.0), showing that the answer was in the negative. That is, most (95%) participants were certain it would not come to their being required to quit the trade on account of EEE toxic effects on them. This ignorance would not help matters.

Forty-eight (48) or 48% of participants strongly disagreed with the speculation that they would not last on EEE business till old age of retirement because of EEE toxic effects on them, 38 disagreed, 12 were not sure, and 2 agreed. None (0%) strongly agreed. The CV (1.68) was less than DV (3.0), showing that the answer was in the negative. That is, most (86%) participants were certain it would not come to their being required to quit the trade before their retirement age because of EEE toxic effects on them. This ignorance is as dangerous as the harm precursor.

Sixty-three (63) or 63% of participants strongly disagreed that they needed to adopt some safety measures for handling EEE, 26 disagreed, 5 were not sure, 4 agreed, and 2 strongly agreed. The CV (1.53) was less than DV (3.0), showing that the answer was in the negative. That is, most (91%) participants were certain that there was no need for adopting some safety measures for handling EEE in order to minimise toxic effects. Again, this ignorance is as dangerous as the harm precursor.

Sixty (60) or 60% of participants strongly disagreed with the safety measures for handling EEE, 31 disagreed, 4 were not sure, 3 agreed, and 2 strongly agreed. The CV (1.56) was less than DV (3.0), showing that the answer was in the negative. That is, most (95%) participants disregarded the safety measures for handling EEE and did not apply them. This attitude was not helpful, but suicidal.

Tests of hypotheses

Questions B1-3 probed the awareness of participants of the toxicity of EEE. Their respective CVs were 1.70, 1.71 and 1.76, each being lower than the DV of 3.0. Therefore, null hypothesis (i) was accepted for each of the cases. That is, marketers and repairers were not aware of EEE toxicity.

Questions B7 and B8 probed the adoption of common safety measures for handling EEE. Their respective CVs were 1.53 and 1.56, each being lower than the DV of 3.0. Therefore, null hypothesis (ii) was accepted for each case. That is, marketers and repairers were not adopting safety measures for handling EEE.

Woolf et al. (2007), shows that lead is a very strong poison. When lead dust is inhaled, some of the poison can stay in the body and cause serious health problems. Worse still, it is more common for lead poisoning to build up slowly over time from repeated exposure to small amounts of lead. In this case, there will be no obvious symptoms. In adults, lead can increase blood pressure and can cause digestive problems, kidney damage, nerve disorders, sleep problems, muscle and joint pain, and mood changes.

Classification of the consequences of lead poisoning by toxicity levels has shown decreased learning, decreased verbal ability, early signs of attention-deficit/hyperactivity disorder (ADHD) and low intelligence quotient (IQ), as general effects. Mild toxicity produces abdominal discomfort, lethargy, mild fatigue, myalgia, and paresthesia. Moderate toxicity produces constipation, irritability, difficulty concentrating, diffuse abdominal pain, mild fatigue, headache, muscular exhaustibility, tremor, vomiting, and weight loss. Severe toxicity produces colic, encephalopathy (seizures, coma, death), lead line on gingival tissue, and paresis or paralysis.

Reports by Karri et al. (2008), Timbrell (2008) and Marshall and Bangert (2008) show that the symptoms and signs of lead poisoning vary according to the individual and the duration of lead exposure. They may be subtle, and someone with elevated lead levels may have no symptoms. Symptoms usually develop over weeks to months as lead builds up in the body during a chronic exposure, but acute symptoms from brief, intense exposures also occur.

The main symptoms in adults are headache, abdominal pain, memory loss, kidney failure, male reproductive problems and weakness, pain, or tingling in the extremities. The classic signs and symptoms in children are loss of appetite, abdominal pain, vomiting, weight loss, constipation, anaemia, kidney failure, irritability, lethargy, learning disabilities, and behavior problems.

In acute poisoning, typical neurological signs are pain, muscle weakness, paraesthesia, and, rarely, symptoms associated with encephalitis. Abdominal pain, nausea, vomiting, diarrhoea, and constipation are other symptoms of acute lead poisoning. Lead's effects on the mouth include astringency and a metallic taste. Gastrointestinal problems, such as constipation, diarrhoea, poor appetite, or weight loss, are common in acute poisoning. Absorption of large amounts of lead over a short time can cause shock (insufficient fluid in the circulatory system) due to loss of water from the gastrointestinal tract. Haemolysis (the rupture of red blood cells) due to acute poisoning can cause anaemia and haemoglobin in the urine. Damage to kidneys can cause changes in urination, such as decreased urine output. People, who survive acute poisoning, often go on to display symptoms of chronic poisoning.

Chronic poisoning usually presents with symptoms affecting multiple systems, but is associated with three main types of symptoms: gastrointestinal, neuromuscular, and neurological. Central nervous system and neuromuscular symptoms usually result from intense exposure, while gastrointestinal symptoms usually result from exposure over longer periods.

Signs of chronic exposure include loss of short-term memory or concentration, depression, nausea, abdominal pain, loss of coordination, and numbness and tingling in the extremities. Fatigue, problems with sleep, headaches, stupor, slurred speech, and anemia are also found in chronic lead poisoning. A "lead hue" of the skin with pallor is another feature. A blue line along the gum, with bluish black edging to the teeth is another indication of chronic lead poisoning. Children with chronic poisoning may refuse to play or may have hyperkinetic or aggressive behaviour disorders.

According to the reports by Hu et al. (2007), about 35-40% of inhaled lead dust is deposited in the lungs, and about 95% of that goes into the bloodstream of adults. Guidotti and Ragain (2007), report that the complications arising from lead poisoning show that lead affects every one of the body's organ systems, especially the nervous system, but also the bones and teeth, the kidneys, and the cardiovascular, immune, and reproductive systems. Hearing loss, tooth decay and cataracts have been linked to lead exposure.

Findings, Conclusions and Recommendations

Summary of results showed that:

1) All participants were completely ignorant of the fact that EEE emit hazardous gases.

2) All participants were completely ignorant of the fact that their inhalation of the hazardous emissions from EEE caused them some health problems.
3) Most (97%) of the participants did not associate any of the common symptoms of ill-health with inhalation of emissions from EEE.
4) Although all participants were aware of some of their colleagues quiting the trade, they were certain that none of such cases had to do with EEE toxic effects on them.
5) Most (95%) participants were certain it would not come to their being required to quit the trade on account of EEE toxic effects on them.
6) Most (86%) participants were certain it would not come to their being required to quit the trade before their retirement age because of EEE toxic effects on them.
7) Most (91%) participants were certain that there was no need for adopting some safety measures for handling EEE in order to minimise toxic effects.
8) Most (95%) participants disregarded the safety measures for handling EEE and did not apply them.

Toxic substances are ubiquitous; no one can run away from them. Their uses are getting increasingly important; people need to use them for personal and society development at various levels. EEE workers are exposed to toxic components of EEE through several pathways (gases emission, dust, water, food contami-nation, and so on). Therefore, exercise of caution is what is required from EEE marketers and repairers. They need to know that their stock contains chemical components that emit hazardous gases and that their inhalation of these emissions leads to adverse health effects that manifest in various symptoms of ill-health. If, through awareness campaign, they are made to know about the hazardousness of WEEE and the need to adopt recommended safety measures, they would exercise caution as they continue with the trade in less risky manners.
It is recommended, therefore, that:

1) EEE shops/workshops be well ventilated;
2) EEE marketers and repairers should take breaks from their shops/workshops;
3) EEE marketers and repairers should take vacations from their jobs;
4) EEE marketers and repairers should wear nose masks and gloves in workplace;
5) The government and non-governmental organisations should conduct awareness creation fora to enlighten the EEE marketers and repairers on proper professional practice to minimise the toxic effects of EEE on them;
6) Policies should be made to control the careless handling of EEE by marketers and repairers, in order to curb health diseases, life tolls and socio-economic loses attendant upon growing EEE sales and services businesses in the era of idiosyncratic adoption of ICTs.

Conflict of Interests

The author(s) have not declared any conflict of interests.

REFERENCES

Blacksmith Institute (2008). *2008 World's Worst Polluted Places Report.* Available at: http://www.worstpolluted.org Accessed 29 August 2010.

Eneh OC, Nkamnebe DA (2011). Gender Gap and Sustainable Human Development in Nigeria: Issues and Strategic Choices. Asian J. Rural Dev. 1(1):41-53.

Gomez R, Ambikar R, Coward C (2009). Libraries, Telecentres and Cybercafes. Performance Measurement Metrics 10(1):33-48.

Hu H, Shih R, Rothenberg S, Schwartz S (2007). The epidemiology of lead toxicity in adults: measuring dose and consideration of other methodologic issues. Environ. Health Perspec. 115(3):455-462.

Karri SK, Saper RB, Kales SN (2008). Lead encephalopathy due to traditional medicines. Current Drug Safety 3 (1): pp.54–59.

Marshall WJ, Bangert SK (2008). Therapeutic drug monitoring and chemical aspects of toxicology. Clin. Chem. 6th ed. Elsevier Health Sciences. p. 366.

Murali BK (2009). Working on e-waste solutions. Chinnai: Express Publication Ltd.

Mutula SM (2005). Peculiarities of the digital divide in sub-Saharan Africa Programme. Electron. Lib. Inform. Syst. 39(2):122-138.

Nkamnebe AD (2010). ICT Consumption and the Challenges of Environmental sustainability in sub-Saharan Africa. Paper presented at the 11th International Conference on Beyond Global Markets organized by the International Society of Markets and Development (ISMD) and hosted by National Economics University, Hanoi, Vietnam 5-8th January 2010.

Osibanjo O, Ogundiran MB (2010). The basics and applications of globally harmonized system (GHS) of chemical classification and labeling. Paper presented at a one-day GHS seminar organized in Onitsha by Institute of Chartered Chemists of Nigeria (ICCON), 17 August 2010.

Osuala E (2007). Introduction to Research Methodology. 3rd ed. Onitsha: Africana-First Publishers Limited.

Slade G (2006). Made to break: technology and obsolescence in America. Washington, DC: Harvard University Press.

Timbrell JA (2008). Biochemical mechanisms of toxicity: specific examples. Principles of Biochemical Toxicology, 4th ed. Informa Health Care.

The absorption and scavenging ability of a bacillus in heavy metal contaminated soils (Pb, Zn and Cr)

TANG Meizhen[1], CHEN Junfeng[2] , SUN Yannan[1], TONG Yanru[1] and LIU Yuling[1]

[1]Key Laboratory of NaSiHu Lake Wetland Ecosystem and Environment Protection, Qufu Normal University, Qufu, Shandong, 273165, China.
[2]College of Resource and Environment, Hunan Agricultural University, Changsha, Hunan, 410128, China.

A bacterial strain, which was able survive Pb, Zn and Cr -heavy metal compound culture was isolated from the soils of Shanjia Villiage, Qufu City, Shandong Province. An observation of the morphological and biochemical characteristics of the compound and the analysis of 16SrDNA sequence revealed that it was *Bacillus*. Standing liquid culture was used to study the tolerance of Pb Zn Cr -heavy metals compound and the ability to absorb and clean the heavy metal compound. Further study was made to compare the absorbing ability of Cu, Cd and Hg.

Key words: *Bacillus*, heavy metal, Pb, Zn, Cr.

INTRODUCTION

With the rapid development of economy and society, the exploitation of coal resources in China increases gradually and environmental problems caused by coal gangue have become increasingly serious (Shao and Cao, 2002). Coal gangue is the main component of coal slag. According to statistics, the cumulative stacked gangue is more than 40 tons (Guo, 2007), but the annual growth rate is still 1-2 million tons. There are heavy metals such as Cr, Pb, Zn, Cd, Cu and other heavy metal elements in the gangue. When there is accumulation of harmful heavy metals in the soil to a certain degree, it will have toxic effects on soil - plant system, which not only causes soil degradation, decline in crop yields and quality, but also pollutes the surface water and groundwater through runoff and leaching, deterioration of water environment, and direct contact with food chain and other ways to endanger human health and life. In the toxic heavy metals, pollution in soil ecological system is long-term and irreversible (Liu, 2008); but the problem is not obvious. Therefore, soil ecosystem metal, heavy metal pollution and prevention especially in the international arena have been a difficult and hot research topic (Chi, 2006).

Microbial remediation is the use of microorganisms (bacteria, fungi, indigenous alien gene engineering bacteria) on the metabolism and transformation of pollutants, degradation of pollutants, mainly used for degradation of organic pollutants in soil. Microbial remediation technology does not destroy the soil environment for plant growth, does not form secondary pollution and does not transfer the pollutants to residual

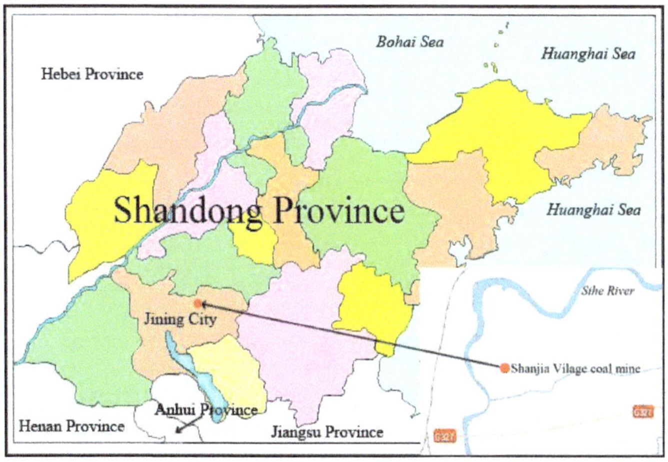

Figure 1. The location map of sampling point.

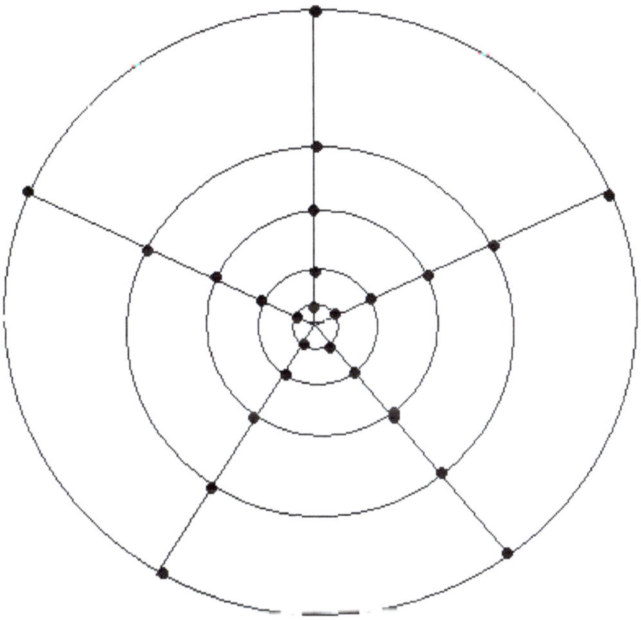

Figure 2. Soil sampling distribution.

Problems (Zhang and Xia, 2000). It can maximize the removal of pollutants or make contaminants harmless. It interferes less in surrounding environment and has the advantages of good environmental and social benefits (Chen et al., 1997). Bioremediation technology, the rapid rise in recent years, has become a green field of scientific research (Ebbs, 1997; Weon, 2011).

At present, domestic and overseas scholars are launching a lot of research on the coal mine area and soil heavy metal management (Li, 2009; Liu, 2009; Murzaeva, 2004). Domestic and foreign experts have adopted the toxic improver method, using local traditional and chemical methods to solve the problem of heavy metal contamination in soil. However, because of their own limitations, they have failed to get the ideal method for soil heavy metal management (Tang, 1996; Wang, 1996). In recent years, there are a many researches about heavy metal resistant bacteria, which are mainly concentrated in heavy metals such as Cd, Cu and As. Therefore, the composite metal reports is even less.

This work studied the soil polluted by coal gangue, based on the theory of microbial remediation technology. Isolated strains can fight a variety of heavy metals. This work at the same time fills the gaps in the domestic research of heavy metal pollution in China caused by coal gangue, and shows how to prevent the pollution of the gangue based on background and scientific information.

This test on coal gangue from mountain soil separated strains that have the ability to absorb compositing metals-Pb, Zn and Cr. After the physiological and biochemical reaction of the strains was identified as well as 16SrDNA, an auxiliary research was done for the bacteria to explore their resistance to Cu, Cd and Hg and their ability to remove degradation.

MATERIALS AND METHODS

The source of the bacteria

The experiment soil samples are taken from Shanjia Vilage Coal Mine in Qufu City (Figure 1), Shandong Province.

The distribution point of the samples is shown in Figure 2; with distance of 2, 5, 10, 15, 20 and 25 m. After retrieving six samples, they were mixed fully and set aside for pretreatment, with quartiles.

The screening culture of the strains

The preparation of the soil solution

The mixed sample was collected into laboratory cultures domesticated for seven days. 1 g of the mixed sample was used, which was diluted with 99 mL of sterile water. After, it was oscillated for 20 min, and stood still for 20 min. The upper clear liquid of 0.5 mL was taken. 4.5 mL sterile water was added into a test tube and the soil diluents concentration of 10^{-3} was obtained. In accordance with the above method, one can get diluted soil concentration gradient of 10^{-4}, 10^{-5}, 10^{-6} and 10^{-7}.

The preparation of the culture media

Beef extract-peptone medium: 5 g beef extract; 10 g peptone; 5 g NaCl; 15-20 g agar; 1000 mL H_2O; pH 7.2-7.4. Potato medium: 200 g selected high quality potatoes; 20 g glucose; 15-20 g agar; 1000 mL H_2O; natural pH. GAO 1st synthetic culture medium: 20 g soluble starch; 1 g KNO_3; 0.5 g $K_2HPO_4{\cdot}3H_2O$; 0.5 g NaCl; 0.5 g $MgSO_4{\cdot}7H_2O$; 15-20 g agar; 0.01 g $FeSO_4$; pH 7.2-7.4.

The culture medium components were configured for preparation and the autoclave was used for sterilization for 20 min under 121°C. After the medium was cooled to about 50°C, 60 mg Pb $(NO_3)_2$, 50 mg $ZnSO_4$ and 80 mg $K_2Cr_2O_7$ were added (every three parts separately) to the medium. It was oscillated and shaken to have a dissolved mix. All the operations were done on bench top.

Table 1. The specific microorganisms culture conditions.

Temperature (°C)	Microbiota	Culture medium	Soil solution concentration	Incubation time
37	Germ	Beef extract-peptone medium contains Pb, Zn and Cr	10^{-5}, 10^{-6}, 10^{-7}	24 h
28	Fungus	Potato medium contains Pb, Zn and Cr	10^{-5}, 10^{-6}, 10^{-7}	48 h
28	Actinomycetes	GAO 1th synthetic culture medium contains Pb, Zn and Cr	10^{-4}, 10^{-5}, 10^{-6}	5 day

The screening of the strains

The strains were kept in 9 cm plate for development, using the pouring plate method. The specific culture conditions are in Table 1. Each concentration has three parallel groups, and the growth of the strains was recorded.

Domestication of the strains

Good growth was obtained by training the resistant strains colony with a cross, using the methods of purification. Purification of culture medium and training methods are used to screen the same species. All strains were purified for three times. At last, the individual colony was purified. All the strains were kept in the refrigerator (4°C) for preservation.

The study of the ability of the strains for removing degradation

Heavy metal processing method

The medium component is 1.2.2, but agar was not added. The four concentrations were done at the same time. The concentrations of the bacteria are screened as a benchmark concentration. The concentrations were set ordinarily for five, ten and twenty times and vaccination training was done. The liquid stalling training method was taken for study. The bacteria culture conditions are given in Table 1; the growing strains were not added to the sune-vector.

Research methods

The liquid static method was used for training; and the liquid static training of the bacteria cultures is centrifuged by high-speed freezer in 13000 r/min and 4°C. The concentration of heavy metal was measured from the centrifugal machine supernatant by the atomic absorption photometer.

Identification of the strains

The bacteria strains with the best degradation ability were selected for identification.

Morphological observation of strains

The morphology of the bacteria colony in the beef was observed specially in a flat plate. Mycelial morphology was observed under oil lens by using gram staining, and the stained spore was added for observation.

Physiological and biochemical reactions of the strains

Glucose oxidation fermentation test, oxidase test, starch hydrolysis test, Acetyl methyl methanol test, methyl red (MR) test, catalase test, producing H_2S test, indole test, nitrate reduction test, and so on were done for the selected bacterium.

PCR amplification and sequence analysis of 16SrDNA

16SrDNA sequence analysis : total DNA strain extracted was cultured , as a template, using primers 8f 5'AGAGTTTGATCCTGGCTCAG 3 '20 bp and 1492r 5'GGTTACCTTGTTACGACTT 3 '19 bp; the PCR product was cloned by the Shanghai Biological Engineering Co., Ltd. for purification identification.

Resistance of the strains to Cu, Cd and Hg

Based on the optimal growth conditions, the screening strains can degrade Cu, Cd and Hg as well as composite metals- Pb, Zn and Cr, using the same methods. The benchmark concentration is 0.5 mg/L. To calculate its removal efficiency of Cu, Cd and Hg as compared to the initial removal efficiency experiments, more comprehensive experimental results are drawn.

RESULTS

Eight strains were obtained after screening: three bacteria, three fungi and two actinomyces. The study of the ability of degradation and removal showed that the strain with most powerful removal and degradation ability was bacteria. The strains with the highest percentage of degradation and absorption are three selected bacteria.

Strains identification

Morphological observation of strains

The selected bacteria are round, neat at the edge, ecru white, opaque, gram-positive and are found in aspen. The form is microscopically shown in Figure 3.

Physiological biochemical reactions of the strains

The results of physiological biochemical reactions of the strains are seen in Table 2.

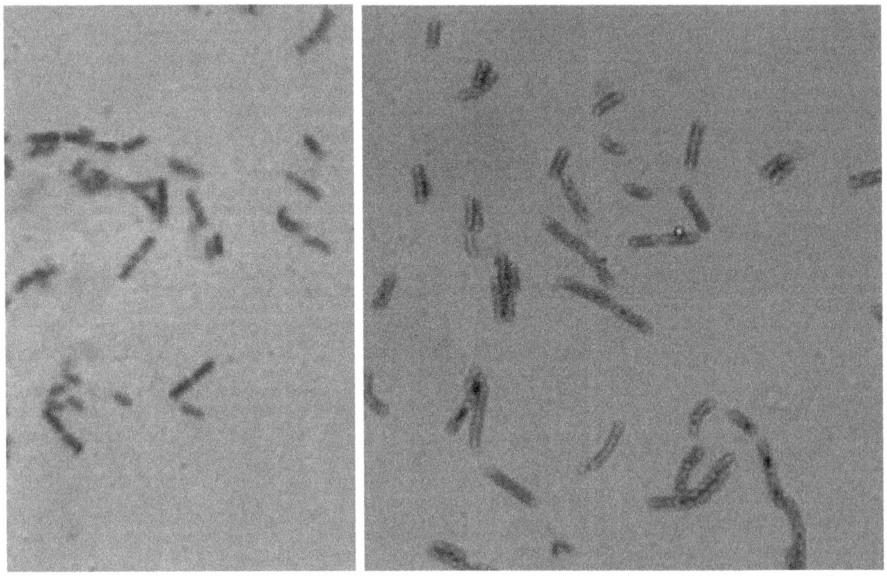

Figure 3. The form observations microscopically.

Table 2. Results of physiological and biochemical reactions of the strains.

Physiological biochemical reactions	Result
Oxidase	-
Catalase	+
Amylolysis	+
Oxidation fermentation of glucose	+
Indole test	-
Nitrate reducing reaction	+
Methyl red test (MR)	+
Acctyl methyl methanol test	+
Produce H_2S test	-

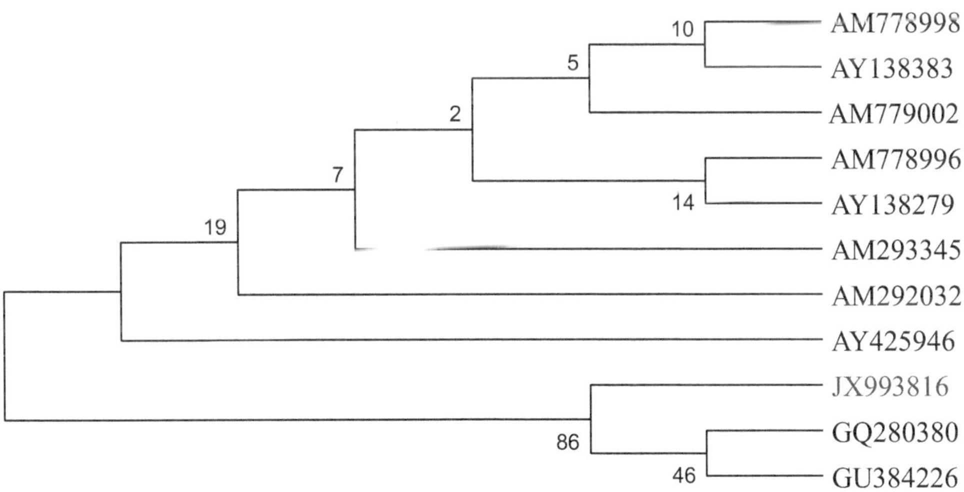

Figure 4. Evolutionary tree.

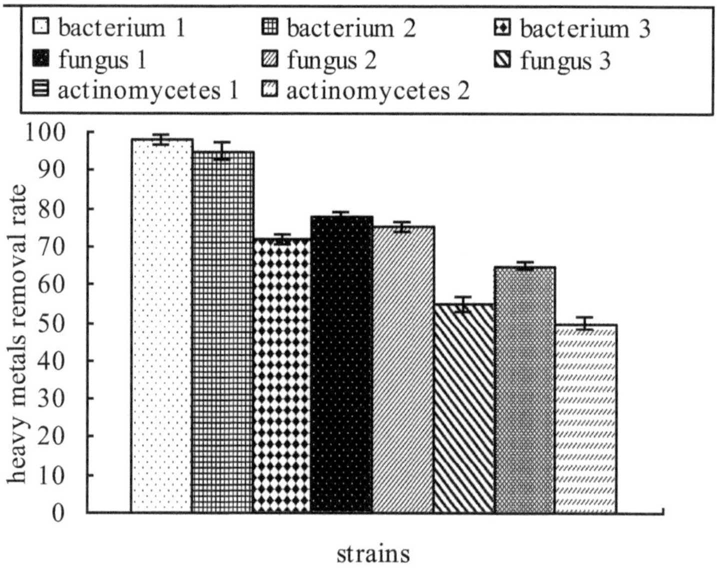

Figure 5. Different rates of strains for heavy metals removal. Annotations: This compound metal concentration is of the benchmark concentration (namely 1 times concentration).

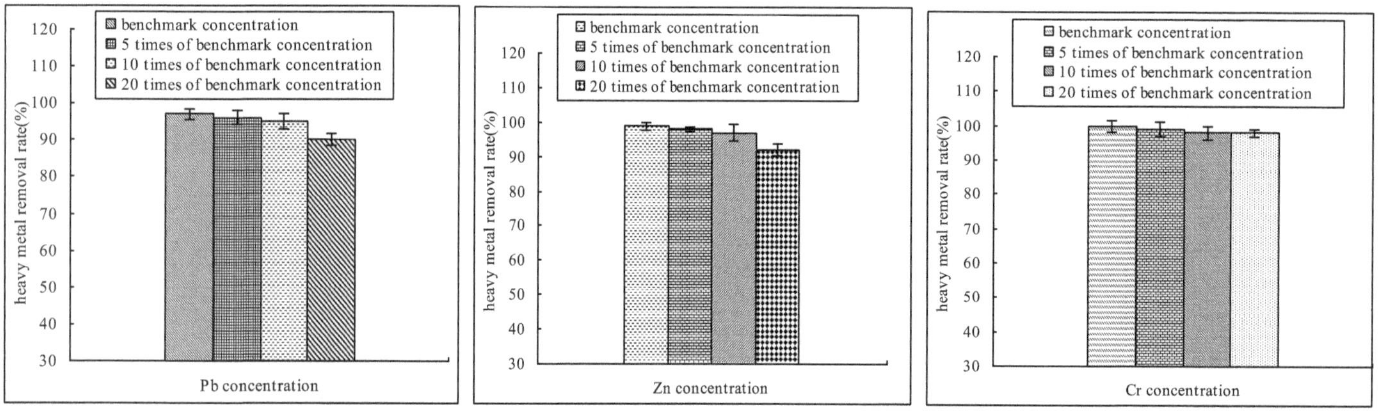

Figure 6. The strain away rate of Pb, Zn and Cr.

Sequence analysis of 16SrDNA

16SrDNA sequence specific to the Shanghai SANGON Biological Engineering Technology Service Co. Ltd selected strain is 1640 bp (GenBank accession number JX993816). BLAST searches for the series, using Clustal W and PHYLIP software to analyze and build the phylogenetic tree (Figure 4).

From the results of the analysis, selected strain and Bacillus homology was 100%. Generally, 16S rDNA sequence homology of less than 98% can be considered to belong to different types; homology less than 93~95% can be considered to belong to different genera (Devereux et al., 1990; Fry et al., 1991). Thus, combining the selected strains based on their morphological, physiological and biochemical reaction characteristics, *Bacillus cereus* is identified.

The ability to degrade and remove Pb, Zn and Cr

The analysis of the ability to degrade and remove different strains of composite metals- Pb, Zn and Cr

Figure 5 shows clearly: (1) Different microorganisms have different ability to degrade and remove heavy metals; the overall performance is: *Actinomyces* > *fungi* > *bacteria*. (2) Different strains (bacterium 1, 2, 3) of the same kind of microorganism (such as bacteria) have different degradation and removal ability.

The ability to degrade and remove heavy metal compound

In Figure 6, the separated bacilli all have certain resistance to Cu, Cd and Hg, and have certain Degradation removal ability. With the increase of metal concentration, the removal rate drops down. From the figure, we can also see that the *bacilli* remove three kinds of metal in the benchmark concentration.

DISCUSSION

In this work, samples are taken from coal gangue mountain soil; and the microbe's resistance to the composite metals- Pb Zn Cr is studied. Bacteria extracted from the soil solution were configured. From the experiment, we can draw the following conclusions:

1) Eight strains are obtained after screening three bacteria, three fungi and two actinomyces. After their morphological, physiological and biochemical reaction observation as well as 16 SrDNA sequence analysis, it is shown that the bacteria are gram positive bacilli. There are different types of media due to differences in nutrients for enrichment and separation of different species of microbes. It does not only Improve the separation efficiency of strains and selective enhancement, but also guarantees the reliability and feasibility of the experiment. So it is the premise of the experiment to be carried out smoothly.
2) Experimental results show that different strains that degrade and eliminate heavy metals exhibit different processing capabilities. The ability to process heavy metals is strongest among bacteria, followed by fungi and actinomycetes. In addition, different strains of the same bacteria species also exhibit different processing capabilities. The results show that with increased concentration of heavy metals, their ability to remove those decreases.
3) Further research proves that bacillus also has certain ability to remove Cu Cd Hg.

Conflict of Interests

The author(s) have not declared any conflict of interests.

ACKNOWLEDGEMENT

This study was supported by Shandong Provincial Natural Science Foundation (ZR2013EEQ009), China; Project of Qufu Normal University (No.XJZ200825).

REFERENCES

Chen YS, Sun QJ, Chen J, Zhuang YY et al. (1997). Research on technology of biosorption of heavy metals. Adv. Environ. Sci. 5(6):34-35.
Chi YH (2006). Application of soil heavy metal pollution and its bioremediation technology. Fujian Agric. Sci. Technol. 3:62-64
Ebbs SD, Lasat MM, Brady DJ, Cornish J, Gordon R, Kochian LV (1997). Phytoextraction of cadmium and zinc from a contaminated soil. J. Environ. Quality. *26*(5):1424-1430.
Guo XD (2007). The study of comprehensive utilization for coal gangue technical. Zhongzhou Coal. (2):19-20.
Li DY, Yao XM, Zuo JP (2009). The report of management, industrial and the investigation about the British soil heavy metal pollution. J. Hunan Environment-biological polytechnic. 12:40-45.
Liu JJ (2008). Current situation and countermeasures of soil contamination by heavy metals. China Environ. Manag.1:36-38.
Liu L, Xiao YB (2009). Advance in the study on recovery and remediation of heavy metals pollution in soil. J. Changchun Inst. Technol. (Social Science Edition). 10(3):73-78.
Murzaeva SV (2004). Effect of Heavy Metals on Wheat Seedlings: Activation of Antioxidant Enzymes. Appl. Biochem. Microbial. 40(1):114-119
Shao J, Cao SY (2002). The current status of environment pollution in coal mine area and its countermeasures. Ind. Safety Environ. Protect. 28(10):26-28.
Tang SR, Huang CY, Zhu ZX (1996). Phytoremediation on Contaminated Soils and Its Research Progress. Shanghai Environmental Science. 15(12):37-39 (In Chinese).
Wang BJ, Yang HF (1996).Interaction of Microorganisms with Heavy Metals. Chongqing Environ. Sc. 18(1):35-38. (In Chinese)
Weon B (2011). Genetic engineering of Escherichia coli for enhanced uptake and bioaccumulation of mercury. Appl. Environ. Microbiol. 67(11):5335-5338.
Zhang C, Xia LJ (2000). Remediation technology. Beijing: China. Environ. Sci. Press. 1-4.

Characterization of Dabagi clay deposit for its ceramics potential

I. Abubakar[1], U. A. Birnin Yauri[2] , U. Z. Faruq[2], S. S. Noma[3] and N. Sharif[4]

[1]Department of Remedial and General Studies, Waziri Umaru Federal Polytechnic, Birnin Kebbi, Nigeria.
[2]Department of Pure and Applied Chemistry, Usmanu Danfodiyo University, Sokoto, Nigeria.
[3]Department of Soil Sciences, Usmanu Danfodiyo University, Sokoto, Nigeria.
[4]Department of Chemistry, Zamfara College of Art and Science, Gusau, Nigeria.

A clay deposit in Gwandu Town of Kebbi State, Nigeria has been evaluated for its potential as industrial raw material. The chemical analysis was carried out using X-ray florescence spectrometery while physical property tests such as porosity, bulk density, linear shrinkage, thermal shock resistance, cold crushing strength and refractoriness were done using standard techniques. The result of chemical analysis indicated that the clay was composed of silica (SiO_2), 64.50%; alumina (Al_2O_3), 16.30%; iron oxide (Fe_2O_3), 14.20%; calcium oxide (CaO), 0.2%; potassium oxide (K_2O), 0.74%; titanium oxide (TiO_2), 1.71% and other oxides in traces. The chemical analysis suggests that the clay deposit is mainly made of kaolinite and free quartz. Result of the physical tests shows that the clay has an apparent porosity of 28.46%, bulk density of 1.81 g/cm^3, linear shrinkage of 6.80%, thermal shock resistance of seven cycles, loss on ignition test (L.O.I) of 4.46%, cold crushing strength of 14138 Nm^{-2} and estimated refractoriness of 1349°C.

Key words: Dabagi clay deposit, characterization, x-ray fluorescence and technological tests, Kaolinite, ceramic applications.

INTRODUCTION

Clay is a natural earthy, fine - grained material that acquires plasticity on being mixed with limited quantity of water (Velde, 1992; Idenyi and Nwajagu, 2003). From a chemical or mineralogical standpoint, clay is a complex aluminosilicate compounds containing attached water molecules, which have their origin in the chemical and mechanical disintegration of rocks, such as granites (Nwajagu, 2005). Clay occurs most abundantly in nature in solids, sediments, sedimentary rocks and hydro-thermal deposits (Velde, 1992). One basic property of or other polar ions into their structure. This is called the swelling property. Based on this property, Velde (1992) broadly classified all clays into swelling and non-swelling type. Swelling clays are called smectites. The important property according to Velde (1992) is the basic composition and structure of the clays; and this is used to further classify the clay minerals into: Kaolinite group ($Al_2Si_2O_{10}$ (OH_6)); this group has one silica and one alumina unit stacked in alternating fashion (1:1 lattice type). No ion or water molecules can enter the adjacent layers, and only the external surfaces determine their colloidal properties.

Owing to its relatively large particles and low specific surface, kaolinite exhibits less plasticity, cohesion and swelling as compared to other clay minerals. Kaolin is an important and widely used industrial mineral which is refined from kaolinite. It is a naturally occurring minerals of the clay family and may contain a number of impurities such as quartz, feldspar, tourmaline, limestone, zircon, etc. which were derived from the parent rock. It is a weathering product of silicate rocks which is whitish, earthy to dull with plastic touch.

The characteristics and chemical composition of a kaolin deposit usually determines its industrial utilization. Kaolin is one of the most valuable of the industrial clays which is used in most manufactured products. Prominent uses include paper filling and coating; paint, plastic, adhesive and ink pigment; rubber reinforcing agent, ceramic raw materials for porcelain, dinner ware, tiles and enamels, catalyst for petroleum cracking and auto exhaust emission catalytic control devices; cosmetics base; and digestive coating remedy (RMRDC, 2010); Bentonite: bentonite belongs to the group of clays whose technical properties are controlled by the proportion of montmorillonite, a sub-group within the smectitic clays. It is clay derived from deposits of weathered volcanic ash. Bentonites are hydrated aluminosilicates, which composed predominantly of the clay mineral montmorillonite. They are composed of a 3-tier structure with alumina silica sheets' sandwiched between tetrahedral silica units. A simplified formula for montmorillonte is Al_2O_3. $4SiO_2.H_2O$, the other minerals that could be found in bentonite in small content are chrystobalite, biotite, chalcedony, calcite, pyrite, dolomite and plagioclase. There are three principal types of bentonite namely: natural sodium bentonite or sodium montmorillonite, natural calcium bentonites or calcium montmorillonite; and sodium activated bentonites or sodium activated montmorillonites.

Natural sodium bentonite as the name suggests, occurs with sodium as the predominant exchange cation. They are characterized by high swelling, high liquid limit and high thermal durability. It is usually used for drilling mud. The vast majority of the montmorillonites occurring in abundance worldwide is of the calcium type and is referred to as calcium bentonite. Much lower swelling and liquid limit values compared to natural sodium bentonite, characterize them.

Calcium bentonite is used as a bleaching agent in cooking oil industries, bleaching agent in lubricant oil recycling, as a catalyst, absorber, filler, etc.

Bentonite has a wide range of industrial uses. The physical and chemical properties of bentonite make it an important industrial mineral, which has widespread application in various industrial sectors such as foundry sand bond in iron and steel foundries and in iron ore pelletizing in metallurgy; this is probably the largest use for bentonite, as insulator in civil engineering as an efficient materials for drilling mud (because the gel-like suspension it forms in water), as bleaching clay in oil refining, clarifying and decolourising, filtering agent for clarifying wine, beer and treating waste water, Ingredient in cosmetics, animals feeds and pharmaceutical (RMRDC, 2010).

Glass sand/quartz: silica sand/quartz constitutes one of the most readily available geological materials used in industries and factories such as glass manufacturing companies. Silica sand/quartz are said to consist of high optimal percentage of silicon oxide (SiO_2) which is a very good chemically stable element and it remains almost the same no matter the series of cycles it may have gone through, either in transportation or re-deposition. Quartz is silica occurring alone in pure state. Silica/glass sand on the other hand are products of weathering, erosion and transportation by rivers or/and the sea. Naturally occurring silica sands may contain some undesirable impurities like accessory haematite, rutile and dolomite etc. The glass sand is used in the production of various glass products: which include sheet glasses, for windows, bottles, mirrors, optical instruments, chemical apparatus, electrical insulation and condensers, pipe, doors, crucibles, automobile and aircraft bodies, filters and building blocks. They are also used for making abrasives and for gravel parking in the petroleum industries (Nwoye, 2009).

Clay products are versatile industrial material that have amazing variety of uses and applications in ceramics porcelain, dinner wares, glasses, refractory bricks, burnt bricks and architectural tiles and ceramic (Akudinobi, 2006). Important uses have also been found for clays in the rubber industry where its reinforcing potential has been exploited since the early part of 21th century, particularly those with tolerable presence of silica (which serve as reinforcing component in rubber compounding) (Akudinobi, 2006).

The important properties of clay are plasticity, colour, strength, drying and firing shrinkages. The percentage of the minerals oxide (Fe_2O_3, MgO, CaO, Na_2O etc.) in the clay ultimately determine the areas of applications of the clay such as in bricks, floor tiles, paper etc. while the quantity of the alkali metal oxides (Na_2O, K_2O, CaO etc) indicates their suitability for ceramic products Nnuka and Agbo (2000).

Nigeria has appreciable distribution of industries engaged in metal production and other process industries, and hence, there is need for raw - material to

Table 1. Chemical composition of Dabagi clay compared with standard clay for industrial applications: Chester (1973) and Grimshow (1971).

Composition	Dabagi clay	Ceramics	Refractory brick	High melting clay	Glass	paper	Paint
SiO_2	64.50	60.50	51.70	53-73	80-95	45.0-45.8	45.3-47.9
Al_2O_3	16.30	26.50	25-44	16-29	12-17	33.5-36.1	37.9-38.4
Fe_2O_3	14.26	0.5-1.2	0.5-2.4	1-9	2-3	0.3-0.6	13.4-13.7
CaO	0.26	0.18-3	0.10-20	0.5-2.6	4-5	0.03-0.60	0.03-0.60
K_2O	0.74	-	-	-	-	-	-
TiO_2	0.17	-	-	-	-	-	-
CuO	0.03	-	-	-	-	-	-
V_2O	0.08	-	-	-	-	-	-
LOI	4.46.	8.18	8-18	5-14	-	-	-

support their growth. Clay products such as ceramic wares, burnt bricks and floor tiles are cheaper and more durable building materials than cement particularly in Nigeria.

There are vast deposits of clay spread across every region in Nigeria, though their properties differ from one site to another on account of geological differences. The present economic state imposes the need for sourcing of local material to meet the increasing demands. This study therefore is to ascertain the chemical composition and mechanical/physical properties of Dabagi clay deposit in Gwandu town of Kebbi State of Nigeria so as to highlight its ceramic value, economic potential and encourage its immediate industrial application/ uses.

MATERIALS AND METHODS

Sample preparation

Stratified random sampling was used. 10 m sample area was divided into ten smaller units and from each unit three samples were collected randomly at a depth of 10 cm, mixed and homogenized. Cone and quartered method was employed until a required (representative) sample was obtained. The coning and quartering means pouring of the clay sample on a flat surface such that it formed a cone which was divided into four equal parts by using a straight edge.

The two alternate quarters are mixed again for further quartering. The representative fractions for the sample was later crushed and grounded to produce 60 mesh (BS). The crushed sample was used for analysis.

Chemical analysis

The chemical composition of the clay was determined using X-ray florescent (XRF). 1 g of clay sample was added to 10 g of anhydrous lithium tetra borate ($Li_2B_4O_7$) acting as a fluxing agent. The constituent were mixed inside a cleansed crucible and heated up to 500°C for 8 min and allowed to cool at room temperature to obtain the fused sample which was used for the analysis. The chemical composition of the sample is presented in Table 1.

Loss on ignition test (L.O.I)

This is the quantity of chemically combined H_2O (and sometimes organic matter content) in inorganic materials (Udochukwu, 2007). Fifty grams of the clay sample was dried in an oven at 110°C and cooled in the desiccator. A porcelain crucible was cleaned, dried and weighed (M_1) to nearest 0.001 g. The dried sample was introduced into the crucible and crucible together with the clay sample ions weighed (M_2) to an accuracy of 0.001 g. The crucible containing the sample was placed in a muffle furnace and heated to a temperature of 900°C for 3 h, were cooled in a desiccator and then weighed (M_3). The loss on ignition was calculated from Equation 1):

$$\text{LOI} = \frac{M2-M1}{M2-M3} \quad (1)$$

Apparent porosity test

Apparent porosity is the ability of the clay materials to be impervious to gasses and liquids. Pores are formed as water and gasses are given off during firing process (Nwajagu, 2005). Specimen measuring 5 x 5 x 4 cm was prepared from a conveniently made clay brick; it was dried in an oven at 110°C and fired in a furnace at interval of 100°C for every 10 min till the temperature of 900°C was attained. The fired specimen was cooled and then transferred into a desiccator and dried weight (D) was recorded. The specimen was then transferred into a 250 ml beaker; water was then introduced into the beaker until the tested specimen was completely immersed in the water. The specimen was allowed to soak in boiled water for 30 min being agitated from time to time to assist to release trapped air bubbles. The specimen was then transferred into empty desiccator to cool. The soaked weight (W) was recorded. The specimen was weighed suspended in water using beaker place on balance. This gave suspended weights (S). The apparent porosity was calculated using Equation 2:

$$\text{Apparent porosity} = \frac{W-D}{W-S} \times 100\% \quad (2)$$

Bulk density test

Bulk density is the mass per unit volume of the clay ignoring the

Table 2. Physical properties of Dabagi Clay compared with standard clay for industrial applications (Omowumi, 2000).

Sample description	Apparent porosity (%)	Bulk density (g/cm^3)	Thermal shock - resistance (cycles)	Cold crushing strength (KN/M^2)	Refractoriness (°C)	Linear shrinkage (%)
Dabagi clay	28.46	1.81	7	14138	1349	6.80
Fire clay	20-30	2.30	20-30	15000	1500-1700	4-10
Siliceous fire clay	23.7	2.0	1	15000	1500-1600	7-10
Ceramics	10-30	2.30	20-30	15000	1430-1717	-
Refractory brick	10-30	2.-30	20-30	15000	1430-1717	-

volume occupied by pores (Idenyi and Nwajagu, 2003). The molded brick of the specimen measuring 5 x5 x4 cm was also prepared and dried for 24 h and then dried in an oven at 110°C cooled in a desiccator and weighed (dried weight) after which the specimen was transferred to a beaker containing water and heated for 30 min to assist in releasing air. The specimen was cooled and soaked weight (W) taken. The specimen was then suspended in water using beaker place on balance. The suspended weight (S) was taken. The bulk density was calculated from Equation 3:

$$\text{Bulk Density} = \frac{\rho W}{W - S} \quad (3)$$

Where, ρw = Density of water.

Thermal shock resistance test

Thermal shock resistance is the ability of the clay materials to withstand heating and cooling several times before a deep crack appears (Lawal, 2005). The test piece of brick was air dried for 24 h and oven at 110°C for 3 h. The specimen was then fired in the furnace at 900°C for 3 h and allowed to cool. The prepared specimen was then inserted in a furnace which has been maintained at 900°C.

The specimen was removed from the furnace with a pair of tongs and allowed cool for 10 min on firebricks. The specimen was returned to the furnace for further 10 min. The process was continued until the test piece was cracked. The number of cycles of heating and cooling before cracking of the specimen was recorded as its thermal shock resistance.

Linear shrinkage test

Linear shrinkage is a property of the clay which makes it to undergo least structural changes and disintegration while being heated (Lawal, 2005). The clay was moistened with water to the point of wedging (which implies that the moistened clay materials remain packed into a ball in hand until intentionally vibration causes the mixture to flow). The wedge sample was cast into fabricated wood box.

The clay bar was prepared and the original/initial length was marked on the sample. The bar was fired at temperature of 900°C for 3 h. The final length after the clay had been fired was recorded. The linear shrinkage was then calculated using the Equation 4:

$$\text{Linear shrinkage} = \frac{A - B}{A} \times 100\% \quad (4)$$

Where, A = Initial or original length; B = Final length.

Cold crushing strengths test

Cold crushing strength is the ability of the clay to bear load (Udochukwu, 2007). A cube measuring 50 mm on a flat surface was made from the refractory brick the test piece was fired in a furnace at 900°C and the temperature maintained for 5 h. The sample was then cooled to room temperature. The specimen was placed on a compressive tester and the load was applied axially by turning the hand wheel at a uniform rate until failure occurred. The manometer readings were recorded. Cold crushing strength (CCS) was calculated from Equation 5:

$$CCS = \frac{Maximum\ \ load\ (KN)}{cross\ section\ area\ (M2)} \quad (5)$$

Refractoriness test

This is the resistance of the clay to fusion and softening at high working temperatures. It is the maximum temperature clay can withstand with no load applied (Nnuka, 2003). The clay sample was formed into simple cones of 1.27 cm base diameter and height 3.81 cm. The cone was mounted on refractory base along with several other standard cones of the same dimensions and standard compositions, the cones were heated at a rate of 5°C/min in the furnace until the test cone bent over its own weight. After cooling, the test cone was compared with the standard cones. The test material was taken to have the pyrometric cone equivalent (PCE) of the standard cone whose behaviour most resembled that of the test cone.

RESULTS AND DISCUSSION

The results from the experimental work are given in Tables 1 to 2. The chemical results of Dabagi clay show high silica (SiO_2) content of 64.50% which satisfies the standard for the manufacture of refractory bricks, ceramics as well as high melting clay with values 51.70, 60.50 and 53-73% respectively but below the range for glass formulation. The alumina content (16.30%) of Dabagi clay is short of the standard required for ceramics (26.50%), refractory bricks (25-44%), paper (33.5-36.1%) and paint (37.9-38.4%) manufacturing industries as reported by

(Chester, 1973). However, it can still be used in the manufacture of high melting clay and alumino silicate and fibreglasses because they required only 16-29% and 12-17% of aluminum oxide respectively as reported by (Chester, 1973). The alumina content of clay is a strong indicator for its refractoriness. The higher the amount of alumina, the higher is the refractoriness of the clay. The iron oxide content of 14.26% is higher than the standard for refractory bricks (0.5-2.4%). Such level of iron oxide usually imparts reddish colour to clay when fired, so making it attractive as a ceramic raw material Nnuka and Agbo (2000). The high iron oxide content also affects the high temperature characteristics of the clay, such as fired strength. The loss on ignition of Dabagi clay is below the range of 8-18% for ceramic and refractory brick production as reported by Chester (1973) and 5-14% for high melting clay. Loss on ignition values are often required to be low (Omowumi, 2000). This is because of its effect on the porosity of material especially refractory bricks.

The physical tests results of Dabagi Clay show an apparent porosity of 28.46%, which is within the standard for the production of fire clay and siliceous fireclay with 20-30% and 23.7% respectively (Omowumi, 2000). The bulk density of 1.81 g/cm^3 is below the value required for the manufacture of fireclay and siliceous fireclay as reported by Omowumi (2000).

The linear shrinkage of the sample after drying and firing fall within the acceptable value of 4 - 10% for fireclay, High Shrinkage values may result in warping and cracking of the clay and this may cause loss of heat in the finished products.

The thermal shock of the clay sample is below the acceptable values of 20-30 cycles as compared in Table 2. The practical implication of this is that their use IS restricted to lining of cables slag pots.

The refractoriness or temperature reached for the sample was 1349°C. This is lower than the recommended range for the manufacture of fireclay, siliceous fireclay and ceramics industries and this may be due to the low amount of Al_2O_3 obtained for Dabagi Clay.

The cold crushing value obtained for Dabagi Clay was 14138 KN/M^2 which is below the specification for the manufacture of fireclay, siliceous fireclay and ceramics industries.

Conclusion

An experimental study was conducted to investigate the suitability of Dabagi clay as an industrial raw material in view of its chemical and physical properties. The result of the chemical analysis shows that Dabagi clay con-tains aluminum oxide (Al_2O_3) and silica (SiO_2) as major constituent. The apparent porosity and linear shrinkage of the clay are moderately high while the bulk density and cold crushing strength values are low.

The clay is therefore found to be a source of raw material for the production of refractory bricks, ceramics and high melting clay materials.

Recommendation

Dabagi clay deposit is one of the unidentified clay in Nigeria. This probably IS the first work on this clay deposit, thus, if this clay is exploited and harnessed, it will no doubt provide internal sourcing raw material used in ceramic, paper, high melting clay, refractory bricks etc. there is also the need for a geological survey to determine the extent of the deposit.

Some additives such as kaolin clay or limestone and rice husk ash should also be added so as to improve the properties of the clay.

Conflict of Interests

The author(s) have not declared any conflict of interests.

REFERENCES

Akudinobi BEB (2006). Aspect of chemical and mineralogical Assessment of Ukpor Clay, Nnewi South Local Government Area, Nigeria. Nig. J. Raw Mate. Res. 3:56-67.

Chester JH (1973) Refractories, production and properties, iron and steel institute, London. p. 4.

Grimshow RW (1971). The Chemistry and Physics of clay and Alhed ceramics materials, 4th Edition Revised New York: Wiley in Conscience. p. 15.

Idenyi NE, Nwajagu, C.O. (2003) Non - metallic materials Technology, Enugu: Olicon Publications. pp. 1-47.

Lawal GI, Amuda MOH Adeosun SO (2005). The effects of Cow Dung and Graphite on the Properties of Local Refractory Clays NJERD.1-8.

Nnuka EE, Agbo JE (2000). Evaluation of the refractory characteristics of Kwa clay Deposit in Plateu State, N.S.E. Technical Transaction. pp. 32-59.

Nnuka EE, Ogo DUI, Oseni M (2003). Characterization and application of some indigenous clay in Nigeria. J. Agric. Sci. Technol.

Nwajagu CO (2005). Foundry Theory and Practical (A bridged Edition) Enugu: Olico Publications. pp. 59-76.

Nwoye CI, Mbah, C.N. (2009). Effect of Porosity on the Shrinkage Behaviour of Olokoro and Otamiri Clay Sinters. JMME. p. 3.

Omowumi OJ (2000). Characterization of some Nigerian Clay as refractory materials for furnace linig, Nigerian found of Engineering management. pp. 1-4.

Raw Materials Research and Development Council (RMRDC, 2010). Non-metallic mineral endowment in Nigeria. pp. 12-17.

Udochukwu M (2007). Investigation of some refractory properties of some kaolite clays from four different location in Abia State, Nigeria.

Velde B (1992). Introduction to clay minerals Chemistry, origins, uses and environmental significance, Chapman and Hall. London p. 12.

24

Bahir Dar tannery effluent characterization and its impact on the head of Blue Nile River

Assefa Wosnie[1] and Ayalew Wondie [2]

[1]Biology Department, Dilla University, P O Box 419, Dilla, Ethiopia.
[2]Biology Department, Bahir Dar University, P.O Box 79, Bahir Dar, Ethiopia.

A study was conducted to characterize Bahir Dar tannery effluent and determine its impact on the head of Blue Nile River using selected physicochemical parameters. Samples were taken from the direct effluent stream and four sampling sites (one upstream and three downstream) were selected along the river to determine its impact on the river. Samples were collected from October to March, 2010/11. Temperature, pH, conductivity and total dissolved solids (TDS) were measured *in situ* using a combined meter. The samples for the rest physicochemical parameters were collected from the sites using the appropriate method. In the laboratory, BOD_5 and COD were measured according to standard methods. Total nitrogen, total phosphorous, chloride and sulphide, were determined with Hach nutrient analysis kits and a Hach spectrophotometer. Total suspended solids were determined photometrically. The heavy metal, chromium (as Cr total), was determined using atomic absorption spectrophotometry according to standard methods. The results show that the impact of the effluent from biological oxygen demand (BOD), chemical oxygen demand (COD), NH_3-N, total nitrogen, chlorides, sulphides and chromium was significant with concentrations of 342±52.5, 850.75±96.2, 288±75.8, 462.5±130, 1408.13±405.3, 16.05±3.04 and 3.54±0.55, respectively and most of the effluent characteristics were beyond the provisional discharge limit set out by the Ethiopian Environmental Protection Authority. Analysis of variance indicated that all the physicochemical parameters except temperature, pH and total phosphorous significantly varied among sampling sites ($p<0.05$); the reference or upstream site having lower value than downstream sites. The result shows the pollution load of the effluent on the river and the urgent need for measures to be taken.

Key words: Tannery effluent, physicochemical parameters, Blue Nile River.

INTRODUCTION

Industrial waste is the most common source of water pollution in the present day (Ogedengbe and Akinbile, 2004) and it increases every year because most countries are getting industrialized. Worldwide, it is estimated that the industry is responsible for dumping 300-400 million tons of heavy metals, solvents, toxic sludge, and other wastes into waters each year (UNEP, 2010). Thus, the environment is under increasing pressure from wastes emanating from such industrial activities. As compared to other industries, leather tanning is one of the most polluting activities (Khan et al., 1999) as it consume huge amount of water in several stages, generating an enormous amount of liquid effluents (Farenzena et al., 2005) which are hazardous to the environment to which they are discharged.

Tannery wastewater is highly polluted in terms of biological oxygen demand (BOD), chemical oxygen demand (COD), suspended solids (SS), nitrogen, conductivity, sulphate, sulphide and chromium (Mondal et al., 2005) and in most developing countries tannery effluents are discharged directly into sewers or water bodies without treatment (Verheijen et al., 1996; Favazzi, 2002). The high BOD_5 content of the effluent will affect the survival of gill breathing animals of the receiving water body and high COD value indicate toxic state of the wastewater along with presence of biologically resistant organic substances. The high level of ammonia-N is toxic to aquatic organism and nitrogen may cause eutrophic condition. The high salinity and TDS of the effluent may result in physiologically stressful conditions for some species of aquatic organisms due to alterations in osmotic conditions. Studies show that increase in salinity causes shifts in biotic communities, limit biodiversity, exclude less tolerant species and cause acute or chronic effects at specific life stages. Changes in the ionic composition of water can also exclude some species while promoting population growth of others (Weber-Scannell and Duffy, 2007). The pollutants are poisonous to man and aquatic life resulting in food contamination.

In Ethiopia, although the use of leather and leather products goes back to prehistoric times, tanning hides and skins of animals into leather has been practiced as industrial activity since the last decades (EEPA, 2003). Presently, Ethiopia's leather industry is in the forefront of the leather sector development within the Eastern and Southern Africa region (UN, 2002). Currently, there are 20 operational tanneries that turnout wastes directly into the nearby water bodies like the other industrial activities. This makes industrial and chemical pollution to become the third major problem in the country and one of the great environmental concerns (Zinabu and Zerihun, 2002). This is becoming evident through the pollution of water bodies and human habitat in the major cities, rivers and lakes. Similarly, Bahir Dar Tannery is discharging its effluent into Blue Nile River. In view of the negative impact of this effluent on the environment, the present study aimed at determining the levels of physicochemical pollutants in effluent samples from the tannery and assessing its impact on Blue Nile River.

MATERIALS AND METHODS

Description of the study area

This study was conducted in Bahir Dar, the capital city of Amhara Region which is situated on the southern shore of Lake Tana, the source of Blue Nile (Abay) River. Bahir Dar Textile factory and Tannery are the most important industries in the city. Both the textile factory and the tannery discharged their effluent directly into the Blue Nile River. The downstream part of the river is used for domestic activities including drinking, irrigation and recreation (swimming and bathing). The use of the river in this way may lead to bioaccumulation of toxic pollutants like chromium which is hazardous to human beings as well as livestock.

Sampling

The study was conducted from October, 2010 to March, 2011. This time was selected to sample from both dry and wet periods so as to include possible seasonal effect. Direct physicochemical samples were taken from the effluent stream so as to compare with the country's discharge limit and four sampling sites (one upstream and three downstream) were established along the river length to assess the impact of the effluent on the river. One site directly below the source of pollution and two other sites at different intervals from the point of effluent discharge were chosen. The reference site was established directly above the effluent discharge where there is little disturbance. The sampling sites were designated as S_1 to S_4. The study area was mapped using Geographical Information System (GIS) using information obtained by Global Positioning System (GPS) (Figure 1).

Data collection

The physicochemical parameters include pH, temperature, BOD_5, COD, ammonia-nitrogen (NH_3-N), total nitrogen, total phosphorous, TDS, conductivity, total suspended solids (TSS), total chromium, chlorides and sulphides. These parameters were selected because they are considered to be deleterious on the receiving environment and they were included in the discharge limit. Temperature, pH, conductivity and TDS were measured *in situ* using combined pH/T°/TDS and conductivity meter. The sample for the rest physicochemical parameters were collected from the sites using a sampler which allows sampling from discrete depth and the sample was transferred into the storage bottle without agitation or aeration (Lind, 1979). Prior to sampling, the polyethylene bottle was cleaned with nitric acid and then washed and rinsed with distilled water. Then, the samples were taken to laboratory. In the laboratory, BOD_5 and COD were measured according to standard methods (APHA, 1998). Total nitrogen, total phosphorous, chloride, and sulphide, were determined with Hach nutrient analysis kits and a Hach spectrophotometer (DR010 Hach Co., Loveland, Colorado, USA). Total suspended solids were determined photometrically. The heavy metal chromium (as Cr total), was determined using atomic absorption spectrophotometer (Buck Scientific Model 210 VGP, USA) according to standard methods (APHA, 1998).

Data analysis

Descriptive statistics was used to analyze the physicochemical data. One-way ANOVA was used to compare the difference in physicochemical data among the sampling sites and means were separated using Tukey HSD.

RESULTS

The mean values for each parameter of the wastewater revealed that, most of them were beyond the standard provisional limit set by Ethiopian Environmental Protection Authority (EEPA, 2003) (Table 1).

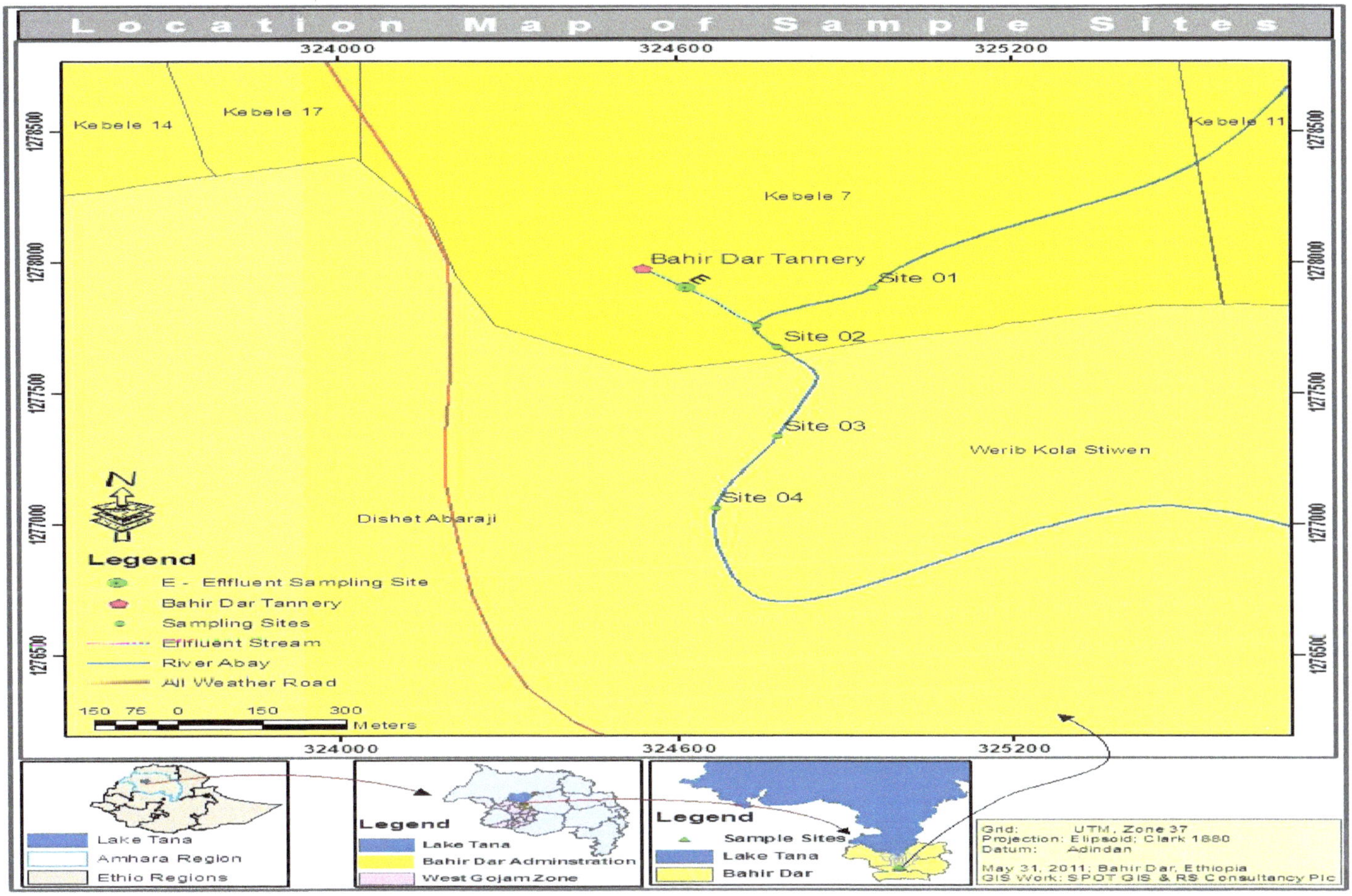

Figure 1. Map showing sampling sites on the effluent stream and along the Blue Nile River.

A detailed characterization of downstream water samples was also carried out to determine the pollution load of the effluent on Blue Nile River. The mean concentrations of pollutants and their variation along the sampling sites are presented in Table 2.

As shown in Table 2, the mean value of temperature and pH at downstream sites ranged from 20 to 24 and 7.13 to 7.15, respectively. Their value did not show significant variation among sampling sites (F=2.22, p=0.16, R^2=0.69 and F=2.04, p=0.18, R^2=0.8), respectively. The pollution profile for BOD_5 and COD along the downstream sites ranged from 34.9±9.05 to 73.4±13.2 mg/l and 107.3±26.7 to 206±32.6 mg/l, respectively. According to the ANOVA, the concentrations of these two parameters significantly varied among sampling sites (F=20.1, P<0.0001, R^2=0.91 and F=26.23, P<0.001, R^2=0.93), respectively. BOD_5 and COD values at the upstream site were significantly lower than the two downstream sites (S_2 and S_3). In addition, their concentrations at the last downstream site (S_4) were significantly lower than that of S_2, but there was no significant variation in their concentration between S_3 and S_4.

The levels of ammonia-N and total nitrogen in the water samples were in the range of 14.3±2.37 to 50±8.11 mg/l and 23±7.86 to 67.5±8.7 mg/l, respectively. Their values differed significantly among sampling sites (F=26.81, P<0.0001, R^2 =0.91and F=40.12, P<0.0001, R^2 =0.94), respectively. The upstream site had significantly lower concentration of ammonia-N than S_2 and S_3. But there was no significant difference in ammonia-N concentration between the last downstream site (S_4) and the upstream site.

Ammonia-N concentration at S_4 was also significantly lower than S_2, but there was no significant difference between S_3 and S_4. The concentration of total nitrogen significantly varied among all sampling sites; the upstream site having lower concentration (Table 2).

The level of phosphorous in the downstream water samples ranged from 5.3±0.98 to 6.5±1.1 mg/l while concentration of phosphorous at the reference site was 6.3±1.7 mg/l. This value did not show significant difference among sampling sites (F=0.12, P=0.95, R^2 =0.15). Total dissolved solids, conductivity and TSS also vary in their values along the sites. Their values at

Table 1. Physicochemical characteristics of Bahir Dar Tannery effluent.

Physicochemical parameter	Mean ± SE	Range	Discharge Limit
Temperature (°C)	25.5 ± 2.2	20.3-30.5	40°C
pH	7.15 ± 0.09	7.13-7.16	6 - 9 pH units
BOD_5 (mg/l)	342 ± 52.5*	214-452	>90% Removal or 200 mg/l
COD (mg/l)	850.75 ± 96.2*	651-1023	500 mg/l
Total ammonia (as N) (mg/l)	228 ± 75.8*	96-420	30 mg/l
Total nitrogen (as N) (mg/l)	462.5 ± 130.8*	89-692	>80% Removal or 60 mg/l
Total phosphorus (as P) (mg/l)	11.5 ± 4.8*	1-24	>80% Removal or 10 mg/l
TDS (mg/l)	2003.25±74.5	1832-2193	-
Conductivity (μs/cm)	3953.25±150.3	3668-4374	-
TSS (mg/l)	339±68,6*	204-525	50 mg/l
Chromium (as total Cr) (mg/l)	3.535 ± 0.55*	1.98-4.51	2 mg/l
Chloride (as Cl) (mg/l)	1408.125 ± 405.3*	613.5-2517	1000 mg/l
Sulphide (as S) (mg/l)	16.05 ± 3.04	8.95-23.39	0.1 mg/l

*Means above the EEPA Discharge Limit.

Table 2. Variation in physicochemical characteristics at the sampling sites along the Blue Nile River (Temperature in °C, pH in pH units, conductivity in μs/cm and the rest in mg/l) in 2010/2011.

Site	pH	Temperature	BOD_5	COD	Ammonia-N	Total nitrogen
Upstream / Reference (S_1)	7.16^a	24.63^a	11.4^c	27.4^c	0.2^c	1.1^d
Just below the effluent discharge (S_2)	7.15^a	24.56^a	73.4^a	206.0^a	50.0^a	67.5^a
200 meters below S_1 (S_3)	7.13^a	20.93^a	55.0^{ab}	164.0^{ab}	32.0^{ab}	45.0^b
400 meters below S_1 (S_4)	7.14^a	23.18^a	34.9^{bc}	107.3^b	14.3^{bc}	23.0^c

Table 2. Contd.

Site	Total phosphorous	TDS	Conductivity	TSS	Chloride	Sulphide
Upstream / Reference (S_1)	6.3^a	150.2^c	307.0^c	41.0^b	34.0^c	0.03^b
Just below the effluent discharge (S_2)	6.0^a	1517.0^a	3051.0^a	222.0^a	299.6^a	0.55^a
200 m below S_1 (S_3)	5.3^a	1167.0^{ab}	2274.0b	157.5^{ab}	213.8^b	0.04^b
400 m below S_1 (S_4)	6.5^a	903.8^b	1793.7^b	108.8^{ab}	174.0^b	0.01^b

Means within a column followed by the same letter are not significantly different from each other according to Tukey HSD (p<0.05).

downstream sites ranged from 903.8±116.1 to 1517±196.8 mg/l, 1793.7±252.5 to 3051±397.2 μs/cm and 108.8±25.1 to 222±46 mg/l, respectively. The values of TDS significantly varied among sampling sites (F=51.18, P<0.0001, R^2=0.95). The upstream site had significantly lower TDS concentration than the three downstream sites. There was also significant difference between S_3 and S_4. Similarly, the values of conductivity showed significant difference among sampling sites (F=54.3, P<0.0001, R^2 =0.96). The upstream site had significantly lower value than all downstream sites and its values at S_3 and S_4 were also significantly lower than S_2. Total suspended solids concentration showed significant variation among sampling sites (F=6.09, P=0.02, R^2 =0.7); the upstream site having significantly lower value than the site just below the effluent discharge (Table 2).

The concentration of chromium at the two downstream sites was 0.165±0.11 (S_2) and 0.142±0.097 mg/l (S_3) while the concentration at the other downstream site (S_4) was not detected. Here, the concentration of Cr at the

Table 3. Literature on concentration of tannery effluents (pH in pH unit, Temperature in °C and the rest in mg/l).

Parameter	Present study	Literature	Reference	Discharge limit
Temperature	25.5	21.2±0.8	Seyoum et al., 2003	40
pH	7.15	4.74-8.66	Deepali et al., 2009	6-9
BOD_5	342	2982.5 840-1860 585-639 1425-1500	Seyoum Leta et al., 2003 Haydar et al., 2007 Akan et al., 2009 Deepali et al., 2009	>90% or 200 mg/l
COD	850.75	11123 1320-54000 2389-3784 1916-27810	Seyoum et al., 2003 Haydar et al., 2007 Seyoum et al., 2003 Deepali et al., 2009	500
NH_3-N	228	1330	Seyoum et al., 2003	30
Nitrogen	462.5	122.2	Seyoum et al., 2003	60 mg/l
Total phosphorous	11.5	16.65-19.93	Seyoum et al., 2003; Seyoum et al., 2003	10 mg/l
TDS	2003.25	42716.33	Deepali et al., 2009	-
TSS	339	3491.9-9485.33	Deepali et al., 2009	50
Total chromium	3.54	8-32.2	Seyoum et al., 2003; Akan et al., 2009	2 mg/l
Chlorides	1408.13	630.4-4313	Akan et al., 2009, Deepali et al., 2009	1000 mg/l
Sulphides	16.05	630.4	Seyoum et al., 2003	1 mg/l

reference site was not detected as that of S_4. Chloride and sulphide levels in the downstream samples varied between 174±66.46 to 299.6±98.13 mg/l and 0.008±0.002 to 0.55±0.23 mg/l, respectively. Chloride values at the downstream samples varied significantly among sampling sites ($F=45.85$, $P<0.0001$, $R^2=0.95$) with the reference site having significantly lower value than the downstream sites. Sulphides showed significant variation among sampling sites ($F=26.1$, $P<0.0001$, $R^2=0.9$). The site just below the effluent discharge had significantly higher value than the two downstream and the upstream sites (Table 2).

DISCUSSION

In this study, a detailed characterization of wastewater was carried out on Bahir Dar tannery effluent and the result showed that most of the parameters under investigation were high and well beyond the provisional standard limit set by EEPA (2003).

The values of temperature and pH in this study were within the discharge limit and below what Seyoum et al. (2003) and Deepali et al. (2009) reported (Table 3). The lower value of pH and temperature in this study might be due to variation in the sampling time of the day and the difference in production capacity of the tanneries. BOD_5 and COD mean values were above the discharge limit. But the values in this study were lower than what Seyoum et al. (2003), Haydar et al. (2007), Akan et al. (2009) and Deepali et al. (2009) reported (Table 3). These high levels of BOD_5 and COD values observed in the waste may be due to high amount of organic matter from various chemicals used during the processing of hides and skins. It has been reported that a significant part of chemicals used in the tanning process is not actually absorbed in the process and discharged into the environment (UNIDO, 1991), thereby increasing the levels of BOD_5 in the effluent. The high BOD_5 content of the effluent will affect the survival of gill breathing animals of the receiving water body and high COD value indicate toxic state of the wastewater along with presence of biologically resistant organic substances.

Ammonia-N and total nitrogen concentrations were above the discharge limit as shown in Seyoum et al. (2003) report. These high levels of ammonia-N and nitrogen might be attributed to several components in tannery effluent containing nitrogen as part of the chemical structure and the nitrogen contained in proteinaceous material of the skin (Bosnic et al., 2000). The high level of ammonia-N is toxic to aquatic organism and nitrogen may cause eutrophic condition. The mean phosphorous concentration of the effluent was somewhat similar to what Seyoum et al. (2003) and Akan et al. (2009) reported (Table 3) and showed little deviation from the discharge limit.

Total dissolved solids, conductivity and total suspended solids values in this study were found to be high. But Seyoum et al. (2003) and Deepali et al. (2009) reported values greater than the result of this study (Table 3). This deviation might be due to the variation in size and production capacity of the tanneries under investigation. The total dissolved solids may increase salinity of the water and thus may render it unfit for irrigation and other purposes for the immediate downstream users while the suspended impurities cause turbidity in the receiving system.

Chromium level in this study is 36 times higher than the discharge limit set out by EEPA (2003). Similarly, sulphide and chloride values were above the discharge limit. These high values observed in the effluent may be due to the chemicals used by tanneries. Studies (Table 3) reported higher values of these pollutants and this might be also due to the variation in size and production capacity of the tanneries under investigation. Generally, the result showed that the effluent is rich in organic pollutants, but relatively poor in phosphorous content.

The mean concentration of pollutants along Blue Nile River indicated that the concentration of the various pollutants decrease downstream from the point where the effluent joins the river to the last downstream site (S_4). This is attributed to the dilution capacity of the river, uptake of pollutants by the vegetation along the river gradient and some reactions that can change the pollutants to harmless ones, may take place as the water flows downstream from the point pollution source.

Even though there was no significant variation among sites, the mean temperature value of the upstream site was slightly higher than the last downstream sites and the slight variation might be due to the difference in the sampling time. The pH in all sites was more or less neutral. The result in this study slightly deviates from the work of Seyoum et al. (2003) (Table 3) and this might be due to the size of the river and its dilution capacity.

The concentration of BOD_5 and COD at S_2 is very much reduced by 76-79% from the direct effluent owing to the dilution capacity of the river. The concentrations at downstream sites were well beyond the reference site indicating the organic pollution load of the effluent on the river. This increase in BOD_5 and COD increases rates of biological or chemical decomposition leading to oxygen depletion which produce both acute (mortality) and chronic (reduced growth, fecundity and disease resistance) impacts on aquatic biota (Allison, 1996).

The concentration of NH_3-N and nitrogen at S_2 showed 78-85% reduction from the direct effluent value. This rapid reduction might be due to the size of the river. Seyoum et al. (2003) reported total NH_3-N and nitrogen reduction of 92-95 and 69% respectively (Table 3). The standard provisional limits for NH_3-N and N of tannery wastewater are 30 and 60 mg/l respectively. All the values at downstream sites were beyond the limit and the value at the reference site. This indicated that the tannery wastewater was responsible for these pollutants. The nitrogen in ammonia form is toxic to aquatic organisms. In the environment ammonia-nitrogen is oxidized rapidly to nitrate, creating demand and low dissolved oxygen in the water. Moreover, the high nitrogen may cause eutrophication problem (USEPA, 2008). Eutrophication reduces dissolved oxygen in water and alters stream habitat available for macroinvertebrates, fish eggs and fish both of which are critical for fish and other aquatic life. This may disrupt the ecological cycle of the stream so that certain biological communities experience severe mortality.

The phosphorous content in downstream samples was within the standard provisional limit set by EEPA (2003). Its concentration did not show significant difference between the sites. The reference site had almost similar phosphorous content as that of the downstream sites showing that the tannery effluent was not responsible for this pollutant. This might be due to the addition of municipal wastewater into the river above the reference site. So in this study, phosphorous does not give concern as severe pollutant from the tannery and may not cause eutrophication.

Total dissolved solids, conductivity and total suspended solids at S_2 showed a 25-34% reduction than the direct effluent value. As compared to the other parameters, the values of the above parameters showed slight reduction at downstream sites. This might be due to their initial high concentration which made the dilution process slower. The values at the downstream sites also significantly differ from the reference site indicating the increasing impact of the tannery effluent on downstream water bodies. Studies by Seyoum et al. (2003) and Birnesh (2007) reported the same result (Table 3). High concentrations of total dissolved solids forms a layer on the bottom of water course and covers natural fauna on which aquatic life depends. This can lead to localized depletion of oxygen supplies in the bottom waters. It also reduces light penetration and thus photosynthesis in the water. On the other hand, high conductivity indicates salinity increment in the receiving water. This may result in adverse ecological effects on aquatic biota (Lefebvre and Moletta, 2006).

Chromium concentration at S_2 showed a 95% decrease from the effluent and it was not detected at S_4. The level at the reference site also was not detected indicating the pollution at downstream sites originated from the effluent. Similarly, chloride and sulphide contents at S_2 showed dramatic reduction with 96-98% from the direct discharge owing to the dilution effect of the river. Chloride concentrations at S_2 were significantly higher than the reference and the other downstream sites. But all the downstream sites have values beyond the standard provisional limit set out by EEPA (2003). The values indicated that most of the concentration of chlorides and sulphides at downstream sites were the contribution of the tannery effluent. The presence of these ions increases the salinity of the receiving water body. The discharge of high salinity and TDS effluents into receiving system may result in physiologically stressful conditions for some species of aquatic organisms due to alterations in osmotic conditions. Studies show that increase in salinity causes shifts in biotic communities, limit biodiversity, exclude less tolerant species and cause acute or chronic effects at specific life stages. Changes in the ionic composition of water can also exclude some species while promoting population growth of others (Weber-Scannell and Duffy, 2007).

Conclusion

Even though tanning industries are very important for the country's economy and to improve standards of living of citizens, their waste is directly discharged into the nearby water body without treatment. Most of the physico-chemical parameters investigated in this study showed that almost all the effluent characteristics were above the provisional discharge limit set by the Environmental Protection Authority indicating the poor treatment mechanism employed by the tannery. Most of the physicochemical parameters along the river gradient were also high and beyond the discharge limit. This will create a problem for downstream users as they use it for domestic, agricultural and recreational value. So, this fact must regularly be brought to public awareness by using media or direct contact with the downstream users. The Environmental Protection Authority should also establish environmental protection laws which consider technical and financial capability of the industries so as to control industrial pollution. Not only establishment, the laws should also be enforced and environmental standards with their protocols should be followed with strict and continuous monitoring to safeguard the environment from heavy loads of pollutants and toxic substances.

Conflict of Interests

The author(s) have not declared any conflict of interests.

ACKNOWLEDGEMENTS

The authors thank Amhara National Regional State for financial support and Bahir Dar University for the provision of material and laboratory facilities.

REFERENCES

Akan JC, Abdulrahman FI, Ayodele JT, Ogugbuaja VO (2009). Impact of tannery and textile effluent on the chemical characteristics of Challawa River, Kano State, Nigeria. Elect. J. Environ. Agric. Food Chem. 8(10):1008-1032.

Allison L (1996). Remedial options book, integrated coastal zone management. Atlantic Coastal action program (ACAP). Saint John.

APHA (American Public Health Association) (1998). Standard methods for the examination of water and wastewater. 20th edn Washington, D.C.

Birnesh A (2007). Assessment of downstream pollution profiles of Awassa textile factory effluent along Tikur Wuha River using physicochemical and macroinvertebrate indicators. M.Sc Thesis. School of Graduate Studies, Addis Ababa University

Bosnic M, Buljan J, Daniels RP (2000). Pollutants in tannery effluents. Definitions and environmental impact, limits for discharge into water bodies and sewers. UNIDO. Regional programme for pollution control in the tanning industry in South-East Asia. US/RAS/92/120.

Deepali, Gangwar KK, Joshi BD (2009). Comparative study of physicochemical properties of effluent from tannery industries. Indian J. Environ. Sci. 3(2):49-152.

EEPA (2003). Provisional standards for industrial pollution control in Ethiopia. Prepared under the ecologically sustainable industrial development (ESID) project - US/ETH/99/068/Ehiopia, EPA/UNIDO, Addis Ababa.

Farenzena M, Ferreira L, Trierweiler JO, Aquim PM (2005). Tanneries: from waste to sustainability. Braz. Archive Biol. Technol. 48:281-289.

Favazzi A (2002). Study of the impact of the main policies and environment protection measures in Africa's leather industry. Principal: Assomac Servizi Srl for UNIDO.

Haydar S, Aziz JA, Ahmad MS (2007). Biological treatment of tannery wastewater using activated sludge process. Pak. J. Engine. Appl. Sci.1:61-66.

Khan SR, Khwaja MA, Khan AM, Ghani H, Kazmi S (1999). Environmental impacts and mitigation costs associated with cloth and leather exports from Pakistan. A report on trade and sustainable development submitted by sustainable development policy institute and IUCN-P to IISD Canada for the IISD/IUCN/IDRC project on building capacity for trade and sustainable development in developing countries, Islamabad.

Lefebvre O, Moletta R (2006). Treatment of organic pollution in industrial saline wastewater: A literature review. Water Res. 40:3671-3682.

Lind OT (1979). Handbook of common methods in limnology. Department of biology and institute of environmental studies. Baylor University, Waco C.V. Mosby Company, Texas.

Mondal NC, Saxena VK, Singh VS (2005). Impact of pollution due to tanneries on groundwater regime. Current Science.88(12):25.

Ogedengbe K , Akinbile CO (2004). Impact of industrial pollutants on quality of ground and surface waters at Oluyole Industrial Estate, Ibadan, Nig. J. Technol. Develop. 4(2):139-144.

Seyoum L, Fassil A, Gunnel D (2003). Characterization of tannery wastewater and assessment of downstream pollution profiles along Modjo river in Ethiopia. Ethiopian J. Biol. Sci. 2(2):157-168.

UN (2002). Investment and innovation policy review Ethiopia. United Nations conference on trade and development investment. UNCTAD/ITE/IPC/Misc.4.

UNEP (2010). Clearing the water: A focus on water quality solutions. Nirobi, Kenya.

UNIDO (1991). Leather industry. Conference on ecologically sustainable industrial development. Case studv No. 3. ID/WG.516/10. UNIDO Secretariat, Copenhagen, Denmark.

USEPA (2008). Onsite wastewater treatment technology fact sheet 9, enhanced nitrogen removal - nitrogen. EPA/625/R/00/008 pp. 45-52.

Verheijen LAHM, Wiersema D, Hulshoff PolLW, De Wit J (1996). Livestock and the environment finding a balance: Management of waste from animal product processing, International agriculture centre Wageningen, The Netherlands.

Weber-Scannell PK and Duffy IK (2007). Effects of total dissolved solids on aquatic organisms: A review of literature and recommendation for salmonid species. Am. J. Environ. Sci. 3(1):1-6,

Zinabu G-M and Zerihun D (2002). The chemical compostion of the effluent from Awassa Textile Factory and its effects on aquatic biota. hEthiopian J. Sci. 25(2):263-274.

Wood-cement composites using suitable mix of sawdust and fibres from veins of palm tree leaves

Adolphe D.Tchéhouali[1] , Chakirou A. Toukourou[1], Agapi K. Houanou[1], Edmond Adjovi[1], Amos Foudjet[2], Antoine Vianou[1] and Degan Gérard[1]

[1]Laboratoire d'Energétique et de Mécanique Appliquée - Ecole Polytechnique d'Abomey Calavi, Université d'Abomey Calavi (LEMA/EPAC/UAC), 01 BP 2009 Cotonou, Benin
[2]CRESA Forêt - Bois Université de Dschang, Cameroun.

Some tropical wood species are currently used in wood work and generating an important quantity of wood waste in Benin, Western Africa. Recycling these wastes through building material is a way to solve environmental challenges, but the wood resources are not inexhaustible and vegetable fibers such as *Eleais guineensis* fibers, obtained from veins of palm tree leaves are available in nature. The main concern of this study is to determine the most suitable mixes of *E. guineensis* fibers and five West African hardwood species namely *Diospyros mespiliformis, Khaya grandifoliola, Tectona grandis, Pterocarpus erinaceus and Isoberlina doka* for wood-cement composites. The compatibility with cement of mixes of fibers and sawdust is studied based on measurements of cement heat hydration and compatibility indices. The tensile and compressive strengths of the composites were determined to evaluate the potential of the mixes. The results show that the woody residues inhibit the hydration reactions of cement and greatly more when containing *E. guineensis* fibers. Hydration tests indicated that all the mixes could be rated as having moderate to good compatibility with cement after hot water treatment. Mechanical tests showed that the mixes containing *Isoberlina doka, Tectona grandis* and *Khaya grandifoliola* sawdust provided the highest mechanical strengths.

Key words: Vegetable fibers, sawdust, compatibility, wood-cement composites, mechanical strengths.

INTRODUCTION

In Benin, a West African country, the forest cover is about 68% of the total superficies of 114 763 km², according to a recent World Bank report (2010). The per capita income according to the same report is 690 USD. Overall wealth of this country has an important component that is derived from forest resources. Many tropical species are identified to be hard wood and generate significant volumes of wood commercially viable. Consequently, large quantity of wood waste is generated in this region and will constitute a considerable environmental problem. In most of the West African countries, a very moderate fraction of wood waste is used to provide energy for domestic use while a great proportion of such waste is consumed in the nature releasing carbon dioxide and other gas in the environment.

The re-use of wood waste as building materials is one of the few ways of tackling the problem. But the wood resources are not inexhaustible and it would be suitable to use the mixes of sawdust and vegetable fibers such as those from palm oil veins, available in the nature for wood - cement composites. Ordinary wood-cement composites obtained by mixing sawdust, water and cement have been widely investigated (Dinwoodie and Paxton, 1984; Hachmi and Sesbou, 1990; Hofstrand et al., 1984). The dimensional variations have been studied by Mougel (1992). The major well known problem occurring with regards to this topic is the aptitude of wood species to be used as wood - cement composites. Such a problem is widely discussed recently (Pereira et al., 2003, 2004). Many studies have been undertaken to assess the compatibility of wood with cement (Hachmi and Moslemi, 1989; Hachmi et al., 1990; Sandermann and Kolher, 1964).

To build houses with walls made from wood cement-composite is a way not only to solve environmental problem but to provide many other advantages. It is a very lightweight material which allows saving of money as reported in the thesis of Doko (2013). It is also a material which provides heat insulation.

Today, although a large number of wood species have been investigated such as pin wood (Semple and Evans 2000), Western Australian *malle Eucalyptus* (Semple et al., 2002), *Acacia mangium and Eucalyptus pellita* (Semple et al. 1999), *Amblygonocarpus andongensis, Brachystegia speciformis, Pterocarpus angolensis, Kaya nyasica, Erythrophleum suaveoleuns, Albizia adianthifolia, Sterculia appendiculata, Millettia stuhlmannii, Julbernadia globiflora and Afzelia quanzenzis* (Alberto et al. 2000), few information related to the compatibility of wood species with cement is available. That is the same thing with the mixes containing various tropical fibers. Moreover, it was widely discussed (Alberto et al., 2000) that the nature of wood species considerably influences the hardening of the cement depending on the origin of the tree. Evaluating some physical properties for oil palm stem as alternative biomass resources, Balkis et al. (2012) concluded that the middle and center sections of oil palm stem contained the highest amount of extractives using hot water extraction. The lowest amount of extractives was located at the bottom outer section of the oil palm stem.

Although today, researchers reveal more and more new environmental problems generated by manufacturing wood cement composite; we must keep in mind that the transformation process of the wood residues should avoid the creation of other environmental problems. Indeed, Silva et al. (2013) reported that in Brazil, the wood panel industry is one of the most important forest-based industries. Assessing the life cycle of medium density particleboard (MDP), they found that unlike other regions, Brazilian MDP is produced from dedicated eucalyptus plantations and heavy fuel oil is an important energy source in MDP manufacture with a negative impact on the environment. They suggest the possibility of using alternative production MDP scenarios as substituting heavy fuel oil (HFO) for in-mill wood residues or diesel or finding other sources of wood for MDP production.

The aim of this research initiated in Benin was to determine the most suitable mixes of *Eleais guineensis* fibers and five West African hardwood species for wood-cement composites. The first step in assessing the suitability of different mixes containing wood species and vegetable fibers for cement composites involves testing whether the mix significantly inhibits the hydration reaction of Portland cement. Compatibility with cement of mixes of *E. guineensis* fibers and five wood species based on the measurement of cement hydration temperature and compatibility indices was studied. In the second step, mechanical tests were carried out on moulded blocks to confirm hydration tests as indicated by Andy and Land Zhongli (1986). In the third step, the suitability of mixes for wood-cement composites were evaluated by carrying out tensile and compressive tests on the samples made from mixes of vegetable fibers and sawdust.

MATERIALS AND METHODS

Sawdust taken from sawmills for five wood species available in Benin (Figure 1a) were used in this study: *Diospyros mespiliformis, Khaya grandifoliola, Tectona grandis, Pterocarpus erinaceus* and *Isoberlina doka. Eleais guineensis* fibers (Figure 1b) are vegetable fibers obtained from ground oil palm veins previously dried.

In order to assess the influence on the compatibility indices of the treatment method applied, the mixes (*E. guineensis* fibers and wood sawdust) were pre-treated firstly with cold water and secondly with hot water. Treatment in cold water consisted of drenching during 48 h and washing thoroughly sawdust and fibers in water at ambient temperature. Treatment in hot water consisted of drenching for 48 h and washing sawdust and fibers in water at 80°C because of their great degree of inhibition assumed (Badejo, 1989; Hachmi and Moslemi, 1989; Pereira et al., 2003).

After washing, both the sawdust and *E. guineensis* fibers were dried in a kiln at 105°C within 48 h then separated by sieving in three size classes. Class retained for the tests was 0, 1 - 4 mm. The type of cement used was CPJ-CEMII 35 manufactured in SCB Lafarge Company and locally available in Benin.

Measurement of cement hydration temperature and compatibility indices

The mass ratios used were: Sawdust/*E. guineensis* fibers/cement/water mass ratio of 12:3:200:90.5 g inspired from the work of Weatherwax and Tarkow (1964) in which wood/cement/water mass ratio was 15:200:90.5 g. In this study, the wood is considered as the mix of sawdust (12 g) and *E. guineensis* fibers (3 g). The lower proportion of vegetable fibers in wood (1:5) has been adopted because of its probable great degree of inhibition presumed.

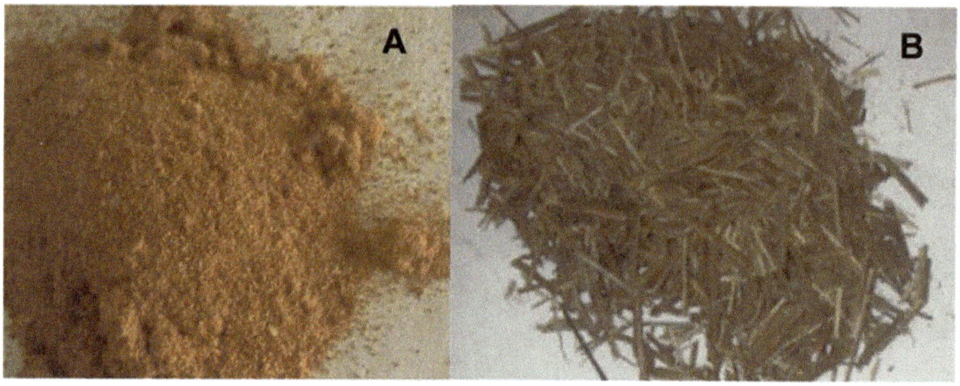

Figure 1. Sawdust (a) and fibres from veins of palm oil leaves (b).

Table 1. Most convenient ratios for wood species.

Wood species in the mixes of sawdust and *Eleais guineensis* fibers	Water/cement/ ratio in weight
Diospyros mespiliformis	00.85
Khaya grandifoliola	0.87
Tectona grandis	0.61
Pterocarpus erinaceus	0.84
Isoberlina doka	1.09

Cement control sample: 200 g of cement mixed with 80 ml of water was used (Weatherwax and Tarkow, 1964).

The samples were thoroughly mixed with water in a sealable polyethylene bag for 2 min. Immediately after mixing, the tip of a temperature thermocouple (type K) was taped to the outside of sample bag and enclosed within the body of the wood-cement mix by folding the bag and contents around it and securing the folded bag with adhesive tape. The bag was then placed in a polystyrene cup and sealed inside a flask. This process was carried out for three samples per test. All experiments were undertaken in a controlled temperature room maintained at 20 ± 1°C. A cement hydration temperature logging apparatus was used to measure the heat of hydration of the three wood-cement samples for 24 h. Temperatures were recorded at 5 min intervals and the curves were smoothed by plotting the progressive averages of every three successive readings. Maximum heat of hydration temperature (T_{max}) and time (*t*) to reach T_{max} were recorded and two wood-cement compatibility indices, C_A - factor and hydration rate (R) were calculated:

$$R = (T_{max} - T_{min})/t.$$

T_{min}: Temperature attained during the time when not more than three repeated values are recorded.

$$C_A = \frac{A_1}{A_2} * 100$$

Where, A_1 : Area under the wood-cement sample hydration curve,

A_2 : Area under the control sample curve.

Control sample is made with cement only. Then, wood species were classified as compatible, moderately compatible or incompatible with cement based on the extent to which they delayed cement hydration (Hachmi et al., 1990).

Mechanical tests on moulded blocks

The evaluation of the aptitude of mixes to wood work by measuring cement hydration temperature and compatibility indices has been completed by mechanical tests on moulded blocks. Thus, sawdust and fibers treated in hot water, cement and water were mixed respecting the following ratio:
sawdust/fibers/cement/water mass ratio of 60/15/1000 /452.5 g according to the ratio adopted in the work of Andy and Land Zhongli (1986) in which total mass of wood was 75 g.

For each batch of mix, twelve cubic specimens (50 x 50 x 50 mm) were used to conduct compressive tests after 28 days of curing in water in room at 20-25°C.

Mechanical tests on wood-cement composites

The sawdust, vegetable fibers, cement and water were mixed according to the most convenient ratio for wood species. For all species, the wood/cement ratio in weight was about 0.80 as found in the work of Tchehouali (2002), but the water/cement ratio in weight varied as indicated in Table 1 with consistency test in fresh mixtures.

The mixes was placed in steel mould and compacted. For each batch of mixture, twelve cubic specimens (50 x 50 x 50 mm) and twelve cylindrical specimens with diameter: 50 mm and height: 100 mm were used to conduct compressive and indirect tensile (split tensile) tests, respectively. After 24 h, the boards were de-clamped and conditioned for 28 days at 20°C and 65% relative humidity

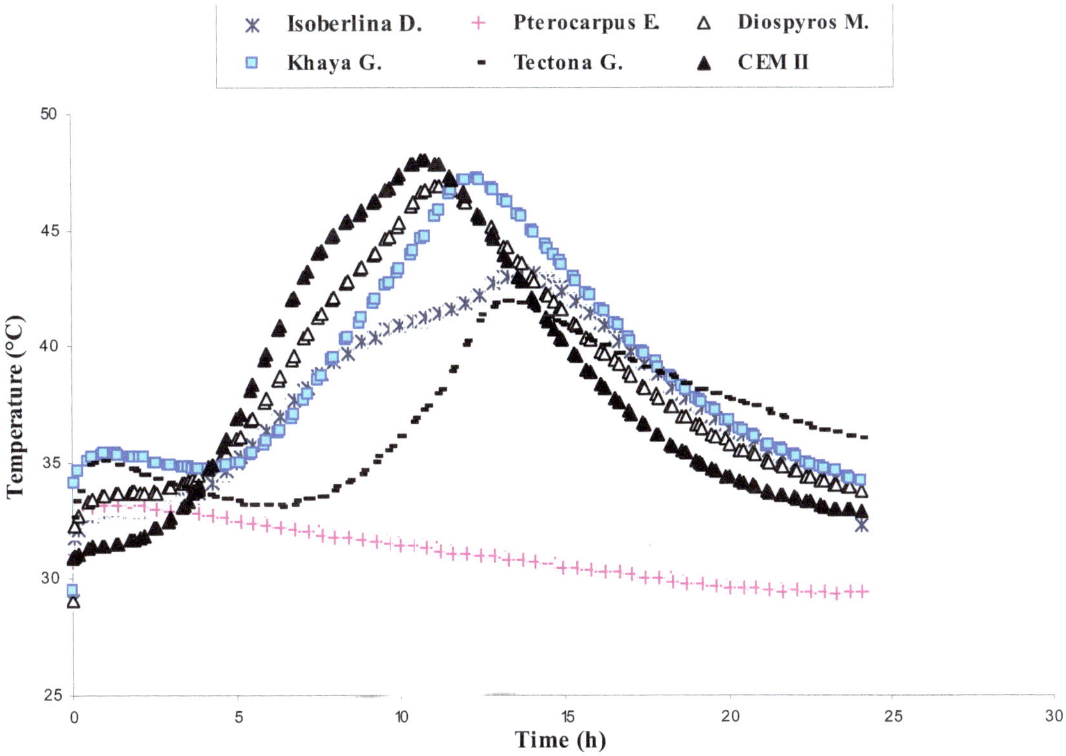

Figure 2. Hydration exotherms (temperature vs. time) for neat cement1 and cement containing tropical wood treated with cold water.

Table 2. Compatibility indices of tropical wood species treated with cold water.

Mix	T_{max} (°C)	T_{min} (°C)	Time (H)	Rate (°C/H)	C_A factor (%)
CEM II	47.8	31.5	8.5	1.80	100
CEMII + *Diospyros mespiliformis sawdust* + *Eleais guineensis* fibers	45.9	34	7.95	1.58	63
CEM II + *Isoberlina doca sawdust* + *Eleais guineensis* fibers	42.8	33.1	10.31	1.0	60
CEM II + K*haya grandifoliola* sawdust+ *Eleais guineensis* fibers	46.3	34.1	7.09	1.72	54
CEM II + Tectona grandis sawdust+ *Eleais guineensis* fibers	41.2	33.0	6.29	1.54	47
CEM II + *Pterocarpus erinaceus* sawdust+ *Eleais guineensis* fibers	0.0	0.0	0.0	0.0	0.0

(RH) to allow the cement to cure and gain strength.

Tensile and compressive tests were carried out directly on half of the samples while others were tested after they were soaked in water at laboratory room temperature (20°C) for 24 h. After soaking, the samples were drained on paper towels for 20 min to remove excess water. Tensile and compressive tests constitute simple and relevant mechanical tests on wood cement - composites to quickly evaluate the aptitude of wood specie for wood work.

RESULTS AND DISCUSSION

Compatibility of wood species with cement based on measurements of cement hydration temperature and compatibility indices.

From Figure 2, when the mixes of sawdust and *E. guineensis* fibers were treated with cold water, the curves indicate that in the mixture of *Isoberlina doka* and *Diospyros melsiformis* species, hydration reactions started very earlier (about 3 hours after adding water) than in other mixtures. In the mixtures of *Tectona grandis* and *Khaya grandifoliola* species, hydration reactions started about 7 and 5 h, respectively. The curve corresponding to *Tectona grandis* species presented a great and favourable tendency in developing heat hydration even over 30 h.

From Table 2, *Isoberlina doka* and *Diospyros melsiformis* species have yielded the highest compatibility indices with C_A - factor equals to 60 and 63%

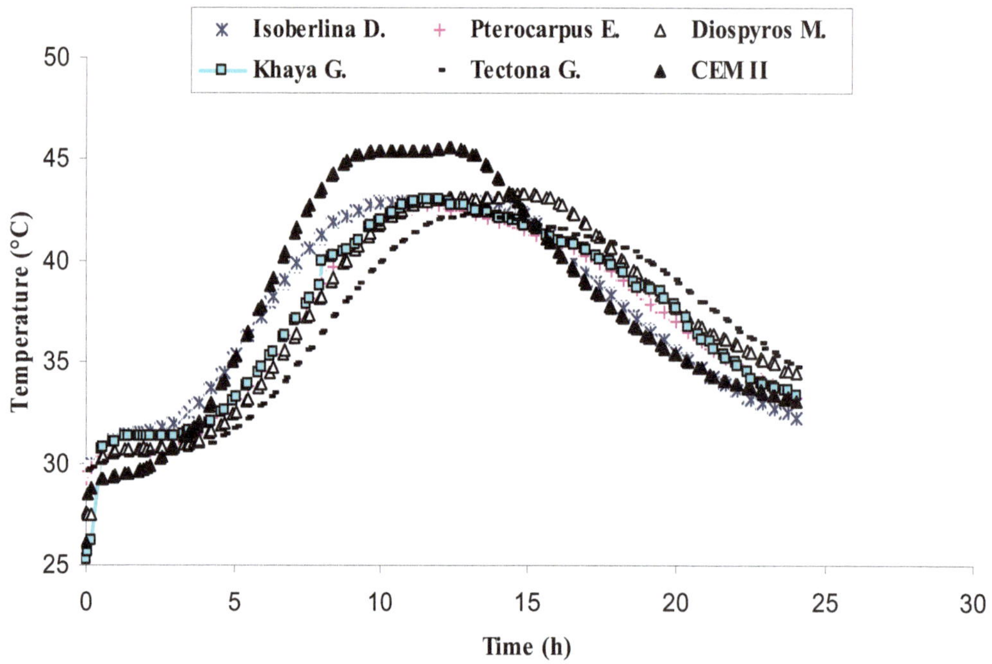

Figure 3. Hydration exotherms (temperature vs. time) for neat cement and cement containing tropical wood and fibres treated with hot water

Table 3. Compatibility indices of tropical wood species treated with hot water.

Mix	T_{max} (°C)	T_{min} (°C)	Time (H)	Rate (°C/H)	C_A factor (%)
CEM II	45.4	31.3	8.5	1.56	100
CEM II + *Isoberlina doca sawdust* + *Eleais guineensis* fibers	42.9	32.1	6.92	1.56	80
CEM II + K*haya grandifoliola* sawdust+ *Eleais guineensis* fibers	41.4	31.8	6.75	1.42	72
CEM II + P*terocarpus erinaceus* sawdust+ *Eleais guineensis* fibers	42.7	31.5	7.26	1.54	70
CEM II +T*ectona grandis* sawdust+ *Eleais guineensis* fibers	42.3	31.3	8.17	1.55	73
CEMII + D*iospyros mespiliformis sawdust*+ *Eleais guineensis* fibers	41.4	31.8	6.67	1.51	66

respectively. *K. grandifolila* and *T. grandis* species had C_A- factors equal to 54 and 47%, respectively. There is no data in the columns corresponding to *Pterocapus erinaceus* species because of its great inhibitory effect on the cement hydration.

From Figure 3, in the case of hot water treatment, the curves of the mixtures of *I. doka* and *K. grandifolila* species show earliest hydration reactions starting from about 2.20 and 3.75 h, respectively after mixing with water. The mixtures of *T. grandis*, *P. erinaceus* and *D. melsiformis* species started their hydration reactions about 4.10, 5.20 and 6.10 h, respectively after mixing with water. In spite of a long delay in starting hydration reactions, *T. grandis* specie had good compatibility indices as indicated in Table 3. *Isoberlina doka*, *Tectona grandis*, *Khaya grandifolila* and *P. erinaceus* species had average C_A - factors equal to 80, 73, 72 and 70%, respectively. The lowest compatibility indices were provided by *Diospyros mespiliformis* (C_A - factor equals to 66%).In general, the results indicate that the treatment of sawdust and fibers in hot water enhance notably the compatibility indices of all species as found in previous work of Alberto et al. (2000).

With simple treatment in cold water, *Isoberlina doka, Diospyros melsiformis* and *K. grandifolila* species had good compatibility indices while compatibility indices of *T. grandis* and *P. erinaceus* species were negatively affected. Alberto et al. (2000) had a similar report on *Khaya* specie which indicated that *P. angolensis* and *K. nyasica* species were compatible with cement after simple treatment in cold or hot water. The contrast of the findings about *P. angolensis* species could be due to the presumed change of the nature of soil where the species has been grown.

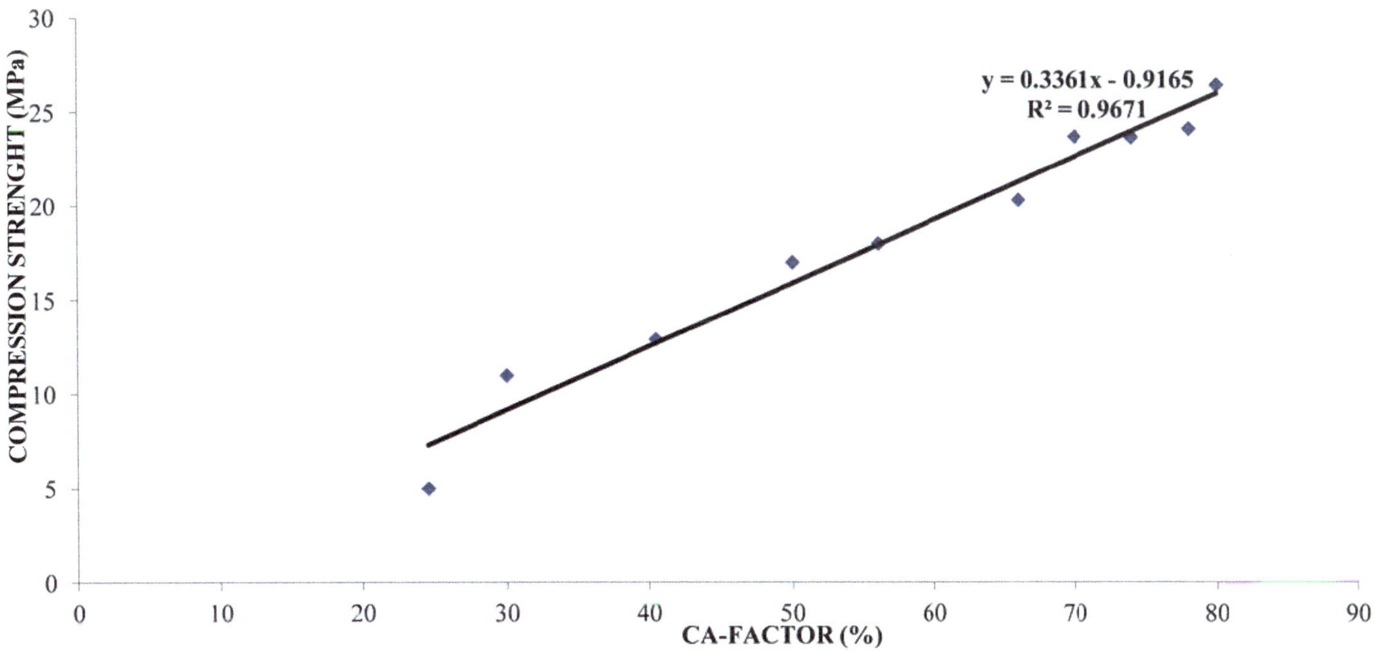

Figure 4. Evolution of compressive strengths of moulded blocks with C_A-factor.

Characteristics of moulded blocks

The blocks have presented an average density of about 1395 kg/m^3. They could not be classified as lightweight concretes which densities vary from 450 to 1200 kg/m^3 (Mougel, 1992). The very slight variation of the densities (minimum value 1310 kg/m^3 corresponding to *Pterocarpus eurinaceus* species and maximum value 1370 kg/m^3 corresponding to *Isoberlina doka* species tallies with the slight variation of the characteristics of the blocks as shown in Figure 4 whereby compressive strengths vary from 21.5 MPa for *P. eurinaceus* species to 24.0 MPa for *I. doka* specie. Moreover, from Figure 4, compressive strengths values were linearly proportional to C_A- factors and such results lead to the conclusion that physical tests could be a real indicator of wood-cement compatibility (Andy and Land Zhongli, 1986). The great proportion of cement in the blocks and the relative high densities obtained constitute a major inconvenience when such material is used as lightweight material with low cost. The compressive tests were carried out on the blocks only to establish the correlation between strengths and C_A - factors.

Mechanical strengths of wood -cement composites

In the case of cold water treatment, the highest mechanical strengths of wood-cement composite were obtained with *I. doka* and *D. mespiliformis* species both in dry and wet conditions as shown in Figure 5. *P. eurinaceus* species provided the lowest mechanical strengths.

In the case of hot water treatment (Figure 6), *I. doka* and *T. grandis* species showed the highest strengths. Samples made from *T. grandis*, *P. eurinaceus* and *K. grandifoliola* species presented good strengths both in dry and wet conditions. *D. mespiliformis* species provided the lowest mechanical strengths. The tensile and compressive strengths of wood-cement composites were very low and decreased drastically in wet conditions (about 40%) especially with *I. doka* and *D. mespiliformis* species.

The term compatibility refers to the degree of cement setting after mixing water, wood and fibers. The performance of specific wood species depends on the chemical composition of their extractives which could be extracted with simple cold water or hot water treatment. Several works (Hachimi et al., 1990; Alberto et al., 2000; Balkis et al., 2012; Weatherwax et al., 1964) indicate that before elaborate wood-cement composite, sawdust should be treated in cold or hot water but treatment in hot water is more efficient than treatment in cold water because hot water act more deeply on the extractives. Thus, some wood species may require a more severe treatment, in NaOH 1% solution for example, in order to overcome their incompatibility with cement. The very low strengths presented in Figures 5 and 6 especially in cases of *P. erinaceus* and *D. mespiliformis* species indicate that more severe treatment should be applied before their utilization as building material in construction. Several other techniques exist to be explored in the next investigations to enhance the mechanical characteristics

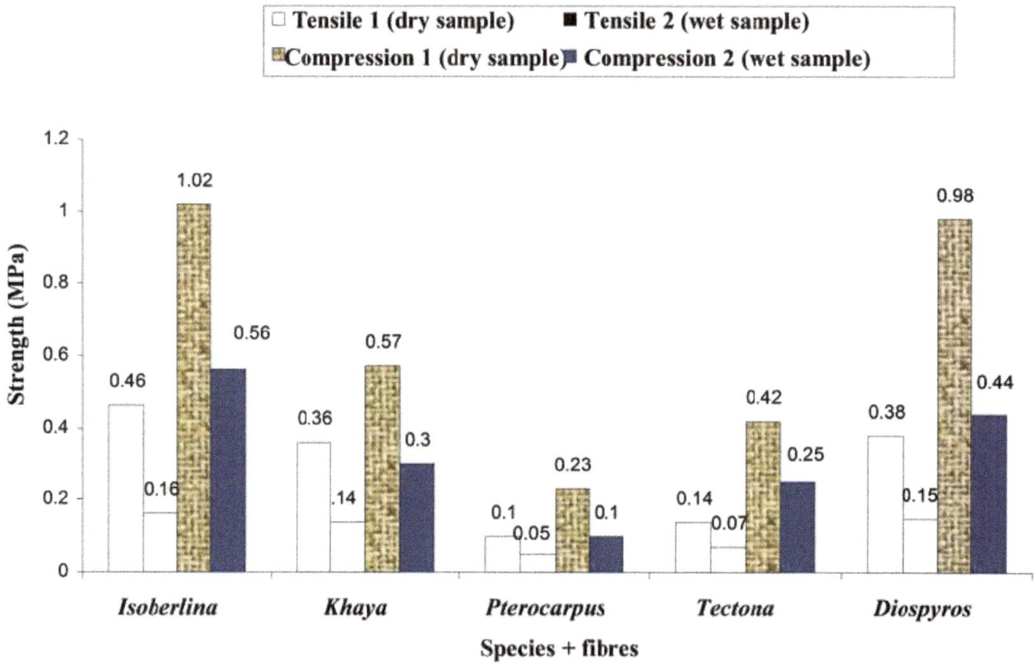

Figure 5. Tensile and compressive strengths of wood cement composites (treatment in cold water).

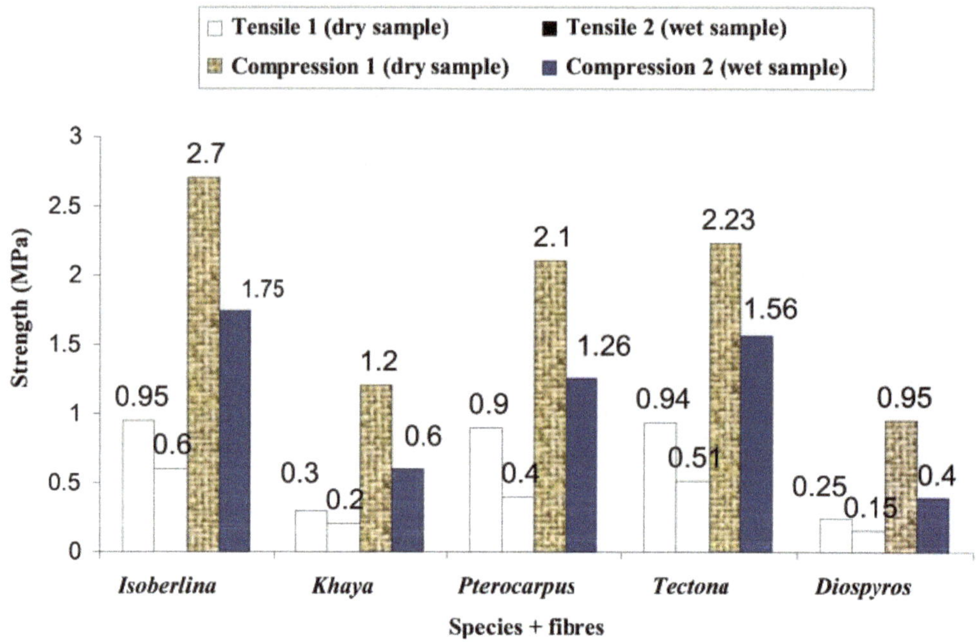

Figure 6. Tensile and compressive strengths of wood-cement composites (treatment in hot water).

of the wood-cement composite made from sawdust and vegetable fibers available in Benin. The new material can be used to manufacture lightweight bricks for walls.

Conclusion

In this study, the sawdust from five West African wood species partially replaced by *E. guineensis* fibers were tested to make wood-cement composites. The compatibility study indicated that all the sawdust and fibers mixes have good aptitude to wood-cement composites. After hot water treatment which enhanced the compatibility of the most mixes, only *Diospyros mespiliformis* species became moderately compatible with cement while the other species namely Isoberlina *doka*, *T.grandis, K. grandifoliola* and *P. erinaceus* species had good compatibility with cement. *T. grandis* and *P. erinaceus* sawdust containing *E. guineensis* species were greatly affected by cold water treatment.

The study confirms that physical and mechanical tests on wood-cement composites could provide quick and satisfactory results to assess the aptitude of tropical hardwood species for wood work. On the basis of the density and compressive strengths of wood-cement composite, the main factor of compatibility of the wood specie with cement could be approximately and quickly evaluated.

In wet conditions, the mechanical strengths of wood-cement composites decrease drastically (about 40%), especially *I. doka* and *D. mespiliformis* species. Such important result must be taken into account when using the composites in a humid environment. Finally, one can suggest the most suitable mixes of fibers and the five West African hardwood species for wood-cement composite in the following order: *I. doka*, *T. grandis, K. grandifoliola*, *P. erinaceus* and *D. mespiliformis* species. Tests will continue to be done in order to find the best way to keep very low the inhibitory effect of the wood species and the fibers on the cement hydration.

Conflict of Interests

The author(s) have not declared any conflict of interests.

REFERENCES

Alberto MM, Mougel E, Zoulalian A (2000). Compatibility of some tropical hardwood species with Portland cement using isothermal calorimetry. For. Prod. J. 50(9): 83-88.

Andy WC, Land Zhongli H (1986). Compressive strength of cylindrical samples as an indicator of wood-cement compatibility. For. Prod. J. 36 (11/12): 87-90.

Badejo S (1989). Influences of pre-treatment temperature and additives concentration on the properties of cement - bonded particleboard from plantation grown tropical hardwoods. Trop. Sci. 29: 285- 296.

Balkis FAB, Paridah Md T, Alinaghi K, Edi SB, Mohd KAU, Adrian CYC (2012). Evaluations of some physical properties for oil palm as alternative biomass resources. Wood Mater. Sci. Eng. 8: 119-128.

Dinwoodie JM, Paxton BH (1984). Wood - cement particle board: a technical assessment. J. Appl. Polym. Sci. 40 : 217- 227.

Doko VK (2013). Formulation et étude comparative des propriétés physiques, mécaniques et thermiques de composites à matrice cimentaire renforcée par des biomasses végétales : cas des fibres de Borassus Aethiopium Mart et des balles de riz. Thèse de Doctorat Unique, Université d'Abomey-Calavi. 196p.

Hachmi M, Moslemi AA (1989). Correlation between wood - cement compatibility and wood extractives. For. Prod. J. 39:55-58.

Hachmi M, Moslemi AA, Campell AG (1990). A new technique to classify the compatibility of wood with cement. Wood Sci. Technol. 24:345-354.

Hachmi M, Sesbou A (1990). Les composites bois - ciment: une nouvelle filière pour la valorisation des produits lignocellulosiques marocains. Annales de la Recherche Forestière au Maroc.

Hofstrand AD, Moslemi AA, Garga JF (1984). Curing characteristics of wood particles from nine northern Rocky Mountain species mixed with Portland cement. For. Prod. J. 34:57-61.

Mougel E (1992). Mise au point d'un composite ciment - bois dont les variations dimensionnelles vis-à-vis des variations d'humidité sont contrôlées. Thèse de Doctorat de l'Université de Nancy I.

Pereira C, Caldeira Jorge F, Ferreira JM (2004). Wood-cement composites: A review. Holz Roh- Werkst 62:370-377.

Pereira C, Caldeira Jorge F, Irle MA, Ferreira JM (2003). Extractive contents of two woods and cork of Portuguese origin and the interaction of these substrates with cement suspensions. For. Prod. J. 53(10):39-43.

Sandermann W, Kohler R (1964). Studies on mineral bonded wood materials VI: A short test of aptitudes of wood for cement - bonded materials. Holzforschung 18:53-59.

Semple KE, Cunningham RB, Evans PD (1999). Cement hydration tests using wood flour may not predict the suitability of Acacia Mangium and Eucalyptus Pellita for the manufacture of wood- wool cement boards. Holzforschung 53(3):327-332.

Semple KE, Cunningham RB, Evans PD (2002). Suitability of five western Australian mallee eucalypt species for wood - cement composites. Ind. Crops Prod. 16:89-100.

Semple KE, Evans PD (2000). Adverse effects of heartwood on the mechanical properties of wood-wool cement boards manufactured from radiata pin wood. Wood Fibre Sci. 32(1):37-43.

Silva DAL, Lahr FAR, Garcia RP, Freire FMCS, Ometto AR (2013). Life Cycle assessment of medium density particleboard (MDP) produced in Brazil. Int. J. Life Cycle Assess. 18(7):1404-1411.

Tchehouali DA (2002). Contribution à la valorisation de bois tropicaux : Mise au point et caractérisation de matériaux ciment-sciure. Thèse de Doctorat Unique, Université d'Abomey Calavi, République du Bénin.147p.

Weatherwax RC, Tarkow H (1964). Effect of wood on setting of Portland cement, Forest Prod. J. 14:567-570.

World Bank report (2010). République du Bénin Analyse environnementale Pays. Rapport n° 58190-BJ. 71p.

26

Removal of azo dye from water by adsorption using waste residue from alum manufacturing process

Haimanot habte[1] , Feleke Zewge[2] and Mesfin Redi[2]

[1]Environmental Science Program, Faculty of Science, Addis Ababa University, P.O. Box 1176, Addis Ababa, Ethiopia.
[2]Department of Chemistry, Faculty of Science, Addis Ababa University, P.O. Box 1176, Addis Ababa, Ethiopia.

Adsorption techniques are widely used to remove certain groups of pollutants such as dyes which are not amenable to biodegradation and environmentally very problematic. Although commercial activated carbon is a preferred sorbent for color removal, its widespread use is restricted due to high cost. As such, alternative non-conventional sorbents should be investigated. It is well-known that natural materials, waste materials from industry and agriculture, and bio sorbents can be obtained without a cost, therefore could be employed as inexpensive sorbents. Therefore the present investigation reports the adsorptive capacity of an industrial residue for the removal of azo dyes. Batch experiments were carried out for the sorption of the dye onto the adsorbent. The influence of contact time, adsorbent dose, pH and initial dye concentration on the adsorption capacity of the material was studied. Kinetics and adsorption isotherms were also studied to evaluate the rate of dye removal and the capacity of the adsorbent. The equilibrium data best fits freundlich isotherm equation. Adsorption capacity (K_f) and intensity of adsorption (n) as determined by fitting the equilibrium data to freundlich isotherm equation are calculated to be 0.26 and 0.82, respectively. The adsorption process was found to undergo via a pseudo-second-order adsorption kinetics with a rate constant of 3.81×10^{-3}(g /mg / min). The removal of dye was not affected for the initial dye concentration range of 15 to 256 mg/L. An adsorption process; requiring an equilibrium time of 30 h with optimum adsorbent dose of 19 g/L for 77.4% dye removal efficiency was observed at near neutral pH. However, as the pH of water is adjusted from 7 to 9, the dye removal efficiency was greater than 90%. The overall result shows that the industrial by product investigated in this study exhibited a high potential for the removal of dye from aqueous solution.

Key words: Dye, adsorbent, adsorption isotherms, batch adsorption, removal.

INTRODUCTION

It is estimated that 10,000 different types of dyes and pigments are produced worldwide annually (Wallace, 2001) out of which a large number of dyes are azo compounds (-N=N-), which are linked by an azo bridge.

Table 1. Percentage composition of the chemical constituents in waste residue.

Chemical composition	Percent (wt. %)	Chemical composition	Percent (wt. %)
Quartz (SiO_2)	40	K_2SO_4	0.005
Kaolin	8.883	$Al_2(SO_4)_3$	1.778
$Al(OH)_3$	0.878	$CaSO_4$	0.194
$Fe_2(SO_4)_3$	0.023	Fe_2O_3	0.001
$MgSO_4$	0.008	Na_2SO_4	0.007

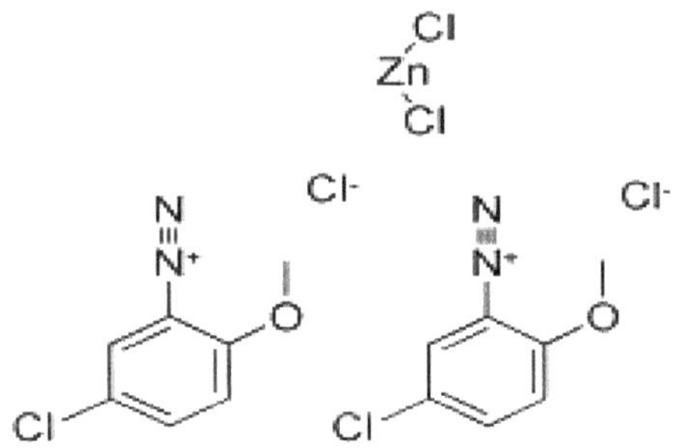

Figure 1. The molecular structure of the dye.

It is estimated that 10,000 different types of dyes and pigments are produced worldwide annually (Wallace, 2001) out of which a large number of dyes are azo compounds (-N=N-), which are linked by an azo bridge. Azo dyes are used by a wide number of industries. While textile mills predominantly use them, azo dyes can also be found in the food, pharmaceutical, paper and printing, leather, and cosmetics industries (Asamudo et al., 2005). Many of these dyes find their way into the environment via effluent discharges. These compounds retain their color and structural integrity under exposure to sunlight, soil and bacteria; they also exhibit a high resistance to microbial degradation in wastewater treatment systems.

Therefore these compounds have become a major environmental concern (Asamudo et al., 2005). In current dyeing processes, as much as 50% of the dye is lost in the wastewater. This can lead to acute effects on exposed organisms due to the toxicity of the dyes, abnormal coloration and reduction in photosynthesis because of the absorbance of light that enters the water (Slokar and Marechal, 1998)

"Currently there are many dye removal methods, out of those methods ion exchange and adsorptions using natural clays are preferable because of cost effectiveness and ease of operation". Activated carbon is the preferred adsorbent widely employed to treat wastewater containing different classes of dyes. Activated carbon adsorption has been cited by the US Environmental Protection Agency as one of the best available control technologies (Bhattacharyya and Sarma, 2003). However, the disadvantage associated with is its high cost (Babel and Kurniawan, 2003). The regeneration of saturated carbon is also expensive, not straightforward, and results in loss of the adsorbent. The use of carbons based on relatively inexpensive starting materials is also unjustified for most pollution control applications (Street et al., 1995). Due to the problems mentioned above, research interest into the production of alternative sorbents to replace the costly activated carbon has been intensified in recent years. Cost is actually an important parameter for comparing the adsorbent materials. A sorbent can be considered low-cost if it requires little processing, is abundant in nature or is a by-product or waste material from another industry. In this paper laboratory investigations were carried out to evaluate the efficiency of the waste residue of aluminum sulfate and sulfuric acid factory to adsorb azo dyes from water.

MATERIALS AND METHODS

The waste residue, generated during the manufacture of aluminum sulfate using kaolin and sulfuric acid, was collected from Awash Melkasa Aluminum Sulfate and Sulfuric Acid Factory, which is located at the central part of the Ethiopian Rift Valley Region. Samples were taken from more than 30 batches of aluminum sulfate production residues and then mixed to maintain homogeneity. The waste directly discarded contains about 52% of solid. The chemical compositions of this industrial by product are given in Table 1. The collected waste residues were sundried for one day and ground to fine powder using mortar and the resulting material is considered as untreated media. The chemical composition of the waste material indicates the absence of any hazardous and carcinogenic substances in the material and hence it is suitable to use as a decolorization agent. The dye used for the investigation of the adsorptive capacity of the waste residue was obtained from Akaki textile industry located in Addis Ababa, Ethiopia. The molecular structure of the dye is shown in Figure 1.

Analytical methods and instrumentation

Uv-vis spectrophotometer is selected as an analytical tool to

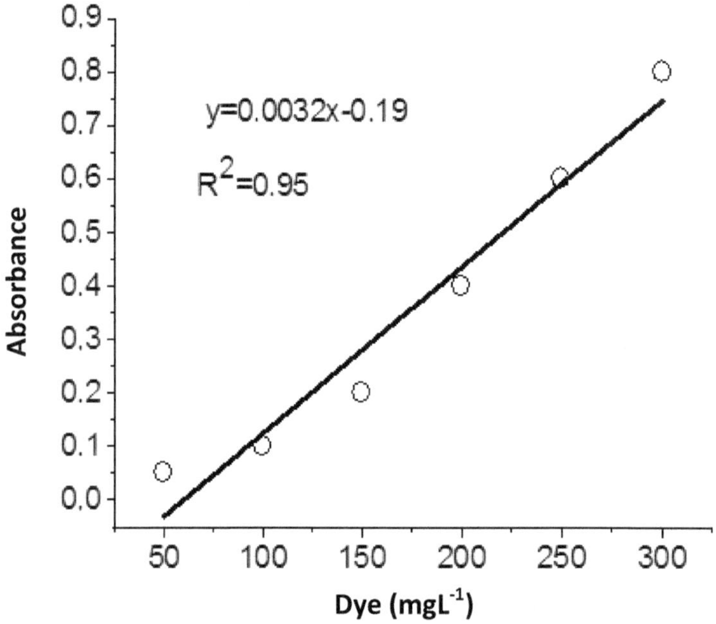

Figure 2. Calibration curve of the dye at its λmax.

determine the concentration of dye solutions. This selection is made based on the fact that most dyes absorb electromagnetic radiation in the Uv-vis region. Concentration of dye solution was calculated from the absorbance measurements using Lambert Beer's law Equation (1)

$$A = \varepsilon cl \quad (1)$$

Where A is absorbance, ε is the molar absorptivity, C is the concentration, and I is the path length.

The dye removal capacity of the adsorbent is expressed as follows using Equation 2:

$$\frac{C_i - C_e}{m} \times V \quad (2)$$

Where C_i and C_e are the initial and final dye concentration (mg/L) in the liquid phase respectively, m is the mass of adsorbent (g) , and V is the volume of experimental solution (mL).

The percentage of dye removal was calculated using Equation 3:

$$\% \text{ dye adsorption} = \frac{C_i - C_e}{C_i} \times 100\% \quad (3)$$

Batch adsorption studies

After the wavelengths of maximum absorbance (λ_{max}= 373 nm) of the solutions is determined calibration curve is constructed by dissolving an amount of dye to make 300 mg/L of dye solution. Then the other series of solutions to make the calibration curve are made by taking the corresponding volume of the previously concentrated dye solutions, diluting using distilled water and measuring their absorbance at their respective λ_{max}. Experimental solutions were prepared by pipetting a known amount of dye into a 1 L conical flask and diluting it with a known amount of distilled water. Batch experiments for decolorization were conducted in 500 mL conical flask containing 300 mL of dye solution at room temperature (22 ± 2°C), to evaluate dye removal efficiency and capacity of the media. The media (waste residue) was placed in the flask and then stirred continuously at a constant slow mixing rate with magnetic stirrer during the experiment. The initial dye concentration selected for batch experiments is within the range recommended in the actual textile effluent (Galster and Hansen, 1998). The effect of dose of the media, pH of water and initial concentration of the dye were investigated by varying any one of the process parameters and keeping the other parameters constant. Investigation of the effect of dose experiments was conducted by varying adsorbent doses (1 to 25 g/L) at constant initial dye concentration of 100 mg/L.

The effect of pH was studied by varying the pH from 7−9 either with 0.1 M NaOH or 0.1 M HCl. To investigate the effect of initial dye concentration, experiments were conducted by varying dye concentrations from 50 to 258 mg/L at constant adsorbent dose of 19 g/L. The kinetic analysis of the adsorption data is based on reaction kinetics of pseudo-first order and pseudo-second-order mechanisms. Adsorption kinetics was determined using constant adsorbent dose of 19 g/L corresponding to the initial dye concentration of 100 mg/L. Isotherm experiments were conducted by changing the adsorbent dose from 1 to 43.3 g/L at initial dye concentration of 100 mg/L. For all experiments, the remaining concentration was determined spectrophotometrically at its corresponding λ_{max} when the equilibrium contact time is reached.

RESULTS AND DISCUSSION

Wave length of maximum absorbance and Calibration curves

The calibration curve (Figure 2) is obtained by preparing dye solution for which the concentration is within the effective concentration range of the dye. The correlation coefficient from the curve shows strong linear relationship between the concentration of the dye solutions and the

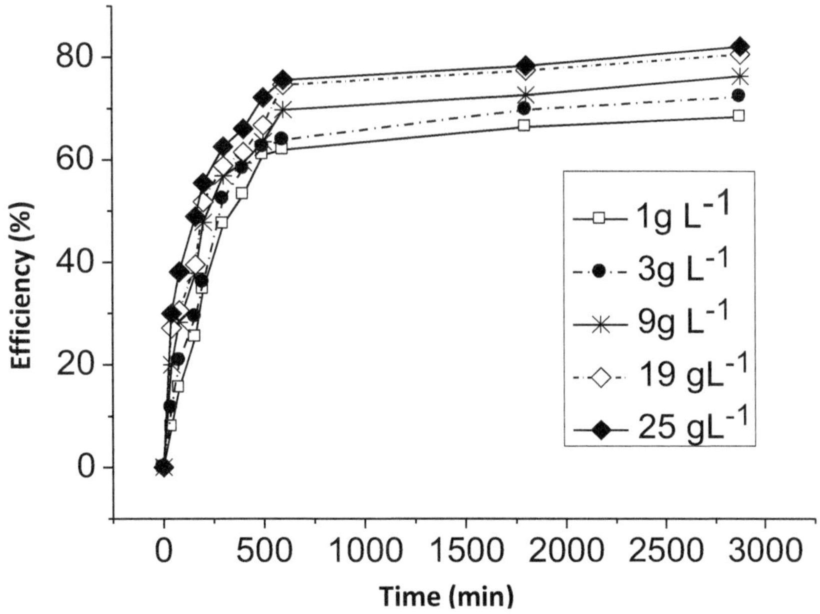

Figure 3. Efficiency of dye removal as a function of time for different doses of untreated media (initial dye concentration = 100 mg /L, pH = 3-4).

absorbance at the dye's wavelength of maximum absorbance.

Effect of dose and contact time

To investigate the effect of dose and contact time, experiments were conducted by varying adsorbent doses (1 to 25 g/L) at constant initial dye concentration of 100 mg/L. Further addition of adsorbent does not provide more increment in the adsorption amount (Figure 3). The results show that the adsorption reached equilibrium after about 30 h. Further increase in contact time did not increase the uptake. In general the decolorization efficiency was increased with dose as reflected by the measured residual dye concentration. The increase in the dye adsorption was due to the increased in availability of dye binding sites resulting from an increase in adsorbent dosage. Higher dose of the media (greater than 19 g/L) will increase the amount of sludge without causing a significant change on the amount of dye removed. On the other hand, the adsorption capacity decreases with increasing dose (Figure 4). To maintain maximum capacity and high removal efficiency, the surface loading (that is the mass ratio of the dye to adsorbent dose) should be lower than the optimum value. A dose of 19 g/L corresponds to the capacity of about 4.74 mg dye /g of adsorbent if there is 90% removal. The surface loading obtained for the dose of 19 g/L in this experiment was 4.1 mg dye /g of adsorbent. Lower capacity of the material is due to the presence of H^+ ions which may cause the surface to be positively charged and hence reduce its interaction with the cationic dye. These experimental results suggest that modification of the surface chemistry of the adsorbent may enhance its capacity.

Effect of pH

About 80% of the dried waste residue is quartz. In quartz [SiO_2], Si and O are structural elements; the Si-O bond has about 50% covalent character (Huang and Ostovic, 1978). The siloxane groups, -SiOSi-, interact with water forming -SiOH (Fan et al., 2003). Any oxide surface attains (positive or negative) charge on its surface. The hydroxyl groups can gain or lose proton, resulting in a surface charge that varies with changing pHs. At lower pH, surface sites are protonated and the surface become positively charged. Maximum removal efficiency at pH 7 to 9 and the removal rate increases with an increase in pH in general. At higher pH, the surface hydroxides lose their protons and the surface becomes anionic. This can be the reason for the higher efficiency of the adsorbent for the cationic dye at higher pH value (Figure 5). The other main constituent of the adsorbent mixture is kaolinite (kaolin), a class of natural clay minerals. Here the adsorption capabilities result from a net negative charge on the surface, which gives clay the capability to adsorb positively charged species. Their sorption properties are also due to their high surface area and high porosity (Alkan et al., 2004). Clay mineral exhibits a strong affinity for both heteroatomic cationic and anionic dyes however; the sorption capacity for basic dyes is much higher than for acid dyes because of the ionic

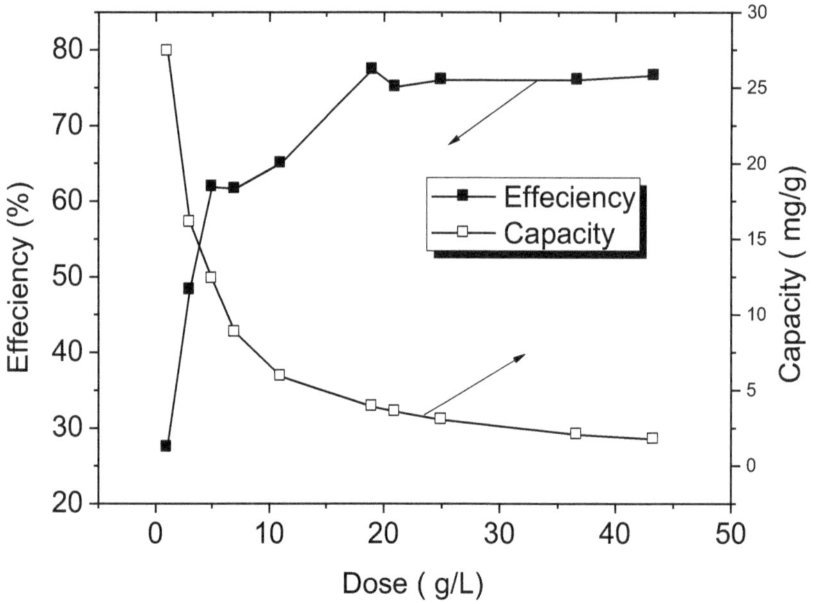

Figure 4. The relationship between adsorbent dose versus percentage of dye removal and adsorbent dose versus removal capacity at fixed dye concentration.

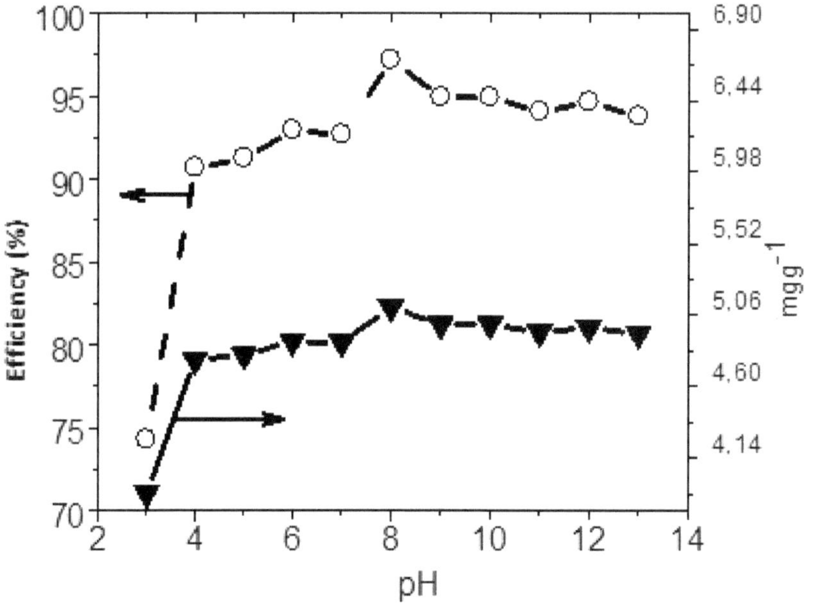

Figure 5. Effect of pH on efficiency and capacity of dye removal at fixed dye concentration and adsorbent dose.

charges on the dyes and character of the clay. Some researchers show that the adsorption of dyes on kaolinite was about 20 times greater than alumina (Harris, 2001). The possible mechanism for the adsorption of the dye for this experimental condition can also be explained based on the theoretical meaning of *pHzpc* (pH of zero point charge). Silica adsorbents have a low pHzpc nearly 2 (Rao and Sridharan, 1984). The measured pH for the dye-adsorbent system is 3 to 4 in this particular experiment, that means the surface of the material will have a slight negative charge even at this lower pH; there is a columbic attraction towards the cationic dye. The other possible mechanism for the adsorption of the dyes by quartz can be attributed to the formation of surface

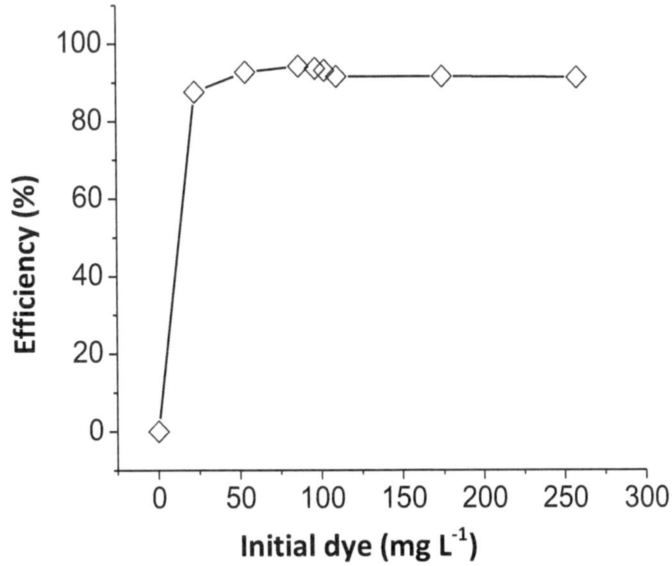

Figure 6. Effect of initial dye concentration on percentage of dye removal at fixed adsorbent dose. (PH = 7 to 9).

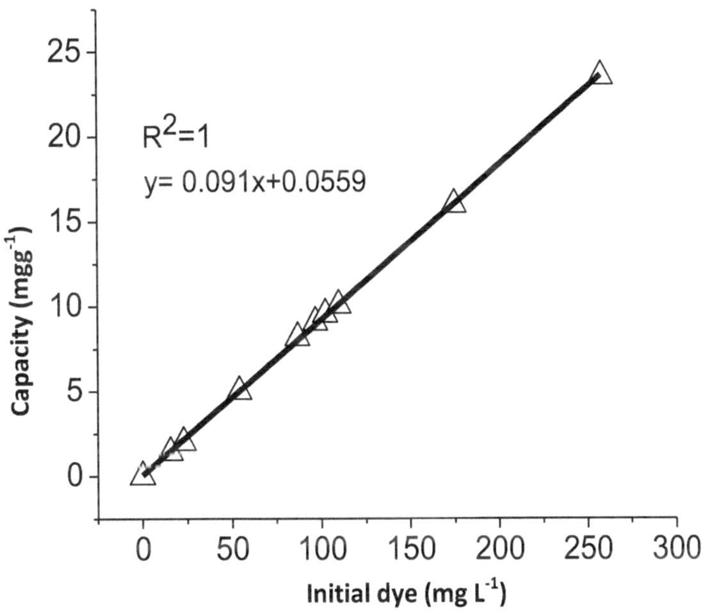

Figure 7. Effect of initial dye concentration on capacity of dye removal at fixed adsorbent dose for (pH – 7 to 9).

hydrogen bonds between the surface hydroxyl groups on the surface of the silica adsorbent and the nitrogen atoms of the dye.

Effect of initial dye concentration

The effect of initial dye concentration at a fixed quantity of the adsorbent dosage is shown in the following figures. The percentage of dye removal at fixed adsorbent dose increased with increasing initial dye concentration (Figure 6). This is expected because in surface adsorption reactions the equilibrium concentration of a given solute in the solid phase is proportional to the equilibrium concentration in the liquid phase. From Figure 7, it is observed that increasing the initial dye concentration

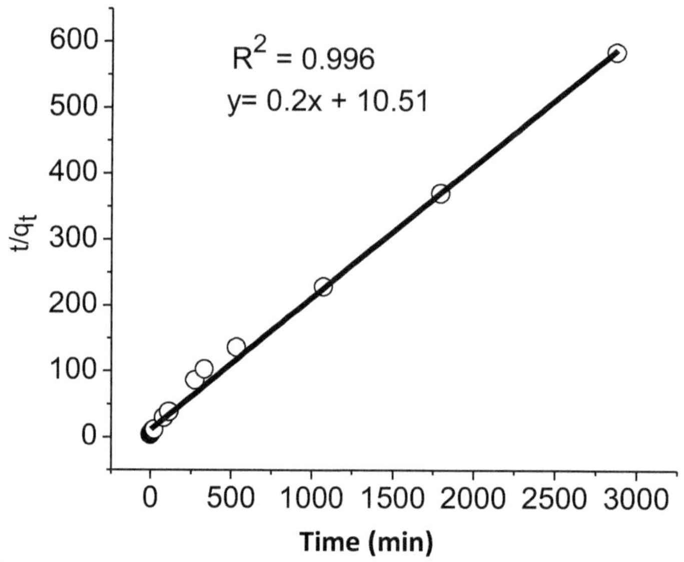

Figure 8. Pseudo-second-order plot of the dye adsorption kinetics on untreated media (pH = 7-9, dose = 10 g/L and an initial dye concentration of 100 mg/L).

increases the removal capacity of the adsorbent. The correlation coefficients of the experiments show that there is a strong linear relationship between the initial concentration of the dye and the removal capacity of the adsorbent. This result indicates that the method is suitable for the treatment of high-strength textile effluent in terms of dye concentration.

Reaction rates

The adsorption kinetics was studied at constant dye and adsorbent concentrations. The kinetic analysis of the adsorption data is based on reaction kinetics of pseudo-first-order and pseudo-second-order mechanisms. The uptake of the dye on untreated adsorbent reached equilibrium in 30 h. The kinetics of adsorption was analyzed by using the Lagergren equation as shown below(Yeheyes et al., 2005):

$$\frac{dqt}{dt} = K_2(qe - qt)^2$$

$$\frac{d(q_e - q_t)}{(q_e - q_t)^2} = -k_2 dt \tag{4}$$

For the boundary conditions t = 0 to t = t and q_t = 0 to q_t = q_t, the integrated form of Equation (5) becomes:

$$\frac{1}{(q_e - q_t)} = \frac{1}{q_e} + k_2 t \tag{5}$$

This is the integrated rate law for a pseudo second-order reaction. Equation (5) can be rearranged to obtain Equation (6), which has a linear form:

$$\frac{t}{q_t} = \frac{1}{k_2 q_e^2} + \frac{1}{q_e}(t) \tag{6}$$

Where, k2 (g /mg /min) is the equilibrium rate constant of second-order sorption. The untreated materials can be described very well by the pseudo-second-order rate equation as shown in Figure 8. The correlation coefficient was found as near to unity. The values of q_e and K_2 are 5 (mg/g) and 3.81 (g /mg /min) respectively which are determined by plotting t/qt against t of Equation (7), and using the slope and y-intercept of the plot. The equilibrium between the adsorbed layer and the solution slightly delayed and this is due to (1) the solute molecules do not diffuse immediately from the bulk of the solution to the interface and (2) the solute molecules do not cross immediately the energy barrier which exists near the interface, and moreover a flow of solute molecules leaving the interface-desorbing-slackens the adsorption.

Adsorption isotherm

The adsorption isotherm of the dye is shown in Figure 9 at a constant temperature (22 ± 2°C). Langmuir (data not shown) and Freundlich isotherms were used to describe the equilibrium nature of dye adsorption by the adsorbent. The values of n and log k_f were calculated

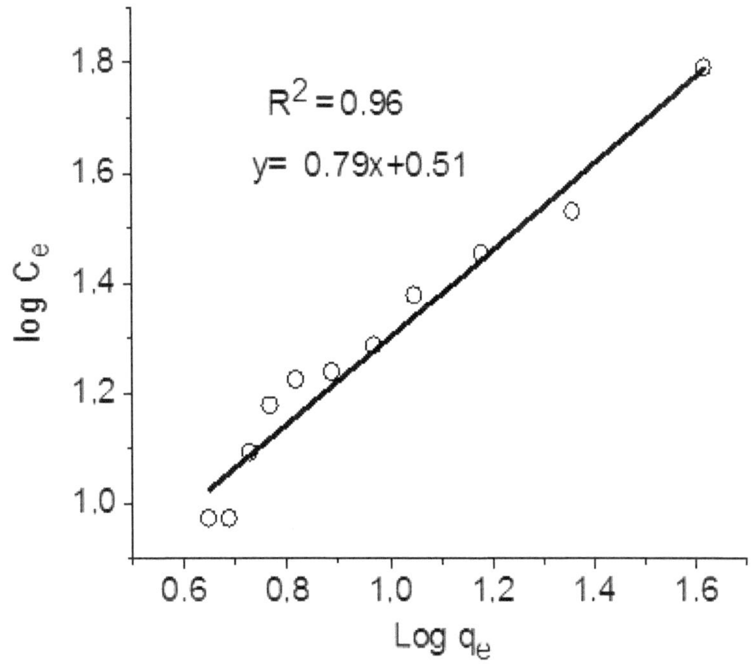

Figure 9. Linearized Freundlich isotherm (initial dye concentration = 103 mg/ L, pH = 7-9, temperature = 24°C).

Table 2. Important adsorption parameters to plot an isotherm for untreated material.

Dose (g/L)	C_e (mg/L)	q_e (mg/g)	$1/q_e$	$1/C_e$	Log q_e	Log C_e
1	61.58	41.53	0.02	0.02	1.62	1.789
3	33.62	23.16	0.04	0.03	1.36	1.527
5	28.25	14.97	0.07	0.04	1.18	1.451
7	23.73	11.34	0.09	0.04	1.05	1.375
9	19.21	9.32	0.11	0.05	0.97	1.284
11	17.23	7.81	0.13	0.06	0.89	1.236
13	16.67	6.65	0.15	0.06	0.82	1.222
15	14.97	5.88	0.17	0.07	0.77	1.175
17	12.43	5.33	0.19	0.08	0.73	1.09
19	9.32	4.94	0.20	0.11	0.69	0.97
25	9.32	3.75	0.27	0.11	0.65	0.97

from the slope and interception of Freundlich plots as listed in Table 2. The adsorption data were analyzed with the help of linear form of Freundlich isotherm. Freundlich model attempts to account for surface heterogeneity (Gupta et al., 1988).

Freundlich isotherm:

$$\text{Log } q_e = \log k_f + \frac{1}{n}\log C_e \quad (7)$$

Where, k_f is roughly a measure of adsorption capacity and 1/n is an indicator of adsorption effectiveness; q_e is the amount of dye adsorbed per unit mass of adsorbent (in mg g^{-1}) and C_e is the equilibrium concentration of dye (in mg l^{-1}). Figure 9 shows a plot of log q_e vs. logC_e. The constants $1/n$ as the slope and k_f as the intercept. The values of Freundlich isotherm parameters along with the kf and n were found as 0.26 and 0.82, respectively. It is found that the related correlation coefficient R_2 value for the Freundlich model is near to unity (0.962) and hence the process of decolorization using untreated material is well fitted to the Freundlich isotherm. The Freundlich coefficient, n should have values in the range of 0 < n <1 for favorable adsorption (Pola et al., 2003). The data did not fit well to the Langmuir isotherm model (data not shown). The condition for the validity of a Freundlich type adsorption model is adsorption on heterogeneous surfaces (Gupta et al., 1988). The increase in equilibrium dye removal capacity with residual dye concentration observed during the investigation of effect of initial dye concentration on capacity of dye removal can support the condition of heterogeneous adsorption.

Conclusions

The ability of locally available industrial by product to remove textile dyes from aqueous solutions was investigated employing an adsorption process. The media could remove about 77.4% of the dye from water at 19 g/L dose in pH range of less than 3-4 in the first 30 h. However, the media has an efficiency of 90% when the pH of the adsorbent and adrobate mixture is between 7 to 9. The removal of dye was not affected by the initial dye concentration range of 15 to 256 mg/L that suggests the industrial by product has high efficiency for high strength textile dye removal. The kinetics of dye adsorption by the industrial by product in the batch study was satisfactorily described by a pseudo second order rate kinetics model, and the experimental data correlated reasonably well with that of the Freundlich adsorption isotherm. The results of these studies indicate that the waste residue is an efficient and low cost adsorbent for dye removal from the aqueous phase.

Conflict of Interests

The author(s) have not declared any conflict of interests.

REFERENCES

Alkan M, Demirbas O, Sermet C, Mehmet D (2004). Sorption of acid red 57 from aqueous solution onto sepiolite. J. Hazard. Mater. B116 (2004) 135-145.

Asamudo NU, Daba AS, Ezeronye OU (2005). Bioremediation of textile effluent using Phanerochaete chrysosporium. J. Biotechnol. 4:1548-1553.

Babel S, Kurniawan TA (2003). Low-cost adsorbents for heavy metals uptake from contaminated water, J. Hazard. Mater. 97:219-243.

Bhattacharyya KG, Sarma A (2003). Adsorption characteristics of the dye brilliant green on neem leaf powder, J. dyes and pigments, pp.57211-222.

Huang CP, Ostovic FB (1978). Cadmium and copper removal by a granular activated carbon. J. Environ. Eng. Div. 5:863-878.

Fan X, Parker DJ, Smith MD (2003). Adsorption kinetics of fluoride on low cost materials, J. Water Res. 37:4929-4937.

Gupta GS, Prasad G, Panday KK, Singh VN (1988). Removal of chrome dye from aqueous solutions by fly ash, J. Water, Air and Soil Pollut. 37:13-24.

Harris RG, Well JD, Johnson BB (2001). Selective adsorption of dyes and other organic molecule to kaolinite and oxide surfaces, Colloid Surf. A: physicochem. Eng. Aspects 180:131-140.

Galster J, Hansen OC (1998). Survey of azocolorants in Denmark: consumption, use, health. Danish EPA.

Yeheyes K, Zewge F, Chandravanshi BS (2005). Dye removal from textile effluents by coagulation using locally produced alum. J. hazard. Mater. Pp.151.

Streat M, Patrick JW, Perez MJC (1995). Sorption of phenol and parachlorophenol from water using conventional and novel activated carbon, J. Wat. Res. 29:467.

Pola A, Tokat E, Erkaya H (2003). Removal of some reactive dyes from textile processing using powdered activated carbon, J. Chem. Eng. Pp.114-122.

Rao SM, Sridharan A (1984). Clays and Clay Minerals, J. clay mineral society, 32:414.

Slokar Ym , Le- marechal A (1998). Methods of decoloration of textile wastewater dyes and Pigments, J. dyes and pigments, 37:335-356.

Strickland AF (1995). Decolorization of continuous dyeing wastewater by ozonation, J. Tex. Chem. Color. 27:11-15.

Wallace TH (2001). Biological treatment of synthetic dye water and an industrial textile wastewater containing azo dye compounds. M.Sc. Thesis, June 18, Black burg, Virginia.

Culture-dependent characterization of hydrocarbon utilizing bacteria in selected crude oil-impacted sites in Bodo, Ogoniland, Nigeria

Chioma Blaise Chikere and Chioma Bertha Ekwuabu

Department of Microbiology, University of Port Harcourt, P.M.B. 5323, Port Harcourt, Rivers State, Nigeria.

This investigation was carried out to characterize microbial communities in selected crude oil polluted sites in Bodo community, Gokana Local Government Area of Rivers State, Nigeria. Total heterotrophic bacterial counts ranged from 0.7 to 1.37x10^{7} cfu/g and 0.2 to 5.9x10^{6} cfu/ml while counts of hydrocarbon utilizing bacteria ranged from 0.1 to 8.0 x 10^{6} and 0.2 to 7.5 x 10^{5} cfu/ml for soil, sediment and water, respectively. Physiochemical parameters of all samples were determined. The ranges obtained were temperature 31-33°C, pH 7.5-8.2, conductivity 1134 - 7680 µs/cm, total nitrogen 792.4 - 886.3 mg/kg, nitrate 36.55 - 42.70 mg/kg, total organic carbon 2.06 - 2.18%, total petroleum hydrocarbon 1007 - 1104 mg/kg, vanadium 0.001 - 0.007 mg/kg, iron 3.772 - 4.889 mg/kg, chromium 52.40 - 66.20 mg/kg, nickel 40.02 - 41.62 mg/kg, lead 17.30 - 19.40 mg/kg and zinc 35.10 - 39.50 mg/kg for soil and sediments while water had total nitrogen 868 mg/l, nitrate 40.6 mg/l, total organic carbon 3.1 mg/l, total petroleum hydrocarbon 768 mg/l, nickel 39.2 mg/l, lead 17.3 mg/l and turbidity 250 NTU. Bacteria isolates characterized belonged to these genera *Bacillus, Proteus, Pseudomonas, Flavobacterium, Corynebacterium, Serratia, Micrococcus, Klebsiella, Enterobacter* and *Azotobacter*. The findings reveal that there is a high population of active indigenous hydrocarbon utilizing bacteria which can be monitored and enhanced to bring about bioremediation in the study area.

Key words: Hydrocarbon pollution, soil, water, sediments, hydrocarbon utilizing bacteria, Bodo, Ogoniland.

INTRODUCTION

Petroleum is at present, Nigeria's and indeed the world's most important derived energy source (Moffat and Linden, 2005). However the growth and activities of petroleum and petroleum associated industries in Nigeria and in other parts of the world has lead to increased oil pollution in our environment. Crude oil, because of its characteristics is one of the most significant pollutants in the environment as it is capable of causing serious damages to humans and the ecosystem (Okpokwasili, 1996).

Oil producing areas of Nigeria especially the Niger-Delta area have experienced the devastating consequences of crude oil spills to both terrestrial and aquatic environments in the past 50 years of crude oil exploration and production (Adati, 2012). One of the major reasons for prolonged negative impact of oil spill on the environment could probably be absence of adequate and qualitative scientific baseline data which is required to provide informed and quick response to emergent environmental challenges (Akinde et al., 2012). The Niger Delta is among the ten most significant wetland and marine ecosystems in the world but unsustainable oil exploration activities has rendered the Niger Delta region one of the five most severely petroleum damaged ecosystems in the world (FME, 2006).

Ogoniland is located in the Niger Delta and oil exploration and production activities have been ongoing in this area since the 1950s. Ogoniland is now characterized by oil fields and installations that have remained dormant for several decades. Past spills, lack of maintenance, oil trapping and damage to oil infrastructures have been a common sight in this region and the environment has been without remediation or partially remediated over the years. UNEP report (2011) concluded that pollution of soil by petroleum hydrocarbons in Ogoniland is extensive in land areas, sediments and swampland. The investigation also showed that the surface water throughout the creeks contains massive hydrocarbons.

Hydrocarbons interact with the environment and micro-organisms determining the fate of the contaminant relative to their chemical nature and microbial degra-dative capabilities respectively. Provided the polluted has requisite values for environmental factors that influence microbial activities and there are no inhibitors of meta-bolism, there is a good chance that there will be a viable and active population of hydrocarbon utilizing micro-organisms in the environment (Chikere et al., 2011, 2012b). Considering the large quantity of oil going into the Niger Delta environment especially farmlands and rivers, the need to cleanup crude oil contaminated sites has become a key environmental issue (Vincent et al., 2011).

Conventional methods such as physical removal are the first response option. It is worthy to note that they do not achieve a complete cleanup of the oil spills. Current mechanical methods typically recovers not more than 10-15% of crude after a major spill and almost always leaves the receiving body in worse conditions (Abu and Dike, 2008).

Due to the abilities of certain microbes to mine-ralize hydrocarbon components into environmentally friendly substances such as carbon dioxide and water, the ability of bacteria in breaking down hydrocarbons has gained growing attention in modern day research (Wackett and Hershberger, 2001; Kadali et al., 2012). Biodegradation by microbes is the key removal process of hydrocarbons which is controlled by hydrocarbon physicochemistry, environmental conditions, bioavailability and the presence of catabolically active microbes (Stroud et al., 2007).

This study was conducted to ascertain the microbial diversity associated with the chronically oil-inundated Bodo community in Ogoniland and to some extent ascertain their natural propensity to utilize petroleum hydrocarbons.

MATERIALS AND METHODS

Sampling

One sample each of crude oil polluted water, soil and sediment were collected under aseptic conditions from Bodo community in Ogoniland, Rivers State in the Niger Delta using appropriate equipments. Soil was collected at 0-15 cm soil auger into sterile polyethylene bags. Sediment was collected with an Eckman grab while water was collected into sterile bottles. Samples were collected at different parts of each site, bulked for homogeneity and thereafter transported to the laboratory at 4°C in ice pack.

Determination of physicochemical parameters of samples

Physicochemical parameters such as pH, moisture content, nitrate, phosphate, total organic carbon (TOC), turbidity, salinity, temperature, conductivity and heavy metals were determined using methods from APHA (2008).

Chromatographic analysis

Residual total petroleum hydrocarbons (TPH) and polycyclic aromatic hydrocarbons (PAHs) were extracted from the samples and quantified using gas chromatograph-flame ionization detector (GC-FID).

Enumeration of total heterotrophic bacteria

Total heterotrophic bacterial (THB) counts were determined using spread plate method on plate count agar (PCA). From each sample 1 g or 1 ml was homogenized in 9 ml of 0.85% normal saline using Heindolph vortexing machine. Decimal dilutions (10-fold) of the suspensions were plated out on agar medium and incubated at 30°C for 24 h. The colony forming units were afterwards enumerated.

Enumeration of hydrocarbon utilizing bacteria

Hydrocarbon utilizing bacteria (HUB) were enumerated by a method adopted from Hamamura et al. (2006) which involved the dilutions of appropriate sample suspensions and plating out on Bushnell-Haas agar (Sigma-Aldrich, USA). Hydrocarbons were supplied through the vapour phase to putative hydrocarbon utilizers by placing sterile Whatman No.1 filter papers impregnated with 5 ml Okono crude oil on the lids of the inverted plates and incubated for 14 days at 30°C.

Purification and characterization of hydrocarbon utilizing bacteria

Discreet colonies of different HUB were randomly picked using a sterile inoculating wire loop and sub cultured for purification by

Table 1. Physicochemical parameters.

Parameter	Sediment	Soil	Water
pH	7.86	5.40	7.66
Electrical conductivity (µS/cm)	7240	1134	3159
Total nitrogen (mg/kg)	886.25	867.20	672.2
Total phosphorous (mg/kg)	11.2	60.4	4.92
Total organic carbon (%)	2.18	5.3	3.1
Nickel (mg/kg)	41.62	41.5	39.2
Lead (mg/kg)	19.40	17.3	17.3
Zinc (mg/kg)	39.5	36.5	32.4
Salinity (PPT)	NA	NA	19.66
Turbidity (NTU)	NA	NA	250
TPH (mg/kg)	1104	1007	768
PAH (mg/kg)	92.6	53.25	85.56

Total 2. Heterotrophic and hydrocarbon utilizing bacterial counts.

Sample	Mean values of THB	Mean values of HUB
Soil (cfu/g)	1.0×10^7	0.5×10^6
Water (cfu/ml)	3.1×10^6	3.9×10^5
Sediment (cfu/g)	2.7×10^6	1.0×10^6

streaking on nutrient agar plates and incubated at 30°C for 24 h. Individual colonies were predominantly identified using biochemical tests as described in Bergy's Manual for Determinative Bacteriology (Holt et al., 1994).

Degradation screening

Representative HUB isolates were further screened for oil degradation capability under aerobic conditions by inoculating a calibrated loop full of 18 h old culture of each hydrocarbon utilizing bacterium into Bushnell Haas Broth containing 1 ml of Okono medium crude oil. Biodegradation was scored by turbidity and emulsification of oil-in-mineral broth medium after 14 days incubation at 30°C (Kostka et al., 2011).

RESULTS AND DISCUSSION

Soil, sediment and water physicochemical parameters are shown in Table 1. The parameters determined indicated that the samples had been exposed to hydrocarbon contamination with traces of other organic and inorganic contaminants (Chikere, 2010, 2012a).

These pollutants cause damages to humans and the ecosystem if not effectively remediated. The contamination may have resulted in the low pH of 5.40 observed in soil as compared to pristine soil while pH in sediments and water were observed to be 7.86 and 7.66, respectively which is neutral to slightly alkaline. Previous studies have demonstrated that the pH range optimal for biodegradation of hydrocarbons is 6-7 (Eweis et al., 1998; Aparna et al., 2010). Nutrients are very important ingredients for successful biodegradation of hydrocarbon pollutants especially nitrogen, phosphorus and in some cases iron. Addition of nutrients is necessary to enhance biodegradation of crude oil pollutants.

The presence of microbial activity was determined by the enumeration of culturable total heterotrophic bacteria and total hydrocarbon utilizing bacteria as presented in Table 2 and soil sample recorded highest cfu count for THB with a mean value of 1.0×10^7 cfu/g and sediment highest in HUB count with a mean value of 1.0×10^6. A similar observation was reported by Ibiene et al. (2011) and Eze and Okpokwasili (2011).

The high counts recorded in soil sample could be attributed to the myriad of nutrients, high organic matter concentration and other ecological factors that influence the survival of soil microorganisms that play important roles in the decomposition and recycling of nutrients. Continuous input of petroleum-based pollutants usually results in an enriched microbial community capable of surviving toxic contamination. The difference between THB and HUB counts was observed to be minimal/-insignificant which suggests that most of the micro-organisms present in the various sample sites are hydrocarbon degraders that can withstand the concentration of crude oil and also use them as source of carbon.

Totally, 47 pure cultures were able to grow on mineral salt medium (Bushnell Haas Agar) with crude oil as carbon source and were identified using phenotypic and biochemical tests. The population of culturable hydrocarbon degraders from soil, water and sediment samples investigated showed that majority of the bacteria were Gram negative belonging to *Gamma proteobacteria* group, this corroborates with the findings of Kaplan and Kitts (2004) although some Gram positive isolates were also observed. A total of 30 isolates had the ability to utilize crude oil and they belonged to the genera *Bacillus, Proteus, Klebsiella, Pseudomonas, Micrococcus, Serratia, Enterobacter, Flavobacterium, Corynebacterium* and *Azotobacter* as presented in Table 3. The flora represents the normal heterotrophic bacteria present in the various samples.

Majority of these organisms isolated were *Proteus, Enterobacter* and *Pseudomonas*. These predominant genera isolated have been shown to contain high numbers of oil degrading species from oil polluted sites, and this gives evidence that these species are probably the active degraders in this environment. These isolates have also been demonstrated by other researchers to be hydrocarbon degraders (Sarma and Sarma, 2010; Ebrahimi et al., 2012). Watanabe (2001) isolated *Micrococcus, Pseudomonas* and *Bacillus* from sediments as marine petroleum hydrocarbon degraders. Chikere et al. (2009) reported the prevalence of *Flavobacterium, Enterobacter, Norcardia* and *Acinetobacter* in a hydrocarbon

Table 3. Characterization of bacterial isolates.

Isolate code	Gram reaction		Tentative identity	Degradative screening
HUBSED1	+	R	*Corynebacterium* sp.	Y
HUBSED2	+	R	*Corynebacterium* sp.	Y
HUBSED3	-	R	*Serratia* sp.	Y
HUBSED4	-	R	*Pseudomonas* sp.	Y
HUBSED5	-	R	*Proteus sp.*	Y
HUBSED6	-	R	*Flavobacterium* sp.	Y
HUBSED7	-	R	*Proteus* sp.	Y
HUBSO1	-	R	*Proteus* sp.	Y
HUBSO2	-	R	*Enterobacter* sp.	Y
HUBSO3	-	R	*Enterobacter* sp.	Y
HUBSO4	-	R	*Flavobacterium* sp.	Y
HUBSO5	-	R	*Pseudomonas* sp.	Y
HUBSO6	+	R	*Bacillus* sp.	Y
HUBW1	+	C	*Micrococcus* sp.	Y
HUBW2	+	R	*Bacillus* sp.	Y
HUBW3	-	R	*Klebsiella* sp.	Y
HUBW4	-	R	*Enterobacter* sp.	Y
HUBW5	-	R	*Serratia* sp.	Y
HUBW6	-	R	*Proteus* sp.	Y
HUBW7	-	R	*Azotobacter* sp.	Y
HUBW8	+	C	*Micrococcus* sp.	Y

Y = Yes, R = rod, C = cocci.

polluted marine sediment undergoing bioremediation. In a related study conducted by Ibiene et al. (2011) at Aluu and Mogho communities in Port Harcourt, Rivers State, Nigeria, *Micrococcus, Bacillus, Corynebacterium, Vibrio, Pseudomonas* and *Flavobacterium,* were isolated from a contaminated soil undergoing bioremediation by natural attenuation. Obire and Nwanbeta (2002) also reported the isolation of *Serratia, Pseudomonas, Proteus, Klebsiella*, *Microccocus* and *Staphylococcus* species from samples collected from petroleum hydrocarbon contaminated soil in Port Harcourt while Eze and Okpokwasili (2010) isolated *Flavobacterium, Proteus, Bacillus, Klebsiella, Lactobacillus* among other bacteria from Okpoka-Woji river sediment serving as a sink for industrial effluents. Table 3 also indicates that the crude oil degradative ability of the individual bacterial isolates was significant as evidenced by turbidity and emulsification of 1 ml of crude oil in 10 ml of Bushnell-Haas broth after 14 days incubation when compared with the test isolates on day zero incubation as presented in Figure 1. Bacteria that are capable of utilizing hydrocarbons as energy and carbon sources in broth culture have been shown to produce bioemulsifiers or bio-surfactants that assist in the transport of hydrocarbons into the cell via efficient uptake systems (Atlas and Philip, 2005; Olga et al., 2008; Satpute et al., 2010; Cho et al., 2011). Previous studies have demonstrated that the bacterial genera characterized in the present investigation contain known hydrocarbon utilizing species (Chaillan et al., 2004; Brito et al., 2006; Olga et al., 2008; Kadali et al., 2012).

Conclusion

One of the major reasons for prolonged negative impact of oil spill on the environment is probably the absence of adequate and qualitative scientific baseline data. These findings have revealed that there is an appreciable population of indigenous hydrocarbon utilizing bacteria in oil-polluted sites in Bodo which can be monitored and enhanced to increase the bioremediation rate in this chronically oil-impacted area. it is pertinent to study the community dynamics of hydrocarbon degrading bacteria in oil-polluted ecosystems using cultivation-independent 16S rRNA-gene-based and functional-gene-based methods in order to fully undertsand the biochemical reactions that underpin hydrocarbon degradation during bioremediation projects (Chikere, 2013).

Conflict of Interests

The author(s) have not declared any conflict of interests.

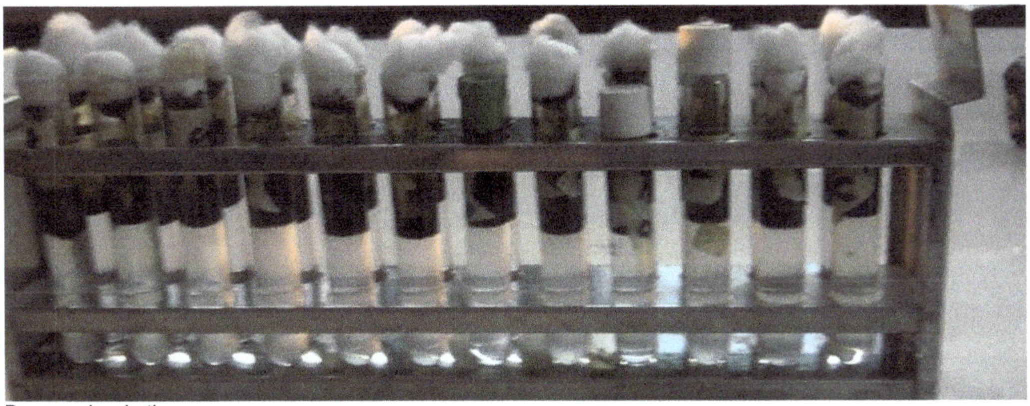

Day zero incubation

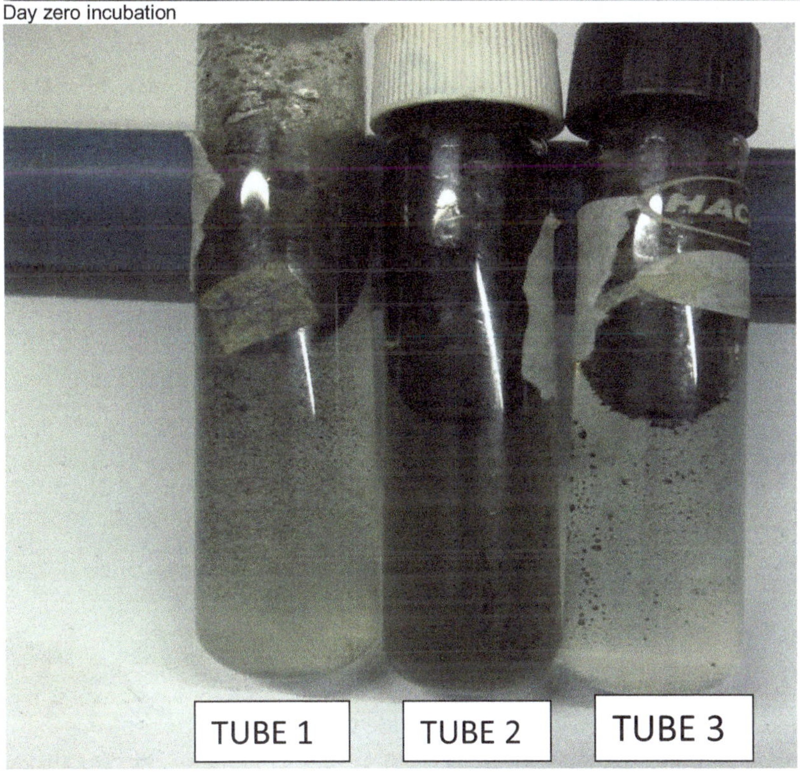

Day 14 incubation

Figure 1. Biodegradation screening of hydrocarbon utilizing bacteria. Day zero incubation of isolates in test tubes in rack above. Day 14 incubation showing crude oil emulsification by specific isolates. Tube 1: Crude oil degradation by HUBSED2 (*Corynebacterium* sp.); Tube 2: Crude oil degradation by HUBSED4 (*Pseudomonas* sp.); Tube 3: crude oil degradation by HUBSO4 (*Flavobacterium* sp.).

REFERENCES

Abu GO, Dike PO (2008). A study of natural attenuation process involved in a microcosm model of a crude oil-impacted wetland sediment in the Niger Delta. Biol. Technol. 4761 - 4767.

Adati AK (2012). Oil exploration and spillage in Niger Delta of Nigeria. Civil. Environ. Res. 2:2222 - 2863.

Akinde SB, Catherine CI, Omokaro O (2012). Alkane degradative potentials of bacteria isolated from deep Atlantic oceans of the Gulf of Guinea. Bioremed. Biodegrad. 3:225 - 328

American Public Health Association (APHA) (2008) Standard Methods for the Examination of Water and Waste Water. 21st ed. Washington DC.

Aparna C, Saitha P, Himabindu V, Alok B, Anjaneyulu Y (2010). Evaluation of bioremediation effectiveness on sediments contaminated with industrial wastes. Int. J. Environ. Sci. 1(4): 607-620.

Atlas RM, Philp J (2005). Bioremediation: Applied Microbial Solutions for Real-World Environmental Cleanup. American Society for Microbiology (ASM) Press, Washington, DC. pp. 78-105.

Brito EMS, Guyoneaud R, Goni-Urriza M, Peyrusr AR, Verbaere A, Crapez MAC, Wasserman JCA, Duran R (2006). Characterization of hydrocarbonoclastic bacterial communities from mangrove sediments in Guanabara Bay, Brazil. Res. Microbiol. 157:752-762.

Chaillan F, Le Fleche A, Bury E, Phantavong Y, Grimont P, Saliot A, Oudot J (2004). Identification and biodegradation potential of tropical aerobic hydrocarbon-degrading microorganisms. Res. Microbiol. 155:587-595.

Chikere CB (2010). Bacterial diversity and community dynamics during the bioremediation of crude oil-polluted soil. Dissertation. Microbiology Department. University of Port Harcourt.

Chikere CB (2012a). Culture-Independent analysis of bacterial community composition during bioremediation of crude oil-polluted soil. British Microbiol. Res. J. 2(3):187-211.

Chikere CB (2013). Application of molecular microbiology techniques in bioremediation of hydrocarbons and other pollutants. Br. Biotechnol. J. 3:90-115.

Chikere CB, Okpokwasili GC Ichiakor O (2009). Characterization of hydrocarbon utilizing bacterial in tropical marine sediments. Afr. J. Biotechnol. 8(11):2541-2544.

Chikere CB, Okpokwasili GC, Chikere BO (2011). Monitoring of microbial hydrocarbon remediation in the soil. 3Biotech. 3:117 - 138.

Chikere CB, Surridge K., Okpokwasili GC, Cloete TE (2012b). Dynamics of indigenous bacterial communities associated with crude oil degradation in soil microcsosms during nutrient enhanced bioremediation. Waste Manag. Res. 30:225-36.

Cho SH, Chae SH, Im WT, Kim SB (2011). *Myroides marinus* sp. nov. a member of the family *Flavobacteriaceae* isolated from seawater. Int. J. Syst. Evol. Microbiol. 61:938-941.

Ebrahimi M, Sarikkhani MR, Fallah R (2012). Assessment of biodegradation efficiency of some isolated bacteria from oil-contaminated sites in solid and liquid media containing oil-compounds. Int. Res. J. Appl. Basic Sci. 3(1):138-147.

Eweis JB, Ergas S J, Chang DPY, Schroeder ED (1998). Bioremediation Principles. McGraw-Hill Companies. Inc. pp. 296

Eze VC, Okpokwasili GC (2010). Microbial and other related changes in Niger Delta River sediment receiving industrial effluents. Cont. J. Microbiol. 4:15-24.

Federal Ministry of Environment (FME) Abuja (2006). Nigerian Conservation Foundation Lagos, WWF UK and CEESP-IUCN

Hamamura N, Olson SH, Ward DM, Inskeep WP (2006). Microbial population dynamics associated with crude oil biodegradation in diverse soils. Appl. Environ. Microbiol. 72:6316 - 6324

Holt JG, Krieg NR, Sneath PHA, Stanley JT, Williams ST (1994). Bergey's Manual of Determinative Bacteriology. 9th Ed. Williams and Wilkins Company, Baltimore, USA. pp. 71-561.

Ibiene AA, Orji FA, Orji-Nwosu EC (2011). Microbial population dynamics in crude oil-polluted soils in the Niger Delta. Nig. J. Agric. Food Environ. 7:8-13.

Kadali KK, Simons KL, Skuza PP, Moore RB, Ball AS (2012). A complementary approach to identifying and accessing the remediation potential of hydrocarbonoclastic bacteria. J. Microbiol. Methods 88:348-355

Kaplan CW, Kitts CL (2004). Bacterial succession in a petroleum land treatment unit. Appl. Environ. Microbiol. 70:1777-1786.

Kostka JE, Prakash O, Overholt WA, Green SJ, Freyer G, Canion A (2011). Hydrocarbon-degrading bacteria and the bacterial community response in Gulf of Mexico beach sands impacted by the deep horizon oil spill. Appl. Environ. Microbiol. 77:7962-74.

Moffat AO, Linden P (2005). Perception and reality. J. Environ. Sci.Technol. 1:283 - 297.

Obire O, Nwanbeta O (2002). Effects of refined petroleum hydrocarbon on soil physicochemical and bacteriological characteristics. J. Appl. Sci. Environ. Manag. 6(1):34 - 44.

Okpokwasili GC (1996) Microbial Degradation of Petroleum Hydrocarbon by Brackish Water Isolates in Nigerian Wetlands. In. The Nigerian Man and the Biosphere, Akapata, T.V.I. and A. Okoli (Eds).

Olga P, Petar K, Jelena M, Srdjan R (2008). Screening method for detection of hydrocarbon oxidizing bacteria in oil-contaminated water and soil specimens. J. Microbiol. Methods 74:110-113

Sarma A, Sarma H (2010). Enhanced biodegradation by some microbial isolates supplemented with heavy metals. Int. J. Bot. 6:441-448

Satpute SK, Banpurkar AG, Dhakephalkar PK, Banat IM, Chopade BA (2010). Methods for investigating biosurfactants and bioemulsifiers: a review. Crit. Rev. Biotechnol. 30:127-144

Stroud JL, Paton GI, Semple, KT (2007). Microbe-aliphatic hydrocarbon interactions in soil; implications for biodegradation and bioremediation. J. Appl. Microbiol. 102:1239 - 1253.

United Nations Environmental Programme (UNEP) 2011. www.unep.org. Accessed 12 June, 2012.

Vincent AO, Felix E, Weltimore MO, Ise-Iyamu OK, Daniel EE (2011). Microbial degradation and its kinetics on crude oil polluted soil. Res. J. Chem. Sci. 1(6):8-14.

Wackett LP, Hershberger CD (2001). Biocatalysis and biodegradation-Microbial transformation of organic compounds. American Society for Microbiology (ASM) Press, Washington DC.

Watanabe K. (2001). Microorganisms relevant to bioremediation. Curr. Opin. Biotechnol. 12:237-241.

Influence of lokpa cattle market wastes on agricultural soil quality

Chinyere G. C, Osuocha K. U. and Imo Chinedu

Department of Biochemistry, Abia State University, Uturu, Nigeria.

This paper examined the influence of Lokpa cattle market waste on soil properties. Soil samples were collected from the Central, 3 and 6 m Northwards, Southwards, Eastwards and Westwards of Lokpa cattle market, Umuneochi Local Government Area of Abia State, Nigeria at a depth of 0 to 25 cm (Sample A) and 26 to 50 cm (Sample B) from each spot while the control samples were collected from an unimpacted area devoid of cattle rearing activities. Results of the physicochemical activities show that cattle waste soil had low acid pH range of 4.02 to 5.83 while soil moisture ranged from 14.90 to 21.58, Organic carbon ranged from 1.92 to 2.83. The enzymatic activities of cattle waste soil were found to be higher ($P<0.05$) than the control. This however could be due to input from the cattle waste leading to increased enzymatic activities. The pH of the cattle waste soils were found to be lower than the control, hence proper care should be taken in the quantity of these waste applied to farms as they can increase soil acidity. The presence of cattle dung could be responsible for the general changes observed.

Key words*:* Cattle, waste, market, soil, quality.

INTRODUCTION

Soil constitutes a dynamic system within which series of changes constantly occur. These changes directly affect the composition, properties and productive potentials of the soil. Oriola and Hammed (2012) reported that soil as a component of landscapes occupies a central position in the landscape balance due to its diverse functions. Soil conditions, constraints on soil quality play an important role on agricultural output and productivity. Livestock production in developing countries has increased rapidly during the last decades (Steinfeld and Chilonda, 2006). In Nigeria and in most developing countries, for animals like cattle and other animals of that type, special markets are kept for them and various activities within the market may affect the soil (Nwaugo et al., 2008). Large quantities of wastes are produced annually in these areas. Oriola and Hammed (2012) reported that the quantity and quality of animal waste are affected by diet composition. These waste materials such as cattle excreta and associated feed losses, wash-water and other materials represent valuable resources that can replace significant amounts of inorganic fertilizers (Leha, 1998).

Animal wastes in the form of manures are valuable

Map 1. Showing sampling location of Umunneochi.

sources of nutrients and organic matter in the maintenance of soil fertility and crop production (Ogbuewu et al., 2012). Tamminga et al. (2000) reported that 55 to 90% of the nitrogen and phosphorous content of animal feed is excreted in faeces and urine, normally used as manure. Animal wastes are routinely applied to cropland to recycle nutrients, build soil quality, and increase crop productivity. Pinamonti and Zorzi (1996) reported that application of feacal waste could be beneficial for soil conservation especially in degraded soils. Animal manures have been effectively used as organic fertilizer. Such manures are valuable fertilizers and soil conditioners when applied under proper conditions at crop nutrient requirements. As reported by Bell (2002), animal wastes contain all essential plant nutrients. This paper examines the effect of Lokpa cattle market wastes on soil physicochemical parameters, soil anions and enzymatic activities and the implication for soil fertility. Findings will assist in adoption of adequate management practices as a panacea for improved utilization of these wastes for sustained agricultural productivity.

MATERIALS AND METHODS

Study area

The study Area Lokpa cattle market is located in Umuneochi Local Government Area of Abia State, Nigeria (Map 1). It is an open surface littered with cow dung, feed fallout and fallen leafs. The market harbours over 48,000 heads of cattle annually as an open feedlot and operates daily throughout the year. Leftover feedstuff and leaves are rarely carted away but left to rot within the market space. The area lies within latitude 055.33° and 06.03° North and longitude 07.10° and 07.29° East.

Soil sample collection

The study area was divided into transects of Central, North, South, East and West. An unpolluted area adjacent to the South was used as control. Samples were collected from the Centre of the market, 30 meters and 60 meters Northwards from the centre. The same was done Southwards, Eastwards and Westwards respectively using plastic auger. Depths of samples collection at each spot were 0 to 25cm and 26 to 50cm, respectively. Samples were transported in plastic bags containing ice packs to the soil Biochemistry laboratory of Abia State University, Uturu for further analysis.

Determination of soil physicochemical parameters

Soil Temperature was determined at the site of soil sample collection using mercury in glass thermometer as described by APHA (1998). Soil pH was measured using fresh soil samples according to the methods described by Bates (1954). Similarly soil moisture content was determined according to the procedure described by APHA (1998). Soil Electrical Conductivity was obtained using conductivity meter in 1:2 soil/water ratio as described by Whitney (1998).

Soil Organic Carbon was obtained according to the procedure presented by Walkely and Black (1934) while soil cation exchange capacity (CEC), sulphur (SO_4^{2-}), phosphate (PO_4^{2-}), nitrate (NO_3) were determined by the method of Dewis and Freitas (1970) and soil calcium carbonate determined by the method of Buuman et al. (1996).

Determination of soil enzymatic activities

The soil enzymes determined were Dehydrogenase, Urease, Hydrogen peroxidase and Alkaline phosphatase. Soil Urease activity was obtained according to the method of Tabatabai and Bremner (1972). Similarly, soil Alkaline Phosphatase, Dehydrogenase and acid Phosphatase activities were determined by the methods described by Tabatabai (1982) while Hydrogen Peroxidase activity was obtained by the method of Alef and Nannipieri (1995).

Statistical analysis

Data collected were subjected to statistical analyses using One Way Analysis of Variance (ANOVA) procedure and difference in means were separated using standard students t - Test. Values were mean ± standard deviation of triplicate determinations. Mean in the same column having different alphabet were statistically significantly ($p<0.05$).

RESULTS AND DISCUSSION

Table 1 and 2 shows the physicochemical properties and soil anions samples analyzed.

Results indicate that cattle waste soils had low acidity than the control. The low acid pH could be due to the presence of cattle dung and urine in the area. This change in pH is due to the release of ammonia following the metabolism of feacal matter which combines with the available moisture to cause change in pH. This agrees with Nwaugo et al. (2008) who observed low acid pH in cattle market waste soil. This pH however is not conducive for plant growth. Joan et al. (2000) reported that cattle manure amendment can increase the pH of soils. However, crop production on acid soil can be improved greatly when soil pH is adjusted to neutrality.

The higher ($P<0.05$) organic carbon content recorded from cattle market waste soils could be due to high input received from cattle waste and feed fallout which are mainly grasses and other weeds; and due to the fact that the initial site preparation does not involve clearing of vegetation hence the organic matter is not lost. This is in consonance with Lakshmikanti and Pramod (2012) who observed high organic carbon in poultry dung amended soils. Olaitan and Lombin (1988) reported that organic matter is a major indicator of soil nutrient due to its colloidal nature. These accumulated organic manures on the surface could be responsible for provision of conducive environment for high enzyme activities.

Jangid et al. (2008) reported that accumulation of organic carbon as a result of manure addition not only results in increased microbial biomass but has also been linked to changes in microbial community, structure and increased functional diversity. Moisture content of cattle waste soil were found to be higher than the control ($P<0.05$). This could be due to the capability of the cattle waste to increase soil infiltration of water. This agrees with Oriola and Hammed (2012) who observed high moisture content in cattle shed soils. Similarly, other nutrients namely calcium carbonate, Phosphate (PO_4^{2-}), Sulphate (SO_4^{2-}) are also higher in the cattle waste soil than the control. This higher values ($P<0.05$) is due to addition through animal manure (Dung and Urine).

Olaitan and Lombin (1988) reported that organic waste is a good and dependable source of these nutrients. Cation Exchange capacity of the cattle waste soil were lower than the control ($P>0.05$). This correlated negatively with the cattle waste soil. This decline could be due to lower concentration of heavy metals in the cattle waste soil since cation exchange capacity is directly related to the capacity of the absorbing metals. This agrees with Oriola and Hammed (2012) who observed that with increase in exchangeable cation, soil acidity may decline. This reduction in cation exchange capacity of the cattle waste soil correlated positively with pH of the control soil.

Table 3 and 4 shows soil enzyme activities of soil samples analyzed. Soil enzymes increase the reaction rate at which plant residues decompose and release plant available nutrients. Soil enzymes play an important role in organic matter decomposition and nutrient cycling. Soil enzymes analysed were found to be higher in the cattle market waste soil than the control. However, the presence of cattle dung which ultimately leads to high microbial activities could be responsible for the general increase in the enzyme activities. This agrees with Oriola and Hammed, (2012) who also observed increased enzyme activities in cattle waste soil. The majority of soil enzymes are extracellular enzymes produced by soil microbes. Similar to microbial biomass, enzymes are very

Table 1. shows soil physicochemical properties of Lokpa cattle market waste soil.

Parameters	Temperature (°C)		pH		Moisture (%)		Organic carbon (%)		C.E.C (cmol/kg)	
Soil Depth	A	B	A	B	A	B	A	B	A	B
Location ↓										
Control	25.01±0.01[a]	26.00±2.00[a]	6.49±0.08[a]	6.23±0.08[a]	11.91±0.01[a]	10.41±0.02[a]	1.29±0.19[a]	0.54±0.39[a]	50.40±0.26[a]	13.20±0.40[a]
C	27.30±0.70[b]	26.20±0.10[a]	5.32±0.38[b]	4.33±0.12[b]	21.58±0.37[b]	11.61±0.03[a]	2.83±0.27[b]	1.42±0.02[b]	17.20±0.26[b]	19.60±0.20[b]
N_1	29.53±0.60[c]	28.00±0.40[c]	5.07±0.03[b]	5.82±0.32[c]	19.47±0.51[c]	12.89±0.07[b]	2.33±0.19[b]	1.51±0.03[b]	9.60±0.02[c]	10.40±0.35[c]
N_2	28.00±1.00[d]	26.40±0.10[a]	4.26±0.28[d]	3.04±0.02[e]	14.81±0.26[d]	11.41±0.01[a]	2.63±0.09[b]	3.51±0.01[d]	14.00±0.44[d]	15.23±0.51[a]
S_1	27.00±0.26[b]	27.00±0.40[b]	4.27±0.22[d]	4.12±0.14[b]	13.72±0.42[d]	15.83±0.034[e]	2.10±0.11[c]	0.94±0.05[a]	12.80±0.20[d]	15.20±0.53[a]
S_2	27.10±0.10[b]	26.33±0.12[a]	4.02±0.01[d]	4.07±0.01[b]	17.63±0.01[f]	15.53±0.01[a]	3.51±0.01[d]	3.22±0.01[d]	12.70±0.20[d]	16.60±0.30[e]
E_1	28.00±0.20[g]	28.33±0.58[c]	4.95±0.13[d]	4.22±0.30[b]	23.42±0.39[g]	10.66±0.42[a]	1.72±0.03[d]	1.62±0.02[b]	23.20±0.20[e]	24.67±0.23[f]
E_2	27.00±0.40[b]	27.00±1.00[b]	4.12±0.01[d]	4.00±0.10[b]	17.81±0.17[f]	11.49±0.05[a]	2.71±0.01[h]	0.18±0.02[a]	11.20±0.36[b]	11.60±0.20[a]
W_1	29.0±1.11[i]	28.80±0.10[c]	4.30±0.01[d]	5.91±0.01[c]	10.88±0.05[e]	8.12±0.02[d]	1.92±0.10[c]	1.68±0.01[b]	26.40±0.36[f]	35.17±0.15[g]
W_2	27.00±0.10[b]	28.00±1.00[c]	4.20±0.01[d]	5.69±0.23[c]	14.90±0.05[d]	10.57±0.31[a]	2.92±0.16[b]	1.74±0.02[h]	21.30±0.60[e]	22.20±0.10[f]

Values are mean ± standard deviation of triplicate determinations. Mean in the same column, having different alphabet are statistically significant ($P<0.05$).
N/B: A = 0-25cm soil depth, B = 26 – 50cm soil depth. N, S, E, and W are North, South, East, and West from discharge point C.

Table 2. Soil anions of Lokpa market waste soil.

Parameters	Calcium carbonate (mg/kg)		PO_4^{3-} (mg/kg)		SO_4^{2-} (mg/kg)		Cl^- (mg/kg)	
Soil depth	A	B	A	B	A	B	A	B
Location ↓								
Control	24.71±0.30[a]	31.22± 0.11[a]	29.50±1.00[a]	27.12±0.01[a]	26.51±0.17[a]	25.90±1.08[a]	163.37±4.37[a]	106.50±2.00[a]
C	23.50±0.31[b]	25.20± 0.02[b]	46.40±0.10[c]	39.13±0.03[b]	1.59±0.62[b]	1.20±0.16[b]	14.20±0.10[b]	14.47±0.31[b]
N_1	21.89±0.32[c]	28.22± 0.26[c]	40.13±0.01[b]	37.29±0.23[f]	0.28±0.08[c]	0.72±0.05[c]	23.27±0.15[c]	85.20±0.60[c]
N_2	26.20±0.27[d]	27.14± 0.02[d]	48.97±0.85[c]	29.61±0.01[d]	1.57±0.30[a]	1.54±0.05[d]	28.40±4.84[f]	49.20±0.20[d]
S_1	24.45±0.09[a]	28.95± 0.10[c]	35.04±0.03[d]	55.62±0.38[c]	0.49±0.04[d]	0.32±0.04[e]	21.30±0.56[e]	63.90±0.61[e]
S_2	25.01±0.01[a]	26.59± 0.06[d]	46.23±0.15[c]	44.26±0.02[f]	0.32±0.01[d]	0.73±0.01[c]	21.50±0.10[e]	51.13±0.15[f]
E_1	27.15±0.14[d]	30.47± 0.38[a]	56.32±0.03[e]	64.44±0.05[g]	1.09±0.38[b]	0.18±0.01[f]	14.20±0.79[b]	42.52±1.99[d]
E_2	22.92±0.03[b]	29.00± 0.13[c]	33.58±0.46[f]	53.14±0.37[c]	0.97±0.07[b]	0.13±0.01[f]	21.30±0.20[e]	78.10±1.13[h]
W_1	23.30±0.29[b]	25.12± 0.01[b]	52.96±0.01[g]	35.47±0.34[i]	0.56±0.02[d]	0.52±0.02[c]	35.50±0.36[d]	60.20±0.20[e]
W_2	21.47±0.32[c]	24.45± 0.30[e]	39.42±0.71[b]	41.02±0.03[j]	1.04±0.02[b]	0.40±0.06[e]	30.40±0.20[f]	85.20±0.90[c]

Values are mean ± standard deviation of triplicate determinations. Mean in the same column, having different alphabet are statistically significant ($P<0.05$).
N/B: A = 0-25cm soil depth, B = 26 – 50cm soil depth. N, S, E, and W are North, South, East, and West from discharge point C.

Table 3. Soil dehydrogenase, acid and alkaline phosphate activities of Lokpa market waste soil.

Parameters	Alkaline phosphatase mg/g/h		Acid phosphatase mg/g/h		Dehydrogenase $mgTPFg^{-1}$ dry-soil $6h^{-1}$	
Soil Depth (cm^2)	**A**	**B**	**A**	**B**	**A**	**B**
Location ↓						
Control	8.69 ± 0.036^{a}	6.72 ± 0.026^{a}	3.71 ± 0.02^{a}	1.08 ± 0.01^{a}	0.34 ± 0.02^{a}	0.36 ± 0.02^{a}
C	7.69 ± 0.060^{b}	6.81 ± 0.015^{b}	14.86 ± 0.06^{b}	8.03 ± 0.02^{b}	0.42 ± 0.01^{b}	0.40 ± 0.01^{b}
N_1	6.90 ± 0.026^{c}	6.24 ± 0.01^{b}	9.65 ± 0.00^{c}	10.40 ± 0.02^{c}	0.88 ± 0.01^{c}	0.32 ± 0.03^{c}
N_2	5.53 ± 0.030^{d}	5.40 ± 0.01^{c}	7.44 ± 0.02^{d}	6.07 ± 0.02^{d}	0.40 ± 0.01^{b}	0.37 ± 0.02^{b}
S_1	6.24 ± 0.053^{c}	6.30 ± 0.08^{b}	17.08 ± 0.02^{e}	3.71 ± 0.00^{e}	0.72 ± 0.01^{d}	0.34 ± 0.04^{c}
S_2	6.18 ± 0.010^{c}	5.27 ± 0.03^{c}	10.04 ± 0.02^{f}	7.04 ± 0.04^{f}	0.77 ± 0.01^{d}	0.53 ± 0.01^{d}
E_1	5.68 ± 0.010^{d}	4.27 ± 0.00^{d}	7.43 ± 0.06^{d}	5.20 ± 0.01^{g}	0.45 ± 0.05^{b}	0.53 ± 0.0^{d}
E_2	5.68 ± 0.030^{d}	7.46 ± 0.05^{e}	5.94 ± 0.03^{e}	9.65 ± 0.02^{h}	0.40 ± 0.01^{b}	0.74 ± 0.01^{e}
W_1	6.09 ± 0.020^{c}	6.02 ± 0.03^{b}	8.01 ± 0.01^{f}	6.05 ± 0.02^{d}	0.87 ± 0.01^{c}	0.77 ± 0.01^{e}
W_2	7.28 ± 0.026^{b}	8.39 ± 0.07^{f}	7.43 ± 0.06^{d}	29.72 ± 0.06^{j}	0.64 ± 0.01^{e}	0.41 ± 0.01^{b}

Values are mean ± standard deviation of triplicate determinations.Mean in the same column, having different alphabet are statistically significant ($P<0.05$). N/B: A = 0-25cm soil depth, B = 26 – 50cm soil depth. N, S, E, and W are North, South, East, and West from discharge point C.

Table 4. Soil Urease and Hydrogen peroxidase activities of Lokpa market waste soil.

Parameter	Urease $mgNH_3$-Ng^{-1}drysoil		Hydrogen peroxidase Ml $0.1Ml^{-1}kmno_4g^{-1}$	
Soil depth(cm^2)	**A**	**B**	**A**	**B**
Location↓				
Control	124.24 ± 3.29^{a}	108.76 ± 0.82^{a}	0.0019 ± 0.0002^{a}	0.0018 ± 0.00020^{a}
C	101.12 ± 0.59^{b}	101.55 ± 0.94^{a}	0.0013 ± 0.0087^{a}	0.0040 ± 0.0010^{b}
N_1	107.70 ± 2.67^{b}	180.39 ± 0.51^{c}	0.0016 ± 0.0012^{a}	0.0023 ± 0.0000^{c}
N_2	166.61 ± 0.65^{c}	156.97 ± 0.01^{d}	0.0018 ± 0.00079^{a}	0.0090 ± 0.0000^{d}
S_1	153.85 ± 0.10^{d}	268.73 ± 5.49^{e}	0.0029 ± 0.00036^{b}	0.0022 ± 0.00030^{c}
S_2	164.81 ± 0.02^{c}	154.01 ± 0.01^{d}	0.0023 ± 0.00010^{b}	0.0019 ± 0.00010^{a}
E_1	190.91 ± 0.07^{e}	120.93 ± 0.05^{e}	0.0032 ± 0.00031^{c}	0.0028 ± 0.00044^{c}
E_2	187.71 ± 0.33^{e}	384.44 ± 29.98^{f}	0.0027 ± 0.00017^{b}	0.0024 ± 0.0002^{c}
W_1	141.30 ± 1.18^{f}	132.65 ± 0.01^{b}	0.0021 ± 0.60053^{b}	0.0014 ± 0.00015^{a}
W_2	186.75 ± 0.02^{e}	125.13 ± 0.105^{e}	0.0029 ± 0.00036^{b}	0.0025 ± 0.00052^{c}

Values are mean ± standard deviation of triplicate determinations. Mean in the same column, having different alphabet are statistically significant ($P<0.05$). N/B: A = 0-25cm soil depth, B = 26 – 50cm soil depth. N, S, E, and W are North, South, East, and West from discharge point C.

responsive to manure availability.

Conclusion

Recycling of these wastes via land application could lead to improvement in agricultural soil properties. Since high concentration of chemical elements, enzyme activities and organic matter determine fertility status of soil, it thus implies that these waste can be used as fertilizers for increased agricultural productivity. The pH of the analyzed soil sample is not conducive for plant growth hence, care should be taken in the quantity of these manure applied to crops as it can increase the pH of soil and have adverse effect on the plants.

REFERENCES

Alef K, Namiperi P (1995). Methods in applied soil Microbiology and Biochemistry. 3rd edition. Academic Press, London. 345-353.

APHA (1998) Standard methods for examination of water works and waste water. American public Health Association, American water works Association and water pollution control federation, 20th edition, Washington DC, USA.

Bates RG (1954). Electronic pH determination. John Wiley and Sons Inc. New York.

Bell DD (2002). Waste management. In: Chicken meat and egg production, 5th edition. Kluwer Academic Publisher, Massachusetts.

Buuman PB, Van L, Vetthorst EJ (1996). Manual for soil and water analysis. Backleys Publishers, Laiden, The Netherlands. 13-17.

Dewis J, Freitas F (1970). Physical and chemical methods of soil and water analysis. Soil Bulletin.

Jangid K, Williams MA, Franzluebbers AJ, Sanderlin JS, Reeves JH, Jenkins MB, Endale DM, Coleman DC, Whitman W.B. (2008). Relative impacts of land-use management intensity and fertilization upon soil microbial community structure in agricultural systems. Soil Biology and Biochemistry. 40: 2843-2853.

Joan KW, Chichang C, George WC, Janna PC (2000).Cattle manure Amendment of soil. American Journal of Soil Science. 64:51-64.

Lakshmikanti B, Pramod CM (2012). Changes in bacterial density, CO_2 evolution and enzyme activities in poultry dung amended soil. Open J. Soil Sci. 2:196-201.

Leha C (1998). Biodigester effluent versus manure, from pigs or cattle, as fertilizer for duckweed livestock res. and rural development: http://www.cipav.org.co/lrrd

Nwaugo VO, Etok CA, Obiekezie SO, Chinyere GC (2008). Evaluation of the effect of Okigwe cattle market wastes on the surrounding agricultural soil parameters. Bio-Research. 6(11):367-370.

Ogbuewu IP, Odoemenam VU, Omede AA, Durunna CS, Emenalom OO, Uchegbu MC, Okoli IC, Iloeje MU (2012). Livestock waste and its impact on the environment. Sci. J. Review. 1(2):17-32

Olaitan, Lombin (1988). Best management practices for poultry manure utilization that enhance agricultural productivity and reduce pollution. Hatfield and B.A Stewart (ed). Ann Arbor Press. 89-123.

Oriola EO, Hammed AT (2012). The influence of cattle wastes on degraded savanna soils of Kwara State Nigeria. Ethiopian Journal of Environmental Studies and Management 5(3):268-275

Pinamonti F, Zorzi G (1996). Experiences of compost use in agriculture and in land reclamation projects. In: De Bertoldi, Editions. U.K 517-527.

Steinfeld H, Chilonda P (2006). Old players, new players, livestock Report, Animal production and health division, Food and Agriculture Organisation of the United Nations, Rome, Italy.

Tabatabai MA (1982). Soil enzymes. In: page Al, methods of soil analysis, part 2.

Tabatabai MA and Bremmeo JM (1972). Use of P-nitrophenyl phosphate for assay of soil phosphate activity. Soil Biology and Biochemistry. 1:301-307.

Tamminga S, Jongbloed AW, Van-Eerdt MM, Aarts HFM, Mandersloot F, Hooggervorst, NJP, Westhoek H (2000). Standard for the excretion of nitrogen by farm animals. Rapport ID-Lelystad editions 71.

Walkely AJ, Black I.A (1934). Estimation of soil organic carbon by the chronic acid titration method. Soil science. *37:29-38*

Whitney DA (1998). Soil salinity. (in JR. Brown Ed.). Recommended chemical soil test procedures for the North Central Region. North central research publications. No 221 (Revised) Missiouri Agricultural Experiment Station and SB wool 1001:59-60.

A simple anaerobic system for onsite treatment of domestic wastewater

Lohani S. P.[1,2] **, Chhetri A.**[2] **and Khanal S. N.**[2]

[1]Department of Mechanical Engineering, Kathmandu University, Nepal
[2]Department of Environment Science and Engineering, Kathmandu University, Nepal POST BOX: 6250, Nepal.

The use of anaerobic process for domestic wastewater treatment would achieve lower carbon footprint as it eliminates aeration and generate methane. Among several anaerobic treatment processes, high rate anaerobic digesters receive great attention due to its high loading capacity and chemical oxygen demand removal rate. Up-flow anaerobic sludge blanket reactor (UASB) is getting wide acceptance among several anaerobic processes. However, its application is still limited to industrial wastewater treatment. There are some unresolved problems to be accepted it in developing countries with lower temperature conditions for domestic wastewater treatment. Several studies have been carried out for the performance of UASB reactor but there is still lack of updated information especially on the issue of low temperature domestic wastewater treatment. Considering the gravity of the issue, an attempt have been made to compile updated information so as to help engineers, researchers and practitioners in the selection of reactors and future prospects of research.

Key words: Wastewater, anaerobic, up-flow anaerobic sludge blanket reactor (UASB), carbon footprint.

INTRODUCTION

Domestic wastewater refers to the wastewater from toilet, bathroom and kitchen of household. Anaerobic treatment of organic material proceeds in the absence of oxygen and the presence of anaerobic microorganisms. It is the consequence of a series of metabolic interactions among various groups of microorganisms which ultimately converts waste into a renewable energy, that is, biogas along with other useful byproducts.

Anaerobic wastewater treatment system is considered to be a sustainable and suitable process for on-site (individual or cluster) treatment due to its low or no energy consumption, low space requirement and relatively simple reactor design (Zeeman et al., 2000). This is partly explained by the elimination of aeration.

The important advantages of anaerobic wastewater treatment are production of methane and bio-fertilizer, reduce carbon dioxide (CO_2) emission and more importantly improve sanitation and public health, while the excess sludge production is low as compared to aerobic processes. Also, the excess sludge is highly stabilized and generally its dewaterability is excellent which eliminates the costs for aeration (Lettinga, 1996;

Hammes et al., 2000; Luostarinen and Rintala, 2005).

A recent study in Nepal shows that an estimated biogas production potential from urban wastewater is about 22 million m^3/year equivalent to 14.5 MWth, at the sametime, the carbon dioxide (CO_2) emission reduction potential is about 2460 tons per year (Lohani et al., 2013). This may directly relate with financial benefits, however, it also has several economic benefits for instance reduction of indoor air pollution, improvement in public health due to improved sanitation and greenhouse gas reduction. This indicates that anaerobic treatment of wastewater could be a sustainable solution due to its vast economic and environment benefits for developing countries like Nepal. Despite enormous benefits, its use is usually limited to high strength industrial wastewater with soluble substrates (Lew et al., 2011).

Domestic wastewater has generally low concentration of chemical oxygen demand (COD) and relatively high concentration of suspended solids with low specific methane yield that requires initial hydrolysis to convert the suspended solids into soluble substrate. Hydrolysis is often the limiting step, especially at low temperature conditions (Lew et al., 2011). Hence, operation of anaerobic reactors for domestic wastewater with a high content of suspended solids is therefore, only feasible at higher ambient temperature or with external heat supply. Nevertheless, standard septic tanks are useful for removal of inert solids, preliminary hydrolysis of particular organic matter, though optimized fermentation and hydrolysis depends on its design (Richard et al., 2005). Therefore, combination of a septic tank, serving as the primary treatment step, combined with a pulse fed UASB reactor as secondary treatment could be a suitable low cost and effective onsite sanitation option. Pulse feeding UASB was found to be simple and effective (Nadais et al., 2005). In addition, it stimulates the development of granular sludge with both improved settling and degradation properties (Franco et al., 2003; Rocktäschel et al., 2013).

For the purpose of anaerobic treatment of domestic wastewater, various technologies have been developed which are core sustainable wastewater treatment (Mahmoud et al., 2003; Luostarinen et al., 2007). Recent reviews mostly focus on anaerobic digestion (Aiyuk et al., 2004; Gomec, 2010) and enhance the start-up and granulation in UASB reactors (Chong et al., 2012). However, there is still lack of information and documentation on the enhancement of treatment efficiency and energy recovery from high rate anaerobic digestion treating domestic wastewater at low temperature climatic condition. Hence, the goal of this study is to compile up to date information of commonly used anaerobic systems with focus on the UASB and septic tank-upflow anaerobic sludge blanket (ST-UASB) reactors, which may serve as background information for perspective research direction.

The information regarding UASB reactor pilot/field operation is given emphasis so that future improvement on reactor process control would be possible.

TECHNOLOGIES FOR ANAEROBIC TREATMENT

Two types of digesters are used in the anaerobic digestion process which are batch and continuous process. In batch process, all the constituent is put in the reactor in the beginning and is removed completely after the reaction is complete, whereas in continuous process fresh influent is fed into the reactor and effluent is regularly drawn out (Ostrem et al., 2004). Most commonly continuous type reactors are used for domestic wastewater treatment.

Septic tank and soak-away system

The most commonly used anaerobic system for pre-treatment of domestic wastewater in on-site application is septic tank. The conventional septic tank is the oldest anaerobic treatment system still widely employed in developing countries and isolated residential locations in developed countries (Coelho et al., 2003). Septic tank has simple design of horizontal flow mode with one-two baffles between the inlet and outlet. But there are certain problems like short circulation in the tank and dissolved oxygen input through inlet (Arceivala and Asolekar, 2007). Moreover, performance of the septic tanks is rather poor due to the horizontal flow mode of the influent sewage despite its lengthy hydraulic retention time (HRT) (Mgana, 2003). About 30-50% COD and about 60% total suspended solids (TSS) removal can be achieved in septic tank treatment system. Thus, septic tank effluent requires further processing in a post-treatment system to meet environmental standards, which would increase the cost and complexity of the system.

Soak-away system is another treatment process in which dispersal of wastewater is done through soil purification process and ground water recharge. As it is basically used for post treatment of septic tank effluent, the performance is dependent on the treatment efficiencies of the septic tank, wastewater distribution and loading to the soil infiltrative surface (Foxon et al., 2004).

Among the anaerobic digesters, high-rate digesters are popularly used in sewage treatment. This is because, unlike the conventional low-rate anaerobic digesters such as anaerobic ponds and septic tanks, high-rate anaerobic reactors are designed to operate at short HRT and long solids retention times (SRT) to incorporate large amounts of high-activity biomass, thus allowing improved sludge stabilization and higher loading capacity (Sperling and et al., 2001). For the treatment process to be considered as high rate, two conditions must be fulfilled which are to retain high concentration of sludge under high loading rate and ensure proper contact between wastewater and retained sludge (Lettinga et al., 1987). Some of the examples of high rate digesters are anaerobic baffle reactor, upflow anaerobic sludge blanket reactor and expanded granular sludge bed reactor as described below.

Anaerobic baffle reactor

Anaerobic baffle reactor (ABR) is a reactor which uses a series of baffles in which the water is forced to rise upward and simultaneously drop downward which ensures good solid retention and more contact between biomass and organic substrate therefore achieving good organic removal rates (Barber and Stuckey, 1999). This technology has been used in the treatment of a variety of wastewater types; however most studies were focused on high strength wastewater with soluble biodegradable material (Barber and Stuckey, 1999). Studies on domestic wastewater have been limited to laboratory scale using synthetic wastewater, or at full scale with limited investigation of internal reactor dynamics (Singh and Viraraghavan, 1998; Foxon, 2004). The ABR functioned as solids retention device where particulate material retained through settling in the first compartment, forming a gel-like matrix and anaerobic conversion to CH_4 and CO_2 reduced the amount of solid (Foxon, 2004). For the smooth operation of ABR, initial loading rates should be kept low to prevent overload of slow growing micro-organisms along with gas and liquid up-flow velocities to encourage flocculent and granular growth (Barber and Stuckey, 1999). According to Henze and Harremoes (1983) approximately 1.2 kg COD/m^{3}d is recommended for start-up period. But for the treatment of dilute wastewater, it is recommended to start-up with high biomass concentration (> 3 gVSS/l) so that sufficiently high sludge blanket and improved gas mixing could be obtained in short time (Barber and Stuckey, 1999).

Expanded granular sludge bed (EGSB) reactor

EGSB is the latest amongst high-rate anaerobic treatment systems which has also become increasingly popular, mainly because of very high loading potential (Lettinga, 2001) with high superficial velocity of 5-10 m/h (Lettinga, 1995). The high surface velocities of the liquid in the reactor are achieved through the application of a high effluent recirculation rate, combined with the use of reactors with a large height/diameter ratio of 20 or more (Kato, 1994; Lettinga, 1995). A practical case of high up-flow velocity in EGSB reactor results in a significant reduction in the area required and this is interesting in the case of treatment of soluble effluents from industries.

The EGSB process use granular anaerobic biomass and have the operation principles as UASB, but differ in terms of geometry and process parameters (Zoutberg and Eker, 1997). In the EGSB process, granular biomass is expanded by the high gas and liquid up flow velocities and hence only granulated sludge is retained at the bottom of the reactor whereas, significant amount of sludge remains in fluidized state at higher levels (Parawira, 2004). Small granule and dispersed biomass are washed out leading to retaining of mainly well structured granular sludge. Advantages like higher OLR up to 40 kgCOD/m^3 is achieved depending on the type of wastewater but drawbacks like granule disintegration, washout of hollow granules and occurrence of fluffy granules are prominent in this reactor (Parawira, 2004). However, in one such study, use of tall reactors with effluent recirculation resulted in EGSB reactor with high superficial velocity and granular sludge bed expansion eliminating dead zones and better sludge-wastewater contact (Man et al., 1986). This prevented the accumulation of excess flocculent between sludge and sludge granules (Van der Last and Lettinga, 1992) with efficient removal of soluble pollutants.

Up-flow anaerobic sludge blanket (UASB) reactor

The UASB reactor was developed by Gazte Lettinga in 1972 during a laboratory study on anaerobic treatment of beet sugar wastewater (Lin and Yang, 1991). Among all high-rate anaerobic digesters UASB reactors have been most widely used (Chong et al., 2012). It is a reactor which operates without any power requirement and is like an inverted septic tank which is more efficient than conventional septic tanks for removal of pollutants and production of biogas (Arceivala and Asolekar, 2007). This has rather become successful due to the presence of solid liquid gas (SLG) seperator on the top for the prevention of solid loss and absence of fixed bed which otherwise would have induced clogging (Foresti et al., 2006). The critical elements of UASB reactor are the influent distribution system, gas-solid separator and effluent withdrawal design (Metcalf and Eddy, 2003). Lin and Yang (1991) concluded UASB to be better performing than other advanced anaerobic systems in terms of having high specific activity (rate of reaction), handling high organic rate, retaining high concentration of biomass due to relatively slow upward flow velocity and good COD removal. The key feature of the reactor that makes it popular high rate anaerobic digester (especially in tropical countries) is the availability of granular or flocculent sludge due to its upflow mode of operation, high COD removal efficiency and the ability to apply high volumetric COD loading rate as compared to other anaerobic processes (Crites and Tchobanoglous, 1998). Due to the granulation and blanketing in a UASB reactor, the solid and hydraulic retention time can be independent as a result, the reduction in treatment time from days to hour is possible (Hickey et al., 1991). The UASB reactor eliminates the need of mixing as rising gas bubbles and upflow mode of operation is enough to provide required mixing, this lowers energy demand for the plant operation.

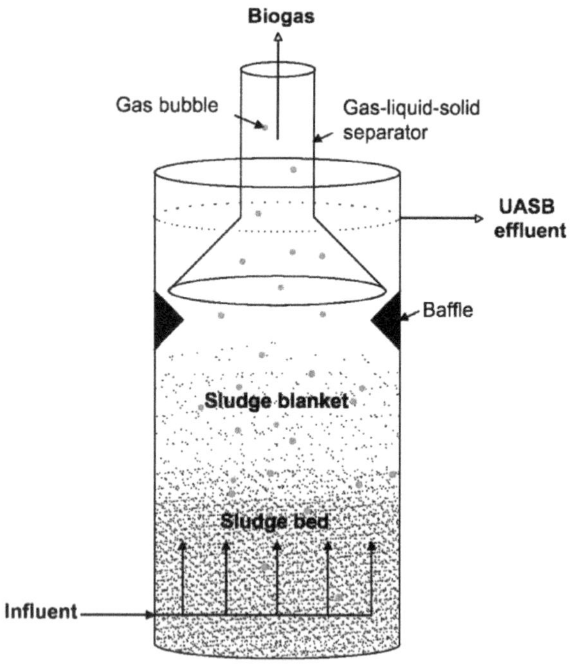

Figure 1. Schematic representation of UASB reactor (Chong et al., 2012).

Working of UASB

A UASB reactor is divided into four components which are sludge bed, sludge blanket, gas-solid separator and settlement compartment as shown in Figure 1. The biomass present at the bottom of the tank is the sludge bed whereas sludge blanket lies above the sludge bed where the suspended particles are found. Gas-solid separator is present at the top for the gas and solids separation and it helps for scum removal as well. The settlement compartment is where settlement of solid into the sludge blanket occurs (Lin and Yang, 1991). The UASB reactor is initially seeded with inoculums such as digested, anaerobic, granular, flocculent and activated sludge. Wastewater is passed from the bottom of the reactor which comes in contact with the inoculums and the biological reaction occurs throughout the sludge bed and sludge blanket (Chong et al., 2012). The baffles prevent wash-out of the viable bacterial matter or floating granular sludge by sliding the settled solids back to the reaction zone. A quiescent zone is created in the settlement zone where heavier sludge settles down and the light and dispersed ones are washed out as effluent. During the upward movement of wastewater, the soluble organic compounds are converted to biogas bubbles consisting of mainly methane and carbon dioxide which are separated by GLS.

Studies show that internal mixing is not sufficient in a pilot-scale UASB reactor treating sewage at temperatures ranging from 4 to 20°C as it may produce dead space in the reactor (Man et al., 1986). Better influent distribution is then required which can be attained by different inlet devices, more inlet feed or high superficial velocities.

Start-up and granulation in UASB

UASB start-up is a time-consuming and delicate process as it begins from the initial feeding to the stage until when stable sludge preferably granular sludge is obtained (Ling and Yang, 1991). This initial start-up is responsible for the overall effectiveness and stability of a UASB reactor which is affected by numerous physical, chemical and biological parameters (Ghangrekar et al., 1996), for instance wastewater constituents, operating conditions, availability and growth of active microbial populations in the inoculum (Chong et al., 2012). Experiences show that the main drawback of UASB reactor is start-up process. It takes long start-up time and is mainly susceptible to temperature and organic loading rates (Lew et al., 2011). The start-up period can take from 2 to several months (Vlyssides et al., 2008) and therefore is a major disadvantage for its applications. An inoculation with seed sludge helps reduce the acclimatization period required before applying the designed organic loading rates. Although a UASB reactor can provide efficient performance without granules, granules formation during start-up would be an advantage for reducing start-up period and has given importance in it (Hulshoff Pol et al., 2004; Liu et al., 2002). Furthermore, the use of septic tank as preliminary treatment step combined with a pulse feed UASB as secondary treatment could shorten the start-up period reducing effect of temperature and granular formation (Richard et al., 2005; Franco et al., 2003; Rocktäschel et al., 2013).

Digested sewage sludge is commonly used as inoculum while other types have also been used (Lin and Yang, 1991). Although, there are many advantages of using UASB, certain processes like interference of particles with flocculation, dead space, bed disruption due to vigorous gas production and speedy inflow rate could hamper the working of UASB (Parawira, 2004).

DOMESTIC WASTEWATER TREATMENT IN UASB REACTOR

The use of UASB reactor for domestic wastewater treatment is usually limited to tropical regions because of low COD concentration and relatively high particulate matter concentration (Lew et al., 2011; Sperling and Oliveira, 2009; Khan et al., 2011?). This needs an initial pre-treatment (hydrolysis) step to convert particulate matter into soluble substrate particularly at low temperature condition which otherwise accumulate in UASB reactor and inhibit anaerobic process. Few studies from various authors on application of UASB reactor for domestic/municipal wastewater treatment are shown in Table 1.

Table 1. Summary of UASB performance results at varied operating conditions.

Temperature (°C)	HRT (h)	OLR (kg COD/m^{3}d)	TSS, removal efficiency (%)	BOD, removal efficiency (%)	COD, removal efficiency (%)	Reference
30	8	1.2	-	-	85	Behling et al. (1997)
20	10-48	-	-	-	60-75	Singh and Viraraghavan (1998)
13-15	10-11	-	75-85	63-73	54-58	Alvarez et al. (2006)
30-35	10	2.7-3.3	-	-	54-72	Mahmoud (2008)
24	48-96	-	56-58	-	81-82	Shayad et al. (2008)
17	48-96	-	51-54	-	74-78	Jamal and Mahmoud (2009)
10-28	6	-	-	-	42-78	Lew et al. (2011)
25-30	9	-	-	77-83	79-81	Rizvi et al. (2014)
8-40	8	0.6- 6.4	65-85	65-85	65-85	Khan et al. (2015)

Behling et al. (1997) examined a pilot scale UASB reactor of 55.5 L capacity for 200 day trial. The equalization tank maintained the OLR 1.21 kgCOD/(m^3.d) with 7.2 l/h flow rate and 7.6 h HRT to feed UASB reactor. The digestion temperature was constant at 30°C. The reactor was fed 10% of the reactor volume (5.5 L) with granulated sludge as inoculum. Results show that the gas production increased steadily until it reached a constant value of about 28 l/d (0.34 m^3 methane per Kg COD removal) and the concentration of effluent COD gradually decreased from 1206 to 60 mg/l (85% COD removal efficiency) which took 60-70 days of plant operation that is when a digestion process is stabilized.

Singh and Viraraghavan (1998) studied start-up and operation of UASB reactors at 20°C for municipal wastewater treatment. They worked on two laboratory scale UASB reactors of 8 L capacity (1 m height and 10 cm internal diameter). Above 50% (4.5 L) digested sludge was seeded as inoculum in both reactors and was operated in a continuous mode. The HRT of the reactor was reduced from 48 to 10 h in about 280 days of operation. The COD concentration of wastewater was about 350-500 mg/l COD (40-50% insoluble) which in initial month of operation COD removal efficiency was in the range of 30-55%. After steady state, the average COD removal efficiency was in the range of 60-75%. But as the insoluble fraction of COD was high in the influent, it appeared that a significant amount of COD removal was due to entrapment of suspended solids in the sludge bed resulting in a significant increase in volatile suspended solids to suspended solids (VSS/SS) ration from 0.5 to 0.8. The high accumulation of particulate matter in the sludge bed seemed to be fine initially because of the fair removal of pollutants but in the long run it, creates disturbance and inhibition of anaerobic digestion in the reactor resulting in diminishing performance of the reactor.

Alvarez et al. (2006) also studied on start-up alternatives and performance of an UASB pilot plant treating diluted municipal wastewater at low temperature. His reactor was a metallic cylinder of total volume 34.9 m^3 (7.1 m height and 2.5 m diameter) and active volume of 25.5 m^3. Three different start-up procedures of an UASB digester were carried out in which start-up A was without inoculum, B was done with inoculation of 10 m^3 digested sludge while C used sludge developed in 50 days of operation of B type digester as its inoculum. Start-up of digester without inoculums (start-up A) was for about 120 day. However, it reached 75-85% TSS removal, 54-58% total chemical oxygen demand (CODt) removal and 63-73% biological oxygen demand (BOD_5) removal at influent concentrations of 240-340 mgCODt/l at temperatures of 13.5-15°C and hydraulic retention times (HRT) of 10-11 h. While in experiment B, start-up period was 75 days, digester efficiencies were 58, 41 and 54% for TSS, CODt and BOD_5 removal, respectively, working at 169 mg CODt/l input at temperature of 14°C and HRT of 11 h. The sludge bed developed and stabilized quickly while using a hydraulically adapted inoculum in experiment C, but CODt and BOD_5 removals remained as low as 43 and 46%, respectively with volatile fatty acids (VFA) accumulation in the effluent.

Luostarinen et al. (2007) studied three pilot scales UASB–septic tank (modified septic tank) reactor of 1.2, 0.2 and 0.2 m^3 capacity used for treating blackwater. The first reactor was inoculated with 100 L granulated methanogenic sludge and operated for 13 years. The flow of the system varied from 1-2 persons for the first year of operation and 3-4 persons in the 13th year. The second reactor was inoculated with 80L sludge obtained from 1.2 m^3 UASB system and maintained at 15°C whereas the third reactor was not inoculated and maintained at 20°C. The performance of the 1.2 m^3 UASB-septic tank was monitored for 52 weeks during the 1st year and for 13 weeks during the 13th year of operation, while others two 0.2m^3 reactor were monitored for 51 (15°C) and 47 weeks (25°C) from the start-up. Comparisons between the gas production in warm and cold season of these reactors showed that higher conversion of COD into methane was in warm season. However, most of the COD washed out as dissolved COD and suspended COD in warm season had been settled in sludge bed in the cold period. This is because with the increase in temperature, hydrolysis of accumulated solids apparently started and dissolved compounds were formed.

Mahmoud (2008) studied high strength sewage treatment in a UASB reactor and an integrated UASB digester system of working volume of 140 and 106 L, respectively. The one-stage UASB reactor was operated at 10 h HRT at ambient temperature for more than a year in order to assess the system response to the Mediterranean climatic seasonal temperature fluctuation. Afterwards, the one-stage UASB reactor was modified to a UASB-digester system by incorporating a continuous stirred tank reactor (CSTR) digester operated at 35°C with continuous mixing at 60 rpm. Pre-treated influent (screens and grit removal) was pumped into a holding tank (200 L plastic container) for about 5 min. The reactor was operated for about a year in which the first 42 days was considered to be start-up. The sludge bed of the UASB reactor was discharged more than two to three times a week to keep its content below 40 cm and that discharge was immediately fed into the digester with the help of peristaltic pump. At that very moment, the digester effluent was pumped out and re-circulated to the lower part of the UASB reactor at 10 cm from the bottom. The whole system operated for 107 days with 57 days as a startup period. The OLR of one stage UASB reactor was 3.35 and 2.73 kgCOD/(m^3.d) for hot and cold six months respectively, whereas it was 2.84 kgCOD/(m^3.d) for the reactor in the UASB digester system. The achieved COD removal efficiency was 54% during the first warm six months of the year, and achieved only 32% over the following cold six months of the year. The modification of the one-stage UASB reactor to a UASB reactor-digester system had remarkably improved the overall performance as the COD removal efficiency of digester alone was 72%.

Shayad and Mahmoud (2008) studied start-up of an UASB-septic tank for community on-site treatment of strong domestic sewage of COD_t concentration of 1189 mg/L. The two pilot scale UASB-septic tank reactors R1 and R2 with two and four days HRT with 0.8 m^3 working volume were operated in parallel for a period of 6 months (October to March) with sewage temperature of 24°C. Domestic sewage was pretreated with screens and grit removal chamber and collected in holding tank (200 L). The UASB reactor was inoculated with anaerobic sludge from cesspit and operated in parallel at ambient temperature. Steady state was considered after 80 days of operation due to poor results in the first 60-80 days. However, methanogenic activity increased with sludge bed development as methane gas began to produce after 30 days. Increase in biogas production was a result of COD conversion likely due to better mixing conditions created by intense gas production along with the bio-conversion. Methane gas production is strongly influenced by the development of sludge bed and the ambient temperature. This experiment resulted in the average removal efficiencies for COD_t and TSS as 56% (1267-56 mg/l) and 81% (623-116 mg/l) for 2 days and 58% (1267-530 mg/l) and 82% (623-113 mg/l) for 4 days HRT which does not have much differences and makes 2 days HRT more adequate and economical.

Jamal and Mahmoud (2009) researched on community onsite treatment of cold strong sewage (average influent COD_t 905 mg/l) in a UASB-septic tank system (up-flow septic tank). They experimented on two pilot scale UASB septic tank reactors R1 and R2 with 2 and 4 days HRT, respectively, with 0.8 m^3 working volume. They were operated in parallel for a period of cold 6 months (October to March) with sewage temperature of 17.3°C. Domestic sewage was pretreated with screens and grit removal chamber and collected in holding tank (200 L). The tank was inoculated with anaerobic sludge from cesspit and operated in parallel at ambient temperature. During the monitoring period, the removal efficiencies in R1 and R2 for COD_t, and TSS were 51%, 74 and 54%, 78, respectively. The difference in the removal efficiencies of those parameters in R1 and R2 is marginal. The sludge filling period of the reactors is expected to be four to seven years.

Sperling and Oliveira (2009) evaluated the comparative performance of full-scale anaerobic and aerobic waste-water treatment processes in Brazil. Evaluation and comparison of 166 full-scale WWTP (both aerobic and anaerobic) operated in Brazil was done which included septic tank anaerobic filter (ST+AF), UASB, UASB-post treatment, facultative pond (FP), anaerobic pond-FP and activated sludge (AS). The results from the statistical tests confirmed that the best performance was achieved by AS followed by UASB with post treatment. These technologies presented better performance than the other processes with regard to the effluent concentration and removal efficiency of almost all constituents. The UASB reactor showed good BOD and COD removal efficiencies but was poor in TSS, FC and nutrients removal.

Lew et al. (2011) studied about an integrated UASB-sludge digester for raw domestic wastewater treatment in temperate climates to improve the performance of UASB reactor treating raw domestic wastewater under such climatic conditions. An experimental UASB reactor made of plexiglass with working volume of 5.3 L was initially seeded with 2 L granular sludge from a full scale UASB reactor treating food wastewater. Domestic wastewater with COD concentration of 1576 (±376) mg/l was fed at temperatures of 28, 15 and 10°C for 2 months each with 6 h HRT. Gas collector was used to collect and measure the gas produced, sludge degradation as mg COD/l was calculated using the theoretical specific methane production per kg COD (350 L CH4 per kg COD). Results show declining removal efficiencies with decreasing temperatures, that is, COD removal decreased from 78% at 28°C to 42% at 10°C with highest methane production of 6.19 L CH_4/day to nil, respectively. This decline was attributed to low hydrolytic activity at lower temperatures that reduced suspended matter degradation resulting in solids accumulation at the top of the sludge blanket. Solids removed from the upper part of the UASB sludge were treated in an anaerobic digester. The anaerobic digester of 2 L working volume was designed for that purpose working in 15°C and increasing temperatures. Result shows gas production increase with rise in temperature reaching up to 4.1 L CH_4/day at 30°C which indicates that non-degradable fraction of sludge decreases with increasing temperature. Maximum utilization of COD was also observed at 30°C which accounted to about 60% (17.9 to7.16gCOD/l) COD removal.

Rizvi et al. (2014) studied start-up of laboratory scale UASB reactor treating municipal wastewater and effect of temperature, HRT on its performance. It was found that the start-up period was about 120 days. The COD and BOD removal efficiencies were in the range of 57-62 and 61-66%, respectively at 17°C and 60 days of operation. However, the removal efficiency increased to the range of 79-81 and 77-83%, respectively at 25-30°C, 9 h HRT and 150 days of operation.

Khan et al. (2015) studied performance evaluation of UASB reactor treating domestic wastewater using 60 L pilot scale reactor. The treatment efficiencies were investigated at 8 h HRT and ambient temperature ranged from 8 to 40°C. TSS, COD and BOD removal efficiencies were found to be in between 65 to 85% under different OLR between 0.57 and 6.35 kgCOD/(m^3.d). The removal of organic matter followed a liner correlation with organic loads.

Reviewing these literatures, it can be concluded that start-up process and temperature are major issues in UASB reactor for the efficient performance of the reactor. There are lack of sufficient information and experience that the UASB reactor has successfully been operated for domestic wastewater in low temperature regions. To bridge this gap, the author has applied simple principle of combining septic tank into UASB reactor for reducing/converting high content of organic particulate matters available in domestic wastewater into soluble solution before supplying to UASB reactor.

Moreover, to get benefit from sufficient mixing and stimulate granulation process (usually takes long time at lower temperature environment), pulse feed strategy was applied to overcome effects of lower temperature environment. The preliminary results and experience is discussed below.

Future prospect (ST-UASB combined system for domestic wastewater treatment)

In tropical countries, UASB reactors are widely accepted for domestic wastewater treatment and there are several pilot and full scale plants in operation in Japan, India, Brazil and Columbia. Studies suggest that COD, BOD and TSS removal efficiencies are in the range of 65 - 75% in these countries at HRT of 6 to 18 h (Seghezzo et al., 1998; Lew et al., 2011; Sperling and Oliveira, 2009; Khan et al., 2011?). But especially in developing countries with lower temperature conditions, there are concerns in pre-removal of particulate matters prior to anaerobic treatment in order to obtain good effluents quality and biogas production, which inhibit the widespread use of this technology (Elmitwalli et al., 2002; Chong et al., 2012). One probable easiest way to improve UASB performance in these countries are to combine it with septic tank or settling tank (usual existing systems in developing countries) to ensure preremoval of suspended solids and allow for hydrolysis of particulate organic matter to generate a feed solution that is appropriate for UASB as described in earlier. Though, there is no published information on how to integrate UASB system into existing infrastructure like septic tanks and the performance of such combined system, it could attract researchers especially from developing countries with fluctuating climatic conditions to investigate on it and find a suitable solution in a local context.

The authors have conducted a preliminary study at Kathmandu University, Nepal on septic tank-UASB combined system which shows very encouraging results. A 250 L pilot scale UASB reactor is fed with septic tank effluents and operated at ambient temperature range of 0 to 30°C. The reactor is fed intermittently 12 times per HRT and the performance is recorded for different HRTs from 18 to 4 h during 1.5 years operation. The start-up period is found to be about 2 to 3 weeks even at low temperature. The average COD hydrolysis and methanogenesis percentage are about 18 to 36%, and 23 to 36%, respectively. About 22 to 37% of influent COD are converted into methane, 12 to 19% accumulated in the reactor while 44 to 62% remained in the effluent. The COD removal efficiency of UASB reactor is in the range of 38 to 56%, while the performance of the combination of septic tank and UASB is 55 to 72% at all HRT except for 4 h. At this HRT, all parameters are significantly lower, because the removal efficiency collapsed. This also shows good pollutant removal efficiencies and short start-up period for reactor operating at fluctuating temperature conditions. Though it needs further long term study and as far as possible parallel study at different countries with similar low temperature conditions to establish the similarity of operation, it can be said that the combined system could be a suitable low cost and effective onsite sanitation option especially for developing countries like Nepal.

CONCLUSION

Start-up of UASB reactor was affected by various factors, most importantly temperature and OLR. In general, a long start-up period of about two months was required, which could even be longer at lower temperature. Treatment efficiency and methane conversion was strongly dependent on temperature, HRT and OLR primarily for the treatment of domestic wastewater. The high organic particulate matter content in domestic wastewater was a major issue for lower temperature conversion and therefore, a septic tank pretreatment with pulse feed UASB (ST-UASB) combined system might be worth attended for prospective research to help solve the concerns. A preliminary study on ST-UASB combined system carried out at Kathmandu University at ambient temperature range of 0 to 30°C. The UASB reactor was fed intermittently 12 times per HRT and the performance was recorded for different HRTs from 18 to 4 h during 1.5 years operation. The start-up period was found to be 2 to 3 weeks even at low temperature conditions and the COD removal efficiency of the combined system was 55 to 72% at all HRT except for 4 h. At this HRT, the removal efficiency collapsed.

Conflict of interests

The authors did not declare any conflict of interest.

REFERENCES

Aiyuk S, Amoako J, Raskin L, van Haandel A, Verstraete W (2004). Removal of carbon and nutrients from domestic wastewater using a low investment, integrated treatment concept. Water Res. 38:3031-3042. http://dx.doi.org/10.1016/j.watres.2004.04.040

Alvarez JA, Ruiz I, Gomez M, Presas J, Soto M (2006). Start-up alternatives and performance of an UASB pilot plant diluted municipal wastewater at low temperature. Bioresour. Technol. 97:1640-1649. http://dx.doi.org/10.1016/j.biortech.2005.07.033

Arceivala SJ, Asolekar SR (2007). Wastewater Treatment for Pollution Control and Reuse, Tata McGraw Hill Education Private Limited.

Barber WP, Stuckey DC (1999). The use of the Anaerobic Baffled Reactor (ABR) for Wastewater Treatment: A Review. Water Res. 33:1559-1578. http://dx.doi.org/10.1016/S0043-1354(98)00371-6

Behling E, Diaz A, Colina G, Herrera M, Gutierrez E, Chacin E, Fernandez N, Forster CF (1997). Domestic Wastewater Treatment Using a UASB reactor. Bioresour. Technol. 61:239-245. http://dx.doi.org/10.1016/S0960-8524(97)00148-X

Chong S, Sen TK, Kayaalp A, Ang HM, (2012). The performance enhancements of upflow anaerobic sludge blanket (UASB) reactors for domestic sludge treatment - A State-of-the-art review. Water Res. 46:3434-3470. http://dx.doi.org/10.1016/j.watres.2012.03.066

Coelho ALSS, Nascimento MBH, Cavalcanti PFF, Haandel AC (2003). The UASB reactor as an alternative for the septic tank for on-site sewage treatment. Water Sci. Technol. 48:221-226.

Crites R, Tchobanoglous G (1998). Small and Decentralized Wastewater Management Systems, McGraw-Hill, USA.

Elmitwalli TA, Oahn KLT, Zeeman G, Lettinga G (2002). Treatment of domestic sewage in a two-step anaerobic filter/anaerobic hybrid system at low temperature. Water Res. 36:2225-2232. http://dx.doi.org/10.1016/S0043-1354(01)00438-9

Foresti E, Zaiat M, Vallero M (2006). Anaerobic processes as the core technology for sustainable domestic wastewater treatment: Consolidated applications, new trends, perspectives, and challenges. Rev. Environ. Sci. Biotechnol. 5:3-19. http://dx.doi.org/10.1007/s11157-005-4630-9

Foxon KM, Pillay S, Lalbahadur T, Rodda N, Holder F, Buckley CA (2004). The anaerobic baffled reactor (ABR): An appropriate technology for on-site sanitation. Water SA 30:592-597.

Franco A, Roca E, Lema J (2003). Improvement of the properties of granular sludge in UASB reactors by flow pulsation. Water Sci. Technol. 48:51-56.

Ghangrekar MM, Asolekar SR, Ranganathan KR, Joshi SG (1996). Experience with UASB reactor start-up under different operating conditions. Water Sci. Technol. 34: 421-428. http://dx.doi.org/10.1016/0273-1223(96)00674-9

Gomec CY (2010). High-rate anaerobic treatment of domestic wastewater at ambient operating temperatures: a review on benefits and drawbacks. J. Environ. Sci. Health 45: 1169-1184. http://dx.doi.org/10.1080/10934529.2010.493774

Hammes F, Kalogo Y, Verstraete W (2000). Anaerobic digestion technologies for closing the domestic water, carbon and nutrient cycles. Water Sci. Technol. 41:203-211.

Henze M, Harremoe SP (1983). Anaerobic treatment of wastewater in fixed film reactors: a literature review. Water Sci. Technol. 15: 1-101.

Hickey RF, Wu WM, Veiga MC, Jones R (1991). Start-up, operation, monitoring and control of high-rate anaerobic treatment systems. Water Sci. Technol. 24: 207-255.

Hulshoff Pol LW, Lopes SI, Lettinga G, Lens PNL (2004). Anaerobic sludge granulation. Water Res. 38: 1376-1389. http://dx.doi.org/10.1016/j.watres.2003.12.002

Jamal AJ, Mahmoud N (2009). Community onsite treatment of cold strong sewage in a UASB-septic tank. Bioresour. Technol. 100: 1061–1068. http://dx.doi.org/10.1016/j.biortech.2008.07.050

Kato MT (1994). The Anaerobic Treatment of Low Strength Soluble Wastewaters, Ph.D thesis, Department of Environment Technology, Wageningen Agricultural University, The Netherlands.

Khan AA, Gaur RZ, Tyagi VK, Khursheed A, Lew B, Kazmi AA, Mehrotra I (2011). Sustainable Options of Post Treatment of UASB Effluent Treating Sewage: A Review. Resour. Conserv. Recycl. 55:1232-1251. http://dx.doi.org/10.1016/j.resconrec.2011.05.017

Khan AA, Mehrotra I, Kazmi AA, (2015). Sludge profiling at varied organic loadings and performance evaluation of UASB reactor treating sewage. Biosyst. Eng. 131:32-40. http://dx.doi.org/10.1016/j.biosystemseng.2014.12.011

Lettinga G (1995). Anaerobic digestion and wastewater treatment systems. Antonie van Leeuwenhoek 67:3-28. http://dx.doi.org/10.1007/BF00872193

Lettinga G (1996). Sustainable integrated biological wastewater treatment. Water Sci. Technol. 33:85-98. http://dx.doi.org/10.1016/0273-1223(96)00303-4

Lettinga G (2001). Digestion and degradation, air for life. Water Sci. Technol. 44:157-176.

Lettinga G, Man AD, Grin P, Hulshoff LP (1987). Anaerobic wastewater treatment as an appropriate technology for developing countries. Tribune Cebedeau 519 (40):21-32.

Lew B, Lustig I, Beliavski M, Tarre S, Green M (2011). An integrated UASB-sludge digester for raw domestic wastewater treatment in temperate climates. Bioresour. Technol. 102:4921-4924. http://dx.doi.org/10.1016/j.biortech.2011.01.030

Lin KC, Yang Z (1991). Technical review on the UASB Process. Int. J. Environ. Stud. 39:203-222. http://dx.doi.org/10.1080/00207239108710695

Liu Y, Xu HL, Yang SF, Tay JH (2003). Mechanisms and models for anaerobic granulation in upflow anaerobic sludge blanket reactor. Water Res. 37:661-673. http://dx.doi.org/10.1016/S0043-1354(02)00351-2

Lohani SP, Adhikari J, Chhetri A, Bakke R (2013). Sustainable Biogas Production Potential from Urban Wastewater in Nepal. Int. J. Environ. Sci. Dev. 4:595-599. http://dx.doi.org/10.7763/IJESD.2013.V4.420

Luostarinen S, Sanders W, Roeleveld KK, Zeeman G (2007). Effect of temperature on anaerobic treatment of black water in UASB-septic tank systems. Bioresour. Technol. 98: 980-986. http://dx.doi.org/10.1016/j.biortech.2006.04.018

Luostarinen SA, Rintala JA (2005). Anaerobic on-site treatment of black water and dairy parlour wastewater in UASB-septic tanks at low temperatures. Water Res. 39:436-448. http://dx.doi.org/10.1016/j.watres.2004.10.006

Mahmoud N (2008). High Strength Sewage Treatment in a UASB reactor and an integrated UASB digester system. Bioresour. Technol. 99:7531-7538. http://dx.doi.org/10.1016/j.biortech.2008.02.021

Mahmoud N, Zeeman G, Gijzen H, Lettinga G (2003). Solids removal in

upflow anaerobic reactors, a review. Bioresour. Technol. 90: 1–9. http://dx.doi.org/10.1016/S0960-8524(03)00095-6
Man AWA, Grin PC, Roersma RE, Grolle KCF, Lettinga G (1986). Anaerobic Treatment of municipal wastewater at low temperatures. Anaerobic treatment: A grown-up technology conference papers (Aquatech 86), Amsterdam. pp. 451-466.
Mgana SM (2003). Towards Sustainable and Robust On-site Domestic Wastewater Treatment for all Citizens, Ph.D. Thesis, Wageningen University, Wageningen, The Netherlands.
Nadais H, Capela I, Arroja L, Duarte A (2005). Treatment of dairy wastewater in UASB reactors inoculated with flocculent biomass. Water SA 19: 603-607.
Ostrem MK, Millrath K, Themelis NJ (2004). Combining Anaerobic Digestion and Waste To Energy, 12th North America Waste To Energy Conference, Columbia University, New York. http://dx.doi.org/10.1115/NAWTEC12-2231
Parawira W (2004). Anaerobic Treatment of Agricultural residues and wastewater, Application of High Rate Reactors. Ph.D. Thesis, Department of Biotechnology, Lund University, Sweden.
Richard L, Weiss M, Pataky K, Jowett C (2005). Re-thinking hydraulic flow in septic tanks. Environmental Science and Engineering Magazine, March 2005.
Rizvi H, Ahmad N, Abbas F, Bukhari HI, Yasar A, Ali S, Yasmeen T, Riaz M (2014). Start-up of UASB reactors treating municipal wastewater and effect of temperature/sludge age and hydraulic retention time (HRT) on its performance. Arabian J. Chem. Article in Press, 2014.
Rocktäschel T, Klarmann C, Helmreich B, Ochoa J, Boisson P, Sørensen KH, Horn H (2013). Comparison of two different anaerobic feeding strategies to establish a stable aerobic granulated sludge bed. Water Res. 47:6423-6431. http://dx.doi.org/10.1016/j.watres.2013.08.014
Seghezzo L, Zeeman G, van Lier JB, Hamelers HVM, Lettinga G, (1998). A review: the anaerobic treatment of sewage in UASB and EGSB reactors. Bioresour. Technol. 65:175-190. http://dx.doi.org/10.1016/S0960-8524(98)00046-7
Shayad MA, Mahmoud N (2008). Start-up of an UASB-septic tank for community on-site treatment of strong domestic sewage. Bioresour. Technol. 99:7758-7766. http://dx.doi.org/10.1016/j.biortech.2008.01.061
Singh KS, Viraraghavan T (1998). Start-up and operation of UASB Reactors at 20°C for Municipal Wastewater Treatment. J. Ferment. Bioeng. 85:609-614. http://dx.doi.org/10.1016/S0922-338X(98)80014-7
Sperling MV, Freire VH, Chernicharo CAL (2001). Performance evaluation of a UASB-activated sludge system treating municipal wastewater. Water Sci. Technol. 43:323-328.
Sperling MV, Oliveira SC (2009). Comparative performance evaluation of full-scale anaerobic and aerobic wastewater treatment processes in Brazil. Water Sci. Technol. 59:15-22. http://dx.doi.org/10.2166/wst.2009.841
Van der Last ARM, Lettinga G (1992). Anaerobic Treatment of Domestic Sewage under Moderate Climatic (Dutch) Conditions Using Upflow Reactors at Increased Superficial Velocities. Water Sci. Technol. 25:167-178.
Vlyssides A, Barampouti EM, Mai S (2008). Granulation mechanism of a UASB reactor supplemented with iron. Anaerobe 14:275-279. http://dx.doi.org/10.1016/j.anaerobe.2008.09.003
Zeeman G, Sanders W, Lettinga G (2000). Feasibility of the on-site treatment of sewage and swill in large buildings. Water Sci. Technol. 41:9-16.
Zoutberg GR, de Been P, (1997). The Biobed_ EGSB (Expanded Granular Sludge Bed) system covers shortcomings of the upflow anaerobic sludge blanket reactor in the chemical industry, Water Science and Technology, 35, 183-187. http://dx.doi.org/10.1016/S0273-1223(97)00268-0

Modeling Jambo wastewater treatment system to predict water re-use options

Kyeyune Simonpeter and Mulamba Peter

Department of Agricultural and Bio-Systems Engineering, Makerere University, P.O. Box 7062, Kampala, Uganda.

The harmful nature of chemicals used in the tannery industries and the disposal of wastes from facilities is well documented. In this study, Jambo tannery which is located in Busia District, (Uganda) with a daily processing capacity of 6.6 tonnes of hides and skin utilises 20 m^3 of water to produce 17 m^3 of wastewater/day. The generated wastewater is treated on site in the wastewater treatment plant whose performance was assessed. The main objective of this study was twofold: (i) to investigate how processes can be efficiently improved and optimized for environmental safety, and (ii) to develop Dev C++ programme to implement Brown's model for determining water quality usage of the effluent of the factory. Specifically, wastewater parameters measured were temperature (T), electro conductivity (EC), pH, total suspended solids (TSS), total dissolved solids (TDS), biological oxygen demand (BOD_5), chemical oxygen demand (COD), total phosphorus (TP), total chromium (TCr), chromium hexavalent (Cr^{6+}) and chromium trivalent (Cr^{3+}). The results of the tannery wastewater were far from the National Environment Management Authority (NEMA) standards in Uganda for COD (100), BOD_5 (50), TP (10), TC (2.0), CH (0.05), TSS (100), TDS (1200), EC (1500) and pH (6.0-8.0). It can therefore be concluded that the performance of the wastewater plant is not satisfactory in the current state. The effluent quality is therefore not currently suitable for release in the environment and re-use for Agricultural and Domestic purposes as per National Environmental Management Authority (NEMA) Standards 1995. However, with improvements in the WWTP unit operations, it could be improved, re-used for irrigation. As such, a model was developed to model wastewater system to predict re-use options especially for agriculture. The study clearly showed that the use of the model and the developed C++ computer programme in predicting the re-use options of the wastewater treatment system was a success and can be applied to other wastewater treatment system as well as improving and managing the existing one.

Key words: Wastewater treatment, tannery effluent, water standards, water quality, environment safety.

INTRODUCTION

In the African continent, because of the erratic rainfall patterns that are dominating the agricultural sector, the possibility of using wastewater such as urban storm water and industrial water is being considered for irrigation, though more studies should be done on the benefits and limitations (Kirkham, 2006; Khambete and Christian, 2014a; Yao et al., 2014). For example, Dikinya and Areola (2010) studied the water quality of treated urban wastewater in terms of heavy metal concentrations that is being used for irrigation of different crops olive, maize, spinach and tomatoes in the Glen Valley near Gaborone City (Botswana). Wang et al. (2007) studied the possibility of recycling treated wastewater for irrigation and the resulting impacts on the soil, crop and the overall environment.

Livestock contributes about 5.2% of Uganda's total gross domestic products (GDP) and agricultural GDP. Small holder farmers in Uganda own about 98% of the cattle herd and 100% of the small ruminant and non-ruminant stock (Muramuzi, 2009). Uganda lacks modern slaughter houses and facilities as such animals are largely slaughtered on open grounds with minimal hygiene and safety standards. It follows that most of the hides and skins from rural slaughter slabs/butchers, slaughter houses and abattoirs pass through wholesale agents. The rest are collected by small traders and then sold to wholesalers/tanners for processing/export. Apart from internal trade in unprocessed hides and skins, there is a substantial informal cross-border trade between the East African (EA) countries. Official statistics show that on average more than 50% of the hides and skins are exported in unprocessed form, though this proportion varies across the EA countries.

Despite the ever growing trade in hides and skins, the effluent quality from tannery industries is still a little studied topic. This study sought to assess the quality of the effluent at the Wastewater Treatment Plant (WWTP) at Jambo Tannery (U) limited. Jambo tannery with a daily capacity of 6.6 tonnes of hides and skins in two loadings, utilize 20 m^3 of water, estimated to be 3.03 m^3 per tonne. The effluent is treated through unit operations that include sedimentation, oxidation, clarification, sludge drying bed, second oxidation pond and maturation for the beamhouse and chromium lines. The tannery uses raw materials in the form of cow hides, goat and sheep skins and a number of imported chemicals. Different types of chemicals are used in leather manufacturing which ranges from common salt to very expensive chrome sulphates. The four major categories of production processes in Jambo tannery include: (i) hide and skin storage and beam house operations, (ii) tarn yard operations, (iii) post tanning operations and (iv) finishing operations.

After the hides and skins are flayed from the carcass at the abattoirs slaughter house, they are delivered to the hide and skin market, directly to the tannery. Where necessary, hides and skins are cured before transport to the tannery in order to prevent them from putrefying. The hides and skins are then sorted, trimmed, cured and stored pending operations in the beam house. The processes involved in the beam house include soaking, unhairing, liming, fleshing and splitting. The tan yard processes include de-liming, bating, pickling, and tanning take place. The tanned hides and skins are tradable intermediate products (wet-blue) as they have been converted to a non-putrescible material (leather). The operations at Jambo tannery end at tanning, with no polishing/finishing operations being done. The Processes typically carried out in post-tanning operations are: summing, setting, splitting, shaving, retaining, dyeing, fat liquoring and drying. The selection of finishing processes depends on the specifications of the final product.

In essence, the current effluent treatment process is quite expensive and time consuming such that the possibility of recycling water is attractive option to consider. After a careful assessment of operations that require water usage, this study proposed to carry out an audit of the prevailing production circumstances. The layout of the wastewater treatment plant (WWTP), the constructed wetland, the factory and the surroundings of the Jambo Tanning Factory in Busia (Uganda) is indicated in Figure 1. The main objective of the study was twofold: (i) to investigate how processes can be efficiently improved and optimized for environmental safety, and (ii) to develop Dev C++ programme to implement Brown' model for determining water quality usage of the effluent of the factory.

Water quality index for categorizing water and wastewater quality

Pollutant yield from the various processing unit operations of the tanning industry are well documented (Bosnic et al., 2000; Ludvik, 2000a,b; Chattha and Shaukat, 1999). Chromium is highly toxic and carcinogenic to human beings, animals, plants and general environment (soil and water), (Jabari et al., 2009; Muramuzi, 2009; Belay, 2010; Afify et al., 2013). There are limits established by Environmental Authorities for these pollutants that are allowable for release into the environment and water reuse as well as water quality monitoring for fish and ecological purposes. Poorly treated effluents from the tannery industry can lead to wide spread contamination of food chains, sharp decline in productivity of food crops, soil, vegetables, livestock including milk production, if used for irrigation.

Wastewater quality physico-chemical parameters estimated in the samples have been used to determine the wastewater quality index (WWQI) which is regarded as a most effective way to communicate wastewater quality parameters (Khambete and Christian, 2014b). There are several ways to assess the quality of water as deemed fit for drinking, irrigation and industrial use. Yogendra and Puttaiah (2008) presented the water quality index (WQI) which is a characteristic number for representing the overall quality of water for public or for any intended use as well as in the pollution abatement programmes and water quality management. The WQI was determined basing on various physico-chemical parameters. WQI provides a single number that expresses overall water quality at a certain location and time, based on these several water quality parameters.

Quality for raw and treated wastewater has also been

Figure 1. Aerial view of Jambo Factory in Busia District, Uganda.

determined by multi-parameter aggregated index (MPAI) using a fuzzy multi-criteria decision making approach (Khambete and Christian, 2014a). Banerjee and Srivastava (2009) successfully applied the WQI for assessment of surface water quality surrounding integrated industrial estate-Pantnagar. WQI approach and statistical tools in fingerprinting of heavy metal pollution and comparison of spatial variability of multiple contaminants have been found effective and a useful tool for assessment of water quality (Boyacioglu, 2012).

WQI has been successfully used to study the water quality of a stream receiving industrial effluents in the Brazilian Northeast where a potential risk to the population was identified (Abrahao and Carvalho, 2009). The application of the WQI classification indicated that the water quality was poor could pose risk. In regards to stream/river water quality flows, Alam and Pathak (2010) successfully carried out the assessment of River Ramganga in using WQI and a computer programme. Additionally, Moscuzza et al. (2007) were able to assess river water quality within the agricultural Pampean plains (Argentina).

There has also been application of WQI in classifying the quality of groundwater (Vasanthavigar et al., 2009; Balakrishnan, 2011). Vasanthavigar et al. (2009) applied the WQI for ground water basin basing on dominant cations (Na^+, $Mg2^+$, Ca^{2+}, K^+) and anions (Cl^-, HCO_3^-, SO_4^{2-}) and was able to show suitability of groundwater for irrigation purposes. Balakrishnan (2011) was able to use WQI values to develop *Isopleth* map that was used for regional decision making geared towards better planning and management of groundwater quality in the Gulburga City (Karnataka State, India).

Model setup equations and tables

In the study, the WQI was used with the NEMA standards using the weighted arithmetic index method (Brown et al., 1970). It incorporates the use of the generated Dev C++ computer language programme with a setup to determine the usage. The overall WQI is calculated by aggregating the quality rating with the unit weight linearly using a model developed by Brown et al. (1970) as in Equation (1) below:

$$WQI = \frac{\sum q_n W_n}{\sum W_n} \quad (1)$$

Where: q_n = Quality rating for the nth water quality parameter; W_n = unit weight for the nth parameters.

Table 1. Comparison of WQI and status of water quality.

Water quality index level	Water quality status
0-25	Excellent water quality
26-50	Good water quality
51-75	Medium water quality
76-100	bad water quality
>100	Very bad water quality

Source: (Chatterjee and Raziuddin, 2002; Yogendra and Puttaiah, 2008).

The quality rating was calculated using Equation (2).

$$q_n = 100 \times \frac{[V_n - V_{io}]}{[S_n - V_{io}]} \quad (2)$$

Where: q_n = Number reflecting the relative value of the parameter in the polluted water with respect to its standard permissible value; V_n = observed value of the nth parameter at a given sampling station; V_{io} = ideal value of the nth parameter in pure water. (That is, 0 for all other parameters except the parameter pH of 7.0 mg/l and dissolved oxygen of 14.6 mg/l; S_n = standard permissible value of the nth parameter.

The unit weight was calculated by a value inversely proportional to the recommended standard value S_n of the corresponding parameter in Equation (3).

$$W_n = \frac{K}{S_n} \quad (3)$$

Where: *K* is a constant for proportionality.

A comparison of water quality index and identification of the status of water quality were made based on the criteria variously used by Yogendra and Puttaiah (2008) and Chatterjee and Raziuddin (2002), which is reproduced in Table 1.

Excessive sodium and salinity concentrations in irrigation water result in sodium hazard, as well as salinity hazard. Sodium ion in water replacing calcium and magnesium ions in soil causes reduced permeability and soil hardening (Zhang et al., 2012). To assess irrigation water quality, the parameters such as percent sodium (Na %) and sodium adsorption ratio (SAR) are calculated based on the chemical variables of water samples (Zhang et al., 2012). Sodium in soil is considered vital for determining groundwater suitability for irrigation purpose because Na reacts with soil to reduce its permeability and support little or no plant growth. The irrigation water assessment indices including:

$$Na\% = \left\{\frac{[Na^+]}{[Na^+] + [Ca^{2+}] + [Mg^{2+}] + [K^+]}\right\} \times 100 \quad (4)$$

$$SAR = \frac{[Na^+]}{\sqrt{\frac{(Ca^{2+} + Mg^{2+})}{2}}} \quad (5)$$

Where, all the ionic concentrations are expressed in millequivalents per liter (meq/L) of the respective ions. The SAR is probably the only one in current use and is generally considered an effective evaluation index for most water used in irrigated agriculture (Zhang et al., 2012).

Water quality assessment programme and prediction of re-use options

The programme was developed using Dev C++ computer language. This programme incorporates the water quality index (WQI) model developed by Brown et al. (1970) and the sodium adsorption ratio equation. The logic flow chart for the programme is shown in Figure 2. The WQI is given by Equation 1. The Quality rating was calculated using Equation 2. The unit weight was calculated by a value inversely proportional to the recommended standard value S_n of the corresponding parameter in Equation 3. To assess irrigation water quality, the parameters such as percent sodium (Na %) and sodium adsorption ratio (SAR) are calculated based on the chemical variables of water samples (Zhang et al., 2012). The irrigation water quality assessment indices were calculated from Equations 4 and 5 (Table 2).

The developed programme determines the usage of effluent water. This commercially available programme has a setup file called KYESP WQIC which runs on the following windows: Windows 95, Windows 98, Windows ME, Windows NT, Windows 2000, Windows XP, Windows vista and Windows 2003 server. After successful installation, it displays an interface where you are required to press numbers one or two depending on what you want to determine. The developed programme prompts the user to decide which usage to test. It does so by asking you either to press 1 or 2 for domestic or irrigation use, respectively.

MATERIALS AND METHODS

Location of the site

Jambo Tannery (U) Limited is a tannery with a capacity to process 250 hides and 3,000 kg skins per day. The tannery is located in

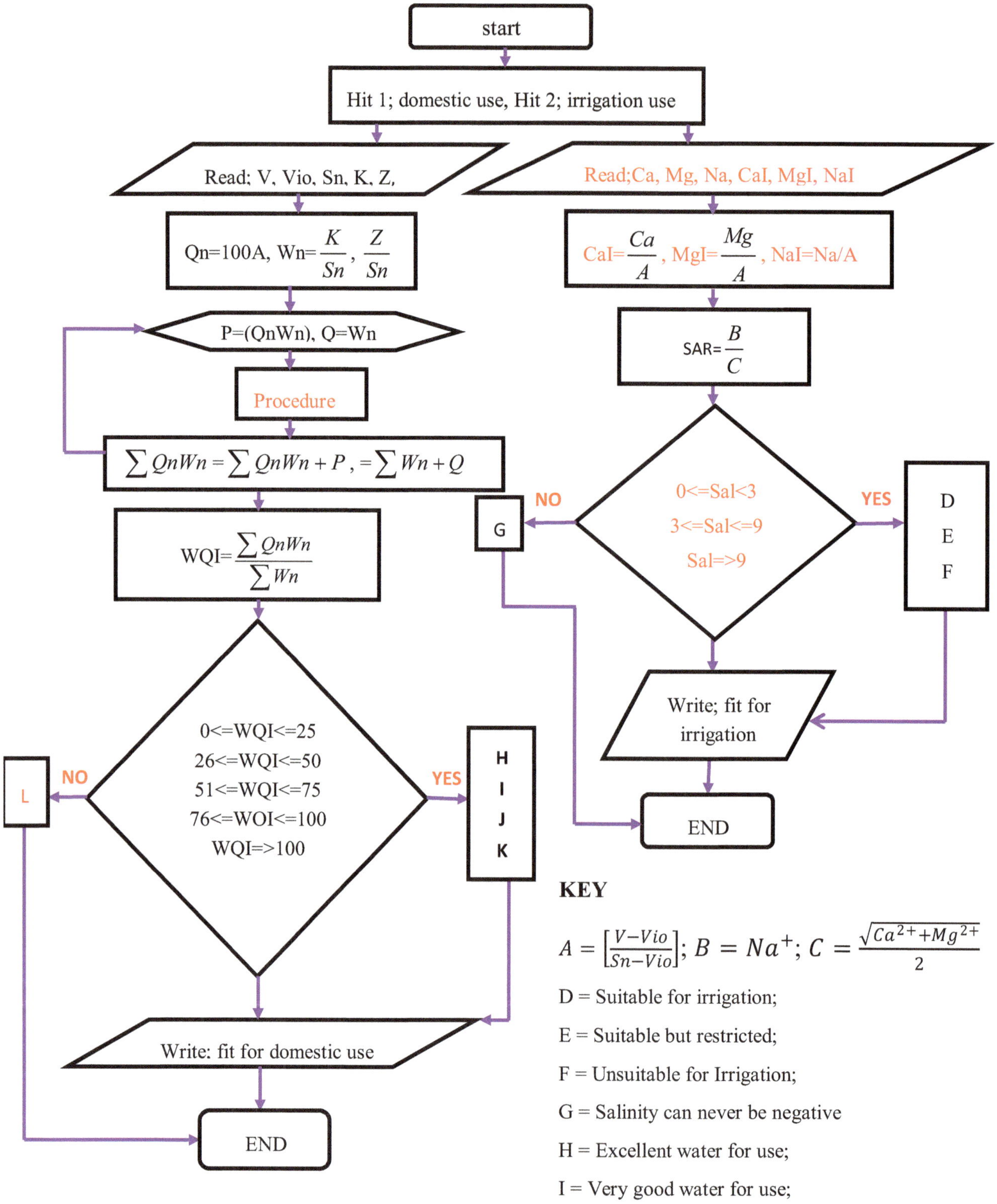

Figure 2. Flow chart of the water quality program.

Solo A village, Busia Municipal Council Industrial Area, in Busia District (Uganda). The tannery is located at GPS coordinates N 00.2840° and E 034.4965°. Busia District developed a master plan which mainly focuses on improving water quality in the area. Currently, the effluent from the wastewater treatment plant at Jambo Tannery (U) Limited flows to Solo stream which is a source

Table 2. Irrigation water quality status.

International SAR standards	Water quality status for irrigation
SAR < 3	Suitable for irrigation
3 ≤ SAR ≤ 9	Used but may be restricted
SAR ≥ 9	Unsuitable for irrigation

of livelihood to communities downstream. The master plan, with a time horizon of 30 to 50 years, requires industrial production units to process their effluents to NEMA standards before disposal.

Sample collection and analysis

Samples were collected in polyethylene plastic bottles which were rinsed out with distilled water and then onsite with sample water. The sampling method followed the spot type but duplicate samples were taken per each site and only one sample was taken to the laboratory for analysis. During sample collection, sample bottles were filled to the top allowing for minimum storage between the time of sampling and delivery to the laboratory, samples were chilled and kept in an insulated container with ice bricks. The water samples were collected at five sample sites. Water samples were analyzed in the Medical School Molecular Laboratory at Makerere University for the following phsico-chemical water quality parameters: Total suspended solids (TSS), pH, electro-conductivity (EC), total dissolved solids (TDS), total phosphorus (TP), ammonium-N, total chromium (TCr), chromium trivalent (CrT), chromium hexavalent (CrH), biological oxygen demand (BOD_5), chemical oxygen demand (COD), sulphide, chlorides and sulphates.

Total dissolved solids (TDS)

TDS were determined by the evaporation method which involved evaporating the waste samples to dryness. In this method, 50 ml of sample were transferred to a weighed evaporating dish, and evaporated to dryness by heating for 1-2 hours at 180°C to a constant weight. Total dissolved solids were calculated using Equation 6.

$$TDS = \frac{1000\, mg\ of\ residue}{ml\ of\ sample} \quad (6)$$

Total suspended solids (TSS)

TSS was determined using the gravimetric method as described by Punmia (1998). Cleaned dish was taken and ignited to constant weight (W_1). 25 ml of well mixed sample was transferred to the above dish. The sample was evaporated to dryness at 103°C for 24 h, in constant temperature oven. It was then cooled in a desiccator and weight was determined (W_2). The dish was ignited at 600°C in furnace for 30 min. The dish was cooled in a desiccator a gain and its weight was determined as W_3. The total solids content was calculated from Equation 7.

$$TSS = \frac{(W_2 - W_1)mg/l \times 100}{V} \quad (7)$$

Where: W_1 = Weight of empty dish; W_2 = weight of the dish after evaporation and V = volume of the sample.

pH and electro-conductivity (EC)

The pH and EC of the wastewater were measured in the field using the HI-98129 type combination meter.

Total phosphorus (TP)

TP was determined following the method of Digestion. 25 ml of the sample was acidified with 1 ml of H_2SO_4, 0.04M, and then 5 ml digestion reagent was added and mixed well. A blank (25 ml distilled water) and phosphate standard was prepared by taking 25 ml of known standard concentration. Both the blank and phosphate standard were treated in the same way as the samples. They were heated for 30 min in an autoclave at 120°C and cooled to room temperature. The color reaction was made in the destruction bottles by adding 3 ml of combined reagent which comprised of 50 ml of 5 N H_2SO_4, 5 ml potassium antimonyl tartrate, 15 ml ammonium molybdate and mixed well. 1 ml of ascorbic acid was added to each sample and the samples were allowed to stand for 20 min for blue color development. The concentration in mg/l at 880 nm wavelength using the spectrophotometer CECIL 1000 series was determined and the reading was multiplied with the dilution factor. To derive the concentration in mg/l of TP in the sample, the formula in Equation 8 was used.

$$TP\left(\frac{mg}{l}\right) = \frac{(mg)\ TP \times 1000}{Sample\ volume\ (ml)} \quad (8)$$

Ammonium nitrogen (ammonium-N)

Ammonium-N was determined by the Nesslerisation method. 25 ml of sample was measured and poured into the nessler tube. 25 ml of the blank was prepared from distilled water. 1 ml of nessler reagent was pipetted into each sample in the nessler tubes, the stopper was placed in each case and tubes inverted several times to mix. They were allowed a minute for the reaction. Each sample was poured into respective blank and the prepared cells. The concentration was measured in mg/l at 425 mm using DREL spootrophotometer. The reading was multiplied by the dilution factor. To derive the concentration in mg/l of ammonium-N in the sample, the formula in Equation 9 was used.

$$Ammonia - N\left(\frac{mg}{l}\right) = \frac{(mg)NH_3 \times 1000}{Sample\ volume\ (ml)} \quad (9)$$

Total chromium (TCr)

TCr was determined using colorimetric method. 0.25 ml (5 drops) H_3PO_4 was added. 0.2 N H_2SO_4 and pH meter to adjust solution to pH of ± 0.5 was used. A solution was transferred to a 100 ml volumetric flask, diluted to 100 ml and mix. 2 ml diphenyl carbazide solution was added, mixed and let to stand 5 to 10 min for full color development. An appropriate portion was transferred to 1 cm absorption cell and its absorbance measured at 540 nm, using reagent water as reference. From the corrected absorbance, chromium micrograms were determined by reference to the calibration curve. A calibration curve was constructed by plotting corrected absorbance values against micrograms of chromium in102 ml final volume. TCr was determined using Equation 10.

$$mg\frac{Cr}{l}=\frac{mg\ Cr\ (in\ 102\ ml\ Final\ Volume)\times 100}{A} \quad (10)$$

Where: A = ml of original sample

Chemical oxygen demand (COD)

The COD was determined according to the method reported by Rand et al. (1975). 10 ml of the sample were taken in a 100 ml bottle then 5 ml of concentrated H_2SO_4 was added and about 1 g of copper sulphates ($CuSO_4$) also added. Then 3 ml of prepared N/40 $KMnO_4$ solution was added and the bottle immersed in boiling water for 30 min while keeping the surface of the boiling water at the higher level than the surface of the sample. Then 3 ml prepared N/40 sodium oxalate (Na_2C2O_4) was added and immediately titrated with N/40 potassium permanganate ($KMnO_4$) until violet color appeared then repeated for the blank separately under same condition using 10 ml of distilled water instead of 10 ml of sample.

$$COD\ as\ mg\ of\frac{O_2}{l}=\frac{\left(\frac{1}{40}\right)\times 8000\times (A-B)}{ml\ of\ sample} \quad (11)$$

Where: A = ml of $KMnO_4$ used for sample; B = ml of $KMnO_4$ used for blank; (1/40 = molarity of $KMnO_4$; and 8000 is the Mill-equivalent weight of oxygen × 1000 ml/l.

Biological oxygen demand (BOD_5)

The BOD_5 was determined using Winkler method as described by Rand et al. (1975). Two 100 ml bottles were obtained with lid and cleaned well. 25 ml sample was taken in each bottle and 75 ml of distilled water was added to the two bottles. The two bottles were closed well. One bottle was kept in the incubator at 20-22°C for 5 days. 10 ml of manganese sulphates solution and 2 ml of alkali-iodide solution were added to the other bottle well below the surface of the liquid by using a syringe. The bottle was closed and its content mixed by inverting the bottle several times. When the precipitate settles leaving a clear supernatant above the precipitate, it was shaken again slowly by inverting the bottle. When the solution setting has produced at least 50 ml supernatant, 8 ml of concentrated H_2SO_4 were added. The bottle was closed and mixed by gentle inversion until dissolution was completed. 100 ml of the sample was titrated with 0.05 M Na_2SO_3 solution until a pale yellow solution is reached. 2 ml of freshly prepared starch solution was added and titration was continued until a blue color appeared. The procedure was then repeated using 100 ml distilled water (blank). This procedure was repeated for incubated sample after 5 days. The BOD_5 was calculated using Equation 12.

$$BOD\ as\ mg\frac{O_2}{l}=16(V_1-V_2) \quad (12)$$

Where: V_1 = ml of Na_2SO_3 used for the sample before incubation; and V_2 = ml of Na_2SO_3 used for the sample after incubation.

RESULTS AND DISCUSSION

Physio-chemical experimental data results and discussions

The physico-chemical wastewater quality parameters analyzed included pH, EC, TDS, TSS, TP, NH_3-N, total chromium, trivalent chromium, hexavalent chromium, BOD_5, COD, sulphide, chlorides, sulphates, and oil and grease. The results from the laboratory analyses for the physico-chemical wastewater quality parameters for the five sampling sites at Jambo Factory (Busia District, Uganda) are summarized in Table 3.

The process involves use of sodium chlorides which acts as a preservative; however, this could have led to high values of EC of 7920 and 27170 µsm and chlorides of 22.5 and 20 mg/l content in the wastewater on average for both S1 and S2, respectively as indicated in Table 3. Higher levels of chlorides are also attributed to the use of constant common salt in prickling process. The high levels of BOD_5 and COD values observed at all the sampling points in tannery treatment system could have been due to high amount of organic matter from various chemicals used during the soaking, tanning and post tanning processing of hides and skins. It has been reported that only about 20% of the large number of chemicals used in the tanning process is absorbed by leather, the rest is released as waste (UNIDO, 2005), thereby increasing the levels of BOD_5 in the effluent. More still, these processes yielded higher values of TDS in the wastewater as observed in Table 3.

In unhairing and liming processes, calcium hydroxide and sodium sulphide as reducing agents are introduced to dissolve and remove the hair. This could have caused the presence of high levels of TDS for the beam and chrome lines respectively in the wastewater as observed from Table 3, higher values of TDS are also attributed to the fleshings from the fleshing process, hence contributing to pollution load. The concentration of different parameters were higher for the chrome linemost likely due to addition of solids from runoff water, mixing of sewage and industrial waste, and the oxidation processes occurring within the chrome treatment line (Table 3). According to Table 3, varying higher values of sulphates are recorded in the wastewater. This could be due to excessive ammonium sulphate which is added to de-lime the hides/skin. A trend of lower pH was due to the dilution of alkaline substances and high turbidity, whereas high pH could be attributed to high temperatures, which enhance microbial activity, causing excessive production of carbon dioxide.

Implementation of the developed programme and the results

Successful implementation of the developed programme depends on the knowledge of the input water physio-chemical quality parameters indicated in Table 3. The typical sample collection and analyses were carried out at five sites. With these typical values in mind from the facilities and the developed programme using Dev C++ computer language, inputs were made into the programme.

Table 3. Wastewater analysis at the sampling sites S1, S2, S3, S4 and solo stream.

Parameters	Units	Sample S1		Sample S2	Solo stream	Sample S3		Sample S4	
		Untreated	Treated	Treated		Untreated	Treated	Untreated	Treated
pH		10.2	8.4	8.3	8.5	11.2	7.31	3.77	8.47
EC	µs/cm	16600	9640	35400	32900	15000	4200	69000	18940
TDS	mg/l	10624	6170	22656	21056	9600	2814	44160	12122
TSS	mg/l	330	46	17	24	708	145	333	16
TP	mg/l	29.08	11	7	1.12	27.6	10.3	33.8	5.9
NH_3-N	mg/l	182.6	23	13	2.6	122	87	300	33
TCr	mg/l	9.5	2	0.9	0.006	10.1	1.6	22.7	0.7
Cr^{3+}	mg/l	3.1	0.7	0.4	0.001	2.93	0.4	7	0.1
Cr^{6+}	mg/l	5.6	0.9	0.4	0.004	6.3	0.8	13.8	0.4
BOD_5	mg/l	869	426	300	266				
COD	mg/l	8864	2986	9008	1860				
S^{2-}	mg/l	0.61	0.48	0.01	0.01	0.7	0.4	1.11	0.13
Cl^-	mg/l	50	42	24	16	35	3	127	16
SO_4^{2-}	mg/l	72	58	173	39	675	125	2525	1475
Oil and grease	mg/l	9.8	3.3	0.64	0.1				

EC = Electro-conductivity; TDS = total dissolved solids; TSS = total suspended solids; TP = total phosphorus; NH_3-N = ammonium nitrogen; TCr = total chromium; Cr^{3+} = trivalent chromium;Cr^{6+} = hexavalent chromium; BOD_5 = biological oxygen demand after 5 days; COD = chemical oxygen demand; S^{2-} = sulphides, Cl^- = chlorides; SO_4^{2-} = sulphates; mg/l = milligrams per litre; S1, S2, S3, S4 are sample sites.

The resulting output from the programme implementation yielded the results shown in Figure 3 for the four samples S1, S2, S3 and S4

For the first sample analysis, the following parameters were chosen to test the quality of water using the developed programme for both samples S1 and S2; EC, TDS, TSS, TP, ammonium-N, TC, BOD_5, COD, chlorides, sulphates, and oil and grease. These parameters yielded a water quality index of 239.568 and 246.962 for sample S1 and S2, respectively, as shown in Figures 3a and b. According to the comparison table incorporated within the programme, the water is characterized as unsuitable for use because the calculated index is greater than 100. The second sample analysis, the following parameters were tested in the programme for both sample S3 and S4; EC, TDS, TSS, TP, ammonium-N, TC, sulphide, chlorides, and sulphates. The parameters yielded a quality index of 274.742 and 152.297 for both S3 and S4 as shown in Figures 3c and d. The water is characterized as unsuitable for use because the calculated index is greater than 100.

However, the option for re-use of effluent for the factory should be considered after improving on the efficiency of the treatment system through (i) separating both the Beamhouse line and Chrome line; (ii) incorporating screens to remove biodegradables like hair and fleshings or where possible, hair should be recovered by using hair-shaving methods to reduce on the pollution loads; and (iii) possible re-use of wastes such as hair in the watershould be explored for making clothes and carpets. The factory should explore alternative use of chemicals to reduce on the bulky use of different chemicals in the process of tanning. The alternative use to be considered should include: free soak (soaking enzyme), Microdep C in unhairing and liming and unhairing assist.

Conclusion

The analyses of the physicochemical parameters indicated that the wastewater has the potential of causing eutrophication due to the high BOD_5 and nitrate concentrations. The levels of COD and BOD_5 in the tannery were found to be higher than the optimum range for the discharge of tannery. A relatively higher concentration of chlorides, phosphates and sulphates indicate the unsuitability of water for domestic use since they were higher than the NEMA limits for the discharge of industrial effluent into the environment. The application of water quality index in Dev C++ computer language programme for the overall assessment of the water quality of Jambo Tannery wastewater treatment system was consistent with the measured values of the physico-chemical water quality parameters and was found to be useful tool for assessing the effluent from the WWTP. The water quality index obtained for the tannery on two different days of the study period showed poor quality of water. This water quality rating study clearly showed that, the status of the tannery wastewater is eutrophic and unsuitable for the human, industrial and agricultural use. This clearly showed that the use of the model and the developed C++ computer programme in predicting the re-use options of the

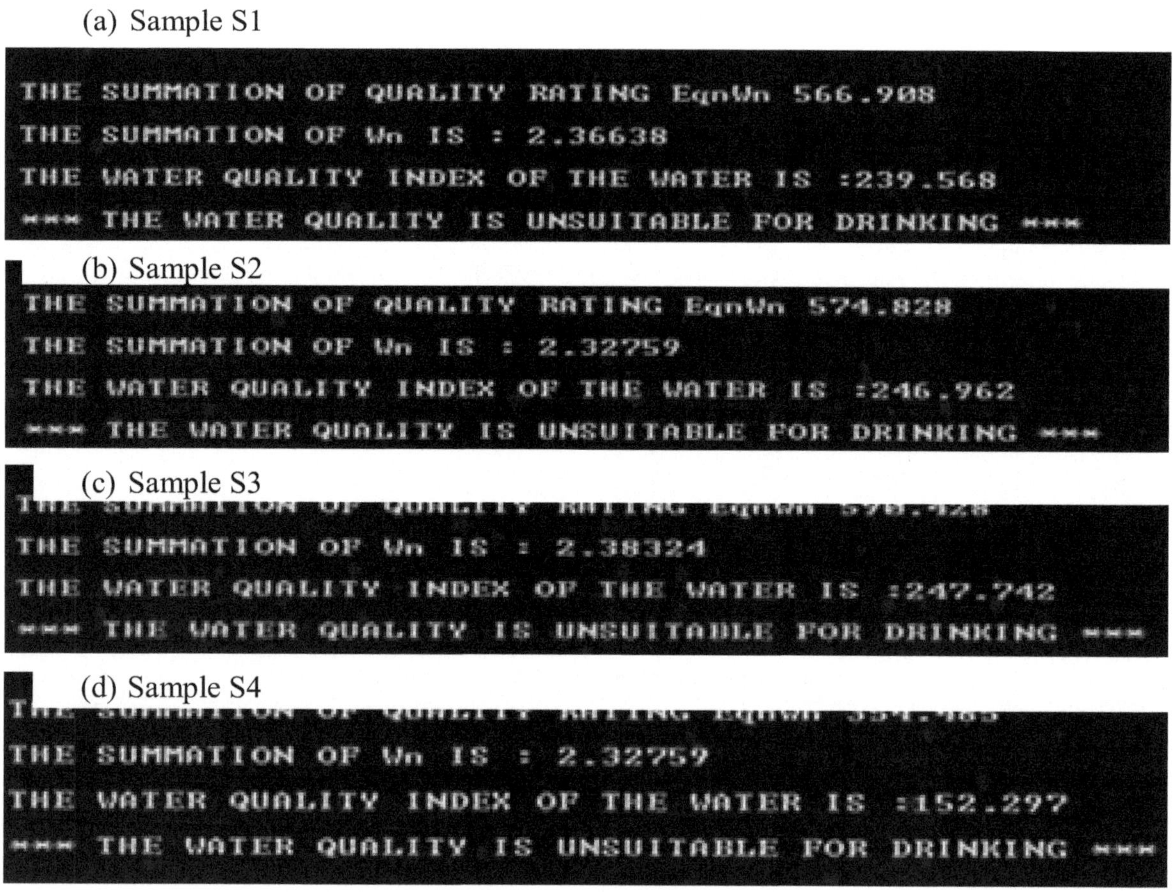

Figure 3. Water quality index programme results for different samples.

wastewater treatment system was a success and can be applied to other wastewater treatment system.

REFERENCES

Afify M, Arafa A, Maksoud NAE (2013). Evaluation of some health hazards among egyptian leather tannery workers. J. Appl. Sci. Res. 9(4):2959-2964.

Alam M, Pathak JK (2010) Rapid Assessment of Water Quality Index of Ramganga River, Western Uttar Pradesh (India) Using a Computer Programme. Nature Sci. 8(11):1-8

Banerjee T, Srivastava RK (2009). Application of water quality index for assessment of surface water quality surrounding Integrated Industrial Estate-Pantnagar. Water Sci, Technol. 60 (8): 2041-2053

Belay A (2010). Impacts of Chromium from Tannery Effluent and Evaluation of Alternative Treatment Options: J. Environ. Protect. (1): 53-58.

Bosnic M, Buljan J, Daniels RP (2000). Pollutants in tannery effluents: Definitions and environmental impact, and Limits for discharge into water bodies and sewers. Regional programme for Pollution Control in the Tanning Industry in South-East Asia. United Nations Industrial Development Organization (UNIDO). US/RAS/92/12.

Boyacioglus H (2012). Utilization of environment and index methods as water quality comparative assessment tools focusing on heavy metal content: Achieves of environment protection. 38 (3):17-28.

Brown RM, McClelland NI, Deininger RA, Tozer RG (1970) A water quality index: do we dare?" Water and Sewage Works. *117: 339-343.*

Chatterjee C, Raziuddin M (2002) Determination of water quality index (WQI) of a degraded river in Asansol Industrial area, Raniganj, Burdwan, West Bengal. Nature, Environment and Pollution Technology. 1 (2): 49-59:181-189.

Chattha J, Shaukat M (1999). An assessment of environmental concerns in the leather industry and proposed remedies: a case study of Pakistan. GIK Institute of Engineering Sciences and Technology Topi, Dist. Swabi NWFP, Pakistan.

Dikinya O, Areola O (2010)Comparative analysis of heavy metal concentration in secondary treated wastewater irrigated soils cultivated by different crops. Int. J. Environ. Sci. Tech., 7 (2), 337-346.

Jabari M, Aqua F, Shahin S, Khatib A (2009). Monitoring Chromium Content in Tannery Wastewater: The Journal of the Argentine Chemical Society. 97 (2):77-87.

Kirkham MB (2006). Cadmium in plants on polluted soils: Effects of soil factors, hyper-accumulation and amendments. Geoderma 137 (1-2):19-32.

Khambete AK, Christion RA (2014a). Predicting efficiency of treatment plant by Multi-Parameter Aggregated Index. J. Environ Res. Develop. 8(3):530-539

Khambete AK, Christian RA (2014b). Statistical analysis to identify the main parameters to effecting WWQI of sewage treatment plant and predicting BOD. Int. J. Research in Engineering and Technology (IJRET). 3(01): 186-195

Ludvik J (2000a). Chrome balance in leather processing: Regional Programme for Pollution Control in the Tanning Industry in South-East Asia. UNIDO. US/RAS/92/120/11-51.

Ludvik J (2000b). The Scope for decreasing pollution load in leather processing: Regional Programme for Pollution Control in the Tanning Industry in South-East Asia. UNIDO. US/RAS/92/120/11-51.

Moscuzza C, Volpedo AV, Ojeda C, Cirelli AF (2007) Water quality index as a tool for river assessment in agricultural areas in the

Pampean plains of Argentina. J. Urban Environ. Engr 1(1):18-25.
Muramuzi F (2009). The Leather tannery industry in Uganda: Risks to the environment and human health.
Punmia BC, Jan AK (1998). Environmental Engineering-2: Waste Water Engineering. 2 nd *Ed. EWW-0618-150: 1-555. Laximid Publications (P) Ltd.*
Rand MC, Taras MJ (1975). Standard methods for the examination of water and wastewater. *13th Ed. American Public Health Association, American Water Works Association and Water Pollution Control Federation.*
Vasanthavigar M, Srinivasamoorthy K, Vijayaragavan K, Rajiv Ganthi R, Chidambaram S, Sarama VS, Anandhan P, Manivannan R, Vasudevan S (2009). Hydrogeochemistry of thirumanimuttar basin: an indication of weathering and anthropogenic impact. Int. J. Environ. Res. 3 (4):617-628.
Wang J, Wang G, Wanyan H (2007). Treated wastewater irrigation effect on soil, crop and environment: Wastewater recycling in the loess area of China. J. Environ. Sci. 19 (9):1093-1099.
Yogendra K, Puttaiah ET (2008). Determination of Water Quality Index and Suitability of an Urban Water Body in Shimoga Town, Karnataka. Proceedings of the 12th World Lake Conference, Taal, Published by the International Lake Environment Committee. 342-346.
Yao Jin, XiaoyanQu, Yuan Li, Ruoren Yu, and Keisuke Ikehata (2014). Health Effects Associated with Wastewater Treatment, Reuse, and Disposal. Water Environ. Res. 86(10): 1970-1993.
Zhang B, Song X, Zhang Y, Han D, Tang C, Yu Y, Ma Y (2012). Hydrochemical characteristics and water quality assessment of surface water and groundwater inSongnen plain, Northeast China. Water Research. 46(8):2737-2748.

Urban water pollution by heavy metals and health implication in Onitsha, Nigeria

A. C. C. Ezeabasili[1] , O. L. Anike[2] and B. U. Okoro[3]

[1]Heriot Watt University, UK.
[2]Nnamdi Azikiwe University, Awka, Anambra State, Nigeria.
[3]Department of Civil Engineering, Nnamdi Azikiwe University, Awka, Anambra State, Nigeria.

Studies of common heavy metals were conducted at Onitsha, Anambra State, the most urbanized city in Southeastern Nigeria. It was discovered that both surface and subsurface water were heavily polluted. Seven (7) heavy metals namely: arsenic (As^{+2}), cadmium (Cd^{+2}), lead (Pb^{+2}), mercury (Hg^{+2}), zinc (Zn^{+2}), copper (Cu^{+2}) and iron (Fe^{+2}) were studied. Their major sources include wastes like refuse dumps, effluent from industries, sewage, fuel spills from dumps and mechanic workshops, hospital and pharmaceutical wastes, agricultural materials, fossil fuel combustion, metallurgical industries, electronic components and semi-conductors, batteries, pigments and paints. Further data were collected in ongoing research works in the area and public enlightenment was done in a small scale. Recommendations are made to government to note the impending dangers of further urban environmental pollution by these trace metals.

Key words: Polluted, effluents, research, urban.

INTRODUCTION

Effluent from industries in Onitsha, Anambra State, Nigeria, together with sewerage and waste spills from households and markets has become a nuisance in the state. These vary from nuisance from the odour to obstructions of traffic to more serious problems of land degradation and water pollution. It is feared that some of the adverse consequences are the noticed increase in disease incidence especially cancer, kidney infection and damage, as well as problems of the cardiovascular system and gastroenteritis (Agbozu et al., 2001; Rajappa et al., 2010). Industrial effluents and effluent from sewerage, as well as those from households contain solvents and other chemical ingredients that have key heavy metals like arsenic, lead, cadmium and mercury. Wastes also derived from shops and small-scale industries like dry cleaning, bleaching, breweries and hospitals has become numerous and may contain plastics, batteries and unused drugs. These wastes require special control, treatment and disposal methods); the practice of dumping of refuse and wastes is common

Table 1. Industries in sampling location (within Onitsha metropolis).

Major industry type	Code	Location	No.	Products
Food, beverage and Tobacco	A	Onitsha	25	Beer, starch and other miscellaneous food products, flavouring, soft drinks, floor and grain milling, meat, poultry and fish, tea, coffee and other beverage, dairy products, etc.
Chemical and Pharmaceutical	B	Onitsha	44	Paints, vanishes, soap and detergents, agro-chemicals, pharmaceuticals, dry cell batteries petroleum products. Toiletries and cosmetics, industrial chemicals
Domestic and industrial plastic, rubber and foam	C	Onitsha	26	Rubber products, plastics, foam
Metal, iron and steel	D	Onitsha	26	Foundry, metal manufacturing & fabrications, aluminum producers, etc.
Electrical and electronics	E	Onitsha	4	Electronics, refrigerators, bulbs, power control & distribution equipment
Textiles, carpet, leather and footwear	F	Onitsha	5	Textile materials, leather products ropes, etc.
Wood and wood products	G	Onitsha	3	Furniture
Non-metallic minerals products	H	Onitsha	8	Glass, ceramic, asbestos, chalk, cement packaging.
Motor vehicle and miscellaneous assembly	I	Onitsha	7	Boat fabrication, automobile manufacturers, etc.

Source: MOI, Anambra State.

in both the old urban areas and the new-urban areas adjoining them (Bhaskar et al., 2010). Efficient waste management requires provision of well selected, designed and operated sanitary landfills. Due to the importance of good water quality, monitoring of pollutants has become imperative. The effectiveness of treating waste before they are released to the environment has been recognized worldwide especially in the developed world, where laws are strictly enforced (Agarwal and Manish, 2011). This is important so as to ensure that permissible concentration of a particular chemical is not exceeded.

Location of study area

Onitsha is located in Anambra State In Eastern Nigeria and lies within latitudes 5°22[1] and 6°48[1] and longitudes 6°32[1] and 7°20[1]. Onitsha in Anambra State is located on the east bank of River Niger and covers an area of about 49,000 km^2. It is one of most important commercial centres in sub-Sahara African, as well as a transit city in Nigeria. It has an estimated population of about one million inhabitants. The socioeconomic characteristics of Onitsha consist of about 75% labour force that is engaged in tertiary sector, such as, trading and services. The remaining 25% of labour force is engaged in manufacturing and industrial activities. However, Onitsha is a centre for the production of local goods and services, as well as, a market for the sale of foreign goods. The Onitsha main market, which is reputed as the largest market in sub-Saharan Africa, has increased the tempo of commercial activities in the city in recent times. There are two main seasons: the dry season, October - March and the rainy season (April - September) approximately corresponding to the dry and flood phase, respectively of the hydrological regime. The vegetation is derived guinea savannah (Awachie and Ezenwaji, 1981).

METHODOLOGY

Random sampling technique was employed in choosing water samples from surface and groundwater samples in Onitsha, Anambra State (Table 2). Water samples were collected in a properly washed and sterilized 500 ml screw-capped glass bottles transported in ice pack and analyzed immediately. An experimental method of research was performed to assess the presence or absence of toxic heavy metals in selected samples and the concentration of each heavy metal when present.

The selected heavy metals were arsenic (As^{+2}), cadmium (Cd^{+2}), lead (Pb^{+2}), mercury (Hg^{+2}), zinc (Zn^{+2}), copper (Cu^{+2}) and iron (Fe^{+2}). Analysis of the heavy metal contents in the water samples was done with the use of the atomic absorption spectrophotometer (AAS).

This method is based upon the absorption of radiant energy, usually in the ultra-violet and visible regions by neutral atoms in the gaseous state. Water sample were stored in 50 ml of PVC bottles, analysis was done using AA210FC, Varian instruments, Fast Sequential, Analytical grade chemicals (HNO_3, Sigma chemicals, Australia and standard heavy metal solutions, Varian instruments, Australia) after preserving at 4°C for a short period of time. Samples for heavy metal analysis were digested using nitric acid (65% purity) as described in the APHA methods (APHA, 2005).

RESULTS AND DISCUSSION

Major sources of pollutants in Onitsha

Effluent discharge from industries

Data from the Ministry of Industries, Anambra State were analyzed statistically (Table 1 and Figure 1). Investigation carried out showed that effluent is discharged directly into

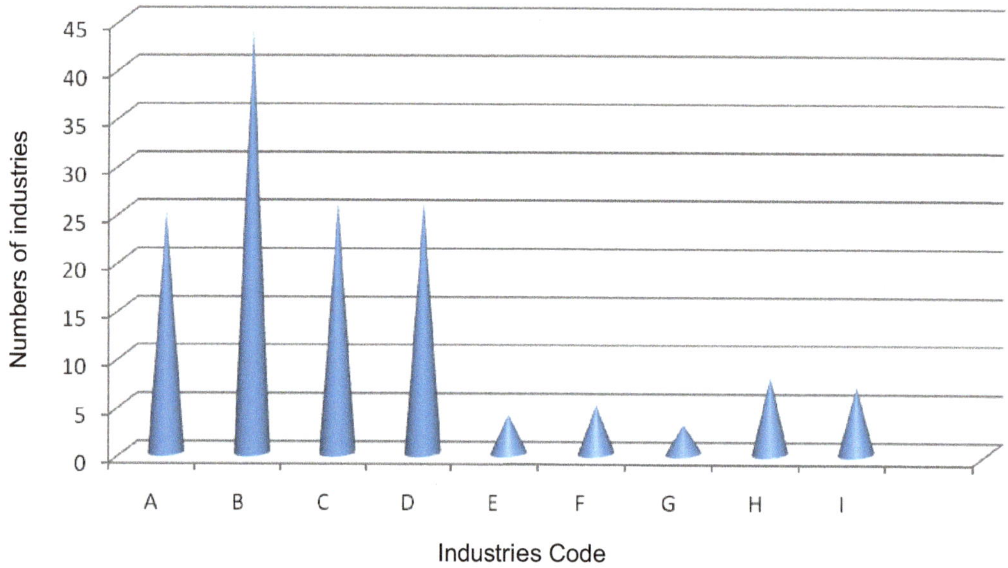

Figure 1. Industrial density in the study locations.

the environment. Motor vehicle and miscellaneous assembly, chemical and pharmaceuticals, domestic and industrial plastic etc, a few of the industrials are located in sampling sites. Onitsha is more prone to pollution because of the vast surface water and the more shallow subsurface water with more permeable soils having little of Imo shale formation. The contributions of these industries are no doubt pronounced in the research results. The skewed distribution of these industries toward the River Niger and Nwangene stream would be responsible for pollution of the rivers. The shallow coastal plain aquifer of the main market environs is also responsible for the ease of pollution of subsurface water in the area (Igwe et al., 2008). None of these industries has any programme for effluent treatment nor are there intense pressures from the government on them to provide for effluent treatment.

Refuse dump sites

Refuse dumps are found in various zones as approved by the Anambra State Environmental Sanitation Agency as well as in zones not so approved. It was discovered that various gully sites are used both by communities, individual households, and the agency for disposal of refuse.

However, the agency has designated two sites each at Omogba phase I and II as final dump sites for the zone. These final dumpsites are not designed landfills and therefore are outlets for the pollution of shallow ground waters around the highly permeable soils of the areas. Other temporary refuse dumps are found at over 20 other sites from where they are emptied in the final dumps at the convenience of the sanitation authority.

Other major dumps are found at Creek road around the Rivers Niger and Nwangene. These dumps constitute the temporary major dumps on land surface of all other dumps in the vicinity. The Ochanga Market and the main markets are also sources of refuse dumps which generate effluent that contain mainly biological and chemical pollutants that contaminate both surface and subsurface water in the coastal area aquifer of the zone.

Fuel (petrol, diesel, kerosene and oil) from filling stations and vehicle exhaust fumes

Fuel filling stations are fastest growing industries in the state. Most of these stations are concentrated along the highways while the rest could be found along major streets of the city. It is therefore easy to identify these stations and their areas of influence where leakages in the underground tanks and spills during discharge pollute the environment. Onitsha has over one hundred and six (106) functional stations. Exhaust fumes of internal combustion engines also contribute to the pollution of soil and water resources in the areas due to the high number of vehicles on the road (Ogbuagu and Ajiwe, 1998).

Previous studies in this area have shown the presence of lead and other heavy metals (Ajiwe, 1996; Ibe, 2000; Okoye, 2004). Although, the values of lead may not have been higher than the WHO standards in many cases, it is feared that the buildup of this metal if not controlled may become major health problems in the next few years. Mechanical workshops, which are spread across Onitsha, are numerous and their contribution to pollution especially lead is obvious. It is also known that car battery and other battery types like those for clocks, torch lights dumped into the system also contribute very much

Table 2. Groundwater and surface water sampled locations.

Sampling points	Code as used in figures 2 and 3
Borehole (Bany water)	1
Borehole (Onitsha South)	2
Borehole (Afasa, Onitsha)	3
Borehole (Life Brewery)	4
Borehole (Metal Industry)	5
Borehole (Otumoye)	6
River Niger (Upstream)	7
Nwagene stream	8
River Niger (Otumoye)	9
River Niger (Central drain)	10

Table 3. Descriptive statistics of heavy metals in sampling points.

Parameter	As	Cd	Pb	Hg	Zn	Cu	Fe
Range	2.61	0.98	1.99	1.91	1.26	0.97	1.24
Minimum	0.00	0.00	0.01	0.00	0.07	0.02	0.12
Maximum	2.61	0.98	2.00	1.91	1.32	0.98	1.36
WHO (2011)	0.01	0.003	0.01	0.006	3.00	2.00	0.30

to the present elevated value of lead in surface and subsurface water (Danquah et al., 2011).

Hospitals

Hospitals are spread over Onitsha and the numbers of private Hospitals have soared because of the confidence of the people and the quick services rendered by the private hospitals. These hospitals, especially the private ones, do not have waste dumps or lined landfills for the special types of pollutants; chemical, biological and even radioactive which emanate from wastes of special drugs and other wastes generated therein. These wastes are either dumped into the popular refuse dumps or buried in shallow dumps. Hospital wastes no matter their quantity remain one type to be given special treatment (Coker et al., 2008). Onitsha metropolis has about 220 hospital/ clinics of different sizes.

Other sources of pollution

Other sources of pollution include soil and gully erosion and flooding that have polluted streams and reservoirs. Presently, soil erosion remains the greatest ecological problem in the state, sewage disposal systems are not adequate and so sewage from septic system, pit latrine also contribute a lot to pollution in the areas as thousands of private and residential buildings have sprung up in the past twenty years. Recent construction works on roads has increased with little attention paid to all the environmental effects of the wastes generated, Agricultural wastes and chemicals are washed down to these cities from the hinterlands. These wastes include nitrate, heavy metals, fluorides and some radioactive wastes that find their way through runoff water to the main rivers of the area and into subsurface water too (Adeyemo and Sangodoyin, 2005).

Appraisal of pollution level in the area of study

Table 3 depicts the concentration of Arsenic level in surface and groundwater samples. Arsenic concentration was between 0.0000 (borehole, Onitsha South) and 2.6058 mg/l (Nwagene stream). Arsenic pollution in these areas are traced to anthropogenic pollution especially industrial pollution, refuse dumps and oil/fuel dumps that get washed down to streams/rivers of these major cities on daily basis (Danquah, 2011). The value of arsenic found in these areas is over and above the world Health Organization standard of 0.01 mg/l in both surface water and the boreholes adjacent and closest to these rivers and streams.

Lead was found in all water samples tested and the average values range from 0.0113 (borehole, Bany water) to 2.0013 mg/l (Nwagene stream). Petrol stations are sited all over the places. Surface spills and leakages from buried tanks that have been operating over 20 years without maintenance pollute water.

Cadmium values as seen in Figure 2 are moderately high with range of 0.0000 to 0.9800 mg/l. As previously discussed, refuse dumps littering everywhere and the reckless disposal of sewage and wastewater, all over the areas are responsible for appreciable high values recorded, with highest values recorded for Nwagene stream.

Apart from the numerous dump sites containing discarded vehicle batteries and small batteries for various uses and some of the manufacturing industrial effluent; appreciable quantities of mercury were detected with

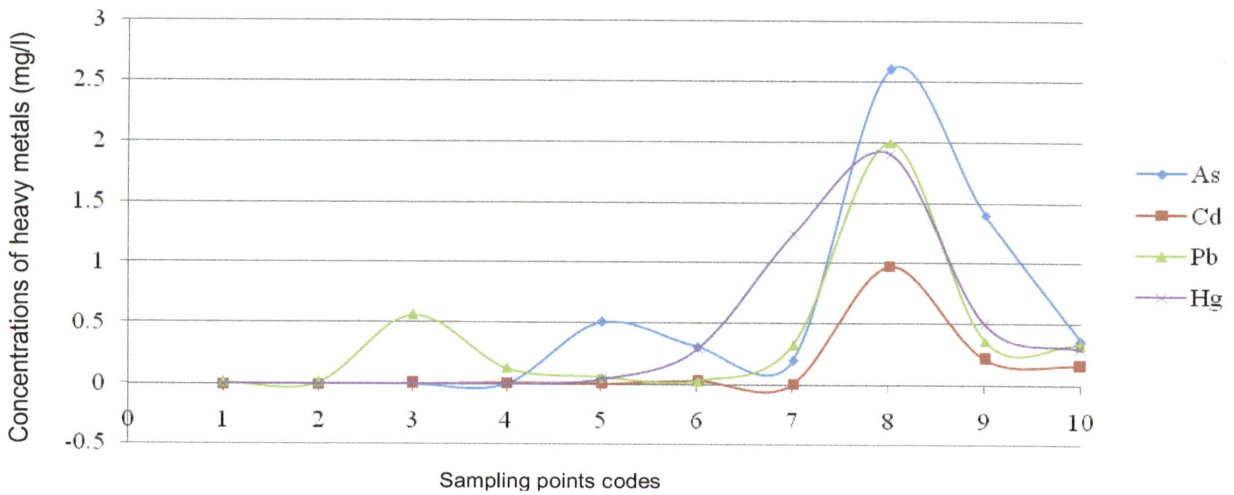

Figure 2. Concentrations of arsenic, cadmium, lead and mercury in the sampling points

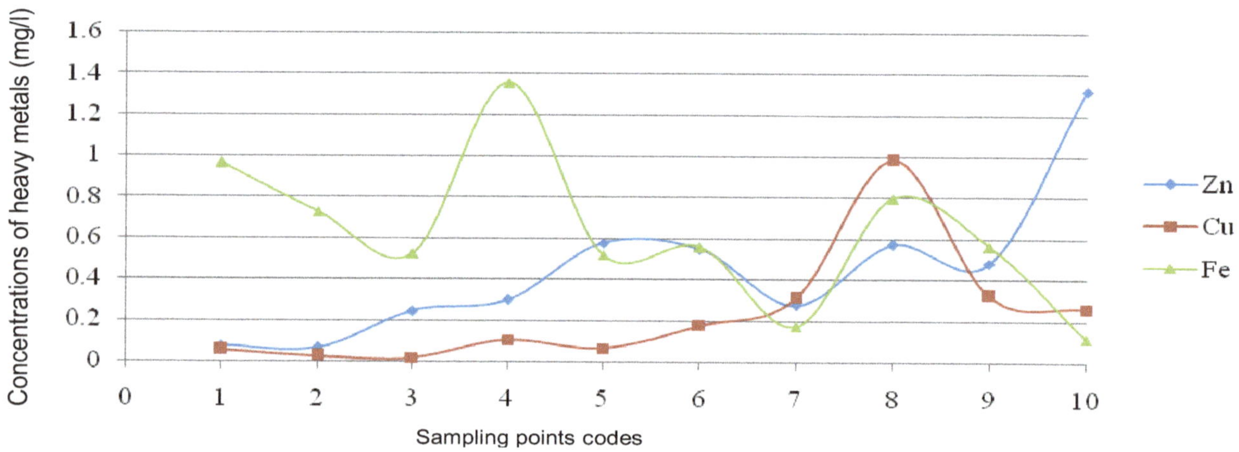

Figure 3. Concentrations of zinc, copper and iron in the sampling points.

values from 0.0000 (borehole Onitsha south) to 1.9076 mg/l (Nwagene stream). Most of the values are far above the World Health Organization standard of 0.006 mg/l.

Zinc is likely to have found its way into the environment from industrial producers in these cities. Galvanized steel is however identified as a major source of zinc in this area. They may come as wire fencing which eventually dissolve in rain and drops to the ground below. Sewage and animal manures has also contributed to the significance presence of zinc in the surface and subsurface water. Onitsha have high values of zinc encountered and this is attributable to the numerous "zinc" roofs that are used for thousands of buildings that have rusted away for up to thirty years in most cases because these are old cities. Fencing with galvanized steel barbed wires and nails are also common contributors. The indiscriminate disposal of sewage and similar wastes from refuse dumps over the areas also contribute to zinc pollution in the area. The standard values for zinc is 3 mg/l for WHO standards. From Figure 3, the highest value of 1.3242 mg/l was recorded at River Niger central drain while the lowest of about 0.0675 mg/l was recorded at borehole (Onitsha South).

Values of copper recorded for both the subsurface water and the surface waters and rivers were generally much lower than the standard set by the WHO (2.0000 mg/l). The highest value of copper recorded in surface water was 0.9829 mg/l at Nwagene stream, and lowest 0.0169 mg/l at borehole (Afasa Onitsha). Copper poisoning is therefore not yet contemplated but the values of copper content remains a pointer to impending problems.

From Table 3, the WHO standard for iron is 0.3 mg/l. The values ranged between 0.1179 (River Niger central drain) and 1.3580 mg/l (borehole, Life brewery). Table 4 reveals that significant interactions exist among

Table 4. Correlation analysis of heavy metals.

	As	Cd	Pb	Hg	Zn	Cu	Fe
As	1						
Cd	0.939**	1					
Pb	0.854**	0.948**	1				
Hg	0.781**	0.811**	0.822**	1			
Zn	0.277	0.274	0.188	0.160	1		
Cu	0.918**	0.957**	0.910**	0.930**	0.308	1	
Fe	-0.003	0.069	0.013	-0.215	-0.524	-0.062	1

**Correlation is significant at the 0.01 level (2-tailed).

heavy metals. At $P<0.01$, positive correlation existed between arsenic and cadmium (r=0.939), lead (r=0.854), mercury (r=0.781), copper (r=0.918); cadmium and lead (r=0.948), mercury (r=0.811), copper (r=0.957), e.t.c as depicted. Result affirms that some indicator parameters influence presence of others.

Impact of heavy metal pollution

Eni et al. (2011) noted that water pollution results to a gradual decline in general economic life of the populace. At Onitsha, the fish industry is affected by gradual poisoning of fish population in areas and periods of heavy pollution loads. Fish production is also reduced because heavy metals at the level of pollution cause stunted growth of the fishes in the River Niger (Ajiwe et al., 2002). As noted by Rish and Rishi (2011), consumption of fishes and other sea foods from the River Niger over long periods could lead to diseases like cancers of the bladder, liver and intestine, clogging of arteries, hypertension, stroke and heart diseases. This therefore, will lead to reduction of the life expectancy of the people. A general hazard in Onitsha includes the presence of odour and colour in the affected waters. Pollution of surface water affects water supply network, resulting in health risks and increased load on water treatment plants, increase in cost of water treatment and therefore obvious health consequences. The leachates and effluents containing heavy metals also contain organic compounds and synthetic material pollutants and contaminants with long half-lives. These remain hazardous for long time within the hydrogeologic environment and are very difficult to remove. High levels of iron bacteria have caused problems in water supplies, coating on piping systems, impellers and motors that reduce flow rates of pumps, total plugging of the wells after prolonged use. It also affects portability of water. Corrosive ground water easily eats off the screens and pumps of boreholes (Okun et al., 1971).

The effect of heavy metals in these areas of study is chronic in that they have a long term effect on health due to frequent exposure to small amounts. The combination or synergistic effects of these metals are suspected to be responsible for notable ailments like cancer, kidney and cardiovascular diseases as attested for by records in the Nnamdi Azikiwe University Teaching Hospital, Anambra State. From the Public Health Department of the same institution, it is also recorded that symptoms of ailment like tiredness syndrome, lethargy, depression and rheumatoid arthritis are common. Other symptoms of heavy metal pollution encountered in treating patients include white lines across the toenails and finger nails, weight loss, nausea, diarrhoea alternating with constipation, damage of gastro-intestinal track and liver, kidney failure, hypertension and loss of hair all ascribed mostly to arsenic, cadmium and mercury (Ezeabasili, 2008).

It is anticipated that if the pollution of water by heavy metals is not abated, the incidence of the disease would increase and new forms of ailment could be experienced in these urban areas as in other parts of the world with similar experiences.

Conclusion and recommendations

Onitsha is generally littered with refuse dumps. Because of the large population density, the level of pollution is getting increasingly serious. The heavy metals in water come from these refuse dumps, industries, drug containers and hospital wastes, various forms of chemical dumps from the market and small scale industries.

In the light of the findings in this work, the author makes the following recommendations; the introduction of modern, sensitive and reliable methods for testing drinking water for physical, chemical and biological parameters is very important and urgent. The development of adequate infrastructure for water testing laboratories is necessary. The presence of AAS equipment in every institution is recommended. This would increase research work on all our water resources like that obtained in most developed countries now. Secondly, the Anambra State Government should strengthen and enforce existing laws bordering on pollution especially for dangerous chemical effluents from industries. The proper monitoring of industries through environmental impact assessment and environmental audit report (EAR) should be enforced by the Ministry of Environment. It would be better if industrial effluents were treated before discharged. Finally, the Anambra State Government should ensure that the concerned officials of the health departments should report results as and when the water quality deviates from the national and international norms, to the government and the concerning public.

The infrastructure of water quality testing laboratories requires improvement. It is noticed, that various research results cannot be reconciled because of faulty equipment. Water quality reporting system needs considerable change to alert concerned officials in inter-disciplinary sectors. As the drinking water testing laboratories are presently governed by the NAFDAC and Urban and Rural Drinking water supply agencies there should be interaction with health professionals/personnel of other sectors. Health education and health department's interaction need to be introduced and promoted at all levels at frequent intervals.

Conflict of interest

The authors did not declare any conflict of interest.

REFERENCES

Adeyemo JA, Sangodoyin AY (2005) Urban Storm Water Quality as affected by Traffic Density and other factors. J. Sci. Technol. Res. Accepted 6th June, 2005.

Agarwal A, Manish S (2011). Assessment of pollution by Physicochemical Water Parameters Using Regression Analysis: A Case Study of Gagan River at Moradabad- India. Adv. Appl. Sci. Res. Vol. 2(2): 185-189.

Agbozu IE, Ekweozor IK, Bamgbose I (2001). Heavy metal pollution by effluent discharges from some industries at Trans Amadi Industrial Layout in Port-Harcourt metropolis. J. Appl. Sci. 4(3): 1951-1968.

Ajiwe VI (1996). Characteristics of groundwater in Anambra State, Nigeria: Policies and Remedies, Book of Abstracts, Stockholm Water Symposium, 4th-9th August Stockholm, pp. 152-158.

Ajiwe VI, Ezirim SO, Ajibola VO (2002). Heavy metal pollution on coastal fishes of the coast of River Niger. J. Appl. Sci. 5(2): 2836-2843.

APHA (2005) American Public Health Association. Standard methods for the examination of water and wastewater. 21st ed. Washington, D.C.

Awachie JB, Ezenwaji H (1981). The fisheries development of the Anambra River Basin, Nigeria. CIFA Technical Paper. 8: 212-224.

Bhaskar CV, Kumar K, Nagendrappa G (2010). Assessment of heavy metals in water samples of certain locations situated around Tumkur, Karnataka, India', viewed 6 April, 2014, Coker A, Sangodoyin AY Sridhar M, Booth C, Olomiolaiye F, Hammonf F (2008). Medical Waste Management in Ibadan, Nigeria: Obstacles and Prospects. Waste Manag. 29:804-811.

Danquah L, Abass K, Nikoi AA (2011). Anthropogenic Pollution of Inland Waters: the Case of the Aboabo River in Kumasi, Ghana, December. J. Sustain. Dev. 4:6. http://dx.doi.org/10.5539/jsd.v4n6p103

Eni DI, Obeifuna JN, Oko C, Ekwok I (2011) Impact of Urbanisation on Sub-surface water quality in Calabar Municipality, Nigeria. Int. J. Humanities Soc. Sci. 1(10): 167-172.

Ezeabasili ACC (2008) Health implication of Heavy Metal Pollution of surface and subsurface water in Nnewi and Onitsha Urban Anambra State, Unpublished Ph.D Thesis, Nnamdi Azikiwe University, Awka.

Ibe KM (2000). The Impact Assessment of Oil on Soils and Shallow Groundwater in the Niger Delta. A Case Study of Ughelli Southwest, Nigeria. World J. Biotechnol. 1(8): 8-16.

Igwe JC, Abia AA, Ibe CA (2008). Adsorption kinetics and intraparticulate diffusivities. Int. J. Environ. Sci. Tech. 5 (1):83-92. http://dx.doi.org/10.1007/BF03326000

Ogbuagu, Ajiwe VI (1998). Industrial Pollution Survey in Anambra. J. Environ. Contamin. Toxicol. Vol. 9(5): 269-275.

Okoye MI (2004). The Pollution of lead in the River Niger at Onitsha Anambra State. Unpublished M.Sc. Thesis, Department of Production Technology, Nnamdi Azikiwe University Awka.

Okun DA, Fair MG, Geyer CJ (1971). Elements of Water Supply and Waste Water Disposal (2nd Ed.) Newyork: John Wiley and Sons. pp. 206-215.

Rajappa B, Manjappa S, Puttaiah ET (2010). Monitoring of heavy metal concentration in groundwater of Hakinaka Taluk, India. Contemporary Engineering Sciences. 3(4): 183-190.

Rishi V, Rishi M (2011) Ecological Effect of Urbanisation on Swan River Watershed District Una, Himachal Pradesh. Int. J. Pharm. Life Science. Vol. 2(5): 723-779.

WHO (2011). Guidelines for drinking-water quality. World Health Organization Geneva Vol.1: 4 ed. ISBN 978924 1548151.

Performance evaluation of constructed wetlands: A review of arid and semi arid climatic region

Mihret Dananto Ulsido

School of Civil Engineering and Architecture, Wuhan University of Technology, 122 Luoshi Road, Hongshan District, Wuhan, Hubei, 430070, P. R. China.

Aiming at environmental pollution control through the use of constructed wetlands systems (CWs) in arid and semi arid climatic region, a detailed review of CWs was undertaken. Given the practical application and simplicity of the technology, principles for building phytotechnology-ecohydrology environment used for wastewater remediation is appropriate. The ability of wetlands to filter, absorb and metabolize suspended and dissolved matter is the basic philosophy behind constructed wetlands. Ecohydrology is the sub-discipline shared by ecological and hydrological sciences that is concerned with the effects of hydrological processes on the distribution, structure and function of ecosystems, and on the effect of biological processes on the elements of the water cycle. It is actually the application of science and engineering to examine problems and provide solutions involving plants. This paper reviewed the efficiency of constructed wetland in arid and semi arid conditions. The use of sub surface flow (SSF) constructed wetland as a post treatment technique have been found as promising technology for wastewater treatment in arid and semi-arid areas.

Key words: Arid, ecohydrology, efficiency, phytotechnology, sub surface flow (SSF), semi-arid.

INTRODUCTION

The threat an environment faces with technological development while using fossil fuels and other inputs for the manufacturing as well as the transportation sector are very immense. Always, nature tries to cope with the changes in the ecosystem, such as contaminants, as long as the adsorption capacity is not surpassed. The rates at which contaminants are released from the sources (point or non-point sources) as compared with the cleaning potential of the components of the eco-system (Biotic or abiotic) differ.

The major contaminant polluting fresh water ecosystem are macro and micro nutrients from agricultural fields (Barrow,1993; Mihret et al., 2013), heavy metals and carbon dioxide from manufacturing and transportation industries, organic and inorganic contaminants from municipalities and brown fields at the outskirt and inner cities (Rehm and Reed, 2000). To curb the problem of contaminants from municipalities at a certain extent, the author of this review suggested urban farming with partially treated wastewater as a feasible option (Mihret, 2013).

Depending on the effluent sources, the natures of individual contaminants vary. Some are biodegradable with limited spatial and temporal dimensions. Others needs more time and space for degradation or adsorbed by living organisms. Another category of contaminants; conservative contaminants, retain their existence but take part in the assimilation and dissimilation processes. The non-conservative contaminants participate in the assimilation and dissimilation process and are liable for chemical changes. Their destiny may be photochemical degradation, chemical or microbial transformation (Figure 1).

The application of constructed wetlands to facilitate

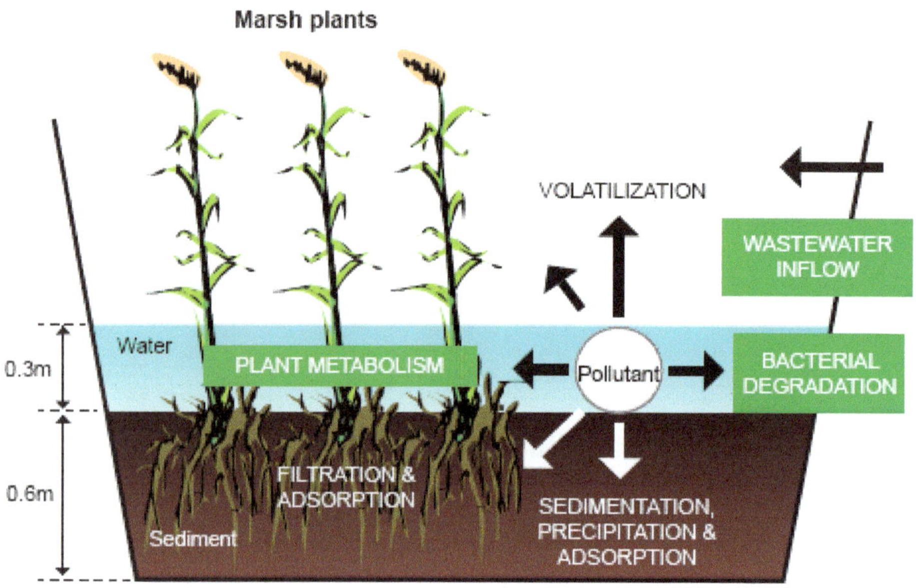

Figure 1. Pollutant removal processes in a constructed wetland system (Wetlands International, 2013).

these processes is a young post treatment technique used in developed countries. How effectively does it purifies contaminated environment? Could be a major question pointed out in relation to constructed wetlands. Competitive efficiency and effectiveness when compared with other techniques leads to the desired sustainability; which calls for evaluation of the existing constructed wetlands in arid and semi arid conditions for proper articulation of future processes.

In the following sections ecohydrology and phytotechnology concepts in the process of sludge treatment are presented, followed by constructed wetland efficiency determination, discussion and finally a conclusion.

Objectives

The plant-water-soil-microorganism interaction plays a major role in the ecohydrology concept. Soil microorganisms and plants connect the lithosphere, hydrosphere and the atmosphere.

The strength of the interaction determines the reliability of the bridge in transferring the contaminated input to environmentally friendly output (Wetland International, 2013). Investigating the efficiency of constructed wetlands in arid and semi arid conditions are the major objective of this paper.

ECOHYDROLOGY AND PHYTOTECHNOLOGY

Ecohydrology is the subdiscipline shared by ecological and hydrological sciences that is concerned with the effects of hydrological processes on the distribution, structure and function of ecosystems, and on the effect of biological processes on the elements of the water cycle (UNEP, 2004). Phytotechnology describes the application of science and engineering to examine problems and provide solutions involving plants (UNEP, 2004). Examples of phytotechnology applications include (UNEP, 2004):

1. The use of plants to reduce or solve pollution problems. Example, the use of natural wetland (W_n) for wastewater treatment.
2. The replication of ecosystems and plant communities to reduce or solve a pollution problem. Examples, constructed wetlands for treatment of wastewater or diffuse pollution sources.
3. The use of plants to facilitate the recovery of ecosystems after significant disturbances. Examples are coal mine reclamation and the restoration of lakes and rivers.
4. The increased use of plants as sinks for carbon dioxide to mitigate the impacts of climate change. Examples of this are reforestation and afforestation.
5. The use of plants to augment the natural capacity of urban areas to mitigate pollution impacts and moderate energy extremes. An example is the use of rooftop vegetation or "green roofs".

Plants possess the ability to absorb carbon dioxide in the process of assimilation from the atmosphere to emit oxygen and thus purify air (Cavalcanti, 2003). Furthermore, in the last three decades, scientists in Europe and America mimicked the way nature coped up polluted environment, and used these naturally pollution adaptive techniques to treat sewage in sewage works, as well as recently to detoxify soil (Rehm and Reed, 2000) and

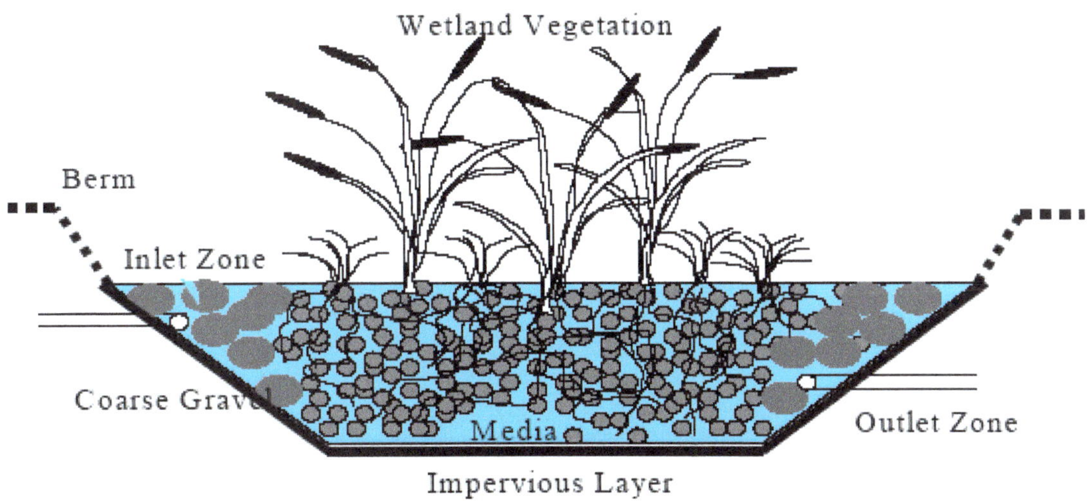

Figure 2. Subsurface flow constructed wet land (Sauer and Kimber, 2001).

sludge using constructed wetlands. Constructed wet lands are among the different techniques used in this field. Constructed wetlands are either free surface flow (FSF) type or sub surface flow type (SSF). The next section deals with the philosophy behind constructed wetlands.

Constructed wetlands

The ability of wetlands to filter, absorb and metabolize suspended and dissolved matter is the basic philosophy behind constructed wetlands. Scientists and engineers are very interested and now working together to mimic these natural systems for handling wastewaters and agricultural run-off (UNEP, 2004). This has prompted the construction of artificial wetlands to cope with the diffuse pollution originating from agriculture, septic tanks, and other sources. In some countries, for instance the U.S., legislation prohibits the drainage of wetlands unless another wetland of the same size is constructed elsewhere (UNEP, 2004).

Constructions of artificial wetlands look like an attractive and cost-effective phytotechnology concept that can be used for controlling various types of wastewater (UNEP, 2004). These wetlands are usually constructed so that water flows primarily over the sediment and through vegetation, or as vegetated submerged bed systems in which water flow is engineered for contact with plant roots.

They are excavated with a shallow gradient in soils of low permeability (Wetland International, 2013; UNEP, 2004) (or lined with an impermeable barrier and then filled with an appropriate soil). They are then either planted or vegetated naturally. Based on the hydrologic scenario in the basin, constructed wet lands are classified into two classes; sub-surface flow and free surface flow types.

Sub-surface flow wetlands

These are shallow excavations (1-1.5 m deep) with a synthetic or clay liner to prevent infiltration of water. They are filled with a media through which the liquid to be purified must flow. All in all, the water flows under the surface of the ground. The filter media can be anything from soil to light, expanded clay aggregate, but 5-10 mm gravel (Wetland International, 2013) is the most common. An inlet zone (Figure 2) made from soil aggregates of larger size ensure the influent liquid is distributed effectively into the media.

A similar outlet zone collects the treated liquid in drainage pipes which pass through the liner into a level control chamber where a simple plastic tube or swivel pipe allows the liquid level in the wetland to be controlled. Area requirement for a sub surface flow depends on the daily BOD_5 loading. The following relation (Equation 1) can be used for calculation (Wetlands International, 2013).

$$A_h = KQ_d (\ln C_o - \ln C_t) \quad (1)$$

Where, A_h = surface area of bed, m^2
K = rate constant, m d^{-1}
Q_d = average daily flow rate of wastewater (m^3 d^{-1})
C_o = average daily BOD_5 of the influent (mg l^{-1})
C_t = required average daily BOD_5 of the effluent (mg l^{-1})

The value of K = 5.2 was derived for a 0.6 m deep bed and operating at a minimum temperature of 8°C

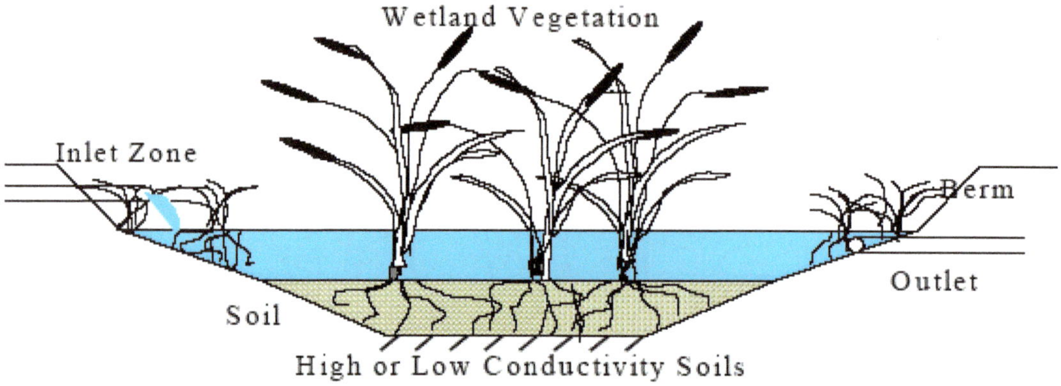

Figure 3. Free surface flow constructed wetland (Sauer and Kimber, 2001)

(Wetlands International, 2013). For less biodegradable wastewater, K-values of up to 15 may be appropriate. Using this formula, a minimum area of 2.2 m^2 $capita^{-1}$ is obtained for the treatment of domestic sewage. In practice, most design systems operate on the basis of 3-5 m^2 cap^{-1} (Wetlands International, 2013).

Free surface flow wetlands

Most free surface flow wetlands are artificial shallow marshy lands or shallow ponds filled with aquatic plants (Wetlands International, 2013). They are very shallow excavations or shallow earth banked lagoons enclosing an area of land demarcated for the purpose. The name free surface comes from the thin free water layers which are formed at the surface (Figure 3).

Soil or some other media such as gravel provides the growing media for the marsh plants. To avoid short-circuiting the surface should be virtually flat with a very gentle slope towards the outlet end (Wetlands International, 2013). The inlet zone distributes the waste-water (Figure 3) over the inlet end of the wetland, and a collection channel collects the treated liquid at the outlet end. The wastewater flows along the surface allowing settlement of solids and coming into contact with the bacterial populations on the surface of the media and plant stems.

These wetlands can serve small communities as natural wetlands and can be incorporated into the treatment systems for larger communities as well (UNEP, 2004). They can also be constructed to treat agricultural runoff or other non-point sources of pollution and are especially well suited for use in surface-mined areas (UNEP, 2004). This reference also mentioned additional ecological advantages, such as nitrogen and saltwater filtering, supply of water and nutrients, production of food and support of endangered species that can possibly increase the economic advantages as compared to conventional wastewater treatment plants.

In general, the effectiveness of phytoremediation technique in constructed wet lands depend on: geology of the site selected, hydrogeology, aquifer characteristics, soil conditions, air quality, climatic conditions, geochemistry, type and distribution of microorganisms, presence and distribution of contaminants and vegetation (UNEP; 2004). Vegetation type and characteristics, microorganism type and distribution, and soil characteristics remains influential through out the life time of constructed wetlands. Better insights of these three parameters are very crucial when dealing with the effectiveness and efficiency of constructed wetlands. In the following sections, some facts about sludge microorganisms, site soil conditions and plants adaptive to highly sludge loaded wastewater and their role in waste water treatment are presented.

Vegetation

Hydrophytes (wetland plants) serve as storage sites for carbon and nutrients and play a role in the movement of gases to and from the hydrosol (wetland soils) (UNEP, 2004). Oxygen is transported from the air through the plant into the rhizome (root zone) and from the soil too. Special types of plants (plants having aerenchym tissue) which are capable of trapping atmospheric oxygen to incorporate it into the soil with their root system are preferred for constructed wetlands. This process ensures that aerobic respiration can be maintained by the non-photosynthetic portion of the hydrophyte tissues; aerenchym tissue; that are buried in the anoxic hydrosol (Wetland international, 2013). Through symbiotic relation-ship some of this oxygen becomes available to the microbes associated with the rhizome.

Various plant species can be used in constructed wetlands (De Sousa et al., 2003). Depending on the nature of the plants they can be used for shallow marsh, marsh and deep marsh lands. Some of this plant species are presented in Table 1.

Table 1. Wetland plant species (Source; Wetland international, 2013).

Planting zone	Common name	Scientific name
Marsh and deep marsh (0.3-1.0 m)	Common Reed	*Phragmites karka*
	Spike Rush	*Eleocharis dulcis*
	Greater Club Rush	*Scirpus grossus*
	Bog Bulrush	*Scirpus mucronatus*
	Tube Sedge	*Lepironia articulata*
	Fan Grass	*Phylidrium lanuginosum*
	Cattail	*Typha angustifolia*
Shallow marsh (0-0.3 m)	Golden Beak Sedge	*Rhynchospora corymbosa*
	Spike Rush	*Eleocharis variegata*
	Sumatran Scleria	*Scleria sumatrana*

Soil nutrients

Hydrophyte growing in the constructed wetland as well as microorganisms in the soil needs optimum level of nutrients in the soil. The levels of essential nutrients control how well the plants (and microbes) grow. For optimum growth, plants require a certain ratio of specific nutrients.

If this nutrient ratio is not balanced, plant and microbial growth is badly affected. Unbalanced nutrient availability affects the growth rate of preferred microbes for waste treatment. Unbalanced ratios change the microbial population and generate smelly metabolites from anaerobic pathways with sulfur and nitrogen containing compounds (Rehm and Reed, 1999).

Microbial activity

Wetland bacteria consume the water soluble organic compounds such as hydrocarbons and BOD (see the reverse mineralization processes in the above reaction). Bacteria contain unique enzymes which allow them to metabolize the suspended organic compounds. In the dissimilation process, the aerobic bacteria produces metabolites which are incorporated into cell mass, used as energy, and/or are converted to nontoxic biological wastes (Rehm and Reed, 1999). Water soluble organic compounds are used as an energy or carbon source by other bacteria, fungi and hydrophytes. Another specific example could be the conversion of ammonium to nitrate by the bacteria nitrosomonas.

$$NH_4^+ + 2O_2 \xrightarrow[\text{Bacteria}]{} NO_3^- + N_2\ H_2O$$

This reaction to get to completion about 4.6 $mg.L^{-1}$ of oxygen is required per gram of ammonia nitrogen (Cheremisinoff et al., 1990). The optimum combined effect of all the parameters stated above gives optimal effectiveness and efficiency for constructed wetlands. Constructed wetland efficiency evaluated in arid and semi arid climatic conditions are presented in the following section.

APPROACH

Analysis on the performance of constructed wetlands was done to evaluate research outputs in arid and semi arid climatic regions. Research outputs based on the numbers of papers published in over 13 journal sources, web pages and papers accepted for publication from presentation on an international conference on constructed wetlands for water pollution control were used for this review.

In view of the current demand of this review, the performance assessment assignment employed a range of tools to gather and analyze data from secondary sources. The review approach considered and used specific indicator parameters important for the environment. The parameters used include:

1. Chemical oxygen demand (COD)
2. Biological oxygen demand (BOD)
3. Total suspended solids (TSS)
4. NH_4-N (ammonia nitrogen)
5. Total Kjeldahl nitrogen (TKN)
6. Total phosphors (TP)
7. Fecal coliforms (FC)

Relevant research documents have been reviewed during data collection and thoroughly reviewed during the preparation of this report. The data collected includes a summary of performance reports, scientific reviews and project completion reports and others.

Performance indicator parameters were computed as a percentage of pollutant reduction as compared to its initial pollutant level. The specific methods used for the study was presented below each case studies investigated.

Findings from a case study held in arid climatic condition (Egypt) and another in semi arid climatic condition are investigated. In addition, a summary of the efficiency of constructed wetlands in tropic and sub-tropic condition are taken from a PhD work of Hristina Bojcevska (Bojcevska, 2010), PhD student in Ecology/Wetland Engineering, Linköping, Sweden. The main intention of this study paper was to find an effective wetland type that can best suit arid and semi arid conditions.

Case study 1: Combining UASB technology and constructed wetland for domestic wastewater reclamation and reuse (El-Khateeb and El-Gohary, 2003)

In this paper the performance efficiency of constructed wetlands in Egyptian conditions are compared. For the study, two treatment schemes consisting of an up-flow anaerobic sludge blanket reactor (UASB) followed by either subsurface flow (SSF) or free surface flow (FSF) constructed wetlands have been investigated. UASB reactor can be briefly described as a system in which substrate passes first through an expanded sludge bed containing a high concentration of biomass (Cavalcanti, 2003; De Sousa et al., 2003).

The sludge in the reactor may exist in granular or flocculent form, but the granular sludge offers advantages over flocculent sludge.

Most of the substrate removal takes place in sludge bed. The remaining portion of the substrate passes through a less dense biomass, called the sludge blanket. The common macrophyte in Egypt *Typha latifolia* (cattail) was used at a planting density of three rhizomes m^{-2}. To evaluate the role of plants in the treatment process, an unplanted gravel bed identical to the SSF unit was operated as control.

During the study period, the wetlands were fed with the UASB effluent at an organic loading rate ranging from 17.3 to 46.8 kg BOD_5/ha·d (55.1 to 134.6 kgCOD.$ha^{-1}\cdot d^{-1}$). Effectiveness of the system for the removal of key constituents (COD, BOD, TSS, nutrients and FC (fecal coli form) has been investigated.

The results show that, the level of COD_{tot}, and TSS in the final effluent of SSF was lower than that of FSF. The possible justifications could be low flow velocity and higher surface area in the SSF media and lower microbial activities as a result of temperature variation in the case of FSF media.

Subsurface flow wetland has demontrated higher overall efficiency than the unplanted control. FC count showed FSF media's superiority over the control and SSF. The reason could be the possible exposure of the bacteria's for UV in the case of FSF. BOD_{tot} showed more or less the same pattern in both SSF and FSF media (Table 1).

Case study 2: River water quality improvement by natural and constructed wetland systems in the tropical semi-arid region of Northeastern Brazil (Ceballos et al., 2001)

In this paper the performance efficiencies of a natural *Typha* spp. wetland (W_n) and constructed wetland (W_c, SSF type) were compared in northeastern Brazil (Paraiba State). The summary of the outcomes from the evaluation are presented in Table 3.

In the natural wetland, removal values were: 75 and 81% BOD_5; 10-53% total phosphorus; 13-55% ammonia; 89-91% FC. Constructed wetland removals increased with HRT with best results on 10 day. Observed removal values were: 74-78% BOD_5; 58-82% ammonia; 94-98% FC. Despite the high remaining values of FC (1.4 x 10^4CFU in 100 ml), the removals were satisfactory and hydraulic retention time (HRT) dependent, suggesting a gradual optimization of the system with time. The W_c exhibited good efficiency for improving water quality from polluted river.

On her PhD work, Bojcevska (2010) did a summary of the research findings of four different researchers who made their researches using different macrophytes other than *Typha* spp. Here this summary is presented to justify the factors which affect the performance efficiency of the two types of constructed wetlands. Different efficiency values are observed as a function of (Tables 2, 3 and 4):

1. Different types of constructed wetlands used
2. Types of hydrophytes used
3. Hydraulic loading rate
4. Hydraulic retention time

CONCLUSION

A Well-designed constructed wetland can perform better than natural wetlands due to easier control and management (Babatunde et al., 2008). Removal efficiencies of different types of constructed wetlands are different.

FSF wetland is better in some aspects (FC removal) than SSF. SSF wetland also showed some superiority over FSF. Even though both of them showed a similarity in BOD_5 removal, SSF wetland showed higher efficiency in COD and TSS removal.

The type and constituents of waste water in the effluent side deter-mines the type of constructed wetland to be used. For waste water rich in fecal coliform and nitrogen, FSF wetland could be a better preference than SSF. For waste water rich in COD, TSS and TP, SSF is a good choice.

Therefore, appropriate decision during wetland construction affects the performance of the system. Performance efficiency can also be dependent on climatic conditions, hydraulic retention time, macrophytes

Table 2. Main characteristics of the sewage measured at the out let of raw sludge conduit, ASB reactor, SSF and FSF wetlands (standard deviation in brackets) (El-Khateeb and El-Gohary, 2003).

Sludge characteristic	Effluent measured at the out let of			
	Raw sludge	UASB Reactor	Constructed wetland type	
			SSF	FSF
COD_{tot} (mg/L)	620.9 (±189)	241 (±66)	53(±19)	3(± 26)
BOD_{tot}(mg/L)	282 (±68)	99 (±27)	22(±(8.5)	20(± (8)
TSS (mg/L)	191 (±68)	59 (±24)	12± (5.5)	26(± 17)
NH_4-N (mg/L)	31 (±5.5)	33 (±5)	24(±6.6)	16(±5)
TKN(mg/L)	61 (±20)	55 (±16)	40(±15)	32(±16)
TP(mg/L)	5.2 (±1.6)	3.4 (±0.7)	2.1(± 0.9)	2.4(±1)
FC (CFU/100 ml)	2.4×10^9 ($\pm 2.18 \times 10^9$)	1.8×10^8 ($\pm 2.5 \times 10^8$)	8.3×10^4 ($\pm 9 \times 10^4$)	7.2×10^4 ($\pm 8.5 \times 10^4$)

Table 3. Summary of the efficiency observed in natural and constructed wetlands (Ceballos et al., 2001).

Effluent characteristic	Natural wetland (W_n) average efficiency (%)	Constructed wetland (W_c) average efficiency (%)	Remark
BOD_5	78	76	Constructed wetland showed relatively modest efficiency over the natural one
TP	24	12	
NH_4-N	27	70	
FC	90	90	
SS	50	66	

Table 4. Performance efficiencies in relative percentage of some constructed wetlands (CW) located in tropical and subtropical regions concerning a selected number of water quality parameters.

CW type	Area (m^2)	HLR (m/d)	HRT (d)	Mean influent concentration (mg/L)		Reduction in concentration (%)		Reference
FSF	300	-	7-12	COD	155.2	C.P	63	Okurut et al. (1999) as cited by Bojcevska, 2010
						P.M	43	
				PO_4	3.71	C.P	16	
						P.M	37	
SSF	n.i	6.48	n.i	SS	104.8		80	Mashauri et al. (2000) as cited by Bojcevska (2010)
				COD	100.8		66	
		55.2	n.i	SS	101.8		50	
				COD	125.8		50	
FSF and SSF	10	0.018	12.8	PO_4-P	2.39		69	Lin et al. (2002) as cited by Bojcevska (2010)
		0.034	6.8		10.45		45	
		0.135	1.7		5.19		35	
SSF	9.36		5	TRP	15,5	C.P	83	Kyambadde et al. (2004) as cited by Bojcevska (2010)
						M.V	48	

n.i = No information; C.P = *Cyperus papyrus*; P.M = *Phragmites mauritianus*; M.V = *Miscanthidium violaceum*; FSF = free surface flow; SSF = sub surface flow, HLR = hydraulic loading rate; HRT = hydraulic retention time.

growth rate and type, root zone maturity and biofilm formation, substrate characteristics and rate of microbial growth and activity. Selecting an appropriate substrate media that suits the type of available hydrophyte species facilitates better removal efficiency. It is essential to avoid the introduction of an invasive hydrophyte species which does not belong to the prevailing wetland ecosystem (personal view). Constructed wetlands are not recommended for treatment of raw waste water. Combining SSF with UASB gives a higher performance efficiency and longer operation time. The different scenarios presented shows that the use of SSF wetland as a post treatment technique after a UASB reactor is a promising technology for wastewater treatment in arid and semi-arid areas.

ACKNOWLEDGEMENTS

The author would like to thank Wuhan University of Technology for providing the necessary facilities to conduct this review. The author is also indebted to the editor and the two anonymous reviewers for their constructive comments, which helped a lot to improve the original manuscript.

REFERENCES

Babatunde AO, Zhao YQ, O'Neill M, O'Sullivan B (2008). Constructed wetlands for environmental pollution control: a review of developments, research and practice in Ireland. Elsevier , Environment International, 34 (1):116-126.

Barrow Chris (1993). Water Resources and Agricultural Development in the Tropics. John Wiley and Sons, Inc., New York, USA.

Bojcevska H (2010). Treatment performance of a free water surface constructed wetland system receiving sugar factory effluents in the Lake Victoria region, PhD thesis. (IFM/Department of Biology, University of Linköping, Linköping, Sweden) online at : http://www.ifm.liu.se/~inuita/researchproposal.html

Cavalcanti PFF (2003). Integrated Application of the UASB Reactor and Ponds for Domestic Sewage Treatment in Tropical Regions. (Thesis; Wageningen University, Wageningen). Online at: http://library.wur.nl/wda/dissertations/dis3368.pdf

Cheremisinoff PN,Trattner RB, Lawson BK, Newberger BI (1990). Introduction to Biotechnology for Waste /Wastewater Treatment. SciTech publishers, Inc., Matawan, NJ, USA.

De Ceballos BSO, Oliveira H, Meira CMBS, Konig A, Guimarães AO, De Souza JT (2003). River water quality improvement by natural and constructed wetland systems in the tropical semi-arid region of Northeastern Brazil. Water Sci. Technol. 44(11-12):599-605.

De Sousa J.T., Haandel A. van, Lima E.P.C and Guimarães A.V.A, (2003). Performance of Constructed Wetland Systems Treating Anaerobic Effluents. Water Science and Technology: 48. Issue 6. pp. 295-299.

El-Khateeb MA, El-Gohary FA (2003). Combining UASB Technology and Constructed Wetland for Domestic Wastewater Reclamation and Reuse. Water Sci. Technol. Water Supply 3(4):201-208.

Mihret DU (2013)). Household Level Wastewater Management: A Pilot Case Study For Hawassa Town, Ethiopia, Int. J. Engine. Res. Technol. (IJERT) , ISSN: 2278-0181 , www.ijert.org. 2:6.

Mihret DU, Ermias AD, Mekonene AG, Adunga EB (2013). Environmental Impacts of Small Scale Irrigation Schemes: Evidence from Ethiopian Rift Valley Lake Basins, Scientific J. Environ. Res. Engine. Manage. ISSN 1392-1649, ISSN (online) 2029-2139. www.erem.ktu.lt. 63:1.

Rehm HJ, Reed G (1999). Biotechnology: Environmental processes I; waste water treatment. Wiley-VCH Verlag GmbH; Weinheim, Germany, 1999.

Rehm HJ, Reed G (2000). Biotechnology: Environmental processes II; Soil Decontamination. Wiley-VCH Verlag GmbH; Weinheim, Germany.

Sauer P, Kimber A (2001). Technical Assessment of Constructed Wetlands for Wastewater Treatment in Iowa. IOWA Energy Center, November, 2001. Online at: *www.iamu.org/main/studies_reports/reports/WetlandReport/SectionIIIntro.HistoryDesign.pdf*

UNEP (2004). Integrated watershed management -Ecohydrology and Phytotechnology Manual. Online at: *www.unep.or.jp/ietc/publications/freshwater/watershed_manual/index.asp*

Wetlands International - Malaysia Office (2013). The Use of Constructed Wetlands for Wastewater Treatment. Selangor, Malaysia.
Online at: http://www.wetlands.org/WatchRead/Currentpublications/tabid/56/ArticleType/ArticleView/ArticleID/1369/PageID/550/Default.aspx

33

Lead uptake of water plants in water stream at Kiteezi landfill site, Kampala (Uganda)

Dana Jill Mugisa[1], Noble Banadda[1] , Nicholas Kiggundu[1] and R. Asuman[2]

[1]Department of Agricultural Bio-Systems Engineering, Makerere University, P. O. Box 7062, Kampala, Uganda.
[2]Chemistry Laboratory, Uganda Industrial Research Institute, P. O. Box 7086, Kampala, Uganda.

The purpose of this study was twofold: (i) to quantify the lead (Pb) uptake by two water plants reeds (*Phragmites australis*) and papyrus (*Cyperus papyrus*) in water stream at Kiteezi landfill site, Kampala (Uganda) and (ii) to compare the two species in Pb uptake downstream. As such, leachate samples were collected at the inlet and outlet of the waste water treatment plant (WWTP) at Kiteezi landfill site. A total of 6 plant samples of both plant species, *P. australis* and *C. papyrus*, were picked from three different sites at intervals of 10, 20 and 30 m taken from the exit point of the WWTP, as the reference point. All samples were taken to the laboratory for analysis in a cool container. The concentration of Pb in the samples was measured using the atomic absorption spectrometer (AAS), Perkin Elmer Model. The obtained data was analyzed using descriptive statistics and two-way Anova. The results showed that there was no significant difference (P > 0.05) in the mean Pb content up taken by both plants (reeds and papyrus). Significant quantities of Pb were present in the plants in the range of 1.68 to 5.46 mg/100 g. The removal efficiency of the plants was found to be 12.4 times higher than WWTP. The highest concentrations of Pb were found downstream at a distance of 30 m away from the reference point. Although, the plants were generalized as having equal uptake levels, the two species had different mechanisms with reeds being accumulators and papyrus being excluders. Therefore, reeds are preferable phytoremediators since when harvested by cutting as practiced by some communities in Uganda, the Pb can easily be removed from the environment.

Key words: Lead-uptake, excluders, accumulators, phytoremediation, leachates and Kiteezi.

INTRODUCTION

Leachate from dumping sites, especially open dumps, contain heavy metals which compromise water quality (Kamarudzaman et al., 2011). Heavy metals are some of the causes of contamination in the environment. They are some trace elements and also defined as elements with metallic properties and atomic number greater than 20. It is a term that can also include light elements such as copper, with an atomic weight of 63. Heavy metals are commonly known for their toxicity. Despite the fact that some heavy metals such as chromium, copper, iron and zinc are essential for life in small quantities, others such as lead and cadmium do not have a vital function and could be harmful to organisms even in small concentrations (Nhapi et al., 2012).

Heavy metal contamination is usually a result of various human activities such as mining and smelting, metalliferous electroplating, internal combustion engine operation, energy and fuel production, fertilizer and pesticide application and the generation of municipal solid waste. Metals enter the municipal solid waste stream from different sources. These include batteries, house dust and paint chips, light bulbs, consumer electronics, ceramics, lead foils such as wine bottle closures, used motor oils, plastics, some inks and also glass (Woodbury, 1992). Movement of these contaminants into non contaminated areas as dust or leachates through the soil is one of the ways that contaminate the ecosystems (Tangahu et al., 2011). Since metals are not easily biodegraded, eliminating them from the environment, necessitates their removal. Currently, not even highly industrialized countries can afford to clean up contaminated sites. For example in Germany, 30% of the soils from contaminated areas are cleaned up in soil remediation facilities, while the rest is stored untreated in waste disposal facilities (Csaba, 2011).

Various methods are being adopted to free the environment from these kinds of contaminants, though majority of them are expensive and not very efficient. Chemical technologies generate large volumetric sludge and thus higher costs. Thermal technologies degrade the valuable component of the soils (Tangahu et al., 2011), hence making the methods not the most efficient. Recent concerns regarding the environmental contamination have led to the discovery of adequate methods to assess the presence and mobility of metals in soil, water and wastewater (Shtangeeva et al., 2004). Currently, phytoremediation, which is the use of special species of plants to remove heavy metals from soil and water, has become an effective and affordable technological solution in the extraction of metals from soil. This is basically because the method is environmentally friendly and cost effective. This method takes advantage of the unique and selective uptake capabilities of plant root systems, together with translocation, bioaccumulation and degradation abilities of the entire plant body. Many species of plants have been successful in absorbing lead, cadmium, arsenic and various radionuclides from soils.

For example Kim et al. (2003) studied heavy metal accumulation in *Polygonum thunbergii* and the soil from Man-Kyung River watershed, Korea, and noted that there was detectable presence of heavy metals in the plant in the order of Zn (2427.3 µg/g) > Cu (863.2 µg/g) > Pb (320.8 µg/g) > Cd (7.4 and 10.1 µg/g) in only the stem and root respectively). The soil samples contained detectable zinc (24.5 µg/g) > lead (17.5 µg/g) > copper (8.4 µg/g) with undetectable cadmium content. This study rendered the plant a very good phyto-extractor. Odong et al. (2013), also studied a range of macrophyte plants to see if anyone was able to clean up waste water and absorb toxic matter from abattoir effluent that is a major source of pollution entering Lake Victoria, in his study, papyrus and weeds were tested and it was found that papyrus was able to remove the highest concentration of four grams of phosphorus per kilogram dry weight from the waste water.

Lead as a metal occurs naturally in the earth, it has many industrial uses and is found in trace amounts everywhere in the environment. It is a very toxic and dangerous metal when ingested in the human body. In humans, exposure to lead can result in biological effects such as problems in the synthesis of gastrointestinal tract, joints, reproductive system and acute chronic damage to the nervous system, depending on the level and duration of the exposure. Research has proven that lead contributes to 0.2% of all deaths and 0.6% of all disability in life globally (WHO, 2009). Developing fetus and infants are more sensitive to the effects than the adults.

Land filling is the primary means of disposal of both residual municipal solid waste and many hazardous wastes in Uganda. According to Ngategize (2000), the current and only landfill site in Kampala is at Kiteezi. It is further situated within a human settlement community, where it has caused social discomfort and environmental pollution. During the process of land filling, waste is subjected to aerobic decomposition which creates social tensions among the communities near the landfills especially odour pollution, flies, vermin and pests (Sabiiti and Katongole, 2012). It is against this background, that this study's main objective was twofold: (i) to quantify the lead uptake by the two water plants, reeds (*Phragmites australis*) and papyrus (*Cyperus papyrus*) in water stream at Kiteezi landfill site, Kampala (Uganda) and (ii) to compare the two species in Pb uptake downstream and as a result suggest which of the species should be promoted to improve the water quality downstream: (i) to quantify the lead (Pb) uptake by the two reeds in the water stream and (ii) to compare the two species in Pb uptake downstream.

MATERIALS AND METHODS

Study area

This study was conducted in Kiteezi landfill site, located at the north of Kampala City, an average distance of 12 km from the city central. Kiteezi landfill site serves the 5 divisions of Kampala, namely, Kampala Central, Nakawa, Lubaga, Makindye and Kawempe. The present access to the site from Kampala City is through Kampala-Gayaza road (about 9 km), then branch off to the left from Mpererwe and follow the road heading to Namulonge for about 4 km. Currently, it is the only landfill site at which Kampala's solid waste is disposed. The neighboring areas of Kiteezi land fill include: Kasangati, Kawempe, Nangabo, Namalere, Bulamoiro, Buye and Kalerwe.

Geographically, Kiteezi is located at latitude: 0° 25' 0" and longitude: 32° 34' 00" as depicted in Map 1. The site was opened in 1996 and it covers an area of 29 acres. By then, KCCA in 2007 acquired an additional six acres to the south of the existing landfill for expansion purposes.

Map 1. Satellite image of Kiteezi landfill location (Source: Google maps).

Experimental procedure

Leachate sample collection and analysis

The leachate samples were collected, preserved and stored in 500 ml plastic bottles. The bottles were first cleaned thoroughly with de-ionized water, then rinsed thrice with the sample leachate before the final collection. The leachate samples were collected by directly inserting 500 ml plastic water bottles in the leachate and picking up the samples. The inlet samples were picked at the inlet collection point, where all the leachate from the landfill is drained to, before it is piped to the water treatment plant. The outlet samples were picked just at the exit of the waste water treatment plant, by also inserting the bottles into the leachate as the leachate is being discharged out of the waste water treatment plant pipes. The samples were then preserved and stored in the 500 ml plastic bottles.

The temperature and pH was measured in the field using portable HACH meters. The leachate was collected from two sampling sites which were, the inlet and outlet of the waste water treatment plant, respectively. The samples were analyzed for lead concentration by digesting them in a solution of 1:3 hydrochloric (50 ml) and nitric acid (150 ml), respectively. Five milliliters of HCl/ HNO_3 (1:3 v/v) was added to the leachate samples to make upto a 25 ml volumetric flask. The solutions were filtered with Whatman filter paper. The actual concentrations of lead were determined using an atomic absorption spectrophotometer (Perkin Elmer Model).

Plant sample collection and analysis

The field survey was conducted to identify which of the two species (papyrus and reeds), accumulated exceptionally high concentrations of lead and the extraction coefficients. Samples were collected from three sites at intervals of 10 m apart, along the stream length, from the start of the stream, at the exit of the treatment plant.

One individual of each plant species was picked from the three different sites of the study area, making a total of six plants. Each plant sample was thoroughly washed in running tap water for 5 min, and with a solution of phosphate free detergent for 15 s, then with tap water for another 15 s. The samples were then carefully rinsed with deionized water twice and separated into shoots, stems and roots. The samples were then oven dried at 105°C for 48 h, ground into fine powder using a blender and sieved through a nylon sieve. Different weights of each plant part sample ranging from 1-5 g, depending on the availability of the sample, were weighed into nickel crucibles that had been initially conditioned and their respective weights recorded. Each plant part were replicated three times, the samples were then carbonized by heating them on a hotplate for 1 h until the powder turned black.

This was followed by dry ashing, which refers to the use of muffle furnace capable of maintaining temperatures of 500 - 600°C. Water and volatiles are vaporized and organic substances are burned in the presence of oxygen in air to carbon dioxide and the oxides of nitrogen. Most minerals are converted to oxides, sulphates, phosphates, chlorides and silicates. Elements such as Fe, Se, Pb and Hg may partially volatilize with this procedure so, other methods must be used if ashing is a preliminary step for specific elemental analysis. Most dry ashing samples need no preparation while fresh vegetables need to be dried prior to ashing.

The carbonised samples were then immediately transferred to the furnace (with the arrangement of the crucibles recorded) and left to ignite for 4-6 h at 550°C, after this period, the furnace was then turned off and not opened until the temperatures dropped to 250°C or below and using safety tongs the samples if well ashed (with no black spots seen), the crucibles were transferred into a desiccator to cool and the weight of sample plus crucible recorded. The concentrations of lead in shoots, stems and roots were determined using the different weights of respective plant samples

Table 1. Lead (Pb) concentration (µg/g) determined in the influent and effluent of the water treatment plant.

Leachate	Concentration (µg/g)
Influent	0.63
Effluent	0.58

Table 2. Mean Pb content (mg/100 g) of the triplicates of shoots, stems and roots of papyrus (P) and reeds (R) determined.

Distance from reference point (m)	P_{roots}	P_{stem}	P_{shoot}	∑Lead*	R_{roots}	R_{stem}	R_{shoot}	∑Lead*
A (10)	1.29	1.71	0.15	3.14	0.58	0.33	0.77	1.68
B (20)	2.45	0.71	0.49	3.65	1.63	1.00	1.71	4.34
C (30)	1.61	1.29	0.82	3.72	1.69	1.35	2.42	5.46

Lead*: the sum of Lead content in the whole plant, P- papyrus, R- reeds.

digested with 5 ml of 20% HCl and the mixture boiled for 5 min on a hot plate. The solution was filtered into 25 ml volumetric flasks and made to the mark using distilled water. The lead concentration was measured using a flame atomic absorption spectrophotometer. Extraction coefficient was calculated as:

$$\text{Extraction coefficient}=\frac{\text{lead concetration in shoot}}{\text{lead concetration in water}} \qquad (1)$$

Equation 1 can be used to evaluate the ability of the plants to accumulate heavy metals.

Determination of waste water treatment plant efficiency

The efficiency of waste water treatment plant was determined using Equation 2.

$$E_0 = \frac{\Delta Pb}{Pb_{in}} \times 100 \qquad (2)$$

Where Pb_{in} is lead in the effluent, E_0 is efficiency of the WWTP

$$E_0 = \frac{(0.63 - 0.58)}{0.63} \times 100 = 7.94\%$$

Data analysis

The data was analyzed using Microsoft Excel package, in which tools like two way-ANOVA (analysis of variance) and other descriptive statistics were used. The data was analyzed in terms of mean and standard deviation. These tools were used to determine the variations of the mean lead concentration in the two species of plants at the three different sites (the two independent variables) and to which level of confidence the variation is if it existed. Using $p > 0.05$, shows non-significant difference in the means, the information helped to check if the determined values conform to a normal distribution.

RESULTS AND DISCUSSION

Table 1 shows the lead concentration in the leachate at Kiteezi landfill site, Kampala, Uganda.The lead at the inlet to the water treatment plant was detected and found to be 0.63 µg/g and that at the outlet to be 0.58 µg/g. The efficiency of the plant was determined from Equation 2 and found to be very poor, 7.94%. This in addition to the build-up of lead in the soil and the stream containing the plants, makes the stream act as a plug flow reactor (where concentrations of pollutants varry along the directions of flow), contributing to the fact that the plants have a higher lead uptake as compared to what the water treatment plant has is depicted in Table 2.

It was noted that the Pb concentration generally increased with distance (10-30 m) away from the reference point (exit point of the WWTP). This trend was attributed to the fact that as distance increased, flow rate reduced, thus making the plants have more contact time with the leachate and hence the increased concentrations in the amounts of lead extracted.

For the reeds, it was found that they accumulated more lead in their shoots than in the roots. This makes the shoot to root ratio > 1 and thus rendering the plant an accumulator (Baker and Whiting, 2002). Also, the stems were seen to have lower concentrations of lead than shoots, and this was attributed to a biological factor that the stems contain vascular bundles (xylem and phloem), that are essential in translocation (xylem), when the transported elements reach the shoots and are subjected to other processes like transpiration, this takes away the water and thus leaving higher concentrations of solids in the shoots.

Papyrus was found to have more lead accumulated in the roots than in the shoots, similar results were reported by Odong et al. (2013). Hence, shoot to root ratio was < than 1 and thus rendering papyrus an excluder (Baker

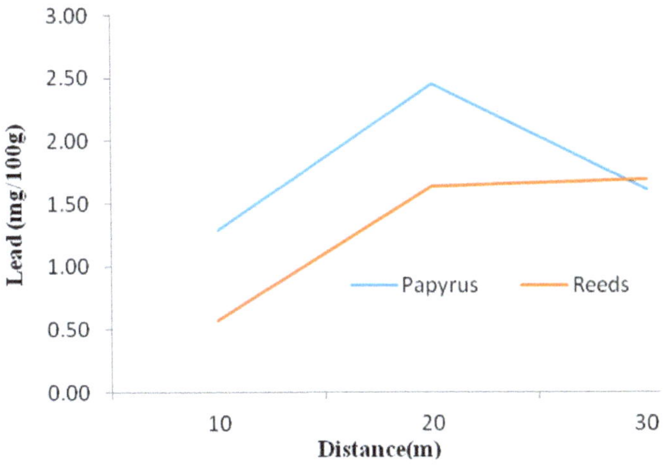

Figure 1. Variation of lead in roots (P> 0.05).

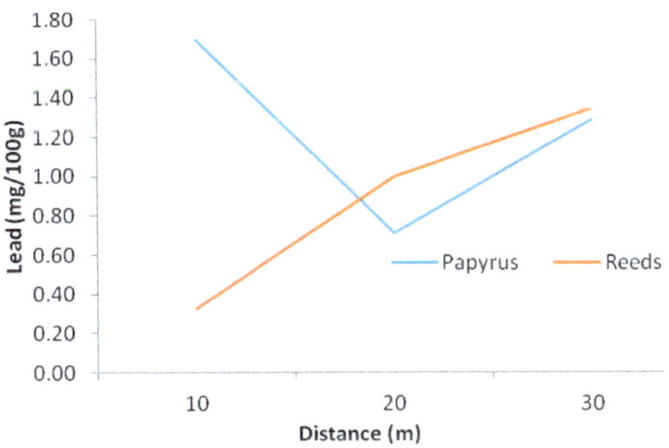

Figure 2. Variation of lead in stems (P> 0.05).

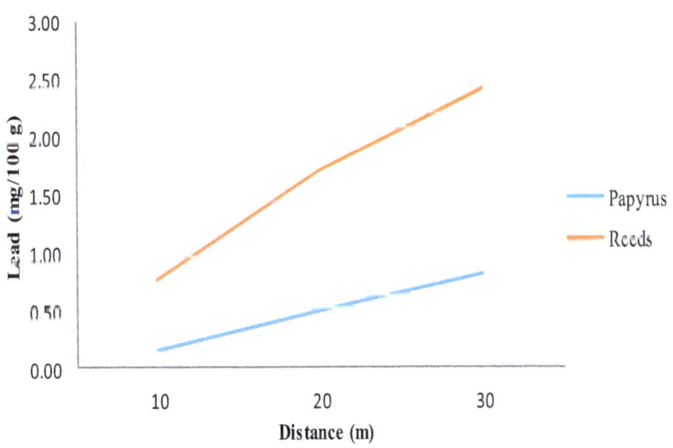

Figure 3. Variation of lead in shoots (P> 0.05).

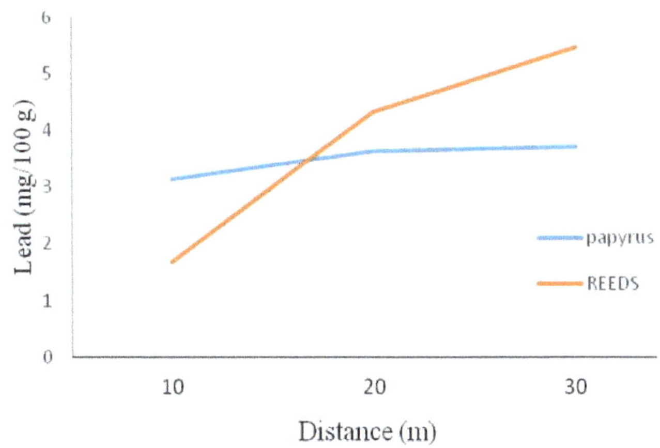

Figure 4. Variation of lead in whole plants (P> 0.05).

and Whiting, 2002). At 20 m, the papyrus roots were found to contain 2.45 mg/100 g which was greater than that at 30 m (1.61 mg/100 g) and this was attributed to the fact that, at the spot where this particular plant was picked, there was a ditch thus making it deeper and having a more root to contact time as compared to the plant at 30 m. On comparing a specific part of both plant species, the variation of lead was as shown in the Figures 1, 2, 3 and 4. Figure 1 shows that papyrus roots had a higher lead concentration as seen above, but for both plants, they increased in the order of C > B > A. Figure 2 shows that the accumulation of lead in the stems increased gradually for reeds in the order of C > B > A, while for papyrus it was in the order of A > C > B. Figure 3 depicts reeds accumulated more lead in the shoots than papyrus but in the order of C > B > A for both plants.

On comparing the results with two-way Anova, it was found that all comparisons showed P > 0.05, this implied that there was no significant difference in variation of the mean lead content of all the parts and the plants as a whole (Figure 4). On further analysis, also the plants were found to be more efficient than the waste water treatment plant as depicted in Table 3.

Conclusions

Significant quantities of lead were present in both the leachate and the plant parts. The waste water treatment plant was found to have a low lead removal efficiency of 7.94%. The removal efficiency of the plants was 12.4 times higher than that of the waste water treatment plant. The plants were generalized as having the same lead uptake levels, since there was no significant difference (P > 0.05) in the comparisons made for the two plant species. On an exhaustive analysis of plant parts, and from the different profiles of the plant parts made, it was also noted that the two species had different uptake mechanisms, with the reeds being rendered accumulators while papyrus are excluders. Therefore, the study proposes that more of the reeds could be propagated

Table 3. The plants' efficiencies (E_{rp} and E_p) at the different distances in comparison with the WTP efficient (E_0).

Distance from reference point (m)	Plant efficiency (%)		Efficiency ratio	
	Reeds (E_r)	Papyrus (E_p)	$(\frac{E_r}{E_0})$	$(\frac{E_P}{E_0})$
10	96.5	98.2	12.2	12.4
20	98.7	98.4	12.4	12.5
30	98.9	98.4	12.4	12.5

due to the fact that they are accumulators, which makes it easier to eliminate the lead (Pb) when they are harvested by cutting them down (Kim et al., 2003) unlike the excluders that keep the lead (Pb) in the roots.

Conflict of interests

The author(s) declare no conflict of interest that may include any of the following:

Funding: Research support for this work (including salaries, equipment, supplies, reimbursement for attending symposia, and other expenses) by any organization that may gain or lose financially through publication of the paper.

Employment: Recent (that is, while engaged in the research project), present or anticipated employment by any organization that may gain or lose financially through publication of the paper.

Personal financial interests: Stocks or shares in companies that may gain or lose financially through publication; consultation fees or other forms of remuneration from organizations that may gain or lose financially; patents or patent applications whose value may be affected by publication.

REFERENCES

Baker AJM, Whiting SN (2002). In search of the Holy Grail-a Further Step in Understanding Metal Hyper Accumulation. New Phytol. 155: 1-7.

Csaba CA (2011). Phytoremediation of Heavy Metals with Cucumber, White poplar and Aquatic plants. Babes-Bolyai University Cluj Napoca. 1-37.

Kamarudzaman NA, Roslaili AZ, Mohd FAJ (2011). Removal of Heavy Metals from Landfill Leachate using Horizontal and Vertical Subsurface Flow Constructed Wetland Planted with Limnocharis flava. Int. J. Civil Environ. Eng. 11(5):85-91.

Kim IS, Kong KH, Green PJ, Lee EJ (2003). Investigation of Heavy Metal Accumulation in Polygonum thunbergii for Phytoremediation. Environ. Pollut. 126:235-2443.

Ngategize P (2000). Draft Strategic Plan for Solid Waste Management for Mpigi District, Environmental Monitoring Association Limited, Kampala Uganda.

Nhapi I, Wali UG, Usanzineza D, Banadda N, Kashaigili JJ, Kimwaga R, Gumindoga W, Sendagi S (2012). Heavy Metal Inflow into Lake Muhazi, Rwanda. Open Environ. Eng. J. 5:56-65.

Odong R, Kansiime F, Omara J, Kyambadde J (2013). The potential of four tropical wetland plants for the treatment of abattoir effluent. Int. J. Environ. Technol. Manage. 16(3):203-222.

Sabiiti EN, Katongole CB (2012). Evaluation of the Urban Solid Waste Management System in Kampala, Uganda.

Shtangeeva I, Laiho JVP, Kahelin H, Gobran GR (2004). Phytoremediation of metal-contaminated soils. Symposia Papers Presented Before the Division of Environmental Chemistry, American Society Anaheim, Calif, USA.

Tangahu BV, Abdullah SRS, Basri H, Idris M, Anuar N, Mukhlisin M (2011). A Review on Heavy Metals (As, Pb, Hg) Uptake by Plants through Phytoremediation. Int. J. Chem. Eng. vol. 2011, Article ID 939161, 31 pages, 2011. doi:10.1155/2011/939161.

WHO. 2009. Regional Office for Europe, Air Quality Guidelines, chapter 6-7, Lead, Copenhagen, Denmark, 2nd edition. Available at: http://www.euro.who.int/document/aiq/6 7Lead.pdf. Accessed 18 November 2013.

Woodbury PB (1992). Potential Effects of Heavy Metals in Municipal Solid Waste Composts on Plants and the Environment. Boyce Thompson Institute for Plant Research at Cornell University. Biomass Energy 3(3):239-259.

Permissions

All chapters in this book were first published in AJEST, by Academic Journals; hereby published with permission under the Creative Commons Attribution License or equivalent. Every chapter published in this book has been scrutinized by our experts. Their significance has been extensively debated. The topics covered herein carry significant findings which will fuel the growth of the discipline. They may even be implemented as practical applications or may be referred to as a beginning point for another development.

The contributors of this book come from diverse backgrounds, making this book a truly international effort. This book will bring forth new frontiers with its revolutionizing research information and detailed analysis of the nascent developments around the world.

We would like to thank all the contributing authors for lending their expertise to make the book truly unique. They have played a crucial role in the development of this book. Without their invaluable contributions this book wouldn't have been possible. They have made vital efforts to compile up to date information on the varied aspects of this subject to make this book a valuable addition to the collection of many professionals and students.

This book was conceptualized with the vision of imparting up-to-date information and advanced data in this field. To ensure the same, a matchless editorial board was set up. Every individual on the board went through rigorous rounds of assessment to prove their worth. After which they invested a large part of their time researching and compiling the most relevant data for our readers.

The editorial board has been involved in producing this book since its inception. They have spent rigorous hours researching and exploring the diverse topics which have resulted in the successful publishing of this book. They have passed on their knowledge of decades through this book. To expedite this challenging task, the publisher supported the team at every step. A small team of assistant editors was also appointed to further simplify the editing procedure and attain best results for the readers.

Apart from the editorial board, the designing team has also invested a significant amount of their time in understanding the subject and creating the most relevant covers. They scrutinized every image to scout for the most suitable representation of the subject and create an appropriate cover for the book.

The publishing team has been an ardent support to the editorial, designing and production team. Their endless efforts to recruit the best for this project, has resulted in the accomplishment of this book. They are a veteran in the field of academics and their pool of knowledge is as vast as their experience in printing. Their expertise and guidance has proved useful at every step. Their uncompromising quality standards have made this book an exceptional effort. Their encouragement from time to time has been an inspiration for everyone.

The publisher and the editorial board hope that this book will prove to be a valuable piece of knowledge for researchers, students, practitioners and scholars across the globe.

List of Contributors

H. W. T. Mapoma
The Polytechnic, University of Malawi, Private Bag 303, Blantyre 3, Malawi

J. J. Zimba
Dairibord Malawi Limited, P.O. Box 30647, Chichiri, Blantyre 3, Malawi

W. Utembe
The Polytechnic, University of Malawi, Private Bag 303, Blantyre 3, Malawi

I. B. M. Kosamu
The Polytechnic, University of Malawi, Private Bag 303, Blantyre 3, Malawi

P. J. K. Rongoei
Egerton University, Department of Environmental Sciences, P.O. Box 536-20115, Egerton, Kenya

J. Kipkemboi
Egerton University, Department of Environmental Sciences, P.O. Box 536-20115, Egerton, Kenya

J. B. Okeyo-Owuor
Chepkoilel University College, Constituent College of Moi University, P.O. Box 1125, Eldoret, Kenya

A. A. van Dam
UNESCO-IHE Institute for Water Education, Department of Water Science and Engineering, P.O. Box 3015, 2601 Da Delft, Netherlands

O. A. Fakankun
Department of Chemical Sciences, Bells University of Technology, Ota, Ogun State, Nigeria

J. O. Babayemi
Department of Chemical Sciences, Bells University of Technology, Ota, Ogun State, Nigeria

J. J. Utiaruk
Department of Chemical Sciences, Bells University of Technology, Ota, Ogun State, Nigeria

O. Otitoju
Faculty of Pure and Applied Sciences, Department of Biochemistry, Federal University, Wukari, Taraba State, Nigeria

G. T. O. Otitoju
Department of Home Science, Nutrition and Dietetics, University of Nigeria, Nsukka, Enugu State, Nigeria

A. N. Mutune
Jomo Kenyatta University of Agriculture and Technology. P.O. Box 62000, 00200. Nairobi, Kenya

M. A. Makobe
Jomo Kenyatta University of Agriculture and Technology. P.O. Box 62000, 00200. Nairobi, Kenya

M. O. O. Abukutsa-Onyango
Jomo Kenyatta University of Agriculture and Technology. P.O. Box 62000, 00200. Nairobi, Kenya

Bintoora Kamukasa Kakurungu Adonia
School of Sciences, Nkumba University, P. O. Box 237, Entebbe, Uganda
Mt Elgon Conservation Area, P. O. Box 135, Mbale, Uganda

N. Pinheiro
Universidade de Cabo Verde, Departamento de Engenharia e Ciências do Mar, CP 163, Mindelo, Cabo Verde

L. Barreira
CIMA, Universidade do Algarve, Campus de Gambelas, 8000 - 117 Faro, Portugal

B. Lopesand
CIMA, Universidade do Algarve, Campus de Gambelas, 8000 - 117 Faro, Portugal

M. J. Bebianno
CIMA, Universidade do Algarve, Campus de Gambelas, 8000 - 117 Faro, Portugal

N. F. Olele
Fisheries Department, Delta State University, P.O. Box 1110, Asaba, Nigeria

B. B. Babatunde
Department of Animal and Environmental Biology, Faculty of Science, University of Port Harcourt, PMB 5323, Choba, Port Harcourt, Rivers State
Lancaster Environment Centre, Lancaster University, UK, LA1 4YQ

I. F. Vincent-Akpu
Department of Animal and Environmental Biology, Faculty of Science, University of Port Harcourt, PMB 5323, Choba, Port Harcourt, Rivers State

G. N. Woke
Department of Animal and Environmental Biology, Faculty of Science, University of Port Harcourt, PMB 5323, Choba, Port Harcourt, Rivers State

E. Atarhinyo
Department of Animal and Environmental Biology, Faculty of Science, University of Port Harcourt, PMB 5323, Choba, Port Harcourt, Rivers State

U. C. Aharanwa
Department of Animal and Environmental Biology, Faculty of Science, University of Port Harcourt, PMB 5323, Choba, Port Harcourt, Rivers State

A. F. Green
Department of Applied and Environmental Biology, Rivers State University of Science and Technology, Nkpolu, Port Harcourt, Rivers State

O. Isaac-Joe
Department of Animal and Environmental Biology, Faculty of Science, University of Port Harcourt, PMB 5323, Choba, Port Harcourt, Rivers State

J. A. Chukwudebelu
Federal Institute of Industrial Research Oshodi, Lagos State

C. C. Igwe
Federal Institute of Industrial Research Oshodi, Lagos State

O. E. Taiwo
Federal Institute of Industrial Research Oshodi, Lagos State

O. B. Tojola
Federal Institute of Industrial Research Oshodi, Lagos State

Francis Attiogbe
Department of Chemical Engineering, Kumasi Polytechnic, P. O. Box 854, Kumasi, Ghana

Akajiaku C. Chukwuocha
Department of Surveying and Geoinformatics, Federal University of Technology, Owerri, Nigeria

Ngozi AC-Chukwuocha
Department of Environmental Technology, Federal University of Technology, Owerri, Nigeria

Osuala, Fidelia Ijeoma
Faculty of Science, Department of Zoology, University of Lagos, P.M.B. 101017, Lagos, Lagos State, Nigeria

Otitoloju,Adebayo Akeemand
Faculty of Science, Department of Zoology, University of Lagos, P.M.B. 101017, Lagos, Lagos State, Nigeria

M. N. Igwo-Ezikpe
Faculty of Basic Medical sciences, Department of Biochemistry, College of Medicine, University of Lagos, P.M.B. 12003, Lagos, Lagos State, Nigeria

N. Abdu
Departments of Soil Science, Faculty of Agriculture, Institute for Agricultural Research, Ahmadu Bello University, Zaria, P.M.B. 1044, Samaru, Kaduna State, Nigeria

A.A Yusuf
Departments of Soil Science, Faculty of Agriculture, Institute for Agricultural Research, Ahmadu Bello University, Zaria, P.M.B. 1044, Samaru, Kaduna State, Nigeria

Prabhat Kumar Rai
Department of Environmental Science, Mizoram University, Tanhril, Aizawl-796004, Mizoram, India

Lalita L. S. Panda
Department of Environmental Science, Mizoram University, Tanhril, Aizawl-796004, Mizoram, India

Biku Moni Chutia
Department of Environmental Science, Mizoram University, Tanhril, Aizawl-796004, Mizoram, India

M. Muni Singh
Department of Environmental Science, Mizoram University, Tanhril, Aizawl-796004, Mizoram, India

S. P. Songca
Faculty of Science, Engineering and Technology, Walter Sisulu University, PO Box 19712, Tecoma, East London, Eastern Cape, South Africa

B. Seleni
Department of Chemistry and Chemical Technology, Walter Sisulu University, Mthatha Campus, Private Bag X1, Mthatha 5117, Eastern Cape, South Africa

O. S. Oluwafemi
Department of Chemistry and Chemical Technology, Walter Sisulu University, Mthatha Campus, Private Bag X1, Mthatha 5117, Eastern Cape, South Africa

AMOS-TAUTUA
Department of Chemical Sciences, Faculty of Science, Niger Delta University, Wilberforce, P.M.B 071, Yenagoa, Bayelsa State, Nigeria

W. Bamidele Martin
Department of Chemical Sciences, Faculty of Science, Niger Delta University, Wilberforce, P.M.B 071, Yenagoa, Bayelsa State, Nigeria

Adebayo O. ONIGBINDE
Chemistry Unit, Basic Sciences Department, Babcock University, Ilishan, Remo, Ogun State, Nigeria

Diepreye ERE
Department of Chemical Sciences, Faculty of Science, Niger Delta University, Wilberforce, P.M.B 071, Yenagoa, Bayelsa State, Nigeria

I. E. Agbozu
Department of Environmental Science, Federal University of Petroleum Resources, P.M.B. 1221, Effurun, Nigeria

F. O. Emoruwa
Department of Environmental Science, Federal University of Petroleum Resources, P.M.B. 1221, Effurun, Nigeria

Angstone Noel J. Thembachako MLANGENI
NR Department, Natural Resources College, Box 143, Lilongwe, Malawi

Shaibu KANANJI
Agriculture Department, Natural Resources College, Box 143, Lilongwe, Malawi

Sosten S. CHIOTHA
Biology Department, Chancellor College, University of Malawi

A. A. Aladesida
Department of Biological Sciences, Federal University of Agriculture, Abeokuta, Ogun State, Nigeria

S. O. Owa
Department of Biological Sciences, Landmark University, Omu-Aran, Kwara State, Nigeria

G. A. Dedeke
Department of Biological Sciences, Federal University of Agriculture, Abeokuta, Ogun State, Nigeria

O. A. Adewoyin
Department of Zoology, University of Ilorin, Kwara State, Nigeria

Onyenekenwa Cyprian Eneh
Institute for Development Studies, Enugu Campus, University of Nigeria, Nsukka, Enugu, Nigeria

TANG Meizhen
Key Laboratory of NaSiHu Lake Wetland Ecosystem and Environment Protection, Qufu Normal University, Qufu, Shandong, 273165, China

CHEN Junfeng
College of Resource and Environment, Hunan Agricultural University, Changsha, Hunan, 410128, China

SUN Yannan
Key Laboratory of NaSiHu Lake Wetland Ecosystem and Environment Protection, Qufu Normal University, Qufu, Shandong, 273165, China

TONG Yanru
Key Laboratory of NaSiHu Lake Wetland Ecosystem and Environment Protection, Qufu Normal University, Qufu, Shandong, 273165, China

LIU Yuling
Key Laboratory of NaSiHu Lake Wetland Ecosystem and Environment Protection, Qufu Normal University, Qufu, Shandong, 273165, China

I. Abubakar
Department of Remedial and General Studies, Waziri Umaru Federal Polytechnic, Birnin Kebbi, Nigeria

U. A. Birnin Yauri
Department of Pure and Applied Chemistry, Usmanu Danfodiyo University, Sokoto, Nigeria

U. Z. Faruq
Department of Pure and Applied Chemistry, Usmanu Danfodiyo University, Sokoto, Nigeria

S. S. Noma
Department of Soil Sciences, Usmanu Danfodiyo University, Sokoto, Nigeria

N. Sharif
Department of Chemistry, Zamfara College of Art and Science, Gusau, Nigeria

Assefa Wosnie
Biology Department, Dilla University, P O Box 419, Dilla, Ethiopia

Ayalew Wondie
Biology Department, Bahir Dar University, P.O Box 79, Bahir Dar, Ethiopia

Adolphe D. Tchehouali
Laboratorise d'Energetic et do Mecanique Applique- Ecole Polytechnique d' Abomey calavi, universitè d' abomey calavi (LEMA/EPAC/UAC), 01 BP 2009 Cotonou, Benin

Chakirou A. toukourou
Laboratorise d'Energetic et do Mecanique Applique- Ecole Polytechnique d' Abomey calavi, universitè d' abomey calavi (LEMA/EPAC/UAC), 01 BP 2009 Cotonou, Benin

Agapi K.Houanou
Laboratorise d'Energetic et do Mecanique Applique- Ecole Polytechnique d' Abomey calavi, universitè d' abomey calavi (LEMA/EPAC/UAC), 01 BP 2009 Cotonou, Benin

Edmond Adjovi
Laboratorise d'Energetic et do Mecanique Applique- Ecole Polytechnique d' Abomey calavi, universitè d' abomey calavi (LEMA/EPAC/UAC), 01 BP 2009 Cotonou, Benin

Amos Foudjet
CRESA foret- Bois Universite de Dschang, Cameroun

Antoine Vianou
Laboratorise d'Energetic et do Mecanique Applique- Ecole Polytechnique d' Abomey calavi, universitè d' abomey calavi (LEMA/EPAC/UAC), 01 BP 2009 Cotonou, Benin

Degan Gerard
Laboratorise d'Energetic et do Mecanique Applique- Ecole Polytechnique d' Abomey calavi, universitè d' abomey calavi (LEMA/EPAC/UAC), 01 BP 2009 Cotonou, Benin

Haimanot habte
Environmental Science Program, Faculty of Science, Addis Ababa University, P.O. Box 1176, Addis Ababa, Ethiopia

Feleke Zewgeand
Department of Chemistry, Faculty of Science, Addis Ababa University, P.O. Box 1176, Addis Ababa, Ethiopia

Mesfin Redi
Department of Chemistry, Faculty of Science, Addis Ababa University, P.O. Box 1176, Addis Ababa, Ethiopia

Chioma Blaise Chikere
Department of Microbiology, University of Port Harcourt, P.M.B. 5323, Port Harcourt, Rivers State, Nigeria

Chioma Bertha Ekwuabu
Department of Microbiology, University of Port Harcourt, P.M.B. 5323, Port Harcourt, Rivers State, Nigeria.

G. C Chinyere
Department of Biochemistry, Abia State University, Uturu, Nigeria

K. U. Osuocha
Department of Biochemistry, Abia State University, Uturu, Nigeria

Imo Chinedu
Department of Biochemistry, Abia State University, Uturu, Nigeria

S. P. Lohani
Department of Mechanical Engineering, Kathmandu University, Nepal

Department of Environment Science and Engineering, Kathmandu University, Nepal POST BOX: 6250, Nepal

A. Chhetri
Department of Environment Science and Engineering, Kathmandu University, Nepal POST BOX: 6250, Nepal

S. N. Khanal
Department of Environment Science and Engineering, Kathmandu University, Nepal POST BOX: 6250, Nepal

Kyeyune Simonpeter
Department of Agricultural and Bio-Systems Engineering, Makerere University, P.O. Box 7062, Kampala, Uganda

Mulamba Peter
Department of Agricultural and Bio-Systems Engineering, Makerere University, P.O. Box 7062, Kampala, Uganda

A. C. C. Ezeabasili
Heriot Watt University, UK

O. L. Anike
Nnamdi Azikiwe University, Awka, Anambra State, Nigeria

B. U. Okoro
Department of Civil Engineering, Nnamdi Azikiwe University, Awka, Anambra State, Nigeria

Mihret Dananto Ulsido
School of Civil Engineering and Architecture, Wuhan University of Technology, 122 Luoshi Road, Hongshan District, Wuhan, Hubei, 430070, P. R. China

Dana Jill Mugisa
Department of Agricultural Bio-Systems Engineering, Makerere University, P. O. Box 7062, Kampala, Uganda

Noble Banadda
Department of Agricultural Bio-Systems Engineering, Makerere University, P. O. Box 7062, Kampala, Uganda

Nicholas Kiggundu
Department of Agricultural Bio-Systems Engineering, Makerere University, P. O. Box 7062, Kampala, Uganda

R. Asuman
Chemistry Laboratory, Uganda Industrial Research Institute, P. O. Box 7086, Kampala, Uganda